OXYGEN TRANSPORT
TO TISSUE XIII

ADVANCES IN EXPERIMENTAL MEDICINE AND BIOLOGY

OXYGEN TRANSPORT TO TISSUE XIII

Edited by

Thomas K. Goldstick
Northwestern University
Evanston, Illinois

Michael McCabe
University of Kuwait
Sulaibikhat, Kuwait

and

David J. Maguire
Griffith University
Brisbane, Queensland, Australia

SPRINGER SCIENCE+BUSINESS MEDIA, LLC

Library of Congress Cataloging-in-Publication Data

Oxygen transport to tissue XIII / edited by Thomas K. Goldstick,
  Michael McCabe, and David J. Maguire.
      p.   cm. -- (Advances in experimental medicine and biology ; v.
  316)
    "Proceedings of the Eighteenth Annual Meeting of International
  Society on Oxygen Transport to Tissue, held July 19-22, 1990, in
  Townsville, Australia."--T.p. verso.
    Includes bibliographical references and indexes.
    ISBN 978-0-306-44342-8   ISBN 978-1-4615-3404-4 (eBook)
    DOI 10.1007/978-1-4615-3404-4
    1. Oxygen--Physiological transport--Congresses. 2. Oxygen in the
  body--Measurement--Congresses.   I. Goldstick, Thomas K.
  II. McCabe, Michael, 1933-   . III. Maguire, David J.
  IV. International Society on Oxygen Transport to Tissue. Meeting
  (8th : 1990 : Townsville, Qld. V. Title: Oxygen transport to tissue
  13. VI. Title: Oxygen transport to tissue thirteen. VII. Series.
    [DNLM: 1. Biological Transport--congresses. 2. Oxygen--Blood-
  -Congresses. 3. Oxygen Consumption--congresses.   W1 AD559 v.316 /
  WF 110 098 1990]
  QP99.3.O909394 1992
  599'.012--dc20
  DNLM/DLC
  for Library of Congress   92-48830
                                                               CIP

Proceedings of the eighteenth annual meeting of the International Society
on Oxygen Transport to Tissue, held July 19-22, 1990,
in Townsville, Australia

ISBN 978-0-306-44342-8

© 1992 Springer Science+Business Media New York
Originally published by Plenum Press, New York in 1992

INTERNATIONAL SOCIETY ON OXYGEN TO TISSUE 1989-90

Officers

President: M. McCabe, Australia
President-Elect: W. Erdmann, The Netherlands
Past President: J. Piiper, Germany
Secretary: N.S. Faithfull, England
Treasurer: S.M. Cain, USA

Executive Committee

D.T. Delpy, England
A.G. Hudetz, USA
I.S. Longmuir, USA
A. Mayersky, Israel
H.P. Metzger, Germany
R.N. Pittman, USA
N. Sato, Japan
Z. Turek, The Netherlands
P. Vaupel, Germany

Townsville Meeting, July 19-22, 1990

Local Organizing Committee

K. Bondeson
D. Maguire
M. McCabe (chairman)
S. McCabe

PREFACE

The International Society on Oxygen Transport to Tissue (ISOTT) was founded in 1973 "to facilitate the exchange of scientific information among those interested in any aspect of the transport and/or utilization of oxygen in tissues". Its members span virtually all disciplines, extending from various branches of clinical medicine such as anesthesiology, ophthalmology and surgery through the basic medical sciences of physiology and biochemistry to most branches of the physical sciences and engineering.

The eighteenth annual meeting of ISOTT was held in 1990 for four days, from July 19 to 22, in the Sheraton Hotel in Townsville, Queensland, Australia. The usual ISOTT format, which was originated in 1985 by Dr. Ian Longmuir, was continued. Almost all presentations were posters with an accompanying, scheduled, brief, slide presentation and discussion. All posters remained in place for the entire four days of the meeting. There were no simultaneous sessions. Essentially all aspects of physiological transport were covered at this meeting with possibly somewhat more emphasis on methods and instrumentation.

The editors gratefully acknowledge the photographic skills of Dr. Jens Höper who took the group picture during the outing to Magnetic Island on July 21. We are also most grateful to Dr. Rod D. Braun of Evanston for his invaluable editorial assistance.

This volume is the thirteenth in the Plenum series Oxygen Transport to Tissue. In many cultures thirteen is considered to be an especially unlucky number. Some hotels and public buildings are even built without a thirteenth floor. As one might expect then, the preparation of this particular volume has been associated with considerable bad luck which has, unfortunately, led to a sizeable delay in publication. The editors sincerely apologize for this delay.

For the editors

Thomas K. Goldstick

CONTENTS

OXYGEN TRANSPORT MODELS

METHODS AND INSTRUMENTATION

TUMORS

SHOCK AND WOUND HEALING

OTHER ORGANS AND TISSUES

OXYGEN TRANSPORT MODELS

MODELING OF OXYGEN TRANSPORT TO SKELETAL MUSCLE:

BLOOD FLOW DISTRIBUTION, SHUNT, AND DIFFUSION

Johannes Piiper

Department of Physiology, Max Planck Institute for Experimental Medicine, Göttingen, F.R.G.

Introduction

The analysis of gas exchange in the lungs has advanced importantly in the last decades. Some generally accepted models have been developed, and they have proved to be useful for understanding gas exchange in normal man and animals, and in particular in patients with pulmonary disease. In this report it will be attempted to apply the models developed for pulmonary gas exchange to analysis of O_2 transfer in tissues, with particular reference to skeletal muscle. It will be shown that application of such models is meaningful, enabling us to identify the factors which may be involved in O_2 delivery to skeletal muscle and to estimate their role.

Gas exchange models: tissue vs. lungs

The highest possible gas exchange efficiency is attained by an "ideal lung" in which alveolar ventilation of every lung element is perfectly matched to its perfusion, and shunt and diffusion limitation are absent (see below). In such a lung model the partial pressure of O_2 in the endcapillary (arterial) blood (Pa) is equal to that in alveolar gas (PA): there exists no alveolar-arterial Po_2 difference. Conventionally (after Riley and Cournand, 1951) the following three factors are considered as capable of producing a gas exchange inefficiency measurable as alveolar-arterial Po_2 difference (Fig. 1, upper half):

(1) unequal distribution of alveolar ventilation (VA) to perfusion (Q),
(2) shunt or venous admixture, and
(3) diffusion limitation.

Each of these factors can give rise to a component of the alveolar-arterial Po_2 difference, and then these components add up to the total alveolar-arterial Po_2 difference.

In tissue models, counterparts of the three factors can be easily conceived as proposed by Piiper (1985) (Fig. 1, lower half).

(1) Unequal distribution of blood flow. The simplest reference for blood flow is tissue mass. Thus unequal distribution of blood flow would mean variance of specific blood flow (= blood flow per unit tissue mass). But the proper reference parameter for blood flow distribution evidently is the local O_2 requirement. A local reduction of blood flow, if of sufficient magnitude, leads to local anoxia, to a decrease of local O_2 uptake, and thereby to a reduction of total O_2 uptake below the total

O_2 requirement. This may occur when the overall blood flow is kept constant because increase of blood flow to other areas would not help to relieve the local deficiency of O_2 supply.

(2) Shunt. Although true anatomical shunts or arterio-venous anastomoses apparently do not occur in muscle, a functional shunting of blood through vessels of relatively large diameter and short length, and thereby much reduced gas exchange capacity, is feasible. Another particular case is the arterio-venous diffusion shunt by diffusion of O_2 from arterial to venous vessels (cf. Piiper, 1988). Such mechanisms contribute to decreased O_2 extraction (= arterial-venous O_2 content difference divided by arterial O_2 content) by admixture of arterialized blood to tissue effluent venous blood.

(3) Diffusison limitation. Also diffusion-limited gas transfer between capillary blood and tissue may lead to decreased O_2 extraction (see below). It is evident that a functional shunt may be considered as due to local diffusion limitation. Indeed, it is difficult and in many cases arbitrary to differentiate between shunt and local diffusion limitation. Transition between shunt and diffusion limitation has also been considered in lungs (Piiper, 1961).

For analysis of pulmonary gas exchange, methods which are specific for one of the factors have been developed. The ventilation/perfusion inhomogeneity can be determined by the method of intravenous infusion of inert gases and measurement of their concentrations in arterial and mixed venous blood (Wagner et al., 1974). Shunt can be estimated from alveolar-arterial P_{O_2} differences in hyperoxia, and diffusion limitation (at least in its equally distributed form) from measurements of P_{O_2} performed in hypoxia (Riley and Cournand, 1951). But all these methods are not applicable to tissues. To estimate the roles of diffusion, shunt and blood flow distribution in O_2 supply to tissues other methods must be used.

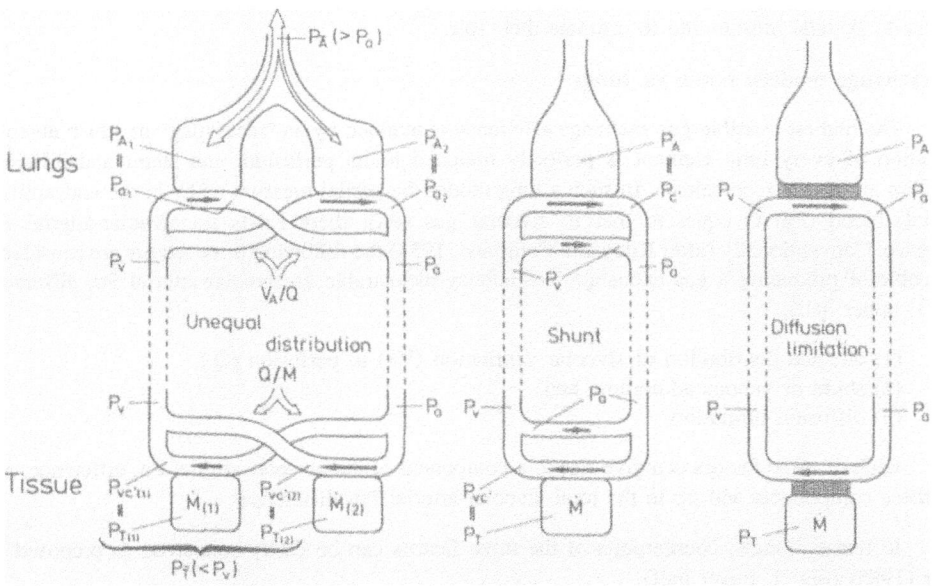

Fig. 1. Mechanisms producing O_2 transfer inefficiency in lungs (upper part) and in tissues (lower part): unequal distribution of blood flow (left), shunt (middle) and diffusion limitation (right). P, O_2 partial pressures; M, O_2 consumption; Q, blood flow; VA, alveolar ventilation; A, a, c', v, vc' and T refer to alveolar, arterial, end-pulmonary capillary, venous, end-systemic capillary and tissue, respectively.

Diffusion limitation of O_2 uptake in skeletal muscle

The fact that muscle venous Po_2 and O_2 content do not approach zero even in conditions of maximal O_2 uptake has been attributed to diffusion limitation (Mercker et al., 1949; Stainsby and Otis, 1964; Stainsby et al., 1988; Hogan et al., 1988, 1989; Roca et al., 1989). In most cases models based on the Krogh cylinder model (Krogh, 1919) have been used to quantify diffusion limitation.

In the Krogh cylinder model (Fig. 2A) there is an axial arterio-venous Po_2 gradient determined by blood flow and O_2 carriage properties of blood, and a radial gradient due to diffusion which is equal in all segments since the O_2 transfer determined by local O_2 consumption is assumed to be equal. At the periphery of the venous end of the cylinder, tissue Po_2 may be assumed to approach zero in critical O_2 supply conditions. In this case the total radial Po_2 drop, ΔPo_2 is equal to venous Po_2.

According to the well-known Krogh-Erlang equation (Krogh, 1919), ΔPo_2 is determined by the specific O_2 consumption, Mo_2/V (O_2 consumption/tissue volume), the Krogh diffusion constant for O_2, Ko_2, the capillary radius, r and the cylinder radius R:

$$\Delta Po_2 = \frac{\dot{M}o_2/V}{4\,Ko_2} \cdot [2R^2 \ln (R/r) - R^2 + r^2] \tag{1}$$

With ΔPo_2 estimated as Pv_{O_2}, measured Mo_2 and V, and assumed Ko_2 and r, the cylinder radius R and therefrom the capillary density can be estimated (Stainsby and Otis, 1964).

The Krogh model may be further simplified. Because of the geometry, most part of the resistance to O_2 uptake resides in the central, pericapillary region. This central resistance is further increased by diffusion and reaction resistance within red cells and plasma, and the resistance in the muscle fibers is reduced by facilitated transport by myoglobin (cf. Groebe and Thews, 1990). Thus the Po_2 drop (ΔPo_2) may be considered to be limited to the intracapillary and pericapillary zone and tissue Po_2 may be viewed as constant within a cylinder segment. According to both models, the O_2 uptake is expected to be proportional to the end-capillary or venous Po_2 in critical O_2 supply conditions.

In rhythmically stimulated dog gastrocnemius muscle, Hogan et al. (1988) found a remarkably good proportionality between O_2 uptake and venous Po_2 as the O_2 delivery in muscle

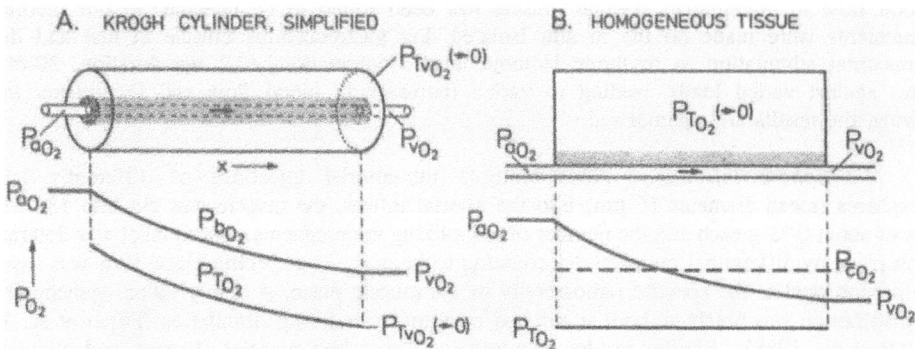

Fig. 2. Models for analysis of diffusion limitation to O_2 uptake in tissue. In both models (A and B) Po_2 in blood drops in the capillary from the arterial to the venous value. In model A, tissue Po_2 drops in parallel with capillary Po_2, but in model B tissue Po_2 is constant. In conditions of critical O_2 supply, Ptv_{O_2} and Pt_{O_2} approach zero.

stimulated to maximum O_2 uptake was progressively reduced by reduction of arterial Po_2 at constant blood flow. There was only a slight increase in the O_2 uptake/Pv_{O_2} ratio as hypoxia progressed. A similar result had been obtained by Stainsby and Otis (1964), but with more pronounced increase of the ratio O_2 uptake/venous Po_2 with decreasing arterial Po_2. Since these results were expected on the basis of the models, the limitation in O_2 extraction was attributed to diffusion, with diffusion conditions remaining constant or slightly improving with progressing hypoxia, to compensate for the decreasing O_2 delivery.

The simplification is further advanced in the homogeneous model (Fig. 2B) in which the tissue is assumed to have everywhere the same Po_2. This may come about by axial facilitated transport of O_2 and by interaction between distributed capillaries. Indeed, Gayeski and Honig (1986, 1988) found by cryophotometric determination of myoglobin O_2 saturation only small axial and radial and longitudinal Po_2 variations in dog gracilis muscle fibers. In this simplified model, blood-capillary O_2 transfer is described by the model conventionally used for analysis of alveolar-capillary diffusion in lungs, with tissue Po_2 replacing alveolar Po_2. The Po_2 difference driving the diffusion is in this case the difference between the integrated mean capillary Po_2 and the tissue Po_2. The mean capillary Po_2 may be determined by the Bohr integration technique in the same manner as for pulmonary capillaries (e.g. Riley and Cournand, 1951). Applications of this relationship to the above-mentioned data of Hogan et al. (1988) yields a reasonably proportional relationship also between mean capillary Po_2 and O_2 uptake.

Some authors (e.g. Mercker et al., 1949) have preferred to use the solid cylinder model (Hill, 1928). In this model O_2 is supplied from the outer surface of the cylinder, and the Po_2 drop from the surface to the center, ΔPo_2, is given by the relationship:

$$\Delta Po_2 = \frac{\dot{M}o_2/V}{4 \ Ko_2} \cdot R^2 \qquad (2)$$

The cylinder may represent a muscle fiber whose radius, R, can be calculated from experimentally determined ΔPo_2, $\dot{M}o_2/V$ and assumed Ko_2. Also in this case O_2 uptake is proportional to ΔPo_2, like in the Krogh model, although the shape of Po_2 gradients differs considerably between the models (cf. Piiper and Scheid, 1986).

Perfusion heterogeneity and shunt: experimental evidence

In a number of studies using various preparations and methods (cf. Piiper, 1990) distribution of blood flow in mammalian skeletal muscle has been found to be unequal. In our laboratory, measurements were made on the in situ isolated dog gastrocnemius muscle at rest and during supramaximal stimulation to rhythmic isotonic tetanic contractions (0.2 sec duration, 30-60 per minute) against varied loads, leading to varied increases in blood flow and O_2 uptake. In the following, the results are summarized.

Microsphere injection. After multiple intraarterial injections of differently labeled microspheres (mean diameter 15 μm) into the arterial inflow, the muscle was cut into 180 to 250 pieces of about 0.75 g each and the number of embolizing microspheres of each label was determined in each piece by differential radioactivity counting techniques. The specific blood flow was assumed to be proportional to the specific radioactivity of the muscle piece. A highly inhomogeneous blood flow distribution was obtained, both at rest and even more so during stimulation (Piiper et al., 1985; Marconi et al., 1988). Similar results were obtained on rabbit muscles (Iversen and Nicolaysen, 1989a; Iversen et al., 1989).

Clearance of locally injected xenon. A small amount of saline containing ^{133}Xe was injected into the muscle at varied sites, and the washout clearance was measured by a counter placed on the injection site. The specific blood flow was calculated from the radioactive decay and the blood/tissue partition coefficient. The ratio of the blood flow calculated from Xe clearance to the directly measured venous outflow averaged 57% and scattered widely (SD/mean = 33%) (Cerretelli et al., 1984). The large scatter may in part be due to methodological problems, but mainly it appears to indicate perfusion inhomogeneity. The low blood flow from Xe clearance may be interpreted as due to a large fraction of functional shunt flow which may have been produced by short residence times, diffusion limitation or veno-arterial back diffusion.

Washout of inert gases. After equilibration of the muscle preparation with one or several inert gases (He, Ar, CH_4, SF_6), washout was performed by switching the perfusion to blood devoid of the inert gas(es). The decay of the inert gas concentration in venous outflow was determined by gas chromatography on spot samples. The washout was markedly non-monoexponential both at rest and during stimulation (Piiper and Meyer, 1984). The simplest explanation is unequal distribution of blood flow to tissue volume. Moreover, a mass balance analysis based on estimated values of the tissue/blood partition coefficient, about 30% of the total blood flow had to be attributed to shunt. Comparison of the washout of test gases with differing diffusivity suggested a role of counter-current veno-arterial back diffusion (Piiper, 1988).

Validity and limitations

All three methods indicate unequal distribution of blood flow to tissue volume in the muscle. Quantitatively, the perfusion inhomogeneity derived from microsphere injection and inert gas washout have been found to be in reasonable agreement (Piiper et al., 1985). Moreover, a functional shunt results from both local xenon clearance and inert gas washout from the whole muscle. Each individual method has its inherent problems and uncertainties. Therefore, the order-of-maganitude agreement in the results obtained by the different methods is of importance.

The xenon clearance method and the microsphere injection method could also be applied to intact running dogs, i.e. to muscles contracting in physiological conditions. The agreement with the results in isolated, supramaximally stimulated muscle was remarkably good (Pendergast et al., 1985). This agreement indicates that the inhomogeneity and functional shunt found in the isolated muscle preparation are not artefactual and are not related to different contraction patterns and mechanics.

For O_2 supply the proper reference for blood flow distribution is O_2 requirement, not tissue volume. It may well be that the O_2 requirement is also unequally distributed within a muscle, and the blood flow may be distributed according to the O_2 requirement. Iversen and Nicolaysen (1989b) measured the distribution of glucose uptake (by the deoxyglucose method) along with microsphere distribution in rabbit limb muscles, but found no correlation between these variables. Thus the use of tissue volume as reference for blood flow distribution may be used as a reasonable approximation in the analysis of O_2 supply.

Unequal distribution of blood flow and shunt: effects on O_2 supply

Calculations were performed on simple models with unequal distribution of blood flow and with shunt, but without diffusion limitation, to show the pattern of O_2 supply limitation as contrasted to the ideal model, i.e. without flow inequality and without shunt. The models were roughly matched to the experimental results. Thus a distribution of blood flow to three compartments was assumed in such a manner that the specific perfusions stood in the ratio 9/3/1. Alternatively, a shunt flow of 33% was assumed. The overall blood flow, the arterial O_2 content and the O_2 requirement were varied, the O_2 uptake (O_2 consumption) and the venous O_2 content were calculated. Examples of calculated results are shown in Fig. 3.

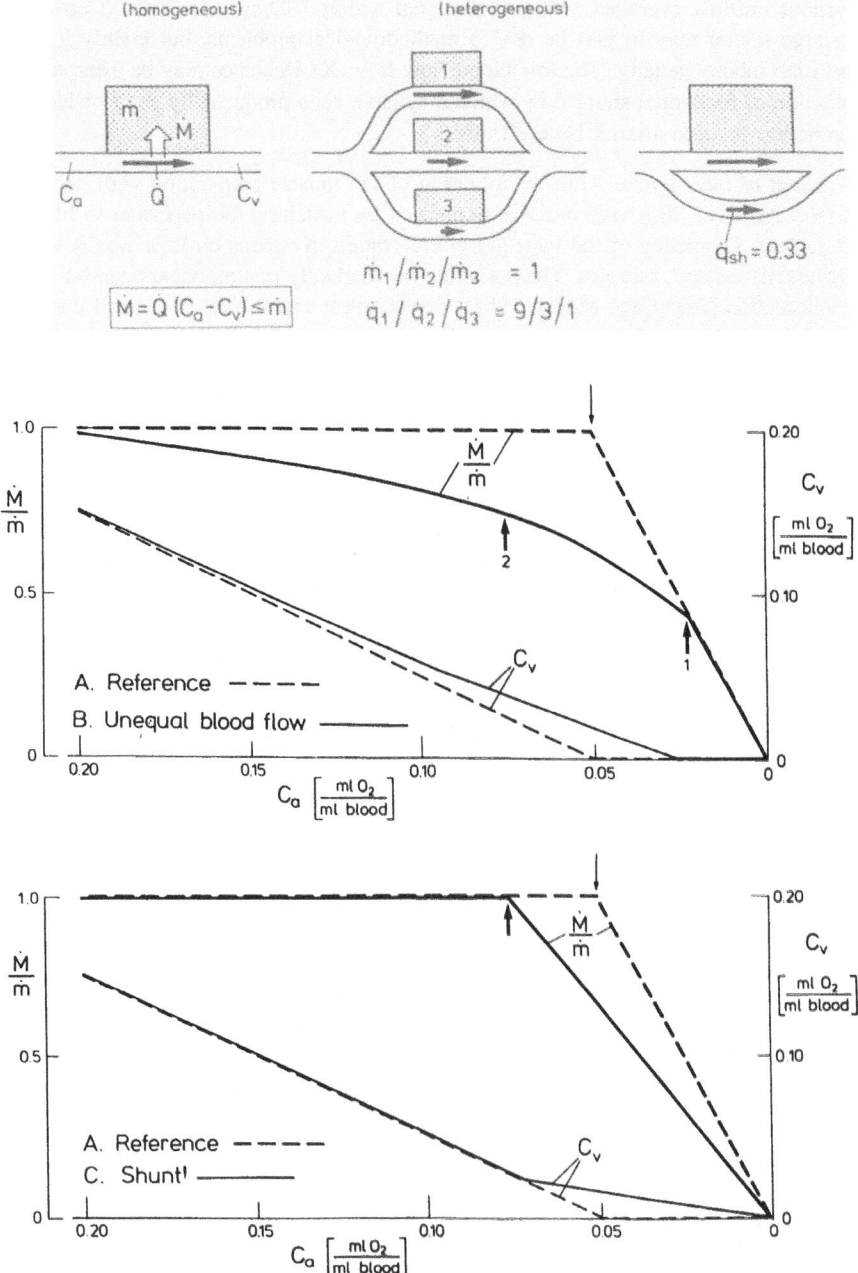

Fig. 3. Models for perfusion-limited O_2 supply and their behavior with decreasing arterial O_2 content (Ca) at constant blood flow (Q) and O_2 requirement (m). Venous O_2 content (Cv) and the O_2 uptake/O_2 requirement ratio (M/m) are shown. The behavior of the reference (homogeneous) model is represented by dashed lines. Arrows mark onset of critical O_2 supply conditions.

In the reference model (homogeneous), as the O_2 delivery (= blood flow x arterial O_2 content) is reduced or the O_2 requirement is increased, the O_2 uptake is equal to O_2 requirement and then starts falling as the ratio O_2 delivery/O_2 requirement ratio drops below unity and venous O_2 reaches zero (critical point). In the model with unequal blood flow distribution, the ratio O_2 uptake/O_2 requirement falls continuously, together with venous O_2 content, as the ratio O_2 delivery/O_2-requirement is reduced. Instead of a critical point, there is a broad critical range. The shunt model behaves in part similarly to the homogeneous model, in part like the distributed flow model.

Of particular interest is the behavior with decreasing arterial O_2 content (Fig. 3). With both unequal blood flow and shunt, both O_2 uptake and venous O_2 content decrease continuously as the arterial O_2 content is reduced. In the case of unequally distributed blood flow, this behaviour is produced by successive attainment of the critical O_2 supply conditions in the individual compartments, as shown by arrows in Fig. 3. With constant shunt, venous O_2 content is proportional to arterial O_2 content when O_2 extraction by tissue is complete. A similar relationship was experimentally found in muscle preparations stimulated to maximum O_2 uptake while reducing arterial O_2 saturation (Stainsby and Otis, 1964; Hogan et al., 1988). This behavior was interpreted as resulting from diffusion limitation, but it may well have been, at least in part, due to unequal blood flow distribution and/or shunt. For a distinction between the roles of blood flow distribution, shunt and diffusion limitation, accurate measurements by simultaneous application of various methods are required.

Summary

By injection of embolizing microspheres, by local radioactive xenon clearance and by inert gas washout in resting and stimulated gastrocnemius dog preparation, experimental evidence for unequal blood flow distribution and for shunt flow has been provided. Model calculations show that in some respect unequal blood flow and shunt produce effects predicted for a homogeneous model with diffusion limitation of O_2 supply. This finding must be taken into account when the role of diffusion limitation to O_2 supply is to be ascertained.

References

Cerretelli, P., C. Marconi, D. Pendergast, M. Meyer, N. Heisler and J. Piiper, 1984, Blood flow in exercising muscles by xenon clearance and by microsphere trapping, *J. Appl. Physiol.*, 56: 24-30.

Gayeski, T.E.J. and C.R. Honig, 1986, O_2 gradients from sarcolemma to cell interior in red muscle at maximal Vo_2, *Am. J. Physiol.*, 259 (*Heart Circ. Physiol.* 20): H 789-H799.

Gayeski, T.E.J. and C.R. Honig, 1988, Intracellular Po_2 in long axis of individual fibers in working dog gracilis muscle, *Am. J. Physiol.*, 254 (*Heart Circ. Physiol.* 23): H 1179-H 1186.

Groebe, K. and G. Thews, 1990, Role of geometry and anisotropic diffusion for modelling Po_2 profiles in working red muscle, *Respir. Physiol.*, 79: 255-278.

Hill, A.V., 1928, The diffusion of oxygen and lactic acid through tissues, *Proc. Royal Soc. London (Biol.)*, 104: 39-96.

Hogan, M.C., J. Roca, P.D. Wagner and J.B. West, 1988, Limitation of maximal O_2 uptake and performance by acute hypoxia in dog muscle in situ, *J. Appl. Physiol.*, 65: 815-821.

Hogan, M.C., J. Roca, J.B. West and P.D. Wagner, 1989, Dissociation of maximal O_2 uptake from O_2 delivery in canine gastrocnemius in situ, *J. Appl. Physiol.*, 66: 1219-1226.

Iversen, P.O. And G. Nicolaysen, 1989a, Heterogeneous blood flow distribution within single skeletal muscles of the rabbit: role of vasomotion, sympathetic nerve activity and effect of vasodilation, *Acta Physiol. Scand.*, 137: 125-133.

Iversen, P.O. and G. Nicolaysen (1989b), Is regional blood flow correlated to regional glucose uptake within single rabbit skeletal muscles? *Proc. IUPS, 17: 168* (Abstracts; International Congress of Physiological Sciences, Helsinki).

Iversen, P.O., M. Standa and G. Nicolaysen, 1989, Marked regional heterogeneity in blood flow within skeletal muscle at rest and during exercise hyperemia in the rabbit, *Acta Physiol. Scand.*, 136: 17-28.

Krogh, A., 1919, The number and distribution of capillaries in muscles with calculations of oxygen pressure head necessary for supplying the tissue, *J. Physiol. (London)*, 52: 409-415.

Marconi, C., N. Heisler, M. Meyer, H. Weitz, D.R. Pendergast, P. Cerretelli and J. Piiper, 1988, Blood flow distribution and its temporal variability in stimulated dog gastrocnemius muscle, *Respir. Physiol*, 74: 1.14.

Mercker, H., B. Ochwadt and W. Schoedel, 1949, Der Einfluss der Erregungsfrequenz und der Belastung auf Durchblutung und Sauerstoffaufnahme des Muskels, *Pflügers Arch.*, 251: 73-82.

Pendergast, D.R., J.A. Krasney, A. Ellis, B. McDonald, C. Marconi and P. Cerretelli, 1985, Cardiac output and muscle blood flow in exercising dogs, *Respir. Physiol.*, 61: 317-326.

Piiper, J., 1961, Unequal distribution of pulmonary diffusing capacity and the alveolar-arterial Po_2 difference: theory, *J. Appl. Physiol.*, 16: 493-498.

Piiper, J., 1985, Mechanisms of functional shunting in mammalian skeletal muscle, in: "Cardiovascular Shunts" (Alfred Benzon Symposium 21), K. Johansen and W.W. Burggren, eds., Munksgaard, Copenhagen, pp. 467-485.

Piiper, J., 1988, Role of diffusin shunt in transfer of inert gases and O_2 in muscle, in: Oxygen Transport to Tissue X (Adv. Exp. Med. Biol. 222), M. Mochizuki, C.R. Honig, T. Koyama, T.K. Goldstick and D.F. Bruley, eds., Plenum Press, New York and London, pp. 55-61.

Piiper, J., 1990, Unequal distribution of blood flow in exercising muscle of the dog, *Respir. Physiol.* 80: 129-136.

Piiper, J. and M. Meyer, 1984, Diffusion-perfusion relationship in skeletal muscle: model and experimental evidence from inert gas washout, in: Oxygen Transport to Tissue V (Adv. Exp. Med. Biol. 169), D.W. Lübbers, H. Acker, E. Lehniger-Follert and T.K. Goldstick, eds., Plenum Press, New York and London, pp. 457-466.

Piiper, J. and P. Scheid, 1986, Cross-sectional Po_2 distributions in Krogh cylinder and solid cylinder models, *Respir. Physiol.*, 64: 241-251.

Piiper, J., D.R. Pendergast, C. Marconi, M. Meyer, N. Heisler and P. Cerretelli, 1985, Blood flow distribution in dog gastrocnemius muscle at rest and during stimulation. *J. Appl. Physiol.*, 64: 241-251.

Riley, R.L. and A. Cournand, 1951, Analysis of factors affecting partial pressures of oxygen and carbon dioxide in gas and blood of the lungs: theory. *J. Appl. Physiol.*, 4: 77-101.

Roca, J., M.C. Hogan, D. Story, D.E. Bebout, P. Haab, R. Gonzalez, O. Ueno and P.D. Wagner, 1989, Evidence for tissue diffusion limitation of Vo_2max in normal humans. *J. Appl. Physiol.*, 67: 291-299.

Stainsby, W.N. and A.B. Otis, 1964, Blood flow, oxygen tension, oxygen uptake, and oxygen transport in skeletal muscle. *Am. J. Physiol.*, 206: 858-866.

Stainsby, W.N., B. Snyder and H.G. Welch, 1988, A pictographic essay on blood and tissue oxygen transport, *Med. Sci. Sports Exercise*, 20: 213-221.

Wagner, P.D., H.A. Saltzman and J.B. West, 1974, Measurement of continuous distributions of ventilation-perfusion ratios: theory, *J. Appl. Physiol.* 36: 588-599.

THE HALDANE EFFECT OF RABBIT BLOOD UNDER DIFFERENT ACID-BASE CONDITIONS

H. Kiwull-Schöne, F. Werkmeister and P. Kiwull

Department of Physiology
Ruhr-University
D-4630 Bochum, FRG

INTRODUCTION

The binding of protons during deoxygenation of hemoglobin (Hb), i.e. the Haldane effect (HE), can be quantified by the difference in plasma pH of oxygenated and deoxygenated blood. For human blood, detailed data concerning the HE are available in a wide range of respiratory and metabolic acid-base conditions (v. Mengden et al., 1969; Siggaard-Andersen, 1974).

Furthermore, by the former authors was shown that the HE-induced pH-difference was linearily related to the logarithm of the bicarbonate concentration ($lgHCO_3^-$), at least within the physiological acid-base range. Hence it follows an empirical correction formula for the pH-value under conditions of incomplete O_2-Hb saturation, which has to be considered e.g. when using equilibration methods (Astrup and Schrøder, 1956) for indirect determination of the CO_2 partial pressure in the blood.

Since there are, however, marked differences between species in the O_2-Hb affinity not only under standard conditions but also as consequences of the Bohr-effect (Kiwull-Schöne et al., 1987), it has to be questioned whether Haldane effect data established so far in human blood can be transferred quantitatively to laboratory animals. Thus, we investigated the HE in rabbit blood (partly also in cat blood) for different shifts of acid-base or CO_2 and compared the results with those reported for dogs (Reeves et al., 1982) and humans (v. Mengden et al., 1969; Siggaard-Andersen, 1974).

METHODS

Blood sampling. The measurements were carried out in arterial blood samples (approximately 25 ml) from 17 air-breathing rabbits, partly awake and partly anaesthetized by pentobarbital sodium for other experimental purpose. During general anaesthesia, the blood was taken from a permanent catheter placed into the right femoral artery, whereas in the awake rabbits the caudal auricular artery was punctured under local anaesthesia of the skin. In order to prevent blood coagulation, per ml blood were added either 1mg of pure ethylenediamine tetraacetic acid (EDTA) or about 10 μl of a 10% solution of potassium bicarbonate and EDTA. These mixtures were kept on ice before measurements. The initial standard bicarbonate concentrations were corrected to about 24 mM by addition of a 0.5 M NaHCO₃ solution, in order to achieve a pH of 7.4 at 5.3 kPa (40 mmHg) PCO₂ and 38°C for standard conditions in each blood sample.

Measurements. Plasma lactate concentration was determined photometrically (Hitachi) by means of the enzymatic UV-method (Boehringer biochemica test combination), the hemoglobin concentration (Hb) was determined photometrically as cyanohemiglobin (Merckotest (R)), and the hematocrit value (Hct) by centrifugation. Plasma pH, PCO₂ and bicarbonate concentrations were measured by the equilibration technique (Astrup and Schrøder, 1956), using a microtonometer (Radiometer), a pH-meter (Knick) and two precision gas mixing pumps (Wösthoff).

Experimental protocol. After correction to standard conditions, the blood samples were divided into several 2 ml-portions and again stored in ice before further treatment. In order to achieve different levels of base excess (BE), these portions were shortly centrifuged and to the supernatant plasma were added 50 μl either of lactic acid in different concentrations (0.2, 0.4, 0.6 M) or of sodium bicarbonate (0.2 M) or of sodium chloride (0.3 M). Subsequently, the treated samples were mixed carefully and small amounts (0.1 ml) were equilibrated by microtonometry in random sequence by 4% and 8% CO₂ in oxygen or nitrogen to get 100% or 0% O₂-Hb saturation, respectively. To exclude a time-dependent acidotic pH-shift due to anaerobic lactic acid production, paired samples were equilibrated simultaneously with oxygen or nitrogen for 5 min each at the same CO₂ partial pressure. Thus, the resulting pH-difference (△pH) at any given PCO₂ is a quantitative measure of the Haldane effect. In this way, from 340 samples 1360 single pH-values were determined and averaged under 10 different respiratory and metabolic acid-base conditions for oxygenated and deoxygenated Hb (Fig. 1).

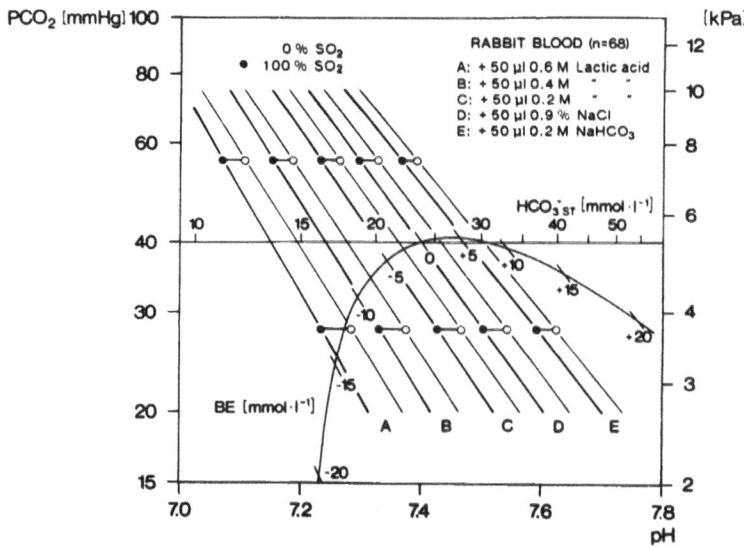

Fig. 1. Dependence of plasma pH on Hb-oxygenation in rabbit blood under different respiratory and metabolic acid-base conditions.
In the lgPCO₂/pH-diagram, the pH-difference between the buffer lines of deoxygenated and oxygenated blood is equivalent to the HALDANE effect. Each value is the mean of 68 single measurements in 17 blood samples.

Table 1. HALDANE effect induced pH-differences, standard bicarbonate
concentrations and slopes of the buffer lines (Fig. 1) for
different levels of base excess in rabbit blood (Means $\pm$SEM).

BE (mM)	$pH_{deox} - pH_{ox}$ PCO_2 = 5.3 kPa	$HCO_3^-{}_{st}$ (mmol·l^{-1})		$\triangle lgPCO_2/\triangle pH$	
		100% SO_2	0% SO_2	100% SO_2	0% SO_2
+ 5	0.028 $\pm$0.002	28.4 $\pm$0.3	30.3 $\pm$0.4	-1.36 $\pm$0.02	-1.32 $\pm$0.02
0	0.036 $\pm$0.003	23.6 $\pm$0.2	25.7 $\pm$0.2	-1.46 $\pm$0.02	-1.41 $\pm$0.01
- 5	0.036 $\pm$0.002	20.1 $\pm$0.3	21.9 $\pm$0.3	-1.53 $\pm$0.02	-1.46 $\pm$0.02
-10	0.039 $\pm$0.003	16.4 $\pm$0.4	18.0 $\pm$0.5	-1.69 $\pm$0.03	-1.59 $\pm$0.03
-15	0.045 $\pm$0.003	13.4 $\pm$0.4	14.8 $\pm$0.5	-1.87 $\pm$0.06	-1.73 $\pm$0.05

RESULTS

1. The HALDANE effect under standard conditions

As can be seen from Fig. 1, the Haldane effect (HE) was quantified
for rabbits by the difference in plasma pH of oxygenated and deoxygenated
blood ($\triangle pH = pH_{deox} - pH_{ox}$). Under standard blood gas (PCO_2 = 5.3 kPa =
40 mmHg) and acid-base conditions (BE = 0 mmol·l^{-1}), and at the average
actual Hb-concentration of 10.6 $\pm$0.6 g/dl, the $\triangle pH$ in rabbit blood was
0.036 $\pm$0.003 pH (mean $\pm$SEM, Table 1). Based on the linear buffer lines in
Fig. 1, it was possible to calculate the HE-induced $\triangle pH$ for any desired
level of PCO_2 in addition to the investigated ranges of base excess. Thus,
the present results in rabbit blood could easily be adapted to the different levels of PCO_2, as being used by other authors (Fig. 2), permitting
better comparison with HE data of humans (Siggaard-Andersen, 1974) and
dogs (Reeves et al., 1982).

Not only PCO_2, but at least two more influences on the HE have to be
considered when performing interspecies comparisons, those of base excess
and of Hb-concentration. Fig. 3 shows the influence of different Hb-con-
centrations, so that even under standard acid-base conditions (HCO_3^- =
24 mM), the HE-induced $\triangle pH$ in human blood recalculated from different
authors differed as much as 0.034 (v.Mengden et al., 1969) and 0.043
(Siggaard-Andersen, 1974). Standardized to 15 g/dl Hb, these values became
0.034 and 0.039, respectively, being distinctly lower than those for
rabbits, cats and dogs, between 0.043 and 0.045 under the same conditions
(Fig. 4).

It is evident that even under standard conditions there are considerable differences of the Haldane effect between species, so that human
blood data should not simply be transferred to laboratory animals. In
particular, the appropriate species-related Haldane-correction would gain
special importance when using the equilibration (Astrup) method at incomplete O_2-Hb-saturation for indirect determination of PCO_2 (see below).

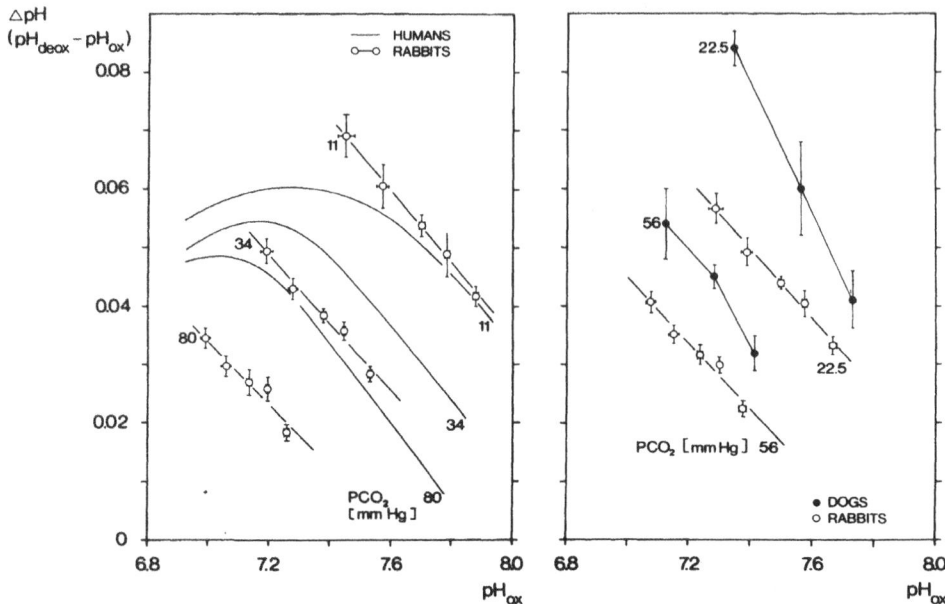

Fig. 2. The HALDANE effect as a function of the pH in fully oxygenated blood.
Left diagram: Comparison of measured values (means ±SEM) in rabbit blood with differential titration curves in human blood (Siggaard-Anderson et al., 1974) at the same levels of PCO_2.
Right diagram: Comparison of mean values ±SEM in rabbit and dog blood (Reeves et al., 1982) at the same level of PCO_2.

2. The HALDANE effect under different acid-base conditions

As can be seen from Fig. 2, the Haldane effect was inversely related to PCO_2 and base excess, increasing with respiratory alkalosis and metabolic acidosis, while decreasing with respiratory acidosis and metabolic alkalosis. Due to the cellular CO_2-buffering of the blood, respiratory acidosis is accompanied by an increase in bicarbonate in virtually the same way as metabolic alkalosis. By inference, respiratory alkalosis and metabolic acidosis both are accompanied by a decrease in HCO_3^-. Based on these considerations, v. Mengden et al., 1969, proposed a valuable parameter reduction by means of the Henderson-Hasselbalch-equation, substituting PCO_2 and pH by HCO_3^-. The linear regression analysis of the HE-induced $\triangle pH$ as a function of $lgHCO_3^-$ yielded rather high coefficients of correlation (Fig. 3, Table 2). Since the actual Hb-concentration was also included in the functional description of the resulting regression lines (see below), the HE-values could easily be adapted to any Hb-level.

Fig. 3 shows the linear regression analysis of HE data in human blood reported by different research groups. The $\triangle pH$-values re-evaluated from the data of v. Mengden et al., 1969, are based on directly measured differences between PCO_2-equilibration lines at either 100% or 0% O_2-Hb saturation, respectively. The $\triangle pH$-values re-evaluated from the data of Siggaard-Andersen, 1974, are based on his Haldane coefficients (Fig. 18, p.75) and converted into plasma pH-differences according to the author (considering a factor 0.114 and the Donnan distribution pH-ratio 1/0.77 between plasma and red cells). Both sets of data yield rather high linear

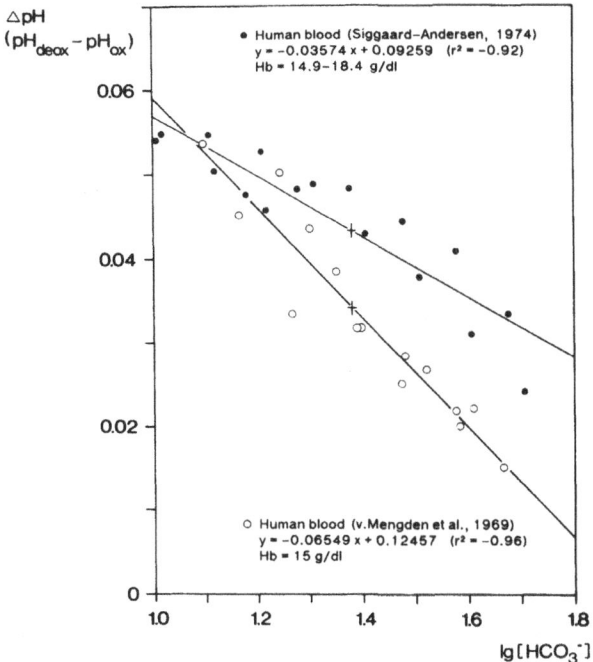

Fig. 3. The HALDANE effect as a function of
plasma bicarbonate concentration.
Linear regression analysis of measured
values in human blood, according to different authors (v.Mengden et al., 1969;
Siggaard-Andersen, 1974).
Note different results for $\triangle$pH even under
standard conditions (+).

correlation coefficients when plotted against lg HCO$_3$$^-$. There are, however, striking differences with regard to slopes and standard values.

Accordingly, the left panel of Fig. 4 additionally shows the linear
regression analysis of the present data determined for rabbit blood, as
well as those for cats and dogs. Based on 10 averaged pairs of $\triangle$pH (= y)
and lgHCO$_3$$^-$ (= x), for rabbit blood, a linear relationship y = ax + b
resulted, the slope being a = –0.06332 and the intercept b = 0.11897. From
these regression coefficients the values of B and A shown by Table 2 were
derived, in order to include the Hb-concentration (g/dl) into the calculation. Thereby, the two factors are defined as B = b/a and A = Hb/a,
yielding the more general expression proposed by v.Mengden et al., 1969

$$\triangle pH = (B - lgHCO_3^-) \ Hb/A.$$

By using this formula, it is possible to calculate the Haldane effect as a
function of bicarbonate for any actual or normalized Hb-concentration
(Fig. 4). As a result, for a given Hb the HE-values in human blood were
generally lower than those in the other species investigated. For quantitative comparison, Table 2 shows the values of B and A calculated by
regression analysis of n pairs of averaged data for different species.

Table 2. Factors describing the relationship $\triangle pH = (B - lgHCO_3^-)\ Hb/A$ between HALDANE effect, acid-base condition and Hb-concentration for different species.

	B	A	r^2	n	Reference
Humans	1.902	225	-0.96	15	v.Mengden et al., 1969
Rabbits	1.879	172	-0.97	10	present data
Cats	1.853	158	-0.69	12	Kiwull-Schöne et al., 1987
Dogs	1.826	150	-0.98	6	Reeves et al., 1982

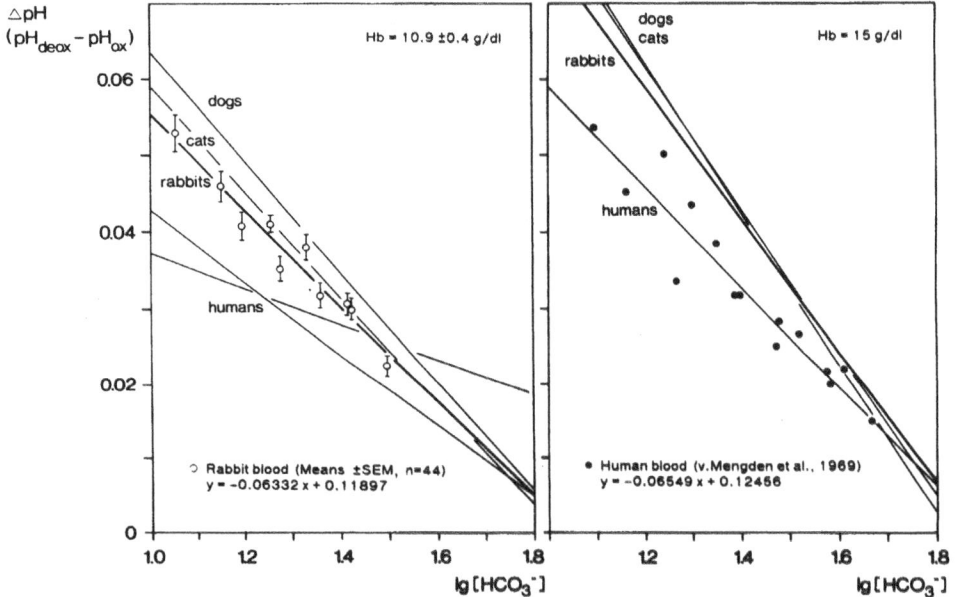

Fig. 4. The HALDANE effect as a function of the plasma bicarbonate concentration.
Left diagram: Linear regression analysis of measured values (means ±SEM) in rabbit blood (r^2 = -0.97, thick line). For comparison are given the regression lines of human blood data (same as in Fig. 3), as well as those from cats (Kiwull-Schöne et al., 1987) and dogs (Reeves et al., 1982), standardized each to the same Hb-concentration as in rabbit blood.
Right diagram: Linear regression analysis of the measured values in human blood (v. Mengden et al., 1969). For comparison are given the regression lines of rabbit, cat and dog blood, standardized each to the same Hb-concentration as in human blood.

DISCUSSION

The present data for rabbits and those for cats (Kiwull-Schöne et al., 1987) and dogs (Reeves et al., 1982), show a generally greater Haldane effect than those for humans (v.Mengden et al., 1969; Siggaard-Andersen, 1974), not only under standard conditions but also during respiratory and metabolic acid-base changes (Fig. 4). The data for human blood reported by the authors just mentioned, slightly differ between $\triangle pH = 0.034$ and 0.039 even at normal PCO_2 and base excess (BE) and standardized to 15 g/dl Hb. According to the calculations of Siggaard-Andersen, 1974, these pH-differences correspond to Haldane factors (f_H) of -0.23 and -0.26, respectively, whereby the former value (v.Mengden et al., 1969) represents the lowest one within the spectrum reported for human blood. There are, however, two more reports in close agreement, Müller et al., 1988, finding $\triangle pH = 0.039$ (i.e. $f_H = -0.26$) and Loeppky et al., 1983, reporting $f_H = -0.28$ at 17.2 g/dl Hb, i.e. $f_H = -0.24$ at 15 g/dl Hb. The only value considerably higher ($f_H = -0.40 \pm 0.025$) was found by Zwart et al., 1984, but, for theoretical reasons (see below), is not compatible with the rather low fixed acid Bohr factor measured in the same laboratory (Kwant et al., 1988).

On the other hand, the corresponding pH-differences of 0.043 for rabbits and of 0.045 for cats and dogs are significantly higher than in humans, the Haldane factors consequently ranging about -0.30. Therefore, differences in the Haldane effect at least between primates and non-primates cannot be excluded entirely. Moreover, they are rather likely when regarding analogously the marked species differences reported for the Bohr-effect (discussed by Kiwull-Schöne et al., 1987). By physico-chemical reasons, the quantitative parameters of the oxygen-linked proton release (Haldane effect) and of the proton-linked O_2 affinity (fixed acid Bohr effect) should be closely related, and under certain conditions even be identical (Siggaard-Andersen, 1974). Thus, it is reasonable to perform interspecies comparisons of proton Bohr factors in much the same way as of Haldane factors. Interestingly, the standardized proton Bohr factors (normocapnia, half-saturation) for cats, rabbits and dogs range about -0.43, -0.44 and -0.49, respectively (Kiwull-Schöne et al., 1987; Reeves et al., 1982), whereas those for humans under the same conditions again are numerically lower, between -0.36 and -0.40 (see measurements and discussion by Kwant et al., 1988).

The exact knowledge of the Haldane effect is of practical importance in not fully oxygenated blood, when determining the PCO_2 indirectly by the Astrup method. By this method $lgPCO_2$/pH equilibration lines normally are established in fully oxygenated blood, so that only under this condition the PCO_2 can be unequivocally concluded from the actual pH (Astrup and Schrøder, 1956). The alkalinic shift due to deoxygenated Hb in hypoxic blood would thus lead to an underestimation of the PCO_2. The tentative position of the $lgPCO_2$/pH equilibration line under hypoxic conditions can be approximated according to Siggaard-Andersen and Engel, 1960. They proposed a HE-dependent base excess correction, whereby $\triangle BE$ $(mmol \cdot l^{-1})$ was assumed to be numerically identical with the actual Hb-concentration (g/dl) multiplied by 0.3. In case of 12 g/dl Hb, this would mean $\triangle BE = 3.6$ $mmol \cdot l^{-1}$, corresponding to a HE-induced pH-difference of about 0.050 under standard conditions and of about 0.065 during the most severe metabolic acidosis which we investigated (BE = -15 $mmol \cdot l^{-1}$). However, the $\triangle pH$ values derived from the present experimental rabbit blood data correspond to only about 70% (Table 1) of this approximation, which as a whole thus represents an overestimation.

In fact, more recently the same author (Siggaard-Andersen, 1974) found a distinctly smaller Haldane effect, as can be judged from the reported Haldane factor (f_H = −0.31), based on differential titration of human blood. This Haldane factor, when standardized to 15 g/dl Hb, would thus correspond to a $\triangle$ pH of only 0.041 instead of 0.070, i.e. about 60% of the originally proposed correction (Siggaard-Andersen and Engel, 1960). Unfortunately, the broad theoretical understanding of the Haldane effect under different blood gas and acid base conditions derived from the differential titration has never been adapted for practical use with the indirect determination of PCO_2. This, however, was performed by v.Mengden et al., 1969, who measured $lgPCO_2$/pH equilibration curves over a wide range of BE for either fully oxygenated or fully deoxygenated human blood. Due to the fact that the HE-induced $\triangle$ pH was inversely related to both PCO_2 and BE, these authors demonstrated a sufficiently linear inverse relationship between $\triangle$ pH and $lgHCO_3{}^-$. This simple relationship is most useful for an appropriate HE-correction, as long as the quantitative differences among species are thouroughly considered, which are shown by the present study (Fig. 4, Table 2).

Fig. 5 demonstrates the order of magnitude of the errors induced by the Haldane effect when using the Astrup-method. As to be expected, the error increases with the degree of hypoxia. Furthermore, the error is greater in the hypercapnic than in the hypocapnic range, as a consequence of the logarithmic dependence of PCO_2 on pH. Finally, there is a growing error with degree of metabolic acidosis, as can be judged from the influence of the base excess on the Haldane effect (Table 1, Fig. 2).

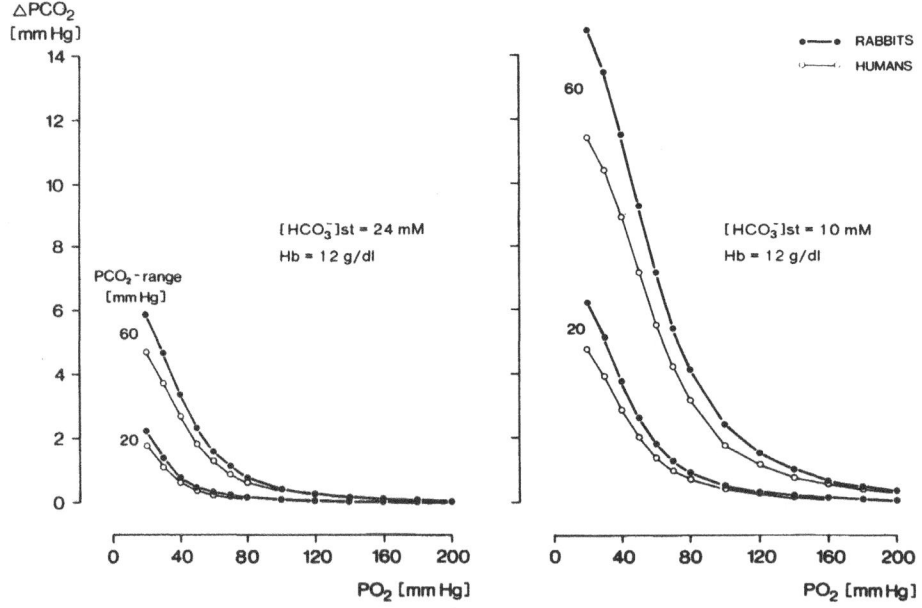

Fig. 5. Errors induced by the HALDANE effect when using the Astrup-method for indirect PCO_2-determination.
Comparison of correction formulas derived either from human blood, according to v. Mengden et al., 1969, or from rabbit blood (for factors see Table 2). The CO_2-difference ($\triangle PCO_2$) as a function of the O_2 partial pressure in different ranges of PCO_2, both under standard metabolic acid-base conditions (left panel) and during strong metabolic acidosis (right panel).

Under most unfavourable conditions, e.g. combined hypercapnia and metabolic acidosis, the error in the indirect PCO_2-determination may reach as much as 15 mmHg (2 kPa). Under these acid-base conditions it is of special importance to use the species-related correction formula (Table 2), considering a higher Haldane effect for rabbits and other common laboratory animals than for humans. The coefficients proposed for humans by v.Mengden et al., 1969, would lead to an underestimation of the indirectly measured PCO_2 values in rabbit blood by about 20 to 25%. In absolute terms, the species-dependent error in the PCO_2-determination would reach up to 3 mmHg (0.4 kPa), which is more than twice as much as the error caused by the inaccuracy of the pH-electrode (Fig. 5, right panel).

SUMMARY

The Haldane effect (HE), i.e. the difference in plasma pH of oxygenated and deoxygenated blood, was investigated in rabbits over a wide range of respiratory (PCO_2 2.7 to 8.0 kPa) and metabolic (BE +5 to -15 mM) acid-base conditions and compared to cats, dogs and humans. Even under standard conditions (PCO_2 = 5.3 kPa, BE = 0 mM) and normalized to the same Hb-concentration, the HE-induced pH-difference was distinctly greater in rabbits, cats and dogs (about 0.045) than in humans (0.034 - 0.039). During respiratory and metabolic acid-base changes, the HE-induced $\triangle$pH was inversely related to PCO_2 and BE. The dependency of the Haldane effect on the acid-base status can be estimated by means of a linear inverse relationship between $\triangle$pH and $lgHCO_3^-$, as originally proposed for human blood by v.Mengden et al., 1969. Correspondingly, the regression analysis of the present experimental Haldane effect data of rabbits, cats and dogs, yielded highly linear correlations as well as characteristic species-related regression coefficients. Additionally considering the influence of the Hb-concentration, a most useful tool for quantitative estimation of the Haldane effect is thus available. This is of practical importance, e.g. when determining the arterial PCO_2 indirectly by the Astrup method. However, Haldane corrections based only on human blood data available so far, would lead to an underestimation of the Haldane effect in common laboratory animals. If the appropriate Haldane shift of pH for a species is not considered, the resulting PCO_2 may be erroneous by up to several 100 Pa, particularly in the range of metabolic acidosis.

ACKNOWLEDGEMENT

We would like to thank Ms. S. Adler for carefully reading the manuscript and, together with Ms. C. Rochna, for her expert help with the drawings.

REFERENCES

Astrup, P., and Schrøder, S., 1956, A simple electrometric technique for the determination of carbon dioxide tension in blood and plasma, total content of carbon dioxide in plasma, and bicarbonate content in "separated" plasma at a fixed carbon dioxide tension (40 mmHg), Scand. J. Clin. Lab. Invest., 8:33.

Kiwull-Schöne, H., Gärtner, B., and Kiwull, P., 1987, The effect of CO_2 and fixed acid on the O_2-Hb affinity of rabbit and cat blood, Pflügers Arch., 408:451.

Kwant, G., Oeseburg, B., Zwart, A., and Zijlstra, W. G., 1988, Human whole-blood O_2 affinity: effect of CO_2, J. Appl. Physiol., 64:2400.

Loeppky, J. A., Luft, U. C., and Fletcher, E. R., 1983, Quantitative
    description of whole blood CO_2 dissociation curve and Haldane effect,
    Respir. Physiol., 51:167.
v.Mengden, H.-J., Schultehinrichs, D., and Thews, G., 1969, Dependence of
    plasma pH on oxygen saturation, Respir. Physiol., 6:151.
Müller, R., Grote, J. und Steinhausen, F., 1988, Die CO_2-Bindungskurve des
    normalen menschlichen Blutes und ihre Beeinflussung durch die Oxyge-
    nation des Hämoglobins, Funktionsanalyse biologischer Systeme, 18:61.
Reeves, R. B., Park, J. S., Lapennas, G. N., and Olszowka, A. J., 1982,
    Oxygen affinity and Bohr coefficients of dog blood, J. Appl.
    Physiol., 53:87.
Siggaard-Andersen, O., 1974, The acid-base status of blood. 4th edn.
    Munksgaard, Copenhagen
Siggaard-Andersen, O., and Engel, K., 1960, A new acid base nomogram,
    Scand. J. Clin. Lab. Invest., 12:177.
Zwart, A., Kwant, G., Oeseburg, B., and Zijlstra, W. G., 1984, Human
    whole-blood oxygen affinity: effect of temperature, J. Appl.
    Physiol., 57:429.

INTERACTION OF BLOOD FLOW, DIFFUSIVE TRANSPORT AND CELL

METABOLISM IN ISOVOLEMIC ANEMIA

C.R. Honig, R.J. Connett and T.E.J. Gayeski

The University of Rochester, School of Medicine and Dentistry
601 Elmwood Avenue
Rochester, NY 14642

A difference in PO_2 (ΔPO_2) between red cells and tissue cells defines a driving force for diffusive O_2 transport. This ΔPO_2 depends on the flux and the over-all tissue conductance for O_2:

$$\dot{V}O_2 = [PcapO_2 - PmbO_2] \times C \qquad\qquad \text{equation 1}$$

In the steady state, $\dot{V}O_2$ (rate of O_2 consumption), equals the transcapillary O_2 flux. $PcapO_2$ can be defined as mean PO_2 in the capillary population. Since diffusive shunting is negligible in red muscle (Honig and Gayeski 1989), effluent venous PO_2 is a lower bound on mean $PcapO_2$. $PmbO_2$ is defined as the PO_2 in equilibrium with myoglobin (Mb). Measurements of $PmbO_2$ at the center of a cell profile, as in the present report, furnish a lower bound on $PmbO_2$. C in equation 1 is a lumped conductance for O_2. This conductance varies inversely and non-linearly with $PmbO_2$.

Recent mathematical models predict that the ΔPO_2 over the short extracellular diffusion path from red cell to sarcolemma is greater than the ΔPO_2 over the long diffusion paths within red myocytes (Federspiel, 1986; Federspiel and Popel, 1986; Groebe, 1990; Hellums, 1977). That prediction is supported by measurements on skeletal and cardiac muscles (Gayeski and Honig, 1986; 1988; Wittenberg and Wittenberg, 1985). Since the extracellular and intracellular fluxes are identical in the steady state, a larger extracellular ΔPO_2 denotes a proportionately lower extracellular conductance for diffusive O_2 transport. This extracellular component of conductance is thought to vary directly with capillary transit time and inversely with O_2 flux density (flux/area) (Honig et al., 1984, 1991; Groebe, 1990).

Isovolemic anemia decreases the aggregate red cell surface area. The associated hyperemia shortens transit times, thereby increasing the O_2 flux per red cell. The result is high O_2 flux density at the capillary "bottleneck". We therefore used isovolemic anemia to further test the concept that the lowest conductance for diffusive O_2 transport is extracellular in working red muscles. To that end we compared normal muscles with anemic ones that had the same $\dot{V}O_2$ and hence the same transcapillary O_2 flux. If current theory is correct the extracellular ΔPO_2 should be greater and intracellular PO_2 should be lower in anemic muscles than in their controls at equal $\dot{V}O_2$.

Oxygen Transport to Tissue XIII, Edited by T.K. Goldstick *et al.*
Plenum Press, New York, 1992

METHODS

Random source dogs were anesthetized with pentobarbital and studied during spontaneous respiration on room air. The gracilis muscle was vascularly isolated, taking care to preserve both sets of supply vessels (Gayeski and Honig, 1986). Muscles were wrapped in saran® and maintained at 37°C. Isometric twitch contraction was induced by stimulating the cut obturator nerve. Gracilis blood flow, the concentration of hemoglobin (Hb), SaO_2 and SvO_2 were determined, and gracilis $\dot{V}O_2$ was calculated from the Fick principle (Gayeski, Connett and Honig, 1987).

In 5 dogs the hematocrit was lowered to about half normal by acute hemorrhage, followed by replacement of an equal volume of plasma expander. Each anemic muscle was paired with a muscle with equal $\dot{V}O_2$ in which the [Hb] was normal. The normal muscles were chosen from among several hundred collected for other purposes and stored in liquid N_2 for retrospective analysis. Criteria for pairing were: 1) $\dot{V}O_2$ of the paired muscles must be within the overall error of the Fick determination. This error was <10% if all the multiplicative errors in flow, [Hb], and Hb saturation were maximal and in the same direction. 2) Arterial O_2 saturation must be within normal limits. Consequently, arterial O_2 content (CaO_2) varied with hematocrit. 3) From the set of normal muscles that met criteria 1 & 2 we chose one that maximized the difference between normal and anemic muscles in respect to O_2 extraction per red cell.

Muscles were frozen in situ by application of a copper block cooled to −196°C in liquid N_2. Samples were trimmed under liquid N_2 and transferred to the cold stage of a microspectrophotometer regulated at −110°C. Mb saturation is stable for at least 5 hours at −110°C.

Mb Saturation and PO_2

Mb saturation was determined with subcellular spatial resolution using a four–wavelength method that takes account of the effect of scattered light (Gayeski, 1981). Comparison with Hb and Mb grundspectra demonstrate that the recorded spectra were free of Hb contamination. The PO_2 in equilibrium with Mb ($PmbO_2$) was calculated from the measured saturation using the Mb P_{50}. Values of this constant determined at 37°C range from 2.5–6.7 torr (Gayeski, 1981); the value determined for canine Mb in our laboratory is 5.3 torr. Mb saturations are reported from which a reader can recalculate $PmbO_2$ for any desired P_{50}. $PmbO_2$ can be interpreted as the PO_2 to which cytochrome a,a_3 is exposed because the ΔPO_2 between cytosol and a mitochondrion is <0.05 torr (Clark et al, 1987).

All Mb saturation determinations were made at the center of a cell profile, near the nadir of the intracellular PO_2 gradient. Each cell chosen was <1000 μm from the surface that had been in contact with the heat sink. At these depths the calculated error in Mb saturation attributable to freezing is <0.1% (Clark and Clark, 1983). Fifty randomly chosen cells were sampled in each muscle, ten from each of the 5 tissue blocks. Results are reported as the 5th percentile, the median, and the interquartile range (25th −75th) percentiles) of cumulative probability distributions.

Metabolic Assays

When spectroscopy was completed the block was removed to an alcohol–dry ice bath. The fascial layer on the muscle surface was removed and the underlying 0.5 mm of muscle was shaved off. Shavings from all blocks scanned for Mb saturation were pooled, weighed, and extracted in perchloric acid. Creatine phosphate (PCr) ATP and Pi were analyzed on the neutralized extract within 4 hours using enzymatic reactions coupled to the production or consumption of reduced pyridine nucleotides.

TABLE 1 - Bulk parameters for anemic muscles and their flux-matched controls.

COL.		1	2	3	4	5	6	7
I.D.	Pair	[Hb] $g \cdot dl^{-1}$	$\dot{V}O_2$	Flow $ml \cdot 100g^{-1} \cdot min^{-1}$	O_2 Offered	SvO_2 %	PvO_2 torr	E $ml \cdot dl^{-1}$
336	1	13.8	8.4	56	10.1	15.9	18	15.0
375		7.0	8.2	120	11.0	27	23	6.8
270	2	15.7	8.1	50	10.6	23.0	18	16.1
321		6.4	8.7	165	12.2	22.4	18	5.4
130	3	14.4	10.2	75	14.6	30.0	21	13.6
381		7.3	10.1	157	15.1	31.1	20	6.4
246	4	17.4	10.6	59	13.2	18.7	15	18.0
380		8.2	9.5	99	10.9	12.1	16	9.6
125	5	17.7	15.5	101	23.5	33.3	24	15.2
288		8.7	16.2	203	22.7	23.1	23	8.2

RESULTS AND DISCUSSION

Role of convective transport

Bulk parameters are compared in Table 1. Blood flow and O_2 flux per red cell were 1.7 to 3-fold higher in anemia. This compensated for low hematocrit, so the O_2 offered to the capillary exchanger, the O_2 extraction per red cell, and the mean end-capillary PO_2 were about the same as in normal control muscles at equal $\dot{V}O_2$; see Table 1, columns 4 and 6. Even though the amount of O_2 convectively transported was normal, $PmbO_2$ was lower in each anemic muscle, indicating less favorable conditions for diffusive transport; see Table 2.

Diffusive Transport from Hb to Mb

Effluent PO_2 (PvO_2) exceeded the 75th percentile for $PmbO_2$ in both the control and anemic muscles, indicating that O_2 in red cells did not

TABLE 2 - Interquartile ranges for Mb saturation or $PmbO_2$ in muscles characterized in Table 1.

I.D.	[Hb] g·dl^{-1}	$\dot{V}O_2$ ml·100g^{-1} min	PERCENTILES OF PROBABILITY						PvO_2–$PmbO_2$ torr
			25th	50th	75th	25th	50th	75th	
			%Mb Saturation			$PmbO_2$, torr			
336	13.8	8.4	42	57	67	3.9	7.0	11	11
375	7.0	8.2	14	27	40	0.9	2.0	3.5	21
270	15.7	8.1	37	45	52	3.1	4.3	5.7	14
321	6.4	8.7	16	21	29	1.0	1.4	2.2	17
130	14.4	10.2	50	64	71	5.3	9.4	13	12
381	7.3	10.1	20	36	59	1.3	3.0	7.6	17
246	17.4	10.6	48	58	67	4.9	7.3	11	8
380	8.2	9.5	14	40	67	0.9	3.4	11	13
125	17.7	15.5	40	56	71	3.5	6.7	13	17
288	8.7	16.2	13	22	38	0.8	1.5	3.2	21

equilibrate with O_2 in myocytes during a capillary transit time. The last column in Table 2 demonstrates that (PvO_2 – $PmbO_2$) was greater in anemia. (PvO_2 – $PmbO_2$) is a lower bound on the average driving force defined in equation 1. Since the total transcapillary flux and PvO_2 were about the same in control and anemic muscles, a larger mean driving force in anemia denotes a lower overall diffusive conductance. Intracellular O_2 conductance was actually greater in anemia because Mb facilitated diffusion is greater at lower $PmbO_2$ (Groebe, 1990; Kreuzer and Hoofd, 1987; Wittenberg and Wittenberg, 1989). Consequently, Tables 1 and 2 confirm theories which predict large PO_2 drops across the extracellular diffusion path (Federspiel, 1986; Federspiel and Popel, 1986; Groebe, 1990; Hellums, 1977). This phenomenon is thought to reflect the high O_2 flux density within a few microns of red cells, and the absence of O_2 carrier in plasma, endothelium, and interstitium.

The role of flux density can be considered from the standpoint of the capillary or the tissue cell. In the former case, shorter red cell transit time increases flux density by increasing O_2 flux per red cell. Transit time depends on volume flow and aggregate capillary cross-sectional area. The latter is nearly maximal at only 10% $\dot{V}O_2$max (Honig, Odoroff, and Frierson, 1982). Consequently, capillary transit time in the anemic muscles should have been shorter than normal by roughly the 2-3 fold higher volume flow; see Table 1. Viewed from the standpoint of a tissue cell, anemia lengthens the plasma gaps between red cells (Honig, unpublished observations). This decreases the effective capillary surface area and increases flux density where the latter is normally greatest.

$PmbO_2$ and Mb-Facilitated Diffusion

The Mb-facilitated O_2 flux increases as $PmbO_2$ falls and the slope of the Mb oxydissociation curve becomes steeper, and more carrier (deoxymyoglobin)

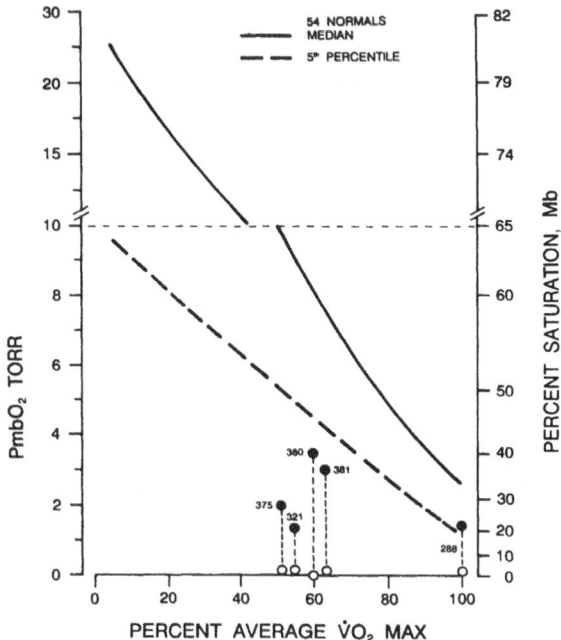

Fig. 1. $PmbO_2$ falls to defend the blood-tissue driving force as $\dot{V}O_2$ increases. Numbered data points represent muscles characterized in Tables. Note scales of ordinates.

becomes available (Kreuzer and Hoofd, 1987; Wittenberg and Wittenberg, 1989). The principal determinants of $PmbO_2$ are the $\dot{V}O_2$, as illustrated in Fig. 1, and O_2 flux per red cell as shown in Tables 1 and 2.

The abscissa is scaled to an average value of $\dot{V}O_2max$ of 16 ml 100 g^{-1} min^{-1} for dog gracilis. The actual $\dot{V}O_2max$ for a particular muscle depends on its mitochondrial capacity; the lower the capacity the greater the O_2 flux per cytochrome at any $\dot{V}O_2$. The scales of the ordinates are compressed above 10 torr and 65% saturation (horizontal dashed line) to accommodate the effect of the hyperbolic Mb oxydissociation curve. The solid and dashed lines represent average data for the 54 normal muscles reported in Connett and Honig, 1989. Median $PmbO_2$ in normal muscles fell from ~38 torr at 5% $\dot{V}O_2max$ to 2 torr at $\dot{V}O_2max$. About 3/4 of this fall was between 5 and 50% $\dot{V}O_2max$, reflecting the shallow slope of the Mb dissociation curve above ~70% saturation. In contrast, the drop in $PmbO_2$ between 50 and 100% $\dot{V}O_2max$ was relatively small due largely to more effective facilitated diffusion at $PmbO_2$ below the Mb P_{50}.

The classical function of Mb-facilitated diffusion is to lower the ΔPO_2 for intracellular O_2 transport (Kreuzer and Hoofd, 1987; Wittenberg and Wittenberg, 1989). In light of the important role of the carrier-free region two additional functions have been proposed: 1) Facilitation should increase the driving force for release of O_2 from Hb by lowering $PmbO_2$ near the sarcolemma (Groebe, 1990; Wittenberg and Wittenberg, 1989). 2) An important transport mechanism was revealed by calculations of Groebe which predict relatively well saturated Mb immediately beneath a red cell where O_2 enters at high flux density. This "shell" of myoplasm is functionally depleted of O_2 carrier (deoxymyoglobin). The carrier-depleted region can be regarded as an extension of the extracellular carrier—free region into the fiber. As $PmbO_2$ falls and the Mb—facilitated flux increases the carrier—depleted region becomes thinner and its conductance increases markedly. This largely accounts for the fact that the over—all O_2 flux can increase 50—100 fold from rest with only a small change in driving force. (Groebe, 1990; Honig, Gayeski, Groebe, 1991). Facilitation is therefore a major reserve of diffusive transport that is recruited automatically as $\dot{V}O_2$ increases and $PmbO_2$ falls. Consequently, the large fall in $PmbO_2$ with $\dot{V}O_2$ shown in Fig. 1 must not be interpreted in a pejorative sense. On the contrary, **low $PmbO_2$ relative to resting muscle is an essential adaptation that allows the transcapillary O_2 flux to match the mitochondrial flux.** The high $\dot{V}O_2$ of red muscle cannot be maintained without it (Cole, 1982).

The dashed curve in Fig. 1 demonstrates that the lowest values of $PmbO_2$ fall proportionately less than the median as $\dot{V}O_2$ increases. For example, at 20% $\dot{V}O_2max$ the difference between the median and 5th percentile was 16 torr; at $\dot{V}O_2max$ the difference was only 1.4 torr. Greater spatial uniformity at low $PmbO_2$ reflects a larger Mb—facilitated O_2 flux, including a flux from myocytes in which $PmbO_2$ is high into those in which it is low (Gayeski and Honig, 1986). In addition, the steep slope of the oxymyoglobin dissociation curve acts as a PO_2 buffer. The filled and open symbols in Fig. 1 denote the median and 5th percentiles respectively for the anemic muscles characterized in Tables 1 and 2. An upper bound on the minimum PO_2 for maximum cytochrome turnover is 0.5 torr in dog gracilis (Gayeski, Connett, and Honig, 1987). Mitochondria that function above this critical PO_2 are considered normoxic; those below 0.5 torr are dysoxic[1]. Median $PmbO_2$ exceeded the dysoxic threshold in every muscle; see Table 2. Nevertheless, each anemic muscle contained a few cells with dysoxic mitochondria, and 3 of the 50 cells sampled in muscle 380 contained some anoxic mitochondria. The anatomical unit of O_2 sufficiency is a mitochondrion, whereas the statistical unit for our probability distributions is a cell. Since all measurement sites were near the nadir of the intracellular PO_2 gradient, the distributions overestimate the fraction of the mitochondrial population that is dysoxic. We conclude that despite anemia and low $PmbO_2$, oxygen per se should not have been a limiting reactant in at least 95% of mitochondria. Nevertheless, low PO_2 relative to flux—matched controls should have had a substantial effect on energy metabolism.

$PmbO_2$ and Cell Metabolism

There are three interactive drives on electron transport and ATP production in mitochondria, namely $[O_2]$ or PO_2, redox, and phosphorylation state. The latter may be defined as the relative fraction of high—energy phosphate bonds in the system (Connett and Honig, 1989). In creatine—containing tissues the phosphorylation state can be measured as the creatine charge, defined as the ratio of phosphocreatine to total creatine (PCr/Crt). If PO_2 falls below the level required to saturate the terminal oxidase at a given O_2 flux **per cytochrome**, the phosphorylation and redox drives on electron transport will be recruited to maintain the electron flux. In effect, changes in phosphorylation and redox states increase the affinity of the terminal oxidase for O_2 as the $[O_2]$ falls (Balaban, 1990; Connett, et. al. 1990; Wilson, Owen, and Erecinska, 1979). Under normal conditions of oxygen delivery the creatine charge decreases linearly from rest to $\dot{V}O_{2max}$ (Connett and Honig, 1989). The creatine charge is coupled to mitochondrial and cytosolic redox at key steps in the Krebs cycle and the glycolytic cascade

---

[1]Dysoxia is an old term resurrected to clarify the meaning of hypoxia (Connett, et al, 1990). Dysoxia is defined as O_2—limited cytochrome turnover. Hypoxia refers to FIO_2 less than normal.

(Balaban, 1990; Connett, et.al, 1990). Consequently, maintenance of $\dot{V}_{O_2}$ at lower cell PO_2 can affect all aspects of energy metabolism. The relative contributions of redox and phosphorylation state depend on the tissue, the flux, mitochondrial capacity, carbon substrate, and other factors.

Role of Mitochondrial Capacity

Table 3 demonstrates that the metabolic effect of a particular $PmbO_2$ depends on the O_2 flux per cytochrome rather than the flux per se. Muscle 321, evidently deconditioned, had an extremely low succinic dehydrogenase (SDH) activity, indicating low mitochrondrial capacity. In contrast, muscle 288 was a well-conditioned labrador with very high SDH activity. Consequently, the turnover rate of each electron transport chain was higher in muscle 321 than in 288, even though its measured $\dot{V}_{O_2}$ was 45% lower. Though both muscles functioned at the same $PmbO_2$, the creatine charge was three-fold higher in muscle 288. Cytosolic redox in red muscle, measured as the lactate to pryuvate ratio (LA/PYR), reflects **aerobic** glycolysis coupled to mitochondrial redox and cytochrome turnover (Connett, 1988; Connett et al, 1990). Note that the LA/PYR ratio was twice as high in muscle 321. Thus, muscle 321 had to recruit a larger fraction of its metabolic reserves to respire at the $PmbO_2$ required to release O_2 from capillaries. The cost of such recruitment is greater susceptibility to fatigue at any $\dot{V}_{O_2}$. Note particularly that metabolic changes thought to be associated with fatigue were observed in muscle 321 even though all its mitochondria should have been functioning above the dysoxic threshold.

TABLE 3

I.D.	$\dot{V}_{O_2}$ $ml \cdot 100g^{-1}$ min	[Hb] $gr \cdot dl^{-1}$	Median $PmbO_2$ torr	$\dfrac{PCr}{Cr_t}$	$\dfrac{Lac}{Pyr}$	SDH nmoles/min per mg Protein
321	8.7	6.4	1.4	0.051	421	51
288	16.2	8.7	1.5	0.154	196	193

Role of System Interactions

The comparisons in Table 3 illustrate the need for multiple measurements of both metabolic and transport parameters for analysis of O_2 - linked stresses. Multivariate analysis is required because energy metabolism and mass transport interact as a system. The fact that transport supports metabolism is obvious. The reverse is also essential: By maintaining $\dot{V}_{O_2}$ over a wide range of $[O_2]$, metabolic

reserves permit $PmbO_2$ to fall in defense of O_2 extraction. These system interactions play an essential role in matching the transcapillary and mitochondrial O_2 fluxes.

SUMMARY

1) High blood flow can compensate for half—normal hematocrit, leaving the rate at which O_2 is offered to the capillaries unchanged. Nevertheless, intracellular PO_2 is lower in anemia, indicating impaired diffusive transport.

2) Anemia increases O_2 flux per red cell and decreases functional capillary surface area. These changes increase flux density and the extracellular component of resistance to diffusive O_2 transport, in accord with current theory (Federspiel and Popel, 1986; Groebe, 1990; Hellums, 1977).

3) Maintenance of diffusive flux in presence of anemia required a larger ΔPO_2 between Hb and Mb, and higher intracellular O_2 conductance brought about by greater Mb—facilitated diffusion. Both compensations depend on lower $PmbO_2$.

4) $PmbO_2$ and creatine charge fall with increasing $\dot{V}O_2$ and ATP demand. These responses, as well as adaptive changes in redox help maintain $\dot{V}O_2$ in the presence of a lower O_2 drive on electron transport.

5) Greater engagement of reserves of both transport and metabolism limits the range of aerobic performance in anemia.

6) The match between the transcapillary and mitochondrial O_2 fluxes depends on interaction of transport and metabolism as a system.

ACKNOWLEDGEMENTS

We thank Dr. Karlfried Groebe for his thoughtful comments on the manuscript. J.L. Frierson, W.R. Reaves and C. Vullo, Jr. furnished skillful technical assistance. Our research is supported by Grants HLB03290 and AR36154 from the National Institutes of Health.

REFERENCES

Balaban, R.S., 1990, Regulation of oxidative phosphorylation in the mammalian cell. _Am. J. Physiol._ 258:C377.

Clark, A., and Clark, P.A.A., 1983, Capture of spatially homogenous chemical reactions in tissue by freezing. _Biophys. J._ 42:25.

Clark, A. Jr., Federspiel, W.J., Clark, P.A.A., and Cokelet, G.R., 1985, Oxygen delivery from red cells. _Biophys. J._ 47:171.

Cole, R.P., 1982, Myoglobin function in exercising skeletal muscle. _Science_, 216:523.

Connett, R.J., 1988, The cytosolic redox is coupled to $\dot{V}O_2$: a working hypothesis. _Adv. Exper. Med. Biol._ 222:133.

Connett, R.J., and Honig, C.R, 1989, Regulation of $\dot{V}O_2$ in red muscle: Do current biochemical hypotheses fit in vivo data? _Am. J. Physiol._ 256:R898.

Connett, R.J., Honig, C.R., Gayeski, T.E.J., and Brooks, G.A., 1990, Defining hypoxia: a systems view of $\dot{V}O_2$, glycolysis, energetics, and intracellular PO_2. _J. Appl. Physiol._ 68:833.

Federspiel, W.J. and Popel, A.S, 1986. A theoretical analysis of the effect of the particulate nature of blood on oxygen release in capillaries. _Microvasc. Res._ 32:164.

Gayeski, T.E.J., 1981, A cryogenic microspectrophotometric method for measuring myoglobin saturation in subcellular volumes; Application to resting dog muscle. (Ph.D. Thesis). Rochester, NY: University of Rochester, 1981. (Univ. Microfilms, No. DA9224720, Ann Arbor, MI).

Gayeski, T.E.J., and Honig, C.R., 1986, O_2 gradients from sarcolemma to cell interior in a red muscle at maximal $\dot{V}O_2$. _Am. J. Physiol._ 251:789-799.

Gayeski, T.E.J., Connett, R.J., and Honig, C.R., 1987, The minimum intracellular PO_2 for maximum cytochrome turnover in red muscle in situ. Am. J. Physiol. 252:H906–H915.

Groebe, K., 1990, A versatile model of steady state O_2 supply to tissue. Application to skeletal muscle. Biophys. J. 57:485.

Groebe, K. and Thews, G, 1988, Effects of red cell spacing and red cell movement upon O_2 release under conditions of maximally working skeletal muscle. Adv. Exper. Med. Biol. 248:175.

Hellums, J.D., 1977, The resistance to oxygen transport in the capillaries relative to that in the surrounding tissue. Microvasc. Res. 13:131–136.

Honig, C.R., and Gayeski, T.E.J., 1989, Precapillary O_2 loss and arteriovenous O_2 diffusion shunt are below limit of detection in myocardium. Adv. Exper. Med. Biol., 247:591.

Honig, C.R., Odoroff, C.L., and Frierson, J.L., 1982, Active and passive capillary control in red muscle at rest and in exercise. Am. J. Physiol. 243:H196.

Honig, C.R., Gayeski, T.E.J., Federspiel, W., Clark, A., and Clark, P., 1984, Muscle O_2 gradients from hemoglobin to cytochrome; new concepts, new complexities. Adv. Exper. Med. Biol. 169:23.

Honig, C.R., Gayeski, T.E.J., and Groebe, K., 1991, Myoglobin and oxygen gradients. in: The Lung: Scientific Foundations. R.G. Crystal, JB West, et. al, Eds. Raven Press, New York.

Klitzman, B. and Duling, B.R., 1979, Microvascular hematocrit and red cell flow in resting and contracting striated muscle. Am. J. Physiol. 237:H481.

Kreuzer, F., and Hoofd., 1987, Facilitated diffusion of oxygen and carbon dioxide, in: "Handbook of Physiology, Section 3: The Respiratory System. Volume IV: Gas Exchange. L.E. Farhi, and S.M. Tenney, eds., Am. Physiol. Soc., Bethesda, Maryland.

Wilson, D.F., Erecinska, M., Drown, C., and Silver, I.A., 1979, The oxygen dependence of cellular energy metabolism. Arch. Biochem Biophys. 195:494.

Wittenberg, B.A. and Wittenberg, J.B., 1985, Oxygen pressure gradients in isolated cardiac myocytes. J. Biol. Chem. 260:6548.

Wittenberg, B.A. and Wittenberg, J.B., 1989, Transport of oxygen in muscle, Ann. Rev. Physiol., 51:857.

THE ROLE OF WALL SHEAR STRESS IN MICROVASCULAR NETWORK ADAPTATION

Antal G. Hudetz and Mohammad F. Kiani[*]

Department of Physiology, Medical College of Wisconsin,
Milwaukee, WI 53226 and [*]Department of Biophysics, University
of Rochester, Rochester, NY 14642, USA

INTRODUCTION

Traditionally, structural adaptation of the microcirculation to
physiological states with increased blood flow has been associated with
changes in oxygen transport or metabolism. Recent findings in the
literature suggest that long term changes in blood flow per se may
influence vessel growth (Hudlicka, 1988). Chronic fluid retention with
elevations of cardiac output result in progressive increases of the
arterial vascular resistance and hypertension. Vascular wall hypertrophy of
the larger arteries and arterioles has been repeatedly demonstrated. More
recently, it has been observed that small arterioles of skeletal muscle
tissue disappear over time (rarefaction) which may also contribute to the
elevations of the total peripheral resistance (Lombard et al, 1989). Clues
to these mechanisms may be found in studies related to long term adaptation
of large arteries (Langille and O'Donnell, 1986) as well as acute dilation
of arterioles in response to increased wall shear stress which has been
suggested by other investigators (Davies, 1989).

For laminar, Newtonian blood flow, wall shear stress in a vessel is
proportional to F/d^3, where F is the volumetric flow rate and d is the
vessel diameter. Thus, relatively small changes in vessel diameter may help
to maintain wall shear stress constant in face of changing flow rate. If
wall shear stress can be detected by the endothelium (Davies, 1989), then
it may serve as a signal for the adjustment of vascular diameter to flow.
The generalization of this idea for anastomosing microvascular networks is,
however, not straightforward. Dilatation or constriction of each
microvessel has an influence on the flow and wall shear stress in other
vessels of the network, therefore, the outcome of local diameter
adjustments at the network level is difficult to predict. A computer
simulation study can help to evaluate the expected network behavior based
on certain assumed local feedback mechanisms.

In the present work, the possible role of wall shear stress in
microvascular network adaptation was investigated by computer simulation.
The hypothesis was tested that adaptation of microvessels to wall shear
stress is responsible for long term adjustment of physiological vascular
diameters in anastomosing vascular networks. Initially, we assumed that
each microvessel controls its diameter to maintain constant local wall
shear stress. An alternative assumption was that only the mean wall shear

stress in the network is maintained constant. The present paper describes the results and conclusions based on a computer simulation study of these hypothesized adaptation schemes.

METHODS

Computer simulation of vascular adaptation was carried out in planar model networks which were topologically similar to the microvascular bed of the mesentery. A series of these networks was generated by computer, simulating neovascularization in a rectangular field (Kiani and Hudetz, 1990). In the simulation, vessels were growing inward from each side of the field and they were bifurcating dichotomously in a probabilistic fashion. The probability of bifurcation was proportional to vessel length, and the angle of bifurcation was randomly drawn from a normal distribution. The vessels were allowed to join and form polygons until the growth of all vessels was terminated by joining or reaching an edge of the growth field. In the present work, networks were generated with a mean bifurcation angle of 50 degrees. These networks contained 117 ± 20 vascular segments and 28 ± 6 polygons. Most of the polygons consisted of 5 to 6 anastomosing segments.

Wall shear stress was calculated in each vessel of the networks using either constant flow or constant pressure boundary conditions. Vessels at the periphery of the network, which served as origins of growth during network generation, were alternately designated as arterioles and venules. An intravascular pressure of 35 mmHg was assigned to arterioles and 25 mmHg to venules. Other peripheral branches were designated as capillaries with an intermediate pressure of 30 mmHg. A first approximation of pressure distribution in the network was calculated assuming constant blood viscosity. The hydraulic resistance of each vessel was estimated from its diameter and length. Equations for flow were written for each vessel segment, and at each branch point the flow balance was written as a second equation. The resulting system of equations was solved for the intravascular pressures at the branch points. A numerical algorithm utilizing the sparseness of the conductance matrix was used for this purpose. Wall shear stress (s) in each vessel was calculated as $s=(P_1-P_2)r/2L$, where P_1 and P_2 are the intravascular pressures at the ends of the vessel segment, r is the radius, and L is the vessel length.

Network adaptation was simulated by the iterative adjustment of vessel diameters to control either the local wall shear stress (local adaptation) or the mean wall shear stress (global adaptation). After each adjustment of vessel diameter, the conductance matrix was updated and the distributions of pressure and wall shear stress in the network were recalculated. In the first approach (local adaptation), vessel diameters were adjusted so that wall shear stress in each vessel approached a common value derived from the initial mean wall shear stress. This procedure was based on the assumption that each vessel controls its diameter to maintain constant local wall shear stress. The initial diameters of all vessels were assumed to be equal in this simulation. In the second approach (global adaptation), vessel diameters were adjusted so that the standard deviation of wall shear stress from its initial mean value was minimized. In this case, initial vessel diameters were obtained from the procedure of network generation. Briefly, the vascular diameter was proportional to the relative age of the vessel.

RESULTS

Local adaptation

Vascular adaptation to local wall shear stress was based on the hypothesis of uniform shear stress in the network. This adaptation scheme either did not converge to a stable network pattern, or it resulted in unrealistic network geometry. No convergence was attained when the computations were performed using constant pressure boundary conditions. With constant flow boundary conditions, convergence was reached for some of the networks, however, the networks rarefied to a significant degree. On the average, the diameter of 60 percent of the vessels approached zero. Furthermore, vascular diameters acquired an unrealistic pattern in that the largest vessels developed in the middle of the network, while vessels around the periphery of the network mostly reduced in size. Note that all vessels had equal diameter before the adaptation. Figure 1 depicts a network after local adaptation as an example.

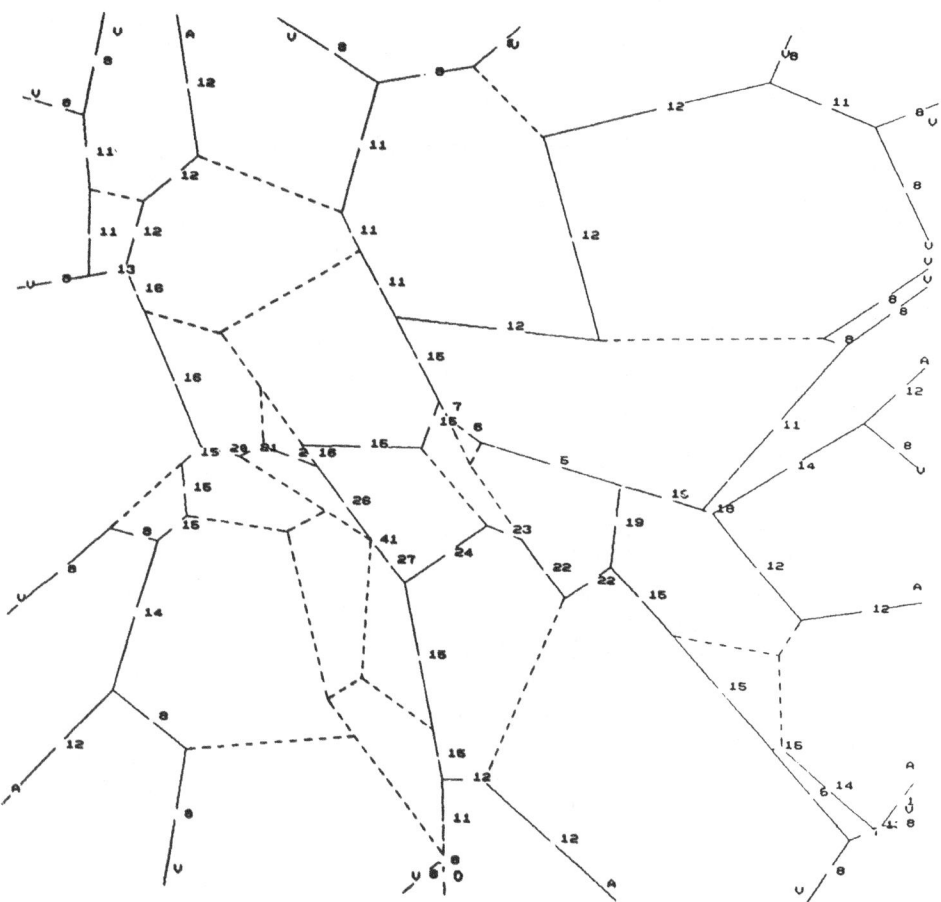

Fig. 1. Network adaptation to local wall shear stress. Dashed lines indicate vessels which were present in the original network but disappeared during adaptation. Numbers indicate diameter of vessel segments in micrometers. "A" denotes arterial, "B" denotes venous connections. Arterial and venous flows were kept constant in this simulation.

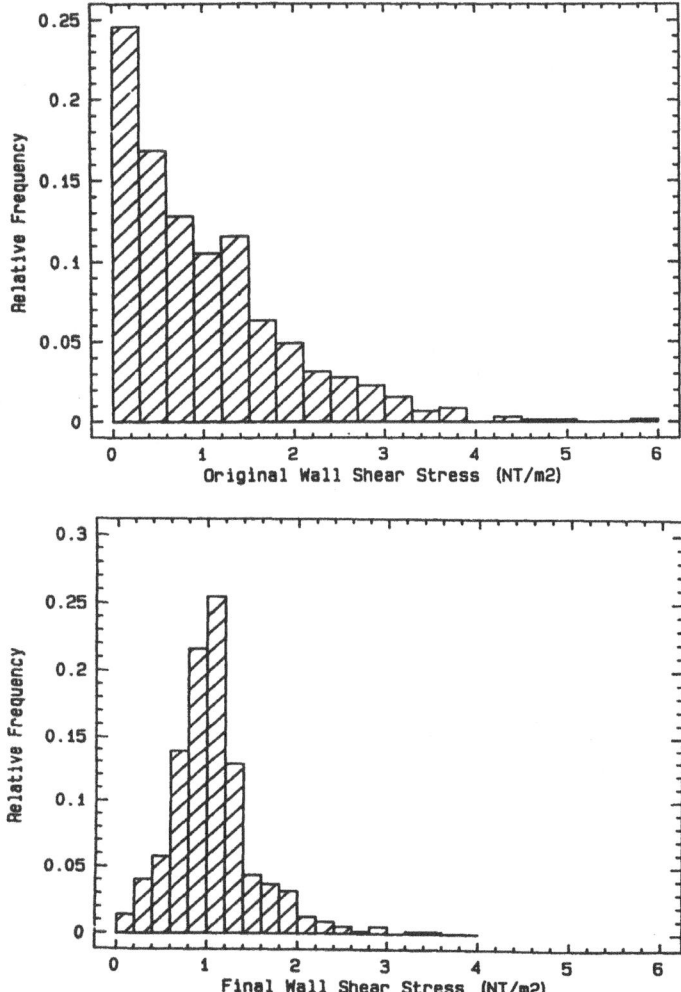

Fig. 2. Frequency distribution of wall shear stress before (top) and after (bottom) adaptation to mean wall shear stress. The standard deviation of shear stress was reduced to about 1/2 of its value before adaptation.

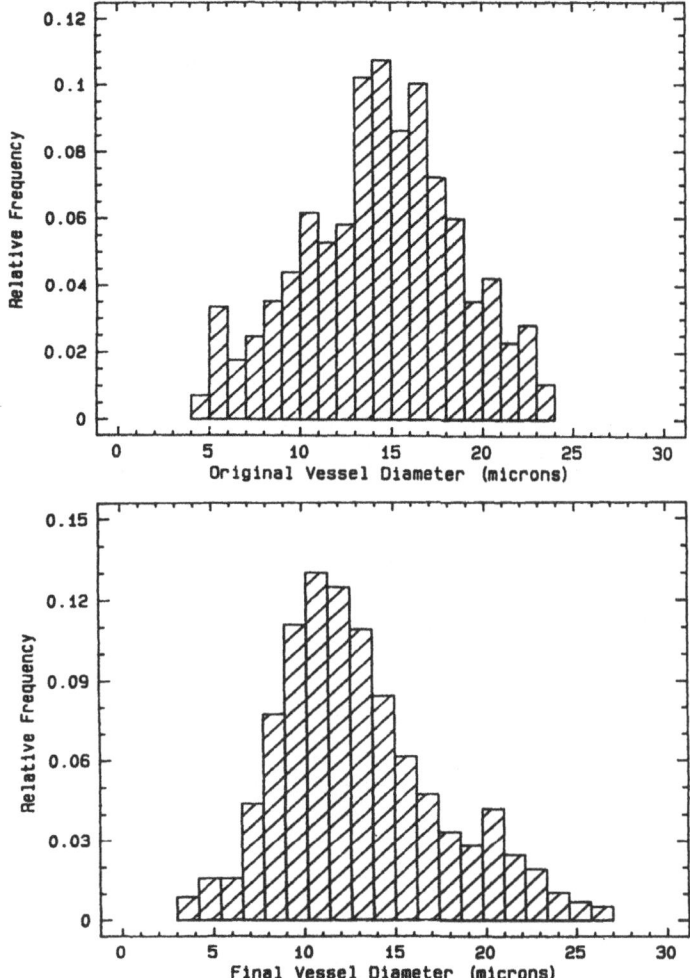

Fig. 3. Frequency distribution of vessel diameter before (top) and after (bottom) adaptation to mean shear stress. Both the mean and the variance of the distribution are unchanged after adaptation.

Global adaptation

This procedure was performed to minimize the variance of wall shear stress in the network without enforcing local shear stresses to approach a common value. Convergence was always reached using this adaptation scheme. From the combined data of 5 networks, mean wall shear stress was 0.62 ± 0.95 before adaptation, and 0.64 ± 0.46 after adaptation at constant flow boundary conditions. The range of wall shear stress was significantly reduced, and its frequency distribution changed from left skewed to near normal (Figure 2). The mean vessel diameter was 12.1 ± 5.4 before, and 11.8 ± 5.1 after adaptation. The frequency distribution of vessel diameter did not change significantly after adaptation (Figure 3). However, the diameter of individual vessels changed by up to 30 percent (Figure 4). No rarefaction was observed in any of the networks. The results from simulations with

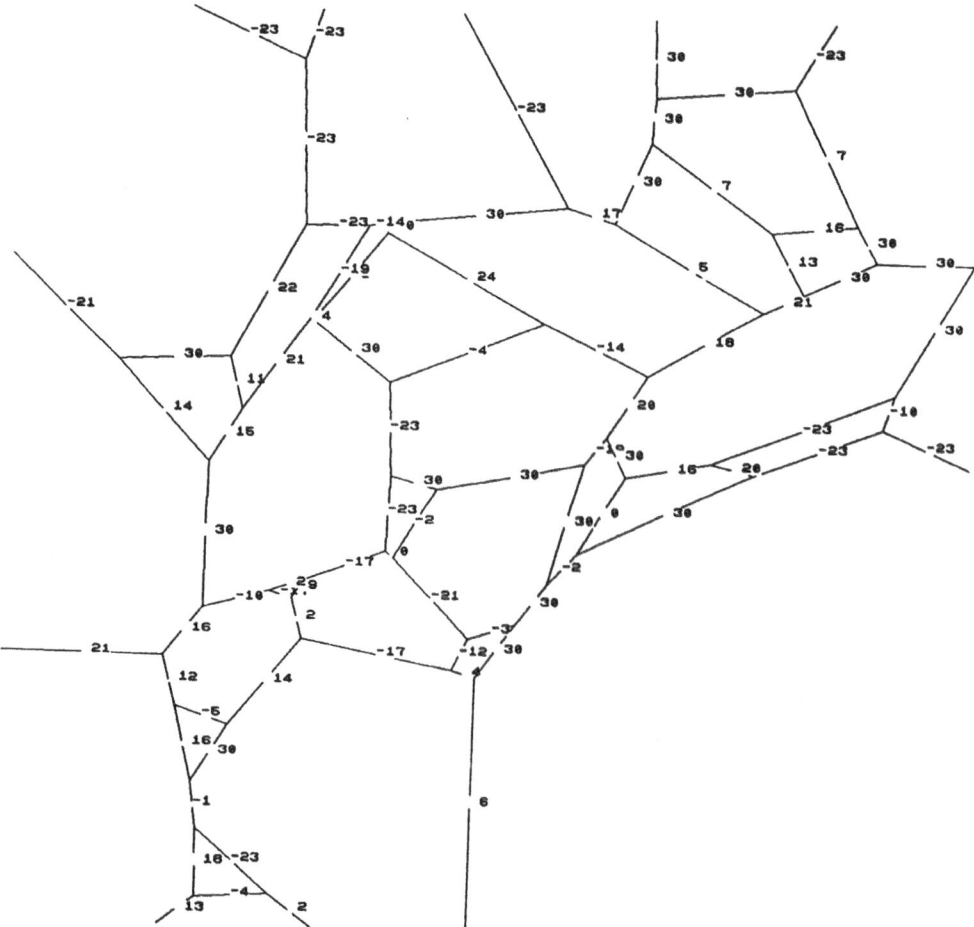

Fig. 4. Network adaptation to mean wall shear stress. Numbers indicate percent change in diameter of vessel segments after adaptation. Note that no rarefaction occurred in this type of adaptation.

constant pressure boundary conditions were quite similar to those obtained with constant flow boundary conditions.

As an additional index of hemodynamic adaptation, the total power dissipation of the networks was calculated at each step of iteration. This parameter was defined as the sum of viscous and metabolic energy costs (Murray, 1926). As seen in Figure 5, power dissipation in the network was reduced monotonously with the standard deviation of shear stress. The maximum decrease in power dissipation due to adaptation was approximately 50 percent in each network.

DISCUSSION

The plausibility of the hypothesis of vascular network adaptation to wall shear stress was investigated in this work. Initially, it was assumed that microvessels control their diameter in order to maintain local wall

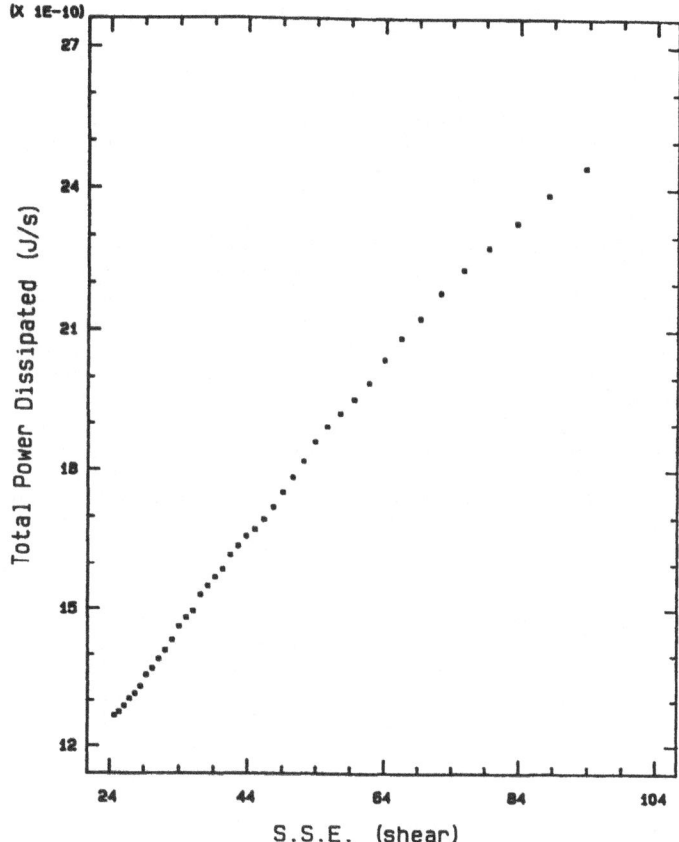

Fig. 5. Change in power dissipation (viscous plus metabolic energy cost) during adaptation to mean shear stress. S.S.E. = sum of squared errors (standard deviation) of local wall shear stress. Plotted values correspond to consecutive steps of iteration progressing from right to left (in the direction of decreasing S.S.E.).

shear stress at a common optimum value. This idea was motivated by the hypothesis of constant shear stress which follows from the principle of optimization of arterial bifurcations (Murray, 1926). From physiological point of view, low wall shear stress is undesirable because it may lead to atherosclerosis (Perktold and Peter, 1990), while high wall shear stress may result in endothelial injury (Hudlicka, 1988). The mechanism of shear stress regulation may be the release of flow dependent vasoactive substances and growth factors by the endothelium, which initiate structural modifications of the vessel wall (Langille and O'Donnell, 1986). Experimental evidence suggests that the value of shear stress is similar in several orders of microvessels (Mayrovitz and Roy, 1983).

The present results of computer simulation support the possible role of wall shear stress in microvascular network adaptation, however, the mechanism of the effect of wall shear stress is not clear. The adaptation scheme designed to equalize local wall shear stress in each vessel did not result in the emergence of realistic network patterns and it resulted in rarefaction of the networks. This finding suggests that local adaptation to wall shear stress alone is unable to govern network adaptation to altered or non-optimal blood flow distributions. Although local wall shear stress may play a role in vascular adaptation, other factors such as, growth factors, tissue metabolism, oxygen tension or a combination of these may influence vascular adaptation. Perhaps, an increased vascular sensitivity to local wall shear stress, relative to the sensitivity to metabolic and growth factors, is partially responsible for the pathological rarefaction of microvessels in chronic hypertension.

Another possibility suggested by this study is that vascular network adaptation is governed by a global measure of wall shear stress, or simply, by mean wall shear stress. The adaptation scheme used to minimize the variance of shear stress was always convergent and yielded more realistic diameter distributions than the local adaptation did. The power dissipation of the networks decreased consistently with the decreasing variance of shear stress. Furthermore, the power dissipation decreased despite the constancy of mean vessel diameter, suggesting that hemodynamic optimization was taking place during adaptation. These results suggest that in addition to local factors, a measure of mean wall shear stress may serve as a control signal for the adaptation of vessel diameter to blood flow in microvascular networks. The mechanism of this adaptation is not known at this time. It is conceivable that propagating vasoactive stimuli may play a role in forming a mean wall shear stress signal.

The mathematical model used in the present analysis was based on several simplifying assumptions, as is usual in computer simulation studies. It is possible that refinement of the simulation by complementing the model with further geometrical and rheological details may improve the convergence of the adaptation process. For example, intrinsic regulation of capillary hematocrit (Papenfuss and Gross, 1981) may help to stabilize local wall shear stress. Unfortunately, little variance was found with the present results when the adaptation simulation was repeated with Non-Newtonian blood flow (Kiani, 1990). Another question was whether the geometry and topology of the model networks were somewhat biased in the study. When the adaptation simulations were repeated in a reconstructed mesenteric microvascular network (Kiani, 1990), identical results to those of the present study were obtained. On the other hand, it may be desirable to repeat the simulations using less arbitrary initial and boundary conditions. This implies that the model be based on measured perfusion pressure and blood flow velocity values of a real microvascular network. Ultimately, the hypothesis of adaptation to shear stress should be tested in a chronic in vivo model.

SUMMARY

The possible role of wall shear stress in microvascular network adaptation was investigated by computer simulation in planar polygonal model networks. Adaptation of vessel diameter to local wall shear stress resulted in false geometry and excessive rarefaction of the networks. Adaptation to mean wall shear stress decreased the range of shear stress, decreased the total power dissipation, and prevented network rarefaction. A measure of mean wall shear stress may serve as a control signal for the adaptation of vessel diameter to blood flow in microvascular networks.

ACKNOWLEDGEMENT

This work was supported in part by grants CBT-8996225 and HL-29587.

REFERENCES

Davies, P. F., 1989, How do vascular endothelial cells respond to flow? NIPS 4:22.

Hudetz, A. .G., and Kiani, M. F., 1989, Hemodynamic optimization of computer generated capillary networks, Int. J. Microcirc. 8(Suppl. 1):S16.

Hudlicka, O., 1988, Capillary growth: Role of mechanical factors, NIPS 3:117.

Kiani, M. F., 1990, "Mathematical Modeling of Microvascular Growth and Adaptation," Ph.D. Dissertation, Louisiana Tech University, Ruston, LA.

Kiani, M. F., and Hudetz, A. G., 1990, Computer simulation of growth of anastomosing vascular networks, J. Theor. Biol. (submitted)

Langille, B. L., and O'Donnell, F., 1986, Reductions in arterial diameter produced by chronic decreases in blood flow are endothelium-dependent, Science 231:405.

Lombard, J. H., Hinojosa-Laborde, C., and Cowley, A. W., 1989, Hemodynamics and microcirculatory alterations in reduced renal mass hypertension, Hypertension 13:128.

Mayrovitz, H. N., and Roy, J., 1983, Microvascular blood flow: evidence indicating a cubic dependence on arteriolar diameter, Am. J. Physiol. 245:H1031.

Murray, C. D., 1926, The physiological principle of minimum work. I. The vascular system and the cost of blood volume, Proc. Natl. Acad. Sci. U.S.A. 12:207.

Papenfuss, H. D., and Gross, J. F., 1981, Microhemodynamics of capillary networks, Biorheology 18:673.

Perktold, K., and Peter, R., 1990, Numerical 3D-simulation of pulsatile wall shear stress in an arterial T-bifurcation model, J. Biomed. Eng. 12:2.

CHANGES IN TISSUE HISTOLOGY ASSOCIATED

WITH ADAPTATION AND ACCLIMATION TO HYPOXIA

Ian S. Longmuir

Department of Biochemistry
North Carolina State University
Raleigh, NC 27695-7622   USA

INTRODUCTION

The terms adaptation and hypoxia are both widely used but not always with quite the same meanings. Now that it is possible to measure the partial pressure of oxygen in every part of the mammalian organism with a fair degree of accuracy and a high degree of spatial resolution it seems appropriate to define hypoxia as a partial pressure of oxygen significantly lower than that normally found in that part of the organism. Thus an inspired Po_2 of much less than 150 Torr and a Lubbers histogram of tissue oxygen tensions shifted to the left would both be defined as hypoxia.

The term adaptation is sometimes used to describe both inborn and acquired mechanisms. More narrowly adaptation is applied to inborn mechanisms only and the terms acclimation and acclimatization are used to define acquired changes produced by artificial and environmental effects respectively.

In this paper some acquired mechanisms will be discussed. Since acclimation, the artificial changes can be more precisely defined and studied than acclimatization only the former will be considered.

Hypoxia produces many changes in tissue but only a few seem to be beneficial. It is some of these that will be described here.

METHODS AND RESULTS

When newborn kittens are exposed to a slowly falling PIo_2 they show oxygen dependent oxygen consumption (Figure 1). That is the rate of oxygen consumption falls with falling PIo_2 and finally becomes zero at a PIo_2 of about 30 Torr and the animals stop breathing and die. If at any point on this curve the PIo_2 is raised oxygen consumption increases again and the animals survive. This facility to reduce oxygen consumption in the face of hypoxia is lost when the kittens are three weeks old (Moore, 1956). Their oxygen consumption rate becomes independent of PIo_2 and when it is lowered to about 60 Torr the animals convulse and die. At 60 Torr the newborn kittens show about a 50% reduction in oxygen consumption and can survive a further 30 Torr fall.

Attempts to explain this difference in response on the basis of changes in regional perfusion failed, and some experiments suggested that tissue oxygen consumption might change in the first few days of life. Since in the newborn the liver consumes more oxygen than any

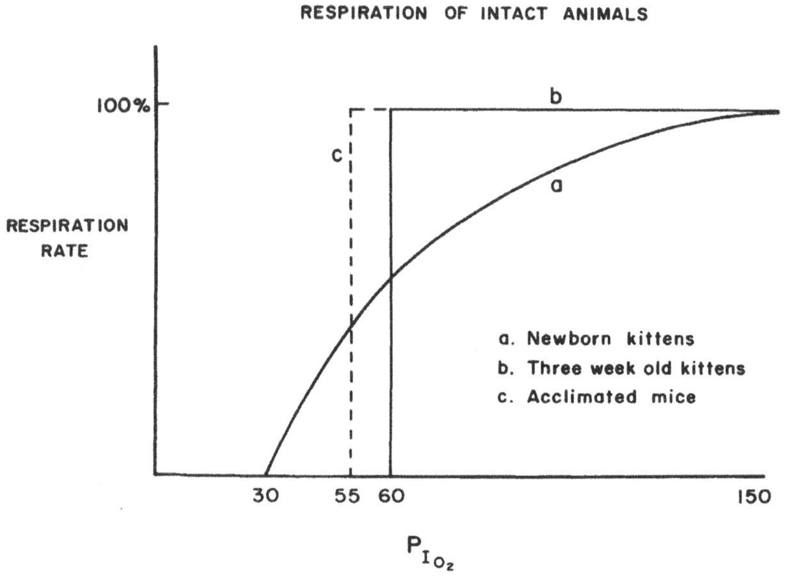

RESPIRATION OF INTACT ANIMALS

Figure 1. Respiration rate vs. Inspired Oxygen Tension. (a) newborn
          kitten, (b) three-week-old kitten, (c) acclimated mouse
          normalized on kitten data.

other organ this tissue seemed worthy of study. However, no difference in the respiratory activity of liver mitochondria or their affinity for oxygen could be found between the newborn and older kittens. We did find a change in the respiration rate of liver tissue slices as a function of ambient Po_2 (Figure 2). Slices from newborn kittens required a 100-200% higher Po_2 to achieve their maximum respiration than did slices from those ten days or more after birth (Longmuir and Moore, 1957). Since there were no differences in the respiratory activities of mitochondria or homogenate it seemed possible that a structural difference might be the basis of the change. The characteristic structure of the liver is the "muralium." This is a branching wall of liver cells. In each side of each wall lie the sinusoids which supply oxygen and other nutrients to the cells. We found that the muraliun of newborn kitten livers is two cells thick. By the tenth day this wall has thinned to one cell (Figure 3). Thus at ten days the maximum distance oxygen has to diffuse is half that at birth.

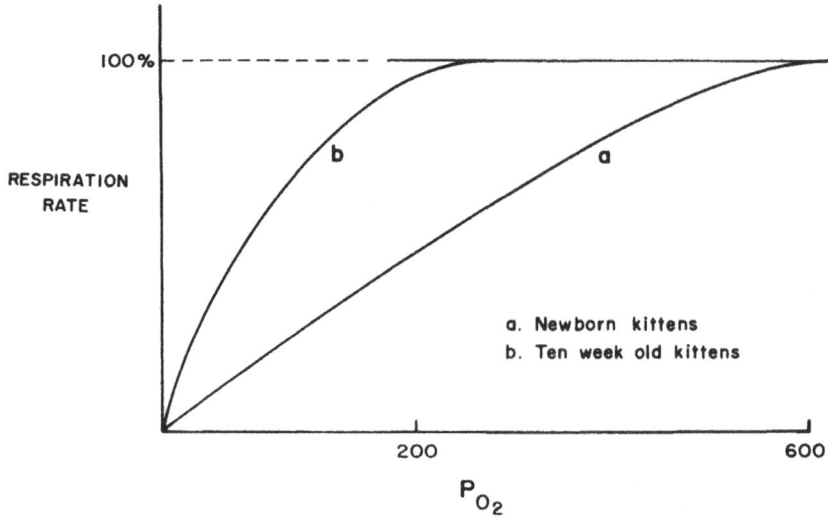

Figure 2. (a) Slices from newborn kitten's liver, (b) Slices from ten-day-old kitten's liver.

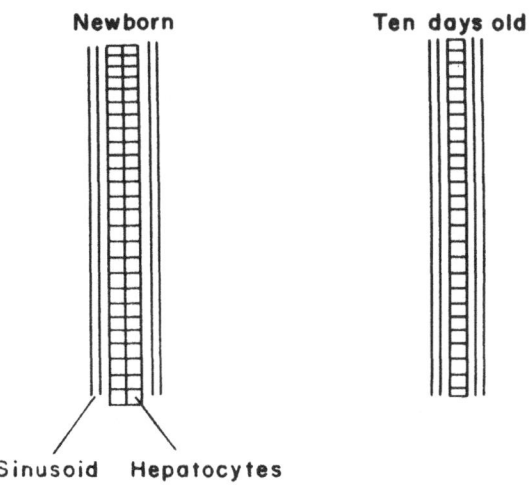

Figure 3. Histological changes with age.

43

The process of acclimation however, is different from adaptation. Acclimation is not a mere return to the newborn condition. Acclimated mammals show oxygen independent respiration but now they can survive to P_{O_2} values below 60 Torr (Figure 1). Thus some change has occurred permitting adequate oxygen supply despite a lowered oxygen delivery potential (hypoxia). The liver is, at rest, a major consumer of oxygen inspite of the fact its major source of oxygen, the portal vein, has a rather low P_{O_2}. Thus it seemed appropriate to look for acclimative changes in this organ. Because of certain political constraints it was not possible to do this work on kittens, so young mice were used instead. Examination by light microscopy of the livers of young mice after six hours exposure of the intact animals to a $P_{I_{O_2}}$ of 100 Torr showed no difference from controls. However, by electron microscopy changes in the endoplasmic reticulum (e.r.) could be seen in the experimental group. Liver is exceptional in that the e.r. constitutes 4% of the liver volume, and it has been shown that oxygen diffuses along

THE LIVER E.R. SCHEMATIC

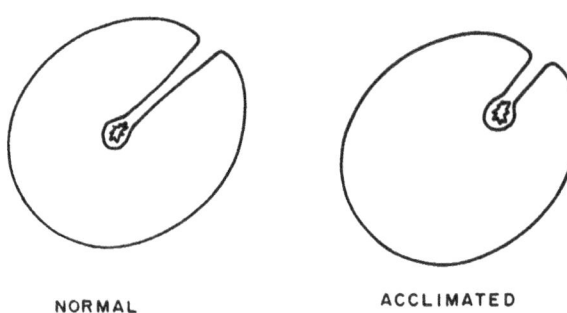

NORMAL   ACCLIMATED

Figure 4. Shortening of major oxygen pathway by acclimation.

the e.r. about ten times as fast as through cytosol (Ho, Ju and Ho, 1986; Vanderkooi and Callis, 1976). Thus, about half the intracellular oxygen transport must be along the e.r. In addition, there is some evidence that the e.r. is continuous with both the plasma membrane and the outer mitochondrial membrane. Thus, the e.r. may constitute a preferential channel for oxygen from the cell surface to the sites of maximum utilization. We considered the possibility that hypoxia might induce change in the e.r. such that oxygen transport could be facilitated. Several hundred electron micrographs (e.r.) of the livers of mice exposed to hypoxia and of controls were examined by curved line sterology (Weibel and Elias, 1967), and in addition, the number of separate elements of e.r. were counted. It was found that hypoxia did not affect the number of elements but halved their total length. Thus, if the e.r. is a major pathway in intracellular oxygen transport in the liver, a shortening of these paths would enable the mitochondria to receive as much oxygen as before despite the lowered oxygen delivery potential.

SUMMARY

Newborn kittens are adapted to hypoxia by having an oxygen
dependent respiration rate. In part, this is due to the structure of
the liver which results in a great reduction of oxygen consumption in
this organ in response to hypoxia.

Older animals possess an oxygen independent respiration rate which
is not lost when they acclimate to hypoxia. Instead, there are
histological changes which permit the liver to consume oxygen at the
same rate as before by facilitating intracellular oxygen transport.

REFERENCES

Ho, C.S., Ju, L. and Ho. C.T. (1986). Measuring oxygen diffusion with
polarographic electrodes. Biotech. and Bioeng. $\underline{28}$, 1086-1092.

Longmuir, I.S. and Moore, R.E. (1957). The change with age in the
critical oxygen concentration of kitten liver slices. J. Physiol. $\underline{138}$,
44.

Moore, R.E. (1956). Loss with age of kittens' adaptation to hypoxia.
J. Physiol. $\underline{133}$, 60-70.

Vanderkooi, J.M. and Callis, J.B. (1976). A probe of lateral diffusion
in the hydrophobic region of membranes. Biochemistry $\underline{13}$, 4000-4004.

Weibel, E.R. and Elias, H. (1967). Introduction to stereologic
principles. In: Quantitative Methods in Morphology. Eds Weibel, E.R.
and Elias, H. Springer-Verlag, Berlin, pp. 89-98.

METHODS AND INSTRUMENTATION

TRANSCUTANEOUS MEASUREMENTS OF SKIN O_2 SUPPLY AND BLOOD GASES

Dietrich W. Lübbers

Max-Planck-Institut für Systemphysiologie
Rheinlanddamm 201
Dortmund 1   FRG

Under steady state conditions in each organ there must be a
balance between O_2 delivery and O_2 consumption. The oxygen is
delivered to the skin by two ways 1) by blood and 2) through
the skin surface by exchanche with the surrounding air. Oxygen
is mainly consumed by the respiratory chain in the mitochondria which are distributed within the tissue to supply the
necessary energy. In the following we will focus mainly on the
O_2 exchange.

O_2 exchange with the surrounding atmosphere

The exchange of oxygen and carbon dioxide through the surface
of the intact human skin was experimentally observed and measured as early as 1851 by Gerlach, a teacher at the Royal Vet-
erinarian School of the University of Berlin. The values he
measured compare well with later results (see 1) In analogy to
lung respiration, the absorption of oxygen and elimination of
CO_2 by the skin surface was called "cutaneous respiration".
Later experiments on gas exchange through the epidermis of the
skin clearly demonstrated that the gas exchange obeys the laws
of diffusion and that no active processes are involved. This
fact has been supported by exchange studies with inert gases
(2, for review see 3)

In infants and adults the O_2 uptake by the skin in air at room
temperature is about 90 ml $O_2/(m^2 \cdot h)$, i.e. about 1 - 2% of the
O_2 uptake by the lung. Larger amounts of oxygen can be exchanged in preterm infants because of the special structure of
their skin (4). The reported values show a large variability.
This is understandable since the exchange rates depend very
much on skin blood flow and diffusion properties of the skin
which are very difficult to standardize.

The O_2 uptake through the skin is directly demonstrated by the
pO_2 profile shown in fig. 1 (5). To measure this profile a
skinfold of the upper thigh was punctured by a needle-elec-
trode directed perpendicularly towards the epidermis of the
opposite side. The position of the epidermis was stabilized by

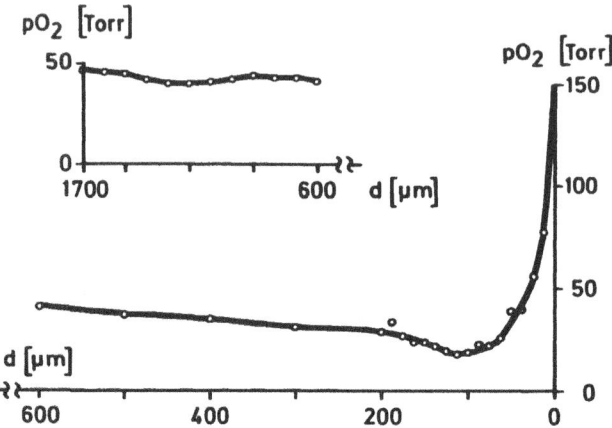

Fig. 1 Oxygen pressure profile in dermis and epidermis. The profile was measured by withdrawing a glass insulated platinum needle electrode (diameter ca 5μm) in steps from the surface of the skin which was in contact with air (left side). Up to a distance of 200 μm the steps were 12.5 μm, then 100 μm. (Data from (5)).

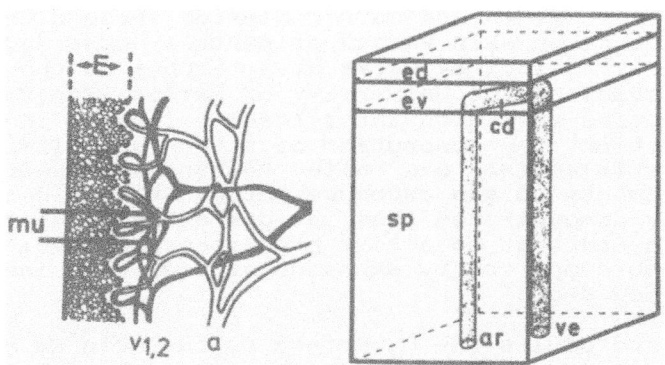

Fig. 2 Schematic drawing of the skin vasculature (left side) and the capillary loop model (right side)
E = epidermis, a = arterial network, v_1, v_2 = venular plexus, mu = microcirculatory unit. The arrow marks a distance of 300 to 500 μm from the epidermis surface. cd, ar, ve = capillary dome, arterial and venous limb of the capillary, ev = viable layer, ed = dead layer.

a plastic ring. Such a stabilization was necessary because of
the high elasticity of the skin. In the outer layer
(epidermis, fig. 1 right side) pO_2 decreases first steeply
from the air value and than levels off at a distance of about
100-150 μm. This pO_2 decrease directly demonstrates that oxygen penetrates into the skin and contributes to its O_2 supply.
Then, a slow pO_2 increase up to a distance of about 500 μm
follows meaning that the deeper skin layers are supplied by
other O_2 sources. In deeper layers only small variations of
pO_2 are observed (fig. 1, insert).

The schematic drawing of the anatomy of the skin (fig. 2)
shows that the epidermis (E) is situated like a membrane between the surrounding atmosphere (fig. 2, left side) and the
blood in the loop-shaped capillaries of the str. papillare of
the dermis. The capillaries receive their blood from the arterial cutaneous network (a) over convex arcades in the middle
dermis. The venous blood is collected in venular plexus ($v_{1,2}$)
and in the venous cutaneous network (see 6). The main diffusion resistance of the epidermis is in the str. corneum as it
was experimentally demonstrated (7, 8). This structure well
explains the measured pO_2 profile with the steep pO_2 decrease
within the epidermis. Smaller branches of arteries and veins
supply the deeper layers of the dermis with oxygen (fig. 1,
insert). The pO_2 decrease towards the epidermis starts at the
level of the arterioles which give off the capillary loops.
The relatively regular structure allows to define a microcirculatory unit (mu) which can be used to study the different
parameters that influence gas exchange between air, tissue and
blood. The microcirculatory unit (fig. 2, right side) consists
out of a capillary loop (ar, cd, ven) with its surrounding
tissue simulating the str. papillare (sp). The str. papillare
is covered by the epidermis consisting out of a viable (ev:
str. germinativum, str. spinosum, str. granulosum) and a dead
(ed: str. corneum) layer (9). To investigate the effect of the
capillary loops on the O_2 supply we assume that from the basal
surface the inflowing blood is the only source of oxygen.

Several other models have been proposed to describe gas exchange through the epidermis (e.g. 8, 10, 11), but these models do not take into account the special vascular structure.
Whang, Quinn, Graves, and Neufeld (12) presented a model which
simulates the vascular network of epidermis and dermis with
different arterial and venous plexus to investigate the contribution of all these vessels to the gas exchange of inert
gases; however, this model neglects the capillary loop. This
may be possible for inert gases, but it seems to be doubtful
for the O_2 exchange.

In fig. 3 a gas exchange, similar to the situation in fig. 1,
is simulated. It shows the pO_2 and pCO_2 profiles along the
arterial limb and across the epidermis (see 9). There is a
steep pO_2 decrease within the str. corneum, a leveling off
followed by an increase towards the arterial inflow. The
measured pO_2 profile in fig. 1 missed the arterial inflow
since it registers the mean pO_2. The pCO_2 profile demonstrates
that skin eliminates CO_2. The pO_2 and pCO_2 profiles clearly
demonstrate the important influence of the atmospheric air on
the O_2 and CO_2 exchange of the skin. As inserts the pO_2 and
pCO_2 values of the diagonal plane of the MU are shown. There
is a large pO_2 gradient between arterial inflow (100 Torr) and

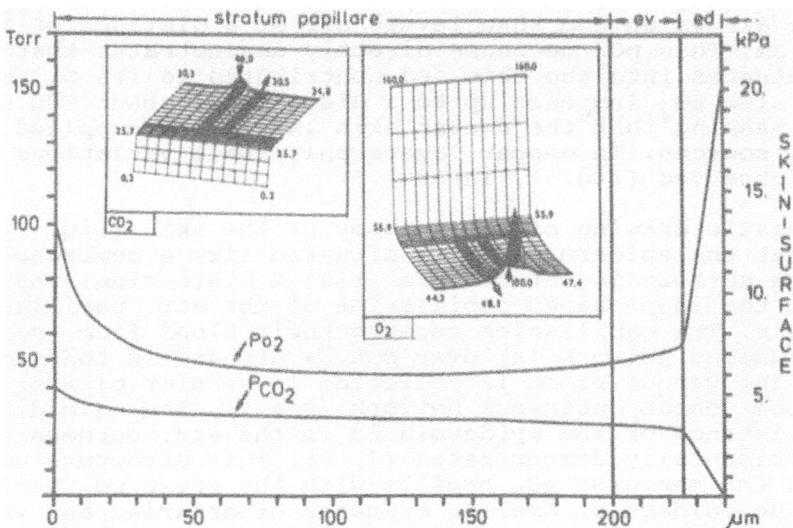

Fig. 3 Simulated pO_2 and pCO_2 profile along the arterial limb
of the capillary and across the epidermis (cross section of a microcircular unit). Inserts: the corresponding pO_2 and pCO_2 fields. Blood flow = 1 ml
$O_2/(100g\ min)$, hemoglobin concentration = 16g/dl, p_{50} =
27 Torr, O_2 consumption = 0.3 ml $O_2/(100g\ min)$. Skin
surface in contact with air.

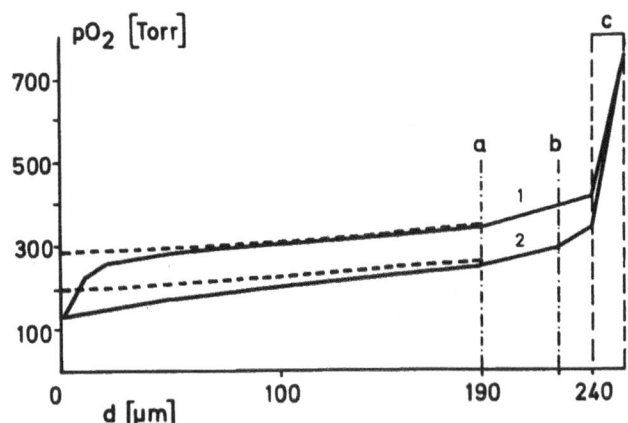

Fig. 4 Simulated pO_2 profiles along the arterial and venous
(broken line) limb of the capillary and across the epidermis. The skin surface is covered by a 12.5 μm
polypropylene membrane (c) which is in contact with
oxygen. $p_{B}O_2$ = 126 Torr, blood flow = 10 (1) and 60 (2)
ml $O_2/(100g\ min)$, a = capillary dome, b = begin of the
str. corneum.

venous outflow (48.3 Torr) of the capillary loop resulting in an O_2 shunt by which a considerable amount of oxygen bypasses the tissue. When at similar conditions skin is covered by a gas impermeable membrane, the pO_2 distribution is distinctly changed. The large pO_2 gradient across the epidermis disappears and because of the relatively small O_2 consumption of the viable layer only a small pO_2 decrease across the epidermis is observed.

The O_2 exchange through the skin depends on several parameters: 1) on the pO_2 on the skin surface, 2) on the O_2 concentration and the pO_2 of the blood, 3) on the blood flow, 4) on the diffusion properties of the tissue and 5) on the geometry of the tissue. By changing these parameters the O_2 exchange can be manipulated. As an example fig. 4 shows simulated pO_2 profiles of the upper skin covered by a 12.5 μm thick propylene membrane. The membrane is in contact with oxygen. Profile 1 corresponds to a flow of 10, profile 2 of 60 ml/(100 g·min). The 6 times larger flow causes a 20% larger O_2 flux into the skin because of the lower tissue pO_2. The lower tissue pO_2 is caused by the about 50% larger O_2 transport of the outflowing venous blood. I think these results clearly underline the complexity of the skin O_2 supply and the difficulty to predict and understand changes without the help of a model. Fig. 4 demonstrates further that the pO_2 difference across the propylene membrane is directly describing the O_2 flux into the skin. This is important because the pO_2 difference across an oxygen permeable membrane can be directly measured by fluorescence based optical sensors as we have shown (13). This is in contrast to polarographic pO_2 measurements of the skin surface pO_2 since in this case the electrode covers the surface and blocks an O_2 exchange with the surrounding air. Therefore such an electrode can only monitor the O_2 supply by the blood.

O_2 consumption of the skin

As in other organs O_2 consumption of the skin, M_{O2} has been determined by measuring the $p_{ss}O_2$ decrease after flow stop if $p_{ss}O_2$ was increased above 150 Torr by respiration of oxygen to eliminate the influence of oxygen chemically bound to hemoglobin (14). Using a membrane covered polarographic platinum pO_2 electrode M_{O2} was found to be 0.278 ml O_2/(100g·min) at 37°C and 0.366 ml O_2/(100g·min) at 45°C (15). Since, however, similar measurements with platinum electrodes of different diameters yielded somewhat inconsistent results (16) we analyzed the parameters by which these results may be influenced by using our capillary loop model .

In a tissue with homogeneously distributed O_2 consumption and oxygen supply the O_2 consumption can be calculated by multiplying the steepness of the pO_2 decrease dpO_2/dt with the O_2 solubility coefficient of the respiring tissue α

$$\alpha \cdot \frac{dpO_2}{dt} = \alpha \cdot s = -M_{O2} \qquad\qquad 1$$

In reality, however the upper skin is not a homogeneous tissue, but it consists out of three compartments (see fig.2): 1) the viable (respiring) compartment A with the O_2 solubility coefficient α_A and the volume V_A 2) the capillary compartment C with α_C and V_C 3) the dead compartment D with α_D and V_D. During the masurement a fourth compartment, the sensor membrane M is added with α_M and V_M. Therewith equation (1) becomes

$$\alpha_A \cdot V_A \cdot s_A + \alpha_C \cdot V_C \cdot s_C + \alpha_D \cdot V_D \cdot s_D + \alpha_M \cdot V_M \cdot s_M = -M_{O2} \cdot V_A \qquad 2$$

Dividing by V_A we obtain the corrected solubility coefficient $\alpha'_i = \alpha_i (V_i/V_A)$. Assuming that in all the compartments the steepness of the pO_2 decrease is the same $(s_A = s_C = s_D = s_M = s)$ it is

$$(\alpha'_A + \alpha'_C + \alpha'_D + \alpha'_M) s = -M_{O2} \qquad 3$$

Eqn. (3) demonstrates that the M_{O2} is determined by the sum of the corrected solubility coefficients and not by the coefficient α_A of the respiring compartment. The possible error depends on the relative contribution of the different coefficients to the total sum. Using the following solubility coefficients (in ml O_2/(100g·atm)) $\alpha_A = 1.8$, $\alpha_C = 2.4$ and $\alpha_D = 0.8$ we obtain for the volume corrected values $\alpha'_A = 1.8$, $\alpha'_C = 0.0271$, $\alpha'_D = 0.0540$ i.e. for the sum of the three compartments of the microcirculatory unit $\alpha'_{mu} = 1.8811$, that is 4.5% larger than the α_A of the viable compartment. With the corrected α'_{mu} 99.7% of the correct value of M_{O2} is found.

Since polarographic as well as optical O_2 sensors add a sensor membrane to the mu we analysed the effect 1) of an added membrane and 2) of a membrane covered polarographic pO_2 electrode.
1) When a 15 μm thick Teflon membrane ($\alpha = 19.0$) is used the corrected α' increases to 3.1638 that is 71.3% larger than α_A. This large increase reduces the steepness of the $p_{ss}O_2$ decrease by 44.1% and therefore leads to a large error of 60.7% if the uncorrected α_A is used to calculate M_{O2}; with volume corrected α' the remaining error amounts to 6%.
2) Table 1 shows the effect of a polarographic $p_{ss}O_2$ measurement on the determination of the O_2 consumption of the skin M_{O2}. During the polarographic measurement all O_2 molecules arriving the cathode surface will be reduced so that pO_2 at the cathode surface is zero. The reduction current is proportional to the amount of the reduced oxygen. The amount of reduced oxygen, the O_2 consumption of the electrode M_{O2} (electrode) is determined by the size of the cathode surface, the diffusion properties of the electrolyte layer and of the membrane and the pO_2 to be measured. In table 1, results are compared that have been obtained with the same electrode but covered by two different membranes. On the left side results with a 12.5 μm thick polypropylene membrane ("polypropylene electrode") and on the right side results with a 15 μm thick teflon membrane ("teflon electrode") are shown. Because of the higher O_2 conductivity of the teflon membrane O_2 consumption of the teflon electrode is 13.5 times larger than that of the polypropylene electrode. At a $p_{ss}O_2$ of 200 Torr the O_2 consumption of the teflon electrode amounts to 62% of the normal O_2 consumption of the skin at 37^OC, i.e. $M_{O2} = 0.3$ ml O_2/(100g min). The results are shown for different O_2 consumptions M_{O2}: 1) 0.0, 2) 0.2, 3) 0.3, 4) 0.4 ml O_2/(100g·min). In the second column the $p_{ss}O_2$ (in Torr) and the time (in s) range are given in which the $p_{ss}O_2$ decrease is about linear.

Table 1. O_2 consumption of the skin

M_{O2}	12.5 μm Polypropylene			15 μm Teflon		
	pO_2	$M_{O2}(\alpha')$	$M_{O2}(\alpha_A')$	pO_2	$M_{O2}(\alpha')$	$M_{O2}(\alpha_A')$
0.0	376 – 332 (1-200 s)	0.0234	0.0197	186 – 142 (9- 67 s)	0.127	0.0724
0.2	258 – 152 (6- 76 s)	0.203 101.5%	0.171 85.5%	140 – 92 (13-46 s)	0.229 114.5%	0.130 65%
0.3	238 – 120 (6 – 51s)	0.289 96.3%	0.244 81.3%	120 – 64 (10-45 s)	0.269 89.6%	0.148 49.3%
0.4	200 – 120 (8- 21 s)	0.375 93.75%	0.316 79%	102 – 52 (10-32 s)	0.312 78%	0.174 43.5%

M [ml O_2/(100g min)], r (electrode) = 15 μm, 37°C, pO_2 [Torr]

As expected the heighth of the starting $p_{ss}O_2$ will be reduced with increasing M_{O2}, but the decrease is much larger at the measurements with the teflon electrode. With the propylene electrode the resulting error in the calculation of M_{O2} remains below 10% (3. column), but the error caused by the large O_2 consumption of the teflon electrode is unacceptable, it is between 10 and 22% (6. column).

O_2 exchange with the capillary blood

The balance equation of the O_2 supply of a tissue states that on steady state conditions the amount of oxygen delivered to the tissue - i.e. the difference between arterial and venous oxygen content ($C_{a'O2}$ - $C_{v'O2}$) times blood flow B - is equal to the amount of O_2 which is consumed by the tissue M_{O2}. Therewith we obtain for $C_{v'O2}$

$$C_{v'O2} = C_{a'O2} - \frac{M_{O2}}{B} \qquad\qquad 4$$

Eqn. (4) shows that there is a hyperbolic relation between $C_{v'O2}$ and blood flow. With increasing flow $C_{v'O2}$ approximates $C_{a'O2}$. At low flow values small changes of flow cause large changes of $C_{v'O2}$ whereas at large flow values - in excessive hyperemia - $C_{v'O2}$ becomes practically independent of flow changes, so that changes in $C_{v'O2}$ mirror changes in $C_{a'O2}$. For the degree of approximation $C_{v'O2}/C_{a'O2}$ we obtain

$$\frac{C_{v'O2}}{C_{a'O2}} = 1 - \frac{M}{C_{a'O2} \cdot B} = 1 - \frac{O_2 \text{ consumption}}{O_2 \text{ offer}} \qquad\qquad 5$$

Eqn. 5 shows that the approximation of $C_{v'O2}$ to $C_{a'O2}$ depends besides on flow also on the amount of oxygen which is physically and chemically bound to hemoglobin. The basic equations 4 and 5 hold not only for O_2 concentration but similarly for oxygen pressure.

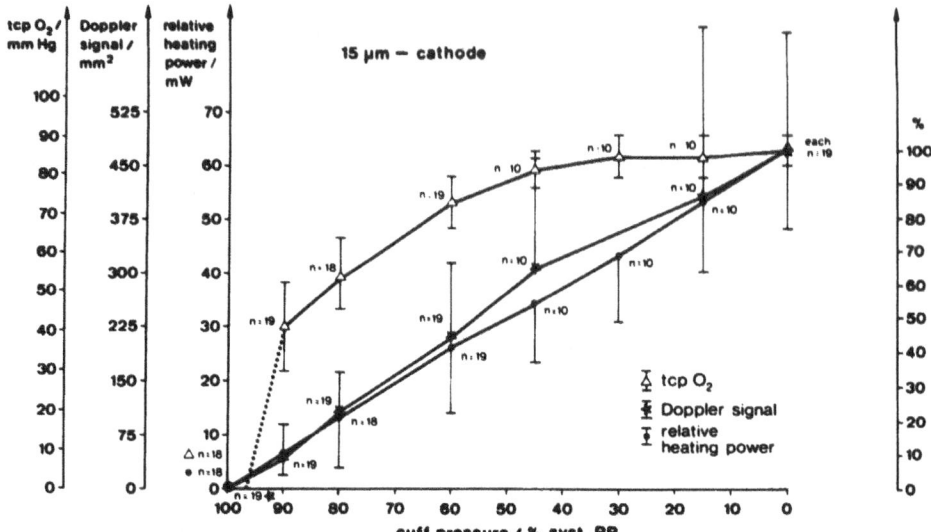

Fig. 5 Relation of transcutaneous pO_2, heating power of the electrode 45°C and Doppler flow signal (relative units/mm^2, measured in the finger tip at 45°C) RR = systolic blood pressure (19).

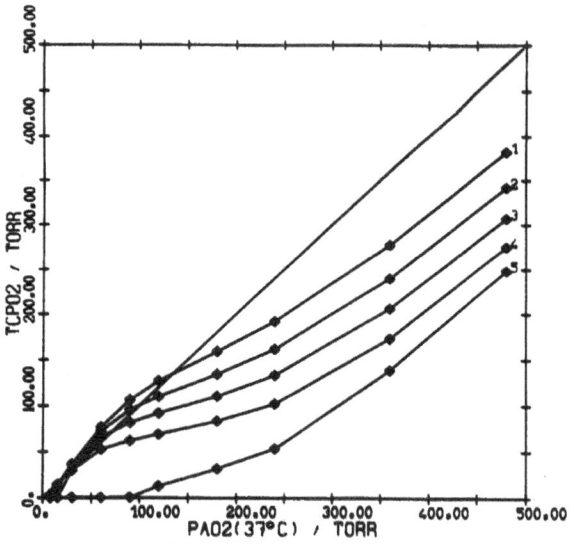

Fig. 6 Transcutaneous pO_2 (43°C) vs arterial pO_2 (37°C) simulated at different blood flow values. M_{O2} = 0.4 ml O_2/(100g min), hemoglobin concentration = 16g/dl, blood flow (in ml O_2/(100g min)): 1) 100.0, 2) 50.0, 3) 25.0, 4)10.0, 5) 1.0.

As already mentioned during the measurement the polarographic
sensor covers the skin so that a gas exchange is only possible
with the blood. Under normal conditions at room temperature
$p_{ss}O_2$ is closed to zero. A large blood flow increase can be
obtained by locally heating up the skin to 43-45°C (see 9). At
this temperature a kind of vasoparalysis is brought about so
that no postocclusive hyperemia can be anymore observed. For a
sufficient heating up the temperature distribution below the
electrode is of great importance (17). Under this condition a
hyperbolic relation between $p_{ss}O_2$ and blood flow is found
(18). Fig. 5 shows such a circulatory hyperbola measured with
a $p_{ss}O_2$ electrode fixed on the intact skin of the volar side
of the forearm and heated up to 45°C (19). Flow was changed by
stepwise inflation of a cuff on the upper arm. As fig. 5 shows
flow could be reduced by 30% without changing $p_{ss}O_2$. This
demonstrates that in this case heating produced an excessive
hyperemia so that skin surface pO_2 reflects - through the skin
as a window - transcutaneously arterial pO_2. Such a measurement of the transcutaneous pO_2, $p_{tc}O_2$ allows a continuous monitoring of arterial pO_2 as long as excessive hyperemia per-
sists. In analogy to the $C_{v,O_2}/C_{a,O_2}$-index (equ. 4) the arterial $p_{tc}O_2$-index, $p_{tc}O_2/p_aO_2$ has been introduced to obtain
information about the state of circulation (20). Fig. 5 shows
that with decreasing flow the arterial $p_{tc}O_2$-index decreases.
Low indices (less than 0.7) can be taken as a sign of disturbed circulation; in this case $p_{tc}O_2$ would monitor flow
changes, whereas at high indices (higher than 0.7) $p_{tc}O_2$ monitors arterial pO_2. However, this simple rule can only give a
rough indication because the actual interdependency between
$p_{tc}O_2$ and p_aO_2 is much more complex.

Since the polarographic sensor consumes oxygen it produces a
pO_2 gradient across the epidermis, especially across the str.
corneum (see table 1). The same holds for the local application of mass spectrometry (e.g. 20). Therefore the measurement
depends very much on the diffusion properties of the str.
corneum. The diffusion resistance of dry str. corneum is very
high, it can be reduced by humidification. It has been observed that at temperatures higher than 40°C structural
changes occur that increase the diffusivity at least by two
order of magnitudes (22). The O_2 conductance of the epidermis
has been determined by using a $p_{tc}O_2$ electrode with two membranes of different diffusional properties (23). Since the
thickness of the epidermis (in adults about 90 μm) is different in different regions of the body and changes with age and
since furthermore local capillarization varies (number of capillary loops/mm^2 between 20-150/mm^2; see 9) such differences
explain that even with excessive hyperemia local and in-
terindividual differences of the $p_{tc}O_2$ are found and that
these differences can be different depending on the properties, particularly on the O_2 consumption of the sensor.

Since the O_2 offer depends on the O_2 content of the blood
(eqn.5) the chemically bound O_2 has a strong effect on the pO_2
distribution in the str. papillare and on the $p_{tc}O_2$. Therefore
according to the O_2 dissociation curve the relation $p_{tc}O_2$ versus p_aO_2 is non linear as it is shown in fig. 6. The $p_{tc}O_2$ is
calculated for a temperature of 43°C. This results in a shift
of the O_2 dissociation curve to the right (5.75%/°C) so that
at the same O_2 concentration p_aO_2 and $p_{tc}O_2$ increase. Since
on the x-axis the p_aO_2 is given for a temperature of 37°C, at

high flow values (fig. 6, curve 1) $p_{tc}O_2$ (43°C) can be higher than p_aO_2 (37°C). This heat induced increase of arterial pO_2 during the pO_2 measurement is the reason that in spite of the pO_2 decrease along the capillary and across the epidermis the approximation of the measured $tcpO_2$ to the arterial pO_2 is relatively good (arterial $p_{tc}O_2$-index larger than 0.7).

This short discussion shows that results obtained with $p_{tc}O_2$ electrodes need a careful interpretation because local tissue geometry and local perfusion is mostly unknown. The advantage of the method is that it monitors continuously and noninvasively the O_2 supply to the upper skin. At low flow it monitors the O_2 offer to the periphery (see eqn. 5), at excessive flow a value which is closely related to the arterial pO_2. Further developments should minimize the influence which the measurements still have on the pO_2 to be measured.

Summary

The data of the O_2 exchange through the surface of the skin show that a part of the normal upper skin can be supplied with O_2 from the surrounding atmosphere. This may be important in pathological situations although probably simultaneous disturbances of the substrate supply may be more serious. The noninvasive continuous registration of skin surface pO_2 (and pCO_2) allows to monitor skin oxygen supply under different conditions. The new optical sensing of $p_{ss}O_2$ and of the local O_2 flux into the skin opens up new promising possibilities for quantification of the skin oxygen supply.

Key words

Skin O_2 uptake, skin O_2 supply, skin O_2 consumption, transcutaneous pO_2, skin O_2 flux.

REFERENCES

1. Fitzgerald, L.R., Cutaneous respiration. Physiol Rev 37: 325-336 (1957).

2. Adamczyk, B.A., Boerboom, J.H., and Kistemaker J., A mass spectrometer for continuous analysis of gaseous compounds excreted by human skin. J Appl Physiol 21: 1903-1906 (1966).

3. Scheuplein, R.J.,and Blank, I.H., Permeability of the skin, Physiol Rev 51: 702-747 (1971).

4. Cartlidge, P.H.T., Rutter, N., Percutaneous oxygen delivery to the preterm infant, The Lancet: 315-317 (1988).

5. Baumgärtl, H., Ehrly, A.M., Saeger-Lorenz, K., and Lübbers, D.W., Initial results of intracutaneous measurements of pO_2 profiles, in: Clinical Oxygen Pressure Measurement, Ehrly A.M., Hauss J., Huch R. eds., pp 121-128, Springer-Verlag Berlin, Heidelberg, New York, London, Paris, Tokyo (1987)

6. Huch, R., Huch, A., and Lübbers, D.W., Transcutaneous pO_2, Georg Thieme Verlag, Stuttgart - New York (1981).

7. Sejrsen, P., Epidermal diffusion barrier to ^{133}Xe in man and studies of clearance of ^{133}Xe by sweat, J Appl Physiol 24: 211-216 (1968).

8. Baumgardner, J.E., Graves, D.J., Neufeld, G.R., and Quinn, J.A., Gas flux through human skin: Effect of temperature, stripping, and inspired tension, J Appl Physiol: 1536-1544 (1985).

9. Lübbers, D.W. and Grossmann, U., Gas exchange through the human epidermis as a basis of $tcpO_2$ and $tcpCO_2$ measurements, in: Continuous transcutaneous blood gas monitoring, Marcel Dekker, Inc., New York (1983).

10. Thunstrom, A.M., Stafford, M.J., and Severinghaus, J.W., A two temperature, two pO_2 method of estimating the determinants of $tcpO_2$, Birth defects: Original Article Series XV (4): 167-182 (1979).

11. Grønlund, J., Evaluation of factors affecting relationship between transcutaneous pO_2 and probe temperature, J Appl Physiol: 1117-1127 (1985).

12. Whang, J.M., Quinn, J.A., Graves, D.J., and Neufeld, G.R., Permeation of inert gases through human skin: Modeling the effect of skin blood flow, J Appl Physiol 67: 1670-1686 (1989).

13. Lübbers, D.W., Völkl, K.-P.,Grossmann, U., and Opitz, N., Lactate measurements with an enzyme optode that uses two oxygen fluorescence indicators to measure the pO_2 gradient directly, in: Progress in Enzyme and Ion-Selective Electrodes, pp 67-73, Lübbers D.W., Acker H., Buck R.P., Eisenman G., Kessler M. and Simon W. eds., Springer-Verlag Berlin - Heidelberg - New York (1981).

14. Evans, N.T.S., and Naylor, P.F.D., The oxygen tension gradient across human dermis, Respir Physiol 3: 38-42 (1967).

15. Severinghaus, J.W., Stafford, M., and Thunstrom, A.M., Estimation of skin metabolism and blood flow with $tcpO_2$ and $tcpCO_2$ electrodes by cuff occlusion of the circulation, Acta Anaesthesiol Scand Suppl 68: 9-15 (1978).

16. Jaszczak, P., Blood flow rate, temperature, oxygen tension and consumption in the skin of adults measured by a heated microcathode oxygen electrode, Lægeforeningens Forlag, Københavns (1988).

17. Tremper, K.K., and Huxtable, R.F., Dermal heat transport analysis for transcutaneous O_2 measurement, Acta Anaesth Scand, Suppl. 68: 4-8 (1978).

18. Wyss, C.R., Matsen, F.A., King, R.V., Simmons, C.W., and Burgess, E.M., Dependence of transcutaneous oxygen tension on local arteriovenous pressure gradient in normal subjects, Clin Sci 60: 499-506 (1981).

19. Steinacker, J.M. and Spittelmeister, W., Dependence of transcutaneous O_2 partial pressure on cutaneous blood flow, J Appl Physiol 64 (1): 21-25 (1988).

20. Tremper, K.K., and Shoemaker, W.C., Transcutaneous oxygen monitoring of critically ill adults, with and without low flow shock, Crit Care Med 9: 706-709 (1981).

21. Targett, R.C., Kocher, O., Muramatsu, K., and McIlroy, M.B., Skin gas tensions and resistance measured by mass spectrometry in adults, J Appl Physiol 56 (5): 1431-1435 (1984).

22. van Duzee, B.F., Thermal analysis of human stratum corneum, J Invest Derm 65: 404-408 (1975).

23. Eberhard, P. and Severinghaus, J.W., Measurement of heated skin O_2 diffusion conductance and pO_2 sensor induced O_2 gradient, Acta Anaesth Scand, Suppl. 68: 1-3 (1978).

RELATING MEASURING SIGNALS FROM P_{O_2} ELECTRODES TO TISSUE P_{O_2}: A THEORETICAL STUDY

K. Groebe

Institut für Physiologie und Pathophysiologie
Johannes Gutenberg Universität Mainz
Saarstr. 21, D-6500 Mainz, West Germany

INTRODUCTION

Organ surface P_{O_2} measurement by oxygen sensitive electrodes has proved to be an efficient tool for monitoring changes in tissue oxygenation status in a number of experimental and clinical situations. A parameter giving more direct information is the P_{O_2} distribution within tissue cells which, however, can only be assessed by more invasive methods. To date, a quantitative relation between P_{O_2} electrode measurements and P_{O_2} in tissue cells has not been established. Part of this problem lies in the fact that any surface electrode P_{O_2} measurement is not confined to tissue cells but rather represents some average over P_{O_2} values in a certain volume containing not only tissue cells but also blood vessels and connective tissue. Such a catchment volume of a typical P_{O_2} surface electrode of 15 μm diameter is thought to be a half sphere of about 25 μm diameter [7], the mean P_{O_2} in which corresponds to the measuring signal. The present study addresses the problem of relating surface P_{O_2} measurements to intracellular P_{O_2}. To that end, the notion of electrode catchment volume is critically evaluated. Its magnitude and errors in electrode P_{O_2} measurement are quantified for a frequently used electrode type. The results are then used to calculate the P_{O_2} which will be measured by the surface electrode in a tissue exhibiting the heterogeneous P_{O_2} distributions typical of working muscle.

ANALYSIS AND DISCUSSION

1. Types of errors in electrode P_{O_2} measurement

P_{O_2} electrodes consist of platinium wires which are connected to the cathode of a voltage source. The anode is linked to the tissue via an indifferent Ag/AgCl electrode. On contact with platinium, O_2 is reduced inducing a current which is in proportion to the oxygen concentration in the measuring medium. GRUNEWALD, who did extensive theoretical studies on P_{O_2} electrodes, distinguishes two types of errors [5]: The diffusion error and the error due to electrode O_2 consumption.

  a. **The diffusion error.** P_{O_2} electrodes continuously reduce oxygen which has to be supplied out of the measuring medium. In an infinite (relative to electrode dimensions) unstirred medium such as tissue, this supply takes place by diffusion. Therefore, the P_{O_2}

near the electrode surface will be lower than remote from it. This P_{O_2} drop constitutes the diffusion error DE. An exact definition of the diffusion error would be $DE = (P_0 - P)/P_0$ where P_0 and P are the P_{O_2}s at the electrode position before and after switching on the electrode voltage. If a large, unstirred medium of the same O_2 conductivity as the measuring medium is used for electrode calibration, this P_{O_2} drop is the same for calibration as for measurement, thus diffusion error has no effect on measured P_{O_2}. GRUNEWALD [5] gives a number of equations which can be used to correct for this error if media for calibration and measurement differ.

b. **The error due to electrode O_2 consumption.** In tissue, the idealized situation described in a. is not exactly true. Oxygen does not enter from an infinitely remote source but is supplied by capillaries throughout the tissue. Thus, the diffusion error becomes smaller than equations in [5] predict. On the other hand, electrode O_2 consumption interferes with tissue O_2 supply in that it renders RBC O_2 extraction larger. This is GRUNEWALD's second type of error which depends on a number of tissue and electrode parameters and which has to be determined in each particular situation separately.

2. General considerations on the catchment volume of P_{O_2} electrodes.

The catchment volume of a P_{O_2} electrode commonly is understood to be the volume of tissue the mean P_{O_2} in which is proportional to the electrode measuring signal. By taking a closer look at the mathematical description of intracellular P_{O_2} distributions without and with a P_{O_2} electrode present, we can obtain some information on the magnitude of this volume.

Quite generally[1], tissue P_{O_2}, P, may be described as the solution of the following boundary value problem:

(i) At any extravascular tissue location $\vec{x}$, P has to satisfy the partial differential equation $K \nabla^2 P(\vec{x}) = -A(\vec{x})$ where K is the tissue O_2 conductivity and $A(\vec{x})$ is the (local) O_2 consumption rate.

(ii) At the interfaces between tissue and blood vessels, P_{O_2} and diffusive O_2 flux have to be continuous and O_2 flux out of the vessels has to be matched with drop in blood O_2 content times flow rate.

(iii) Across the tissue surface there is no flux of O_2, thus a zero flux boundary condition holds, i.e. $\nabla P \cdot \vec{n} = 0$ where $\vec{n}$ is the normal vector on the surface.

An example of such a tissue P_{O_2} distribution is shown in Fig. 1, left panel, where the P_{O_2} in a tissue cross section containing two capillaries is displayed. The tissue surface is facing the viewer. The P_{O_2} at the outer perimeter of the capillary endothelium is prescribed as boundary condition, so no P_{O_2} profile is calculated within capillaries.

If we place a platinium P_{O_2} electrode on the surface of the tissue we obtain a second boundary value problem for the tissue P_{O_2}, in which the only thing that needs to be modified is the boundary condition: As O_2 is reduced instantaneously on contact with the platinium surface,

(iv) underneath the electrode a zero P_{O_2} boundary condition has to be substituted for the zero flux condition (iii).

In Fig. 1, right panel, the P_{O_2} distribution is shown which results if a platinium electrode is positioned in the middle between the two capillaries on the tissue surface of Fig. 1, left panel. This second problem may be solved by adding a suitable function h which is harmonic in the extravascular space (i.e. h satisfies $\nabla^2 h = 0$ in extravascular tissue locations) to P (the solution of the previous boundary value problem), which warrants that the new boundary condition is satisfied. In particular, it is required for this harmonic function that

---

[1]We assume here that the tissue is in steady state and that P_{O_2} is high enough so O_2 consumption rate is not P_{O_2} dependent. For ease of formulation, Mb facilitated diffusion is not addressed explicitly. If Mb is present, similar considerations hold for the effective P_{O_2} (cf. [3]) in place of P.

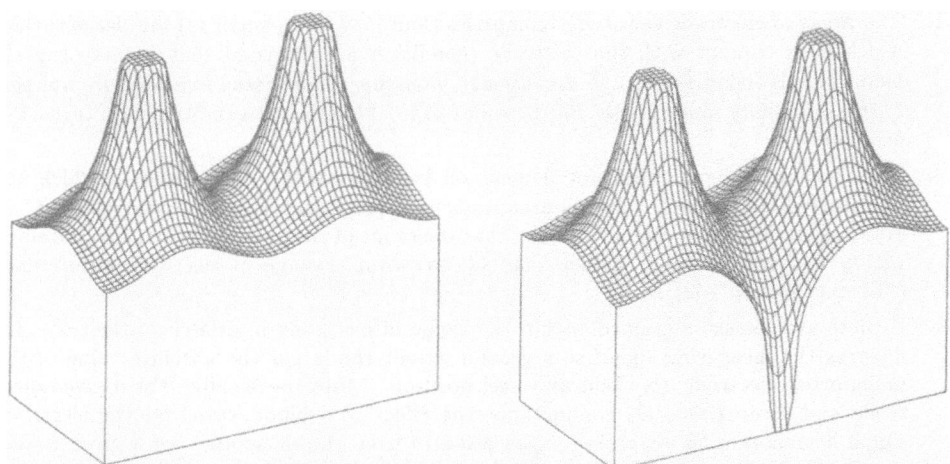

Figure 1, left panel: P_{O_2} distribution in a tissue cross section containing two capillaries. The tissue surface is facing the viewer. The P_{O_2} at the outer perimeter of the capillary endothelium is prescribed as boundary condition, so no P_{O_2} profile is calculated within capillaries. Right panel: P_{O_2} distribution resulting from profile at left if a platinium P_{O_2} electrode is positioned in the middle between the two capillaries on the tissue surface.

a. $h(\vec{x}) = -P(\vec{x})$ at locations $\vec{x}$ beneath the electrode (zero P_{O_2} at electrode),

b. $\nabla h \cdot \vec{n} = 0$ (zero flux) elsewhere in the tissue surface, and

c. P_{O_2} and O_2 flux at vessel-tissue interfaces match. In the special case that electrode O_2 consumption is negligible (i.e. vascular O_2 content and P_{O_2} are practically unchanged by the electrode) it is required that $h(\vec{x})$ approaches 0 as $\vec{x}$ approaches a blood vessel.

In somewhat simplified terms, h describes the P_{O_2} field of the electrode for the particular electrode geometry and tissue conditions considered. $h(\vec{x})$ is less than 0 and rises rapidly towards 0 with increasing distance of $\vec{x}$ from the electrode. As a consequence, interface condition c. is trivially satisfied if the electrode is positioned sufficiently far away from blood vessels. In this case of remote electrode O_2 supply, the electrode P_{O_2} field is not distorted by the presence of blood vessels. Consequently, h and hence the electrode signal (which is proportional to the total O_2 flux into the electrode: $K \int_{\text{electrode surface}} \nabla h \cdot \vec{n}$) exclusively depend on electrode geometry and on the P_{O_2} distribution prevailing at the electrode surface before electrode voltage was switched on (condition a.). We conclude that in this case the catchment volume is zero, more precisely, *it is degenerated to the tissue surface area in contact with the electrode*.

In the general case, there are further highly complex dependencies of the O_2 flux to the electrode and hence of the measuring signal on vascular geometries, O_2 contents, and blood flows (condition c.), rendering the above definition of a catchment volume as the volume of tissue, the mean P_{O_2} in which is reflected in the electrode signal, impractical[2]. However, from the above conditions a.–c. that the harmonic function h has to satisfy, some conclusions may be drawn concerning the region to which the electrode is sensitive:

---

[2]It may be of interest that the notion "catchment volume" has not been introduced in GRUNEWALD's original work [4]. He used the German word "Einzugsbereich" to denote tissue regions in which the P_{O_2} is changed by more than 10 % (or 5 %) of its original value when switching on electrode voltage. Nevertheless, his "Einzugsbereich" has been interpreted to be equivalent to a catchment volume in the above sense. In view of his definition, "region of electrode interference" might be a more appropriate translation than "catchment volume".

1. The range of electrode sensitivity comprises (and may be as small as) the tissue surface which is in contact with the electrode (condition a.). Beyond that, it may contain blood vessels (condition c.). *Extravascular, non-superficial tissue locations are not part of the sensitivity range* unless P_{O_2} becomes so low that O_2 consumption rate turns P_{O_2} dependent.[3]

2. The measuring signal does not depend on locations within the tissue in which the undistorted electrode P_{O_2} field is practically zero (condition c. trivially satisfied). Cum grano salis, this is equivalent to saying that the range of electrode sensitivity is contained within and considerably smaller than GRUNEWALD's range of electrode interference ("Einzugsbereich", [4]).

3. If there are vessels contained within the range of electrode interference, their P_{O_2} influences the measuring signal to a greater extent the larger the absolute value of the undistorted electrode P_{O_2} field at vessel position. More specifically: For a given electrode and given tissue O_2 conductance the effect of a blood vessel on the electrode signal decreases with increasing vessel distance from the electrode. For a given tissue, vessel effects decrease with decreasing electrode O_2 consumption rate.

Concluding we can summarize that an electrode catchment volume in the above straightforward sense is only loosely related to the range of electrode interference defined by GRUNE-WALD and is actually impractical in many situations. The tissue range to which the electrode signal is sensitive to almost never comprises a *volume* of tissue. Rather it consists of the tissue *surface* beneath the electrode and – possibly – some blood vessels.

Small electrode O_2 consumption is desirable as it attenuates both types of GRUNEWALD's errors [5] and distortion of the measuring signal by vessels and by P_{O_2} dependent O_2 consumption rate (both of which would be hard to quantify). Electrode O_2 consumption may be controlled by modifying electrode surface area or by covering the platinium surface with cellophane and/or teflon membranes the O_2 conductivity of which is small compared to tissue O_2 conductivity. Such membranes cut down the O_2 diffusive flux to the electrode, hence its O_2 consumption. However, small electrode O_2 consumption renders electrode sensitivity range shallow. Thus, in the limit of zero consumption, the measured P_{O_2} is an *area weighted average* over tissue surface P_{O_2} rather than a spatial average, the consequences of which will be discussed below.

3. Diffusion error, range of electrode interference, and O_2 consumption rate of P_{O_2} electrodes currently in use.

The diffusion field of a flat circular platinium electrode of radius R_0 has been described by GRUNEWALD [5]. He also showed that this field is well approximated by the diffusion field of a semi-spherical electrode (which has a radius of $r_0 = 2R_0/\pi$) the equations of which are much more practical [4] and derived these equations for electrodes covered by one membrane [5]. In analogy to [5] we find for the semi-spherical P_{O_2} electrode covered by two membranes of thicknesses d_1 and d_2 and of O_2 conductivities K_1 and K_2, respectively, the P_{O_2} distribution

$$
P(r) = \begin{cases}
P_\infty \dfrac{K_2/K_1}{\delta + \eta + \varepsilon}\left(1 - \dfrac{r_0 + d_1}{r} + \dfrac{d_1}{r_0}\right) & r_0 \leq r < r_0 + d_1 \\[2ex]
\dfrac{P_\infty}{\delta + \eta + \varepsilon}\left(1 - \dfrac{r_0 + d_1}{r} + \delta\right) & r_0 + d_1 \leq r < r_0 + d_1 + d_2 \\[2ex]
P_\infty\left(1 - \dfrac{K_2/K}{\delta + \eta + \varepsilon}\cdot\dfrac{r_0 + d_1}{r}\right) & r \geq r_0 + d_1 + d_2
\end{cases}
$$

---

[3]It ought to be kept in mind that to sites in which P_{O_2} is (even massively) changed by the electrode, electrode measuring signal needs not to be sensitive to (and in most cases actually is not). For determining if a tissue location is part of the electrode sensitivity range the crucial question is: Does the P_{O_2} present at this site before switching on electrode voltage exert any influence on the harmonic function h which, in turn, makes up for the electrode signal. From conditions a. and c. it follows that, at the most, surface locations beneath the electrode and vessels can influence h.

where K is the tissue O_2 conductivity, P_∞ is the P_{O_2} remote from the electrode and

$$\delta = \frac{K_2}{K_1} \cdot \frac{d_1}{r_0}, \qquad \eta = \frac{d_2}{r_0 + d_1 + d_2}, \qquad \varepsilon = \frac{r_0 + d_1}{r_0 + d_1 + d_2} \cdot \frac{K_2}{K}.$$

Diffusion error DE and O_2 consumption V_{O_2E} rate of this electrode are

$$DE = \frac{\varepsilon}{\delta + \eta + \varepsilon} \qquad \text{and} \qquad V_{O_2E} = K_1 2\pi r_0^2 P'(r_0) = \frac{K_2 2\pi (r_0 + d_1) P_\infty}{\delta + \eta + \varepsilon}.$$

A frequently used P_{O_2} surface electrode is the multiwire electrode by KESSLER and LÜBBERS (described e.g. in [7]). It consists of 8 platinium wires of 15 μm diameter and a common Ag/AgCl reference electrode, fused into a glass core. The electrodes are covered by a 12 μm cellophane membrane and a 12 μm teflon membrane. GRUNEWALD [5] cites the O_2 conductivities of cellophane and teflon to be $K_1 = 2.6 \cdot 10^{-9} \frac{ml\,O_2}{cm \cdot min \cdot mm\,Hg}$ and $K_2 = 8.7 \cdot 10^{-9} \frac{ml\,O_2}{cm \cdot min \cdot mm\,Hg}$, respectively. Muscle O_2 conductivity for free diffusion (which is a lower bound to effective conductivity) is $K = 1.5 \cdot 10^{-8} \frac{ml\,O_2}{cm \cdot min \cdot mm\,Hg}$ [3]. In the worst case (only free diffusion in muscle, electrode calibration in a medium with infinitely high diffusivity), the diffusion error of this electrode is $DE = 3.7\%$. At effective diffusivities pertinent to muscle cell P_{O_2}s which are typical for working muscle (c.f. [3]), the diffusion error would be about 2 % or less. Consequently, in any case the range of electrode interference (even for the 95 % definition by GRUNEWALD [5]; cf. footnote 2) does not extend beyond the electrode membranes. According to subsection 2, the range of electrode sensitivity is confined to the muscle surface and electrode P_{O_2} readings are rather an area than a volume weighted average.

From the above equations we calculate an O_2 consumption rate of $1.0 \cdot 10^{-11} \frac{ml\,O_2}{min \cdot mm\,Hg}$ for this electrode. In order to estimate the effects of such an additional O_2 consumer on the tissue considered we have to relate it to tissue parameters. For examples, we are going to use two muscles, stimulated at 1 or 6 Hz which have been studied in [2]. Measured median intracellular P_{O_2}s are 13 $mm\,Hg$ and 2 $mm\,Hg$, and capillary densities are 993 mm^{-2} and 738 mm^{-2}, respectively. Assuming that the O_2 flux to the electrode is supplied out of sections of 30 μm length (twice the electrode diameter) of two capillaries, electrode consumption is equivalent to an increase in capillary domain cross sectional area along these sections of 48 μm^2 or of 2 μm^2, respectively. If capillary domains can be modelled as Krogh cylinders this increase would correspond to changes in Krogh radius from 17.9 μm to 18.3 μm or from 20.77 μm to 20.78 μm, respectively. Variations of this magnitude are well within the range of physiological variations in capillary density, so unless electrode application leads to mechanical disturbance in flow or capillary density it should have no appreciable effect on tissue O_2 supply.

4. Electrode P_{O_2} measurements in tissues with heterogeneous P_{O_2} distributions.

For understanding cell metabolism it is essential to know the P_{O_2} present in tissue cells under consideration. At low O_2 consumption rates, intracellular P_{O_2} is largely in equilibrium with capillary P_{O_2}, and P_{O_2} throughout all of the tissue is pretty much constant. Therefore, methods with high spatial resultion (such as cryophotospectroscopy) furnish very similar P_{O_2} values as methods with lower resolution (like P_{O_2} surface electrodes). At high O_2 fluxes, considerable P_{O_2} gradients are required to drive oxygen from capillaries to tissue cell mitochondria rendering tissue P_{O_2} distributions inhomogeneous. In this situation, the P_{O_2} measured by a method which does not resolve the spatial heterogeneity, is the result of some averaging process, and the exact nature of this averaging process is as crucial for the value of the measuring signal as the actual P_{O_2} distribution at the measuring site. In this section an example will be given on how a tissue P_{O_2} distribution in muscle and mean intracellular P_{O_2} relate to the P_{O_2} measured by a surface electrode.

Muscle has been chosen for this study for the following reason: Even though intracellular P_{O_2} in working muscles has been shown to be surprisingly homogeneous [1] the great P_{O_2} differences between capillaries and muscle cells render overall tissue P_{O_2} distribution extremely heterogeneous [3]. In order to assess the effect of such heterogeneities on the P_{O_2} measured by surface electrodes, the measuring process has been modelled in a computer simulation:

1. In a first step, a P_{O_2} distribution for the muscle stimulated at 6 Hz from [2] has been calculated using a mathematical model (the details of which are to be described elsewhere) which considers radial diffusion from capillaries into the tissue. The model takes account of the following characteristics of muscle O_2 supply:

   a. RBC O_2 unloading along the capillary,
   b. effects of the particulate nature of blood,
   c. free and hemoglobin-facilitated O_2 diffusion and reaction kinetics inside RBCs,
   d. free and myoglobin-facilitated O_2 diffusion inside the tissue,
   e. carrier-free region separating RBC and muscle cell,
   f. capillary-to-fiber ratio of 1 or, optionally, of 2.

   The resulting P_{O_2} distribution for stimulation at 6 Hz in a longitudinal section through a capillary domain is displayed in Fig. 2, left panel.

2. In a second step, mean muscle fiber P_{O_2} of the calculated three-dimensional P_{O_2} distribution in the left panel of Fig. 2 has been determined. Furthermore, a muscle cell P_{O_2} histogram has been obtained from the calculated distribution according to the same protocol as the experimental one in [2] ("measuring" sites at least 10 μm away from

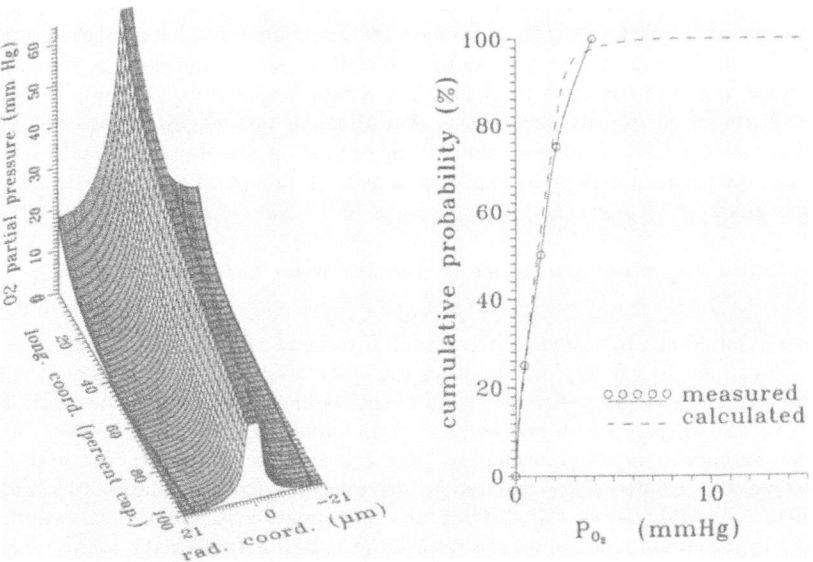

Figure 2, left panel: Longitudinal section through the calculated P_{O_2} distribution in a capillary domain of a muscle stimulated at 6 Hz. Right panel: Measured P_{O_2} histogram for this muscle along with a model histogram which is obtained according to the same protocol as the measured one ("measuring" sites at least 10 μm away from capillaries) from the three-dimensional profile in Fig. 2, left panel.

capillaries). Fig. 2, right panel, shows the measured P_{O_2} histogram for this muscle along with the model histogram. The latter was obtained independently, i.e. was *not* produced to match the experimental histogram. Median P_{O_2}s of both histograms are 1.1 or 1.3 $mm\,Hg$, respectively, and histogram shapes are surprisingly close.

3. In the third step, P_{O_2} electrode measurement is simulated. From the theoretical considerations in section **2.** as well as from electrode data calculated in section **3.** it follows that electrode measuring signal is an area weighted average over muscle surface P_{O_2}. In order to extract an approximation to muscle surface P_{O_2} distribution from the three-dimensional P_{O_2} distribution calculated in 1., morphometric data on muscle surfaces are required. This kind of data has not been obtained yet. Therefore, the present analysis is based on the worst (most heterogeneous) case in which capillaries are located immediately beneath or on the surface. For this case, muscle surface P_{O_2} profile may be assumed to be equal to the two-dimensional P_{O_2} distribution in a longitudinal section through the capillary domain (Fig. 2, left panel) and the mean over this distribution to represent the electrode measuring signal.

Fig. 3 illustrates the resulting mean muscle fiber P_{O_2} (from step 2., dashed horizontal line) and simulated surface electrode P_{O_2} (from step 3., dotted horizontal line) in their relation to a typical radial P_{O_2} distribution (solid) half way down a muscle capillary, with the capillary in its center. The dotted horizontal line labelled "mean surface P_{O_2}" gives the area weighted average over the P_{O_2} profile, while the dashed line labelled "mean fiber P_{O_2}" gives the volume weighted average P_{O_2} in the muscle cells. Simulated electrode P_{O_2} is 7 $mm\,Hg$ while mean muscle cell P_{O_2} is 1.9 $mm\,Hg$ and thus only 27 % of P_{O_2} from simulated electrode measurement.

In spite of these pronounced differences in mean P_{O_2}s, the predicted electrode P_{O_2} is far below P_{O_2} values from actual electrode measurements in electrically stimulated muscles (e.g. [6]). A possible reason (which was not taken into account in the model) could be that capillaries located on the muscle surface have capillary domains the magnitudes of which would be roughly only one half of those located deeper in the tissue (if capillary density does not change with depth into tissue). As a consequence, superficial peri-capillary P_{O_2} drops (which are the predominant cause for tissue P_{O_2} heterogeneities) are less pronounced rendering surface P_{O_2} higher. Other possible reasons are that data for muscles which have been studied by P_{O_2} surface electrodes do not match our present input data: Lower O_2 consumption rates or higher perfusion rates both would render mean surface P_{O_2} higher.

A final problem one has to be aware of when measuring surface P_{O_2} is the presence of

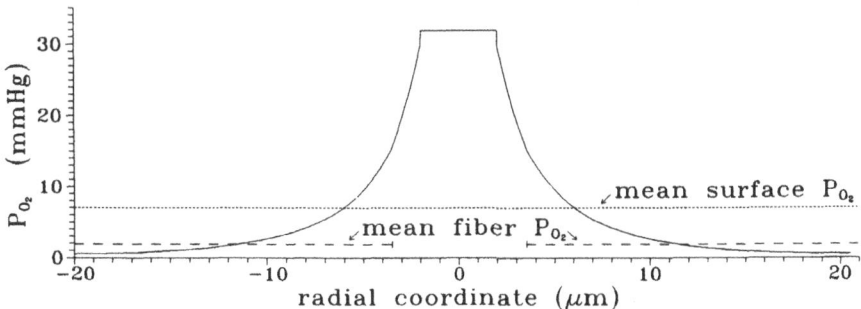

Figure 3: Typical radial P_{O_2} distribution half way down a muscle capillary with the capillary in its center. The dotted horizontal line labelled "mean surface P_{O_2}" gives the area weighted average over this profile, while the dashed line labelled "mean fiber P_{O_2}" gives the volume weighted average P_{O_2} in the muscle cells.

a surface fascia which in dog gracilis typically has a thickness of 500 μm or more [HONIG, personal communication]. LINDBOM [8] reported that superficial terminal muscle arterioles exhibit terminal branches perfusing the fascia. He suggested that these branches could represent functional a-v-shunts (as O_2 consumption in the fascia is negligible). As a consequence, in the fascia P_{O_2} values would be expected which are much higher than muscle cell P_{O_2}s and are rather representative of mean capillary P_{O_2}. This idea is supported by the facts that in the fascia O_2 consumption rate and diffusivity both are low. Thus, if surface P_{O_2} measurement is to reflect tissue conditions (within the limitations discussed above), the fascia needs to be entirely removed. In doing so, care has to be taken not to disrupt the above mentioned terminal arteriolar branches because induced hemorrhage from microlesions can serve as an O_2 source which is large relative to the size of the electrode.

The present study shows quite pronounced differences between mean muscle cell P_{O_2} in heavily working muscles and predictions of P_{O_2} measured by surface electrodes. However, these findings are based on a number of assumptions. Only in the case in which the muscle surface is free of both surface fascia and capillaries, electrode measured surface P_{O_2}s may be expected to represent intracellular P_{O_2}. Any deviation from this arrangement is bound to raise electrode measured surface P_{O_2} above mean intracellular P_{O_2}, possibly by significant amounts. In order to obtain more definite modelling results on differences between muscle cell P_{O_2} and P_{O_2} measured by surface electrodes, quantitative information on vascular morphology in fascia and superficial muscle layers is needed.

SUMMARY AND CONCLUSION

Organ surface P_{O_2} measurement by oxygen sensitive electrodes has proved to be an efficient tool for monitoring changes in tissue oxygenation status. A parameter giving more direct information is the P_{O_2} distribution within tissue cells which, however, can only be assessed by more invasive methods. The present study addresses the problem of relating electrode measured surface P_{O_2} to muscle cell P_{O_2}. To that end, the magnitude of tissue volumes the P_{O_2} in which electrodes are sensitive to, has been reassessed.

It turned out that the measuring signal of current membrane covered P_{O_2} electrodes is only sensitive to the P_{O_2} within the muscle surface and not any deeper, thus rendering the measured P_{O_2} an area weighted average over the surface P_{O_2} rather than a spatial average. The consequences of this finding are illustrated in an example of a maximally working muscle. Under the assumption that there are capillaries located on the muscle surface or immediately beneath it, the P_{O_2} predicted to be measured by a surface electrode is almost four times the average muscle fiber P_{O_2}. However, P_{O_2} values actually measured by surface P_{O_2} electrodes are even higher calling for further explanation, in which superficial blood vessels and surface fascia may play an important role. Quantitative information on vascular morphology near muscle surfaces is needed in order to more definitely determine the importance of mechanisms discussed, for experimental results.

ACKNOWLEDGEMENTS

The author wishes to thank Drs. C. Honig and G. Thews for their many helpful and stimulating comments and suggestions.

References

[1] T.E.J. GAYESKI, C.R. HONIG, O_2 gradients from sarcolemma to cell interior in red muscle at maximal $\dot{V}_{O_2}$, Am.J.Physiol. 251:H789–H799 (1986)

[2] T.E.J. GAYESKI, C.R. HONIG, Intracellular P_{O_2} in long axis of individual fibers in working dog gracilis muscle, *Am.J.Physiol.* 254:H1179–H1186 (1988)

[3] K. GROEBE, A versatile model of steady state O_2 supply to tissue. Application to skeletal muscle, *Biophys.J.* 57:485–498 (1990)

[4] W.A. GRUNEWALD, Zur Theorie der Ausgleichsvorgänge an Pt-Elektroden und ihre mathematischen Grundlagen, Dissertation, Univ. Marburg, 1966

[5] W.A. GRUNEWALD, Diffusionsfehler und Eigenverbrauch der Pt-Elektrode bei P_{O_2}-Messungen im steady state, *Pflügers Arch.* 320:24–44 (1970)

[6] D.K. HARRISON, S. BIRKENHAKE, S.K. KNAUF, AND M. KESSLER, Local oxygen supply and blood flow regulation in contracting muscle in dogs and rabbits, *J.Physiology* 442:227–243 (1990)

[7] M. KESSLER, D.K. HARRISON, AND J. HÖPER, Tissue oxygen measurement techniques, in: "Microciculatory Technology", pp. 391–425, C.H. BAKER, W.L. NASTUK, eds., Academic Press, Orlando, 1986

[8] L. LINDBOM, Distribution patterns of blood flow in the rabbit tenuissimus muscle in response to brief ischemia and muscular contraction, *Microvasc.Res.* 31:143–156 (1986)

A NEW CATHETER FOR QUASI-CONTINUOUS MEASUREMENT OF ARTERIAL PARTIAL

OXYGEN PRESSURE

E.P. Eijking, J.A.H. Bos, W. Schelter*, W. Gumbrecht*, W. Erdmann and B. Lachmann

Dept. of Anaesthesiology, Erasmus University Rotterdam, The Netherlands and
*Siemens AG, Erlangen, FRG

INTRODUCTION

Partial oxygen pressure (PaO_2) and oxygen saturation (SaO_2) of arterial blood are important parameters for evaluation of oxygen delivery from the alveoli to the pulmonary circulation and are used to assess whether the patients' lungs are properly ventilated. Currently, it is routine to take a blood sample from patients whenever hypoxemia is suspected or some adjustment to the ventilator settings is required. The problem with the conventional blood sampling method is that sudden changes of blood gas values cannot be observed (Lam, 1987). Though there have been some developments in continuous monitoring of patients' oxygenation status (e.g. pulse oximetry and transcutaneous PO_2 electrodes; for review see: Parker, 1987), it is not possible to continuously or quasi-continuously measure PaO_2, $PaCO_2$, SaO_2 and Hb concentration at the same time. Recently a method for on-line measurements was introduced for monitoring SaO_2 together with Hb and Ht (Schelter et al, 1990; Eijking et al, 1990). High correlations were found between the new method and standard methods (OSM 2 Hemoximeter, Radiometer, Denmark and a Ht centrifuge).

Recently this catheter has been further developed for quasi-continuous measurements of PaO_2. The purpose of this study was to compare the newly developed measuring cell with a standard method for measuring PaO_2 (ABL 330 Acid-Base Laboratory, Radiometer, Denmark) at different O_2 concentrations.

MATERIALS AND METHODS

Sample taking system

The system comprises a two-lumen catheter with an electro-chemical flow-through cell, a pump unit with the polarographic instrumentation and a personal computer. The flow-through cell is a sandwich arrangement of a chip carrier, a silicon chip with the electrodes and a PMMA slide with an engraved flow-through channel. The working cycle is described elsewhere (Gumbrecht et al, 1990).

The planar sensor arrangement consists of a platinum working-electrode, a platinum counter-electrode and a Ag/AgCl reference-electrode (Fig 1). The area of each electrode is about 25 microns x 1000 microns.

Voltage drops across the working-electrode against reference-electrode is regulated by a potentiostat. The resulting electrode current is current-to-voltage converted and passed to the personal computer. Oxygen tension is a function of the quotient of measuring current divided by calibration current.

The catheter is placed in a blood vessel and every 180 s a blood sample of 10 μl is withdrawn using the computerized pump system. About 30 s after sampling PaO_2 values are available. After this measuring phase, the cell is immediately flushed with a heparinized calibration fluid, which is exposed to air.

Animal study

Two pigs (approximately 14 kg BW) were anesthetized with midazolam 0.5 mg/kg i.m., ketamine 10 mg/kg i.m. and pentobarbital 10 mg/kg i.v., tracheotomized and paralyzed with pancuronium bromide. During the experiments pancuronium bromide 0.5-1.0 mg/kg/h and midazolam 1-3 mg/kg/h were used to maintain anesthesia. The animals were pressure-controlled ventilated with a Servo Ventilator 900 C (Siemens-Elema, Solna, Sweden) at the following ventilator settings: FiO_2 = 0.1-1.0, F = 15/min, peak airway pressure = 15 cm H_2O, PEEP = 3 cm H_2O and inspiratory-expiratory ratio = 1:2. The catheter leading to the measuring cell was placed in the right femoral artery. In the left femoral artery an arterial line was placed from which blood samples were taken for comparison. The amount of oxygen delivered to the lungs was changed by altering the FiO_2 from 0.1 to 1.0 and was monitored by measuring the inspiratory oxygen concentration, which was measured with the Servo Gas Monitor 120 (Siemens-Elema, Solna, Sweden). Approximately 5 min after changing the FiO_2, PaO_2 (as measured by the measuring cell) was recorded and at the same time a blood sample was taken from the left femoral artery and PaO_2 was measured using the ABL 330.

At the end of the study the animals were sacrificed with an overdose of intra-venously administered KCl.

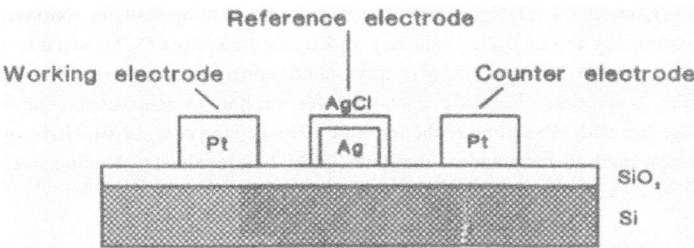

Figure 1.   Diagram of the electrode arrangement.

RESULTS

In both pigs, measurements were performed over a 7-hour monitoring period. Figure 2 shows PaO_2 values from the measuring cell plotted against PaO_2 values measured with the ABL 330 in the range 0-160 mmHg. The PaO_2 values derived from the measuring cell are the result of a function of the quotient of measuring current divided by calibration current and are displayed as percentages; PaO_2 values measured by the ABL 330 are displayed in mmHg. Altogether 83 paired values were acquired. Regression analysis shows a high correlation (r=0.98, p<0.001). Figure 3 shows PaO_2 values of the two methods plotted against each other in the range 160-600 mmHg. In this range 16 paired values were recorded; the correlation is also very high (r=0.99, p<0.001).

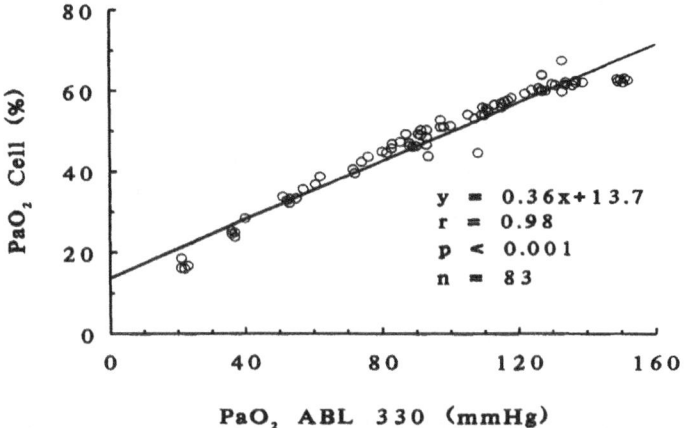

Figure 2. Relationship between PaO_2 measured with the measuring cell and the PaO_2 measured with the ABL 330 in the range 0-160 mmHg.

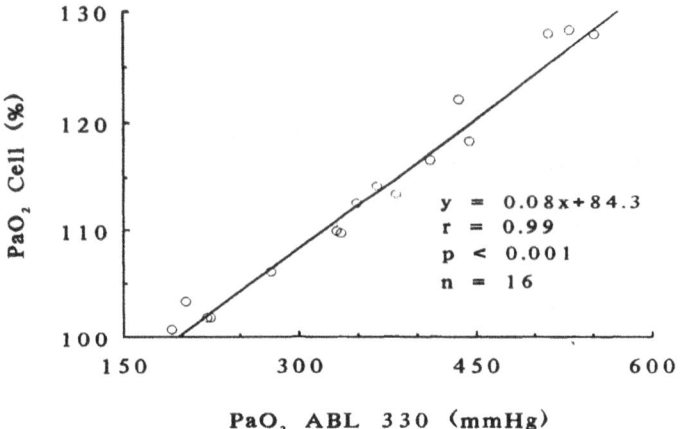

Figure 3. Relationship between PaO_2 measured with the measuring cell and the PaO_2 measured with the ABL 330 in the range 160-600 mmHg.

DISCUSSION

The results of this study show that the newly developed cell is capable to accurately measure PaO_2, both in the range of 0-160 mmHg and 160-600 mmHg. The electrodes behave differently below 160 mmHg compared to higher levels, because the measuring cell is calibrated with a fluid which is exposed to normal breathing air (PO_2 = 160 mmHg). By adjusting the software the PaO_2 values will be displayed in mmHg over the whole range of 0-600 mmHg.

The computerized pump system has been described in detail by Gumbrecht and colleagues (1990). The advantages of this system are that the measuring cell is calibrated after every measuring phase, thus preventing drift of the electrodes and preventing deposition of fibrin in the cell. The duration of the total measuring cycle in this study was 180 s, but can easily be shortened to 120 s. Integration with the SaO_2 and Hb measuring cell as decribed earlier (Eijking et al, 1990; Bos et al, 1990) and the existing system for online measurement of electrolytes (Gumbrecht and Schelter, 1987) is intended.

A system which can continuously or quasi-continuously monitor the total oxygenation status in patients will be of great advantage not only for anesthesia and intensive care of adults, but also for neonates.

SUMMARY

Recently a catheter has been developed based on amperometric measurement principle for in vivo monitoring of PaO_2. A study in pigs was performed to compare the cell with a standard method for measuring PaO_2. The results show that the cell is capable to accurately measure PaO_2.

ACKNOWLEDGEMENT

This work was in part supported by The Dutch Foundation for Medical Research (SFMO).

REFERENCES

Bos, J.A.H., Schelter, W., Gumbrecht, W., Montag, B., Eijking, E.P., Armbruster, S., Erdmann, W. and Lachmann, B., 1990, Development of a micro transmission cell for in vivo measurement of SaO_2 and Hb, Adv. Exp. Med. Biol., (in press).

Eijking, E.P., Bos, J.A.H., Armbruster, S., Schelter, W., Gumbrecht, W., Erdmann, W. and Lachmann, B., 1990, Online measurements of SaO_2, Ht and Hb using a micro transmission cell, Adv. Exp. Med. Biol., (in press).

Gumbrecht, W., Schelter, W. and Montag, B., 1987, A chemFET microcell system for medical and biotechnical online electrolyte monitoring, Third Conference on Sensors and their Applications, Cavendish Laboratory, Cambridge, UK: 155.

Gumbrecht, W., Schelter, W., Montag, B., Rasinski, M. and Pfeiffer, U., 1990, Online blood electrolyte monitoring with a chemFET microcell system, Sensors and Actuators, B1: 477.

Lam, A.M., 1987, Continuous arterial PO_2 monitoring, Can. J. Anaesth., 34:1: 58.

Parker, D., 1987, Sensors for monitoring blood gases in intensive care, J. Phys. E: Sci. Instr., 20: 1103.

Schelter, W., Gumbrecht, W., Montag, B., Bos, J., Eijking, E. and Lachmann, B., 1990, A micro transmission cell for monitoring of oxygen saturation and hemoglobin concentration, Sensors and Actuators, B1: 495.

CONTINUOUS INTRA-ARTERIAL PO$_2$ MONITORING DURING THORACIC SURGERY

M. Fennema, R.J. van Krugten, H.J. de Boer, O. Prakash and W. Erdmann

Department of Anesthesiology, Erasmus University, Rotterdam, The Netherlands

INTRODUCTION

Anesthesia and surgery, especially thoracic surgery, can have negative effects on oxygenation. It is therefore necessary to monitor arterial oxygen partial pressure (PaO$_2$) closely. Conventionally, this is done by intermittent blood gas analysis. This is normally performed at certain intervals so that acute changes in oxygenation are usually not detected immediately. Furthermore, blood gas analysis is often carried out when one expects a change in oxygenation, for instance after one-lung ventilation has commenced. This has the disadvantage of a time delay in results, and possible intervention, of from 3 to 30 minutes. Even within 3 minutes the clinical situation can have completely changed, especially in major surgery or in critically ill patients. It is therefore necessary to monitor PaO$_2$ continuously. Pulse-oximetry has been an improvement as an early warning for hypoxia. Unfortunately this is not always reliable, especially in low perfusion states or during hypothermia (Striebel and Kretz, 1989). Also it is not possible to measure hyperoxia with this sensor because a hemoglobin saturation is being calculated, no partial pressure is measured (Ozorn et al., 1984). Another continuous monitor is transcutaneous PO$_2$ (PtO$_2$) measurements. Although this does reflect arterial PO$_2$, it is only accurate in infants and the newborn and is also unreliable under the same conditions as pulse-oximetry (Green et al.,1987).

In the last 20 years mass spectrometric, gas chromatographic, and optical methods have been developed to monitor PaO$_2$ in a continuous or semi-continuous fashion (Eberhard et al., 1979). However, these methods still appear too cumbersome for routine clinical use, with maybe the exception of the fiberoptic method (Barker et al., 1986). For the last 15 years it has been possible, in experimental situations, to measure arterial PO$_2$ continuously using a Clark-type electrochemical sensor (Holcomb et al., 1976). In recent years a PaO$_2$ sensor system has become commercially available. Several authors have reported on its reliability and its correlation with intermittent blood gas analysis (Rithalia et al, 1981, Pfeifer et al., 1988). Recently the system has been improved, making it easier to insert the sensor, calibrate and correct for temperature.

This study was performed to test the new Continucath PaO$_2$ sensor (Biomedical Sensors, Hilekes, Almere, The Netherlands) under clinical conditions. Thoracic surgery was chosen because radial arterial cannulation is mandatory and large changes in oxygenation are expected, so that a substantial range of PO$_2$ in vivo measurements could be obtained.

MATERIAL AND METHODS

Anesthesia and Surgery

Measurements were made in 20 adults with ages ranging from 20 to 77 years (mean 54.4 $\pm$ 16.7) who underwent thoracic surgery for mostly lobectomy, pneumectomy or pleurectomy. Cardiac surgery was excluded from this study. Depending on the operation 1 or 2 peripheral intravenous lines were inserted. All patients had a central line inserted via the basilic vein. After infiltration with 1% lidocaine, an 18G arterial canula (Surflow) was inserted in the radial artery. The Continucath PaO_2 sensor was inserted in the radial canula as described later. An epidural catheter was inserted between thoracic vertebrae 5 and 6. After a test dose of 3 ml bupivacaine 0.5%, the epidural catheter was topped up with 5 to 6 ml bupivacaine 0.5% and 0.05 to 0.06 mg sufentanil. After receiving 0.002 to 0.005 mg/kg fentanyl i.v., anesthesia was induced by either 3 to 5 mg/kg thiopental i.v. or 0.15 to 0.2 mg/kg etomidate i.v. (depending on the condition of the patient). Muscle relaxation was achieved with administration of 0.08 to 0.1 mg/kg pancuronium and the patient was intubated with a double-lumen endotracheal tube. The patient was ventilated with a Siemens Servo 900 C ventilator.

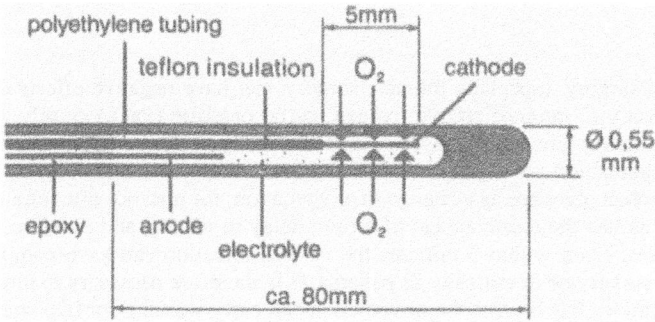

Fig. 1. Schematic drawing of the tip of the Continucath intravascular PaO_2 sensor. For description see the text.

Most patients were ventilated with an O_2-N_2O mixture. Some (3 patients) were ventilated with an O_2-air mixture, depending on the condition of the patient - for instance patients with large bullae did not receive N_2O. Isoflurane was added to the inspiratory gas mixture as required. After surgery had commenced fentanyl and pancuronium was given as required. The following parameters were continuously monitored and registered by a computer system developed by the Thorax Center, University Hospital Dijkzigt, Rotterdam: E.C.G., heart rate (H.R.), arterial blood pressure (A.P.), central venous blood pressure (C.V.P.), inspiratory O_2 concentration (FiO$_2$), expiratory CO_2 concentration (FeCO$_2$), isoflurane concentrations, ventilatory pressures, expiratory minute volume, respiration rate (R.R.), lung compliance and fluid balance. Off-line Arterial blood gas analysis was performed regularly using an automatic blood gas analyzer (Radiometer ABL 300) and PO$_2$, PCO$_2$, pH, bicarbonate, base excess and SaO$_2$ were registered. After surgery residual muscle relaxation was antagonized with neostigmine and the patient was extubated. Extra O_2 was given by mask and the patient was transported to the I.C. unit. Here H.R., A.P., C.V.P., and R.R. were registered continuously. Blood gas analysis was taken 2 hourly until the patient was completely stable. A mixture of bupivacaine 0.1% and sufentanil 0.002 mg/ml was given continuously via the epidural catheter at 3 to

5 ml/hour for 1 to 3 days post-operatively so that no extra pain medication was necessary. When the patient was stable and no extra oxygen was necessary, the arterial canula and C.V.P. line was removed. All patients left the I.C. unit within 30 hours.

The PaO$_2$ sensor

The sensor consists of a flexible, oxygen-permeable, heparin-coated, hydrophobic polymer tubing with a 0.55 mm external diameter (see figure 1.). This tubing contains a silver cathode and a silver-AgCl anode in an electrolyte solution. The sensor is delivered with a protective covering sheath containing an isotonic fluid to keep the sensor sterile. The covering sheath (see figure 2) is removed and the "T" fitting is connected to the hub of the indwelling arterial canula. The sensor is advanced into the artery by pulling on the distal introduction tubing system. It is advanced beyond the tip of the canula by at least 5 cm. The rear screw assembly is tightened to prevent fluid leakage and the distal part of the tubing is cut off. The pressure monitoring port is connected to an arterial pressure transducer. A continuous flush of 4 ml/hour heparinized saline solution is utilized to maintain adequate resolution of the blood pressure waveform. All this can be done aseptically without using any specific

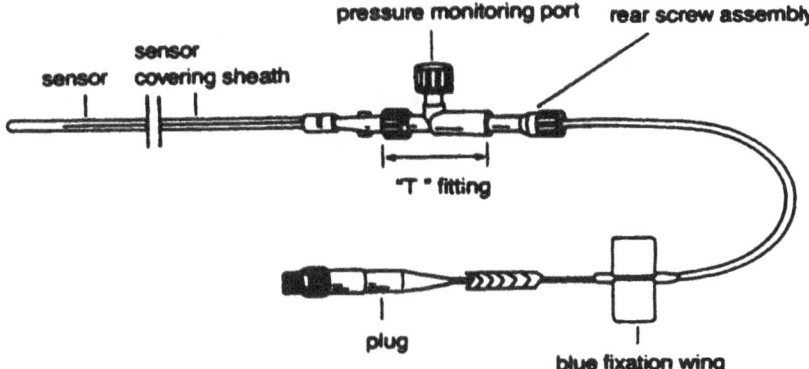

Fig. 2. Schematic drawing of the PaO$_2$ sensor and introduction system. The distal part of the delivery tubing is not shown. For description see text.

sterile procedures - no sterile gloves are needed. With this system it is possible to measure PaO$_2$ and A.P. simultaneously and continuously, while aspirating blood for blood gas analysis intermittently, all with the same arterial canula. The plug of the sensor is connected to a cable which in turn is connected to the PaO$_2$ monitor. The monitor has a digital display of PaO$_2$, scaled in both kilopascals (kPa) and millimeters of mercury (mm Hg). There are variable upper and lower alarm controls with visual and auditory signals and a mute key. There is a key to change mid-plateau polarization voltage from -800 Mv under O$_2$-N$_2$ ventilation to -600 mV when replacing N$_2$ with N$_2$O. This is essential because, as shown in figure 3, the silver cathode is sensitive to N$_2$O at -800 mV and not at -600 mV (Sugioka et al., 1987). Temperature correction keys are present because the sensor
is sensitive to temperature. This is corrected electronically by 4% per degree Celsius. An in vivo calibration is performed electronically in the following way:

After the electrode is connected to the polarization voltage (-600 mV or -800 mV, depending on the situation) the sensor is left to stabilize for 10 minutes. The temperature is adjusted to that of the patient. A button is pressed to store the (uncalibrated) PaO$_2$ readout. Immediately after this blood is aspirated for blood gas analysis. The system is flushed with about 10 ml of heparinized saline solution. When the results of the blood gas analysis return from the laboratory the stored value can be recalled and adjusted to the ABL 300 PO₂ value - all electronically - no calculations are required. Now the system is ready for real-time PaO₂ monitoring. All that is needed is to check that the temperature is adjusted to that of the patient. According to the manufacturer the system should be calibrated once every 24 hours.

All this sounds very complicated but actually it is very easy. The authors could insert and calibrate the sensor system with the aid of the instructions, without having the procedure being demonstrated by a third party. Only in one case was it not possible to insert the sensor because of kinking of the tubing. In retrospect this was due to the fact that the rear screw assembly was not loosened enough. In another case it was not possible to calibrate because of interference caused by diathermy.

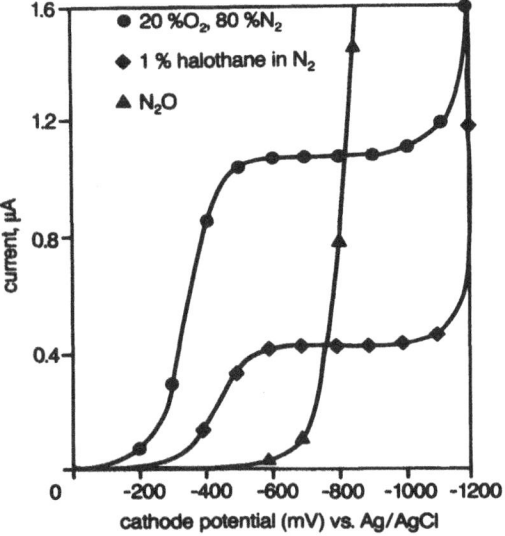

Fig. 3. Polarograms of the PO₂ sensor in presence of oxygen and nitrogen, halothane in nitrogen, or nitrous oxide; measured in vitro in blood stream. Flow 50 cm/sec.

Collection of Results

After induction of anesthesia the monitor was set to -600 mV, except in the 3 cases that the patients were ventilated with O₂-N₂ where the monitor was set to -800 mV. The monitor was adjusted to the rectal temperature of the patient and a calibration was performed as described above. During surgery the read-out of the PaO₂ monitor was recorded. Immediately after this blood gas analysis was performed to compare this read-out with the PO₂ of the ABL 300 blood gas analyzer. After surgery. in the I.C. unit, the sensor was recalibrated with a polarization voltage of -800 mV. The I.C. nurses performed blood gas analysis at there own discretion, always noting the PaO₂ read-out before aspirating blood. The nurses were told to treat these patients in exactly the some way as the other patients. That meant that the sensor could be removed if the system caused any problems. Also the arterial canula (and therefore the sensor system) were removed according to the normal protocol, which was normally about 12 hours after surgery. Problems with the sensor/arterial pressure system were also recorded.

RESULTS AND DISCUSSION

As mentioned before, two patients had to be excluded from the trial because in one case it was impossible to insert the sensor and in the other it was not possible to calibrate because of electrical interference. This interference caused by diathermy also caused problems in other patients but normally readings returned to normal values after cautery ceased. In 4 patients this was not the case and the monitor "froze" so that the monitor had to be switched off and on, and the complete calibration procedure repeated. Rithalia et al. (1981) also had problems with diathermy. As in their study, long burst of diathermy seemed to cause problems. Strangely enough, in many patients diathermy did not affect readings at all. No satisfactory explanation could be found.

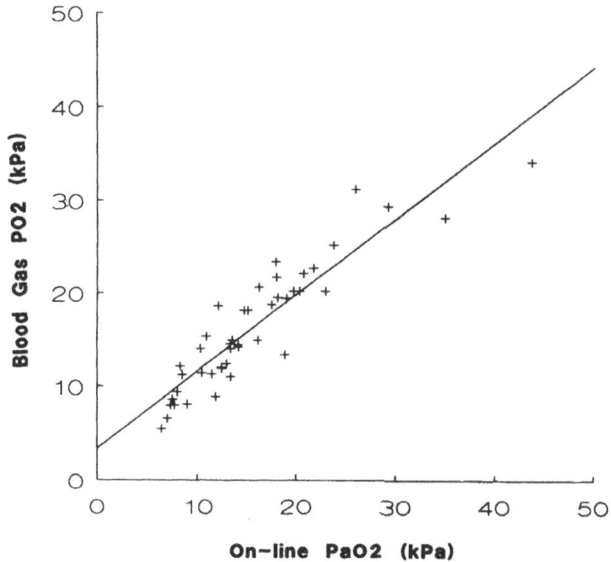

Fig. 4. The relationship between Continucath on-line PaO_2 and simultaneous values obtained from arterial blood gas samples during surgery. Correlation coefficient is 0.92 (n = 44).

Figure 4 shows the correlation between Continucath PaO_2 values and PO_2's obtained by blood gas analysis (PbO_2) in the 18 subjects where it was possible to obtain results. The equation of the regression line is $PbO_2 = 0.82\ PaO_2 + 3.4$ (r = 0.92, p < 0.05, n = 44). The sensor output always responded almost immediately to changes in oxygenation. The 90% response time was 90 seconds. The correlation between PaO_2 and PbO_2 in the post-operative period is shown in figure 5. The equation of the regression line is $PbO_2 = 0.84\ PaO_2 + 3.2$ (r = 0.83, p < 0.05, n = 30). Although significant, this correlation is not as good as during surgery. This could be caused by the different polarization voltage (normally -600 mV during surgery and -800 mV post-operatively) although the work of Green et al. (1987) does not support this as they have a correlation coefficient of 0.91 using -800 mV. When analyzing the results where the polarization voltage was -600 mV, the correlation coefficient was 0.93 (15 patients). Unfortunately, there were only 3 patients where the sensor was supplied with -800 mV during surgery (those patients receiving O_2-N_2) so that it was not possible to test this hypothesis. Another reason for the difference in correlation could be that intra-operative blood gas analysis was performed by the motivated authors, while the post-operative measurements were done by I.C. nurses. It is possible that recording of PaO_2 was not performed just before aspiration of blood. Also the blood samples are not handled as quickly in the post operative period as during surgery.In other words this poor correlation might not reflect inaccuracy of the intravascular electrode, but inaccuracy of blood gas analysis.

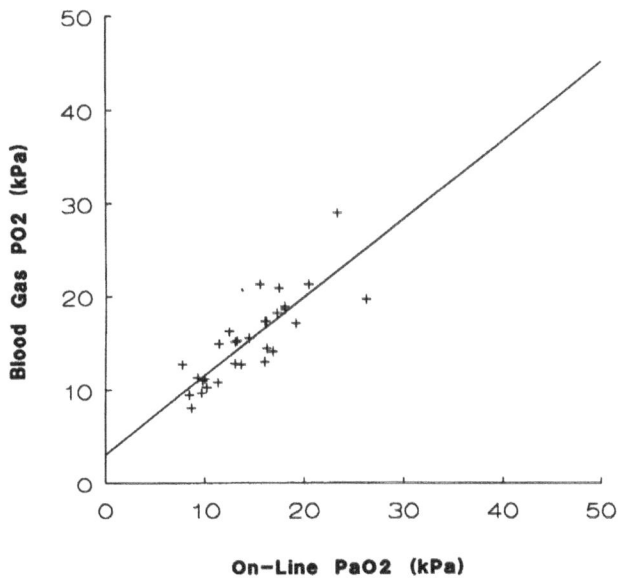

Fig. 5. The relationship between Continucath on-line PaO_2 and simultaneous values obtained from arterial blood samples in the postoperative period. Correlation coefficient is 0.83 (n=30).

In 45% of the patients damping of the arterial curve occurred. This could usually be solved by flushing the arterial line with heparinized saline and/or changing the position of the patients hand. In one patient it become impossible to aspirate blood even though there was a continuous flush and arterial blood pressure could still be measured. As mentioned before, the I.C. nurses were told not to let the indwelling sensor interfere with their work. That meant that if damping of the curve was annoying to the nurses, they could remove the sensor. This occurred in 6 (33%) of the cases, mainly because the low blood pressure alarm kept triggering. Hopefully in the course of time, nurses (and doctors) might be convinced that continuous PaO_2 monitoring could be more important than continuous blood pressure monitoring. Damping of the arterial curve is also described by other workers (Oxorn et al., 1984; Bratanow et al., 1985; Pfeifer et al., 1988). Green et al. (1987) seemed to have no problems. A continuous flush of heparinized saline is very important, as well as an extra flush after aspirating blood. This implies that frequent sampling could traumatize the arterial canula/PaO₂ sensor as indeed propagated by Pfeifer et al. (1988). A larger arterial canula (16 G) might decrease damping (Rithalia et al., 1981) but in this department normally 20G or 18G canulas are used to minimize the risk of arterial damage. Possibly another type arterial canula could solve this problem.

The sensors remained in situ for a maximum of 24 hours. After removal no clotting of the sensor was observed. There were no ischemic problems on examination of the hand after use of these catheters. Comparisons were made up to 17 hours after calibration (in the post-operative period). Figure 6 shows that the duration of electrode placement does not affect reliability. Calculation of drift

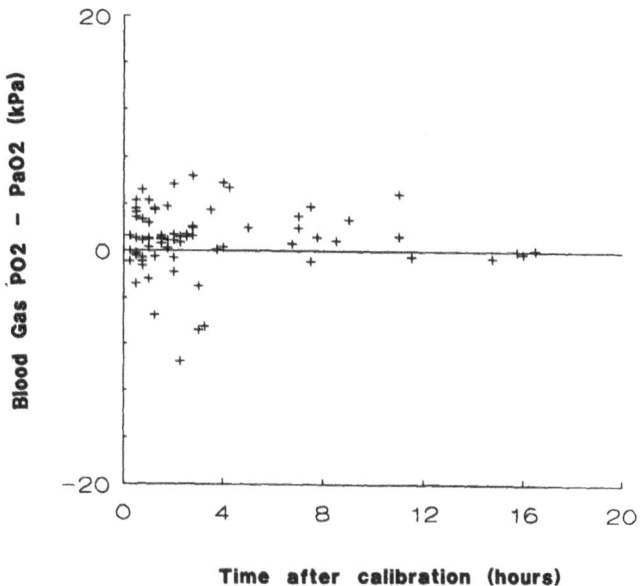

Fig. 6. Differences in blood gas analysis and intravascular PaO₂ in the course of time after calibration.

in vivo is very difficult but Eberhard et al. (1979) using a prototype of this sensor showed a drift in vitro of less than 0.7%/hour. This corresponds very well with the drift of less than 10%/24 hours reported by Bratanow et al. (1985).

The lines of regression in figures 4 and 5 show that on average the blood gas PO₂ is higher than the intravascular PaO₂. This is also shown by other workers (Green et al. 1987). In figure 7, when comparing the difference between PbO₂ and PaO₂ to the average PO₂ ([PaO₂ + PbO₂]/2), one can see that between 5 and 25 kPa there is a linear relationship between the two methods, albeit that on average PbO₂ is indeed higher than PaO₂. This slight underestimation of PO₂ by the intravascular sensor does not outweigh its importance as a trend monitor and early warning sign for hypoxia. Figure 7 also shows that for very high PO₂'s (above 25 kPa) the blood gas analyzer gives much lower readings than the intravascular sensor. This might not be an inaccuracy of the Continucath sensor, but a problem with blood gas handling. As postulated by Eberhard et al. (1979), one of the problems could be that contamination of the blood monster with air could lower the PO_2. This discrepancy between PbO_2 and PaO_2 at higher values is also the reason why the slope of the line of regression is less than 1.

The aim of this study was to compare the accuracy of the Continucath sensor against blood gas analysis. The PaO_2 monitor was not supposed to be used for clinical interpretations. After the authors were convinced of the accuracy of the sensor, and especially of its capability to react immediately to changes in oxygenation, it was obvious that it became an important tool during surgery. Often a decrease in PaO_2 was observed and action was taken, where otherwise one might have waited for the results of blood gas analysis, or even worse, taken no action at all because no analysis would normally have been performed. If the blood pressure was very low, it was noted that the PaO_2 reading

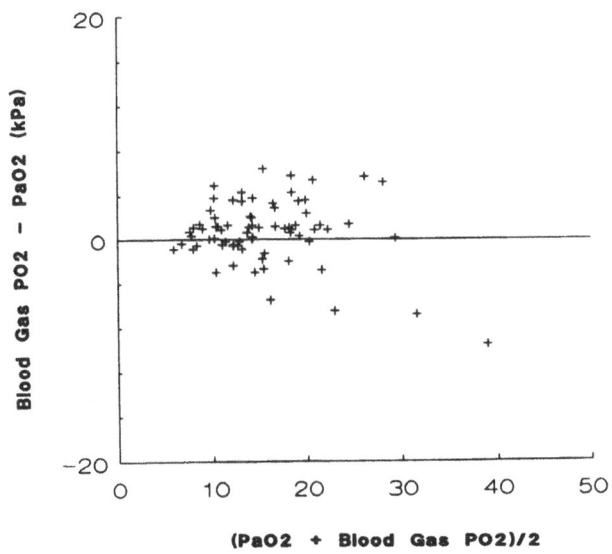

Fig. 7. Scattergram of the (blood gas PO_2 - intravascular PaO_2) differences against the average PO_2 ([PaO_2 + blood gas PO_2] / 2).

dropped. This has been noted by other authors (Rithalia et al. 1981, Pfeifer et al. 1988). This is actually an added safety feature because the drop in PO_2 reading would draw the attention of the therapist to take the required measures. The manufacturers state that the error caused by flow variation is less than 1% above a flow rate of 5 cm/sec.

CONCLUSIONS

This sensor system is a practical method for continuous on-line arterial oxygen partial pressure monitoring. It is suitable in all situation where one would like to be informed about acute changes in oxygenation, especially in situations where a delay cannot be tolerated. This is the case in all major surgery and in seriously ill I.C. patients. In these cases an arterial line is already in situ so that insertion of the PaO_2 sensor does not cause any extra risks or discomfort to the patient. Other examples for its use is in the prevention of retrolental fibroplasia in neonates, or minimizing pulmonary fibrosis in bleomycin treated patients by keeping the FiO_2 as low as possible while

ensuring an adequate PaO_2. Contraindications for its use are the same as those for the use of an arterial canula:
- local infection,
- inadequate circulation distal to the insertion site of the canula,
- serious hemorrhagic disorders,
- Raynaud's disease and diseases where Raynaud's phenomenon can occur.

One advantage of blood gas analysis is that besides PO_2, one is also informed about PCO_2, pH, etc. Hopefully in the near future it will be possible to have continuous on-line monitoring of blood pressure, PaO_2, $PaCO_2$, pH, SaO_2, potassium etc. with the same canula.

SUMMARY

Intermittent blood gas sampling has several disadvantages, the most important being that samples are usually taken at set intervals, or when changes in oxygenation are suspected - when it is too late. Another problem is inaccuracy caused by careless blood sample handling. Continuous intravascular PaO_2 monitoring eliminates these problems. This study shows that the Continucath sensor is an easy-to-use and reliable monitor, with specific early warning capabilities for hypoxia, thereby improving anesthetic and intensive care management. Its characteristics are:
- a stabilization period of 10 minutes,
- a 90% response time of 90 seconds,
- temperature dependence of 4% per degree celsius,
- a flow dependence of less than 1% if the flow is more than 5 cm/sec,
- a drift of less than 0.7% per hour,
- a correlation coefficient of 0.92 when compared to blood gas analysis during surgery.

REFERENCES

Bratanow, N., Polk, K., Bland, R., Kram, H.B., Lee, T.-S., Shoemaker, W.C., 1985, Continuous po-larographic monitoring of intra-arterial oxygen in the perioperative period, *Crit. Care Med.*, 13(10):859-860.

Barker, S.J., Tremper, K.K., Hyatt, J., Zaccari, J., Heitzmann H.A., Holman B.M., Pike, K., Ring, L.S., Teope, M., Thaure, T.B., 1986, Continuous fiberoptic arterial oxygen tension measurements in dogs, *J. Clin. Monit.*, 3(1):48-52.

Eberhard, P., Fehlmann, W., Mindt, W., 1979, An electrochemical sensor for continuous intravascular oxygen monitoring, *Biotelemetry Patient Monitoring*, 6:16-31.

Green, G.E., Hassell, K.T., Mahutte, C.K., 1987, Comparison of arterial blood gas with continuous intra-arterial and transcutaneous PO_2 sensors in adult critically ill patients, *Crit. Care Med.*, 15(5):491-494.

Holcomb, C., Erdmann, W., Corssen, G., 1976, The significance of diffusion hypoxemia, *Southern Medical Journal*, 69(10):1282-1284.

Oxorn, D.C., Chung, D.C., Lam, A.M., 1984, Continuous in-vivo monitoring of arterial oxygen in a patient treated with bleomycin, *Can. Anaesth. Soc. J.*, 31(2):200-205.

Pfeifer, P.M., Pearson, D.T., Clayton, R.H., 1988, Clinical trial of the Continucath intra-arterial oxygen monitor, *Anaesthesia*, 43:677-682.

Rithalia, S.V.S., Bennett, P.J., Tinker, J., 1981, The performance characteristics of an intra-arterial oxygen electrode, *Intensive Care Med.*, 7:305-307.

Striebel, H.W., Kretz, F.J., 1989, Advantages and limitations of pulse oximetry, *in*: "Clinical Aspects of O_2 Transport and Tissue Oxygenation," K. Reinhart, K. Eyrich, ed., Springer-Verlag, Berlin, Heidelberg, New York.

Sugioka, K., Cattermole, R.W., Sebel, P.S., 1987, Arterial oxygen tensions measured continuously in patients breathing 21% oxygen and nitrous oxide or air, *Br. J. Anaesth.*, 59:1548-1553.

CONSTRUCTION, CALIBRATION AND EVALUATION OF pO$_2$ ELECTRODES

FOR CHRONICAL IMPLANTATION IN THE RABBIT BRAIN CORTEX

Koen van Rossem, Herman Vermariën and René Bourgain

Laboratory of Physiology and Physiopathology, University of Brussels
VUB, Laarbeeklaan 103, B-1090 Brussels, Belgium

ABSTRACT

Aiming at continuous polarographic measurement of the mean pO$_2$ in the rabbit brain cortex before, during and after photochemically induced infarction, we designed and constructed monopolar platinum oxygen electrodes of the open type for chronical implantation. The measuring tip (length 1 mm, diameter 0.1 mm) is covered with a homogenous membrane of cellulose acetate. The electrode currents are measured by a four-channel amplifier of proper design; the device permits accurate and stable polarisation, identical for each channel. Moreover, a calibration device has been constructed. It consists of a Buchner funnel filled with Ringer solution and mounted in a temperature-controlled bath. In order to create a specific partial pressure of oxygen in the calibration chamber, predetermined gasmixtures are bubbled through the solution using computer controlled mass flow regulators. The calibration device thus permits the determination of primary and secondary electrode parameters, i.e. linearity, oxygen sensitivity and residual current, and polarisation dependency, temperature dependency, sensitivity to CO$_2$, electrode stability, dynamic behaviour and oxygen consumption.

Three groups, each of them containing ten electrodes, have been tested with regard to electrode parameters : the first group contains bare electrodes, the second and the third group contain membrane covered electrodes, with a membrane thickness of 10 and 20 µm respectively. In order to evaluate acute and long-term effects of implantation on the brain cortical tissue and on the sensors' measuring qualities, electrodes have been implanted for different time periods (51 days, 30 days, 9 days, 5 min). pO$_2$ was recorded regularly and polarograms have been registered. The effects on cortical tissue have been studied with the aid of light microscopy.

INTRODUCTION

The field of cerebral ischemia is one of the main interest in neuroscience. Great efforts are made to extend the knowledge about pathophysiology and possible treatments of stroke and thrombosis. Animal models are widely used in this field of research and cerebral infarcts have been induced in different species applying various methods. Photochemical induction of thrombosis is a recently developed technique which can be applied to induce reproducible cerebrocortical infarctions (Watson et al., 1985). Several investigators have studied morphological, pathological and behavioural changes after such infarction and

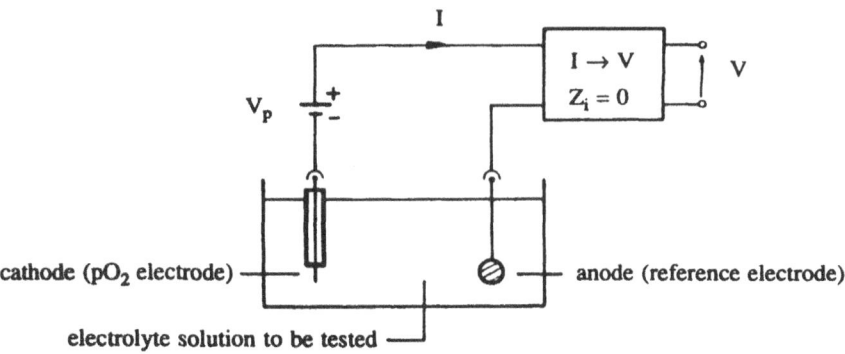

Figure 1 Schematical drawing of a polarographic oxygen sensor and the measuring circuit.

possible beneficial effects of certain treatments. Except for behavioural studies, steady state measurements have been performed including microscopy and autoradiography.

We have developed a method which enables continuous measurement of partial oxygen tension (pO_2) polarographically, before, during and after standardised photochemically induced infarction of rabbit brain cortical tissue (van Rossem et al., 1989). However, in order to obtain more reliable information, we felt the need to evaluate and to improve standardisation of the measuring technique. Standardisation includes reproducible construction and application of pO_2-electrodes, knowledge of their measuring properties and possible influences on cortical tissue after implantation.

At present polarography is the most widely used technique for measurement of oxygen pressure. The measuring principle of the polarographic sensor is based upon the reduction of oxygen molecules at the surface of an electrode which is polarised at a fixed negative voltage (between - 600 and - 800 mV) with respect to a reference electrode. The cathode and the reference electrode must be connected by a suitable ion conduction medium. According to Baumgärtl and Lübbers (1983) four electrons are consumed by each molecule of oxygen :

$$O_2 + 2H_2O + 4e^- \rightarrow 4OH^-$$

As a consequence oxygen reduction results in a current that can be measured in an external circuit (Figure 1). Nobel metals are chosen for cathode material; Ag/AgCl is usually chosen as a reference electrode. With the polarising voltage (V_p) set a value between - 600 and - 800 mV pO_2 is reduced to zero at the cathode; in steady state the current (I) is then limited by oxygen diffusion to the electrode only and hardly influenced by the value of V_p. The latter is expressed as a plateau in the polarogram (I as a function of V_p). Horizontal plateaus are merely theoretical; in practice changes in V_p give rise to variations in current, thus for accurate measurement a stable voltage is a necessity. When V_p exceeds - 800 mV additional chemical reactions occur and current increases steeply.

The oxygen consumption of an adequately polarised electrode can be calculated from the following equation

$$I = nFJ_{O_2} \tag{1}$$

with J_{O_2} representing the oxygen flux to the cathode (mol/s), F Faraday's constant (96.5 kC/Eq) and n the number of electrons exchanged (Eq/mol) (n = 4). An electrode current of 1 nA thus corresponds to an oxygen consumption of 2.59 fmol/s.

Considering a medium where the oxygen tension is held constant at a specific distance from the electrode (e.g. in tissue by capillary blood flow) and taking into account the assumption that the oxygen tension equals zero at the metal surface of the cathode, the relation between the oxygen tension and the flux is described by a diffusion resistance (R_s)

$$J_{O_2} = \frac{pO_2}{R_s} \qquad [2]$$

The latter also implies the existence of a gradient in oxygen tension towards the metal surface. This resistance depends on the shape of the electrode and the distance involved, the oxygen diffusion constant and the solubility of oxygen in the medium and the area of the electrode. The sensitivity (s_{O_2}) of the electrode is defined as

$$s_{O_2} = \frac{I}{pO_2} \qquad [3]$$

In the case of a bare electrode this results in

$$s_{O_2} = \frac{nF}{R_s} \qquad [4]$$

A bare electrode positioned in the medium to be tested cannot make a good oxygen sensor as its sensitivity is determined by the properties of that medium, whereas the ideal sensor should have a sensitivity completely independent of that medium. E.g., when that medium is stirred the existing gradient in oxygen tension will be affected resulting in an effective decrease of diffusion resistance and consequently an increase of sensitivity. The same holds for applications in tissue where perfusion is altered.

In order to eliminate the disadvantages described a membrane is applied to the metal surface. Essentially the membrane stands for an artificial diffusion resistance in series to the one representing the medium. For a flat membrane the diffusion resistance (R_m) is given by

$$R_m = \frac{d}{A_m D_m S_m} \qquad [5]$$

with d representing the thickness of the membrane, A_m the area, D_m the diffusion constant and S_m the solubility of oxygen in the membrane material. In this case the flux is given by

$$J_{O_2} = \frac{pO_2}{R_s + R_m} \qquad [6]$$

The artificial resistance gives rise to a decreased sensitivity

$$s_{O_2} = \frac{nF}{R_s + R_m} \qquad [7]$$

and the pO_2 at the surface of the membrane

$$pO_{2(m)} = \frac{R_m}{R_s + R_m} pO_2 \qquad [8]$$

The sensor can be indicated as ideal if

$$R_m \gg R_s \qquad [9]$$

and in that case

$$s_{O_2} = \frac{nF}{R_m} \quad \text{and} \quad pO_{2(m)} = pO_2 \qquad [10]$$

As one can see, the ideal sensor is unsensitive to stirring, it has a low oxygen consumption and does not influence the oxygen tension distribution in the medium, but, it has a low sensitivity. Temperature dependency and stability of the electrode are thus determined by the membrane material. It should also be noted that response time increases as R_m

increases. According to equation [1] electrode current should be zero when the medium is oxygen free; in practice there is always a small residual current as a result of other electro-chemical reactions.

As contrasted with the bipolar design (Clark, 1956), the monopolar electrode is physically separated from the anode and the conducting medium between cathode and anode is the medium to be measured (e.g. biological tissue). In this case miniaturisation is relatively easy. The membrane material and dimensions (Equation [5]) principally determine the measuring quality; nevertheless secundary aspects are equally important such as mechanical strength, applicability, biocompatibility, stability. Popular membrane materials include polyethylene, propylene, cellulose acetate, teflon, polystyrene and polyurethane. Following electrode parameters are of practical importance concerning pO_2 measurement in general : sensitivity, residual current, linearity, stability, temperature dependency, sensitivity to other gasses, response time and flow dependency; when in vivo measurements are performed additional attention has to be paid to dimensions and implantation requirements, sterilisation, biocompatibility, effects on surrounding tissue (Hale, 1983; Silver, 1966).

In order to evaluate the measuring qualities one has to calibrate the monopolar sensor in an aqueous solution. Bubbling a known gasmixture trough the fluid is relatively easy to apply. The fluid in the calibration chamber should resemble the medium of interest as much as possible (pH, ion concentration, etc.). Nevertheless, as tissue conditions are never completely known (and may vary impredictably), the test medium is never an exact replica of the tissue to be investigated.

Our aim is to monitor mean cortical pO_2 continuously before, during and after photochemically induced infarction in the rabbit's brain cortical tissue. Since pO_2 varies considerably according to depth in the cortex (Baumgärtl and Lübbers, 1983; Fennema et al., 1989) the only applicable type to measure mean cortical pO_2 is the open type with a cylindrical measuring tip which makes direct contact with the tissue over its complete surface (in contrast with the recessed type). As the thickness of the rabbit's cerebral cortex varies between 1.3 and 1.5 mm (Shek et al., 1986), a measuring tip of 1 mm is chosen. Platinum, being a stronger metal than gold, is used as cathode material. With respect to solidity and manageability a wire diameter of 100 μm is chosen. With a cylindrical measuring tip dipcoating is an easy technique for membrane application. Cellulose acetate is selected as membrane material. In order to evaluate the measuring qualities of the electrodes a statistical study has been performed on three groups of membrane thicknesses (0, 10 and 20 μm). Stabilisation time, polarogram slope, linearity, sensitivity, residual current, stability, temperature dependency and sensitivity to CO_2 are studied in vitro in a calibration chamber of proper design. Tests are performed in vivo and the effects of electrode implantation and measurement are evaluated.

ELECTRODE CONSTRUCTION AND MEMBRANE EVALUATION

Miniature monopolar pO_2 electrodes are constructed according to the following procedure (Figure 2). Pure platinum wire (length 15 mm, diameter 0.1 mm) is soldered to a gold connector. After staining with black ink (Edding T25) the wire is insulated with cyanoacrylate glue by dipping. The measuring tip of the electrode is created manually under a microscope. Firstly, the insulation (and the ink) is removed at the free end of the wire over a length of 1 mm by using a scalpel. Except for bare electrodes (group A), the naked tip is then covered with a cellulose acetate membrane by dipping it repeatedly into a cellulose acetate solution (5 % cellulose di-acetate in a 33.3 % ethanol in aceton solution). For our statistical study of electrode qualities three groups are constructed : group A contains bare electrodes, group B and C contain membrane covered electrodes with a membrane thickness of 10 and 20 μm respectively. After selection of the electrode based on membrane evaluation, the soldered connection is covered with shrinking tube which is then hermetically sealed with cyanoacrylate glue.

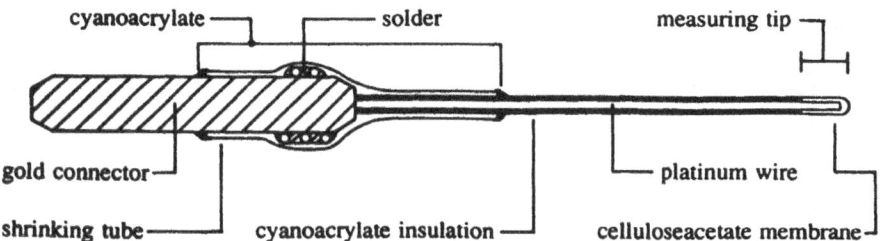

Figure 2 Cross-sectional diagram of the miniature monopolar pO_2 electrode. The measuring tip (diameter 0.1 mm, length 1 mm) is covered with a homogenous cellulose acetate membrane.

The membrane thickness is determined at four equidistant levels of the measuring tip with the aid of a light microscope provided with a measuring ocular. Therefore the tip is positioned on the engraved network of a Burker cel counting chamber. With the tip ending positioned in the middle of a large square of the network (delineated by double lines, 0.2 x 0.2 mm) and the shaft parallel to the engraved lines, one can easily measure the thickness of the shaft at equidistant levels (distance 0.25 mm). Therefore a cross in the ocular is moved from one side of the shaft to the other according to a perpendicular line of the network and the outer diameter can be read from the ocular. As the wire diameter is known, the membrane thickness can be calculated. As such four thickness values (d) are obtained and mean thickness ($\bar{d}$), standard deviation (σ_d) and variation coefficient ($v_d = \sigma_d / \bar{d}$) can be calculated. For establishing a specific group of electrodes limits are set to deviations of the nominal value of the thickness (D) and to the thickness homogeneity. Therefore the paramaters $\bar{d}$ and v_d are used respectively and the following criteria are applied for selection :

$$\left| D - \bar{d} \right| \leq 2.5\mu m \text{ and } v_d \leq 0.2 \qquad [11]$$

with D = 10 μm and 20 μm for group B and C respectively. About 50 % of the electrodes constructed did not meet this criteria and were excluded. Each group contains 10 electrodes. The mean variation coefficient on membrane thickness amounts 0.14 in both groups B and C. The mean thickness per group ± the standard deviation of the mean thicknesses of different electrodes amounts 10.63 ± 1.09 μm in group B and 19.07 ± 0.83 μm in group C.

Electrodes are stored in a medium similar to the one of application : they are kept with their measuring tips in a Ringer solution bath (at room temperature). In order to prevent evaporation and bacterial pollution the bath is hermetically sealed and the solution is changed every week.

MEASUREMENT AND CALIBRATION APPARATUS

A measuring device (FYSPpO$_2$1) allowing pO_2 measurements with four electrodes simultaneously has been developed. The four pO_2 electrodes connected are polarised with a voltage (V_p), either an adjustable stabilized internal voltage (identical for each electrode) or an arbitrary external voltage (single ended). The choice between the internal or the external voltage can be made for each electrode separately. The calibration voltage is continuously displayed (3-digit display). The common reference electrode is connected to ground and may be used as a grounding electrode for additional measurements such as biopotential recordings (in our case EEG and EP). The amplifier is essentially a current-to-voltage converter with zero input impedance; the conversion factor is 10^6 VA^{-1}.

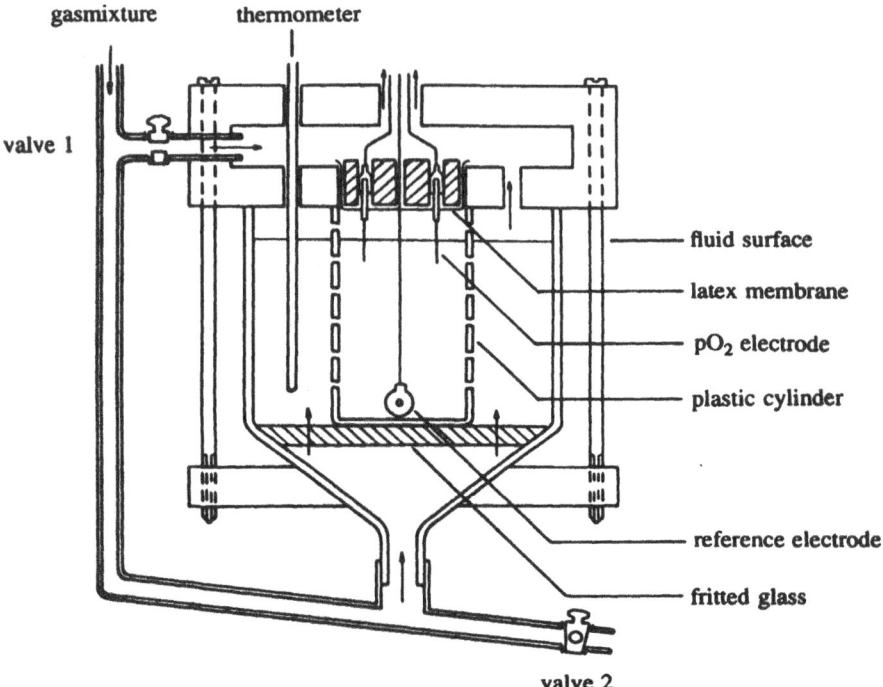

gasmixture thermometer

valve 1

fluid surface

latex membrane

pO$_2$ electrode

plastic cylinder

reference electrode

fritted glass

valve 2

Figure 3 Sketch of the calibration device showing two pO$_2$ electrodes suspended with their connectors, the measuring tip plunged into Ringer solution saturated with a predetermined gasmixture. Arrows indicate the direction of the gasflow through the fritted glass and the fluid. Valve 1 can be used to adjust the part of the total gasflow which is sent through the fluid. Holes in the cover enable insertion of a thermometer or serve as a gas outlet.

On the basis of a model presented by Proctor and Bohlen (1979) and taking into account requirements for accurate calibration (Hitchman, 1983), a calibration device has been developed in order to determine electrode parameters in vitro (Figure 3). A Buchner funnel (Schott Duran 17D3) containing a fritted glass bottom plate (porosity factor 3) forms the heart of the device. Ringer solution is poured into the funnel and predetermined gasmixtures can bubble through the fluid after passing the fritted glass. In this way the fluid gets quickly saturated with the gasmixture so that the pO$_2$ in the fluid and in the gasmixture are identical. Electrodes are suspended with their connectors in a stopper which is mounted into the central hole of the cover. Their measuring tips as well as an Ag/AgCl reference electrode are plunged into the Ringer solution. A latex membrane is fixed to the stopper in order to prevent condensation of water vapor on the connectors. A plastic cylinder, perforated by many small holes, surrounds the electrodes preventing direct contact with gasbubbles, but still allowing convective mixing of the fluid.

The calibrating chamber is positioned in a thermostatic bath. The temperature of the solution is measured using a digital thermometer (Comark 6 900). Only a part of the gas flow is sent through the fluid. The other part adjusted by valve 1 (Figure 3) is flown through the space between the upper and the lower cover plate and improves the stability of the equilibrium between fluid and gas. The flow through the fluid is set at 300 ml/min. Applying this flow the dimensions of the holes in the cover serving as gas outlet (diameter

5 mm, length 10 mm) are adequate to prevent backdiffusion of gas (calculated according to Hitchman (1983)). Pure nitrogen (purity 99.9 %) and an ultra-precise mixture of 15 % O_2 in N_2 are used to provide gasmixtures with a predetermined O_2 fraction (f_{O_2}). The pO_2 in the fluid is thus calculated as follows

$$pO_2 = (P_a - P_v^T)f_{O_2} \qquad [12]$$

with P_a representing atmospheric pressure and P_v^T water vapor pressure at temperature T (thermostatic bath). The flow of each gas is controlled by mass flow regulators (UCAR MFTV-530) which have been calibrated by the manufacturer according to the applied gas or gasmixture. Each flow regulator can be adjusted manually or computer controlled (precision range 0.5 ml/min). The total flow is kept at 500 ml/min and by changing the portion of each gas various fractions of O_2 in the final gasmixture can be obtained at a high precision level. 15 % has been chosen as a maximum as cortical pO_2 remains below 15 kPa in physiological conditions.

CALIBRATION AND EVALUATION OF ELECTRODES IN VITRO

For the statistical study of the three groups of electrodes measurements are performed according to a strict time schedule :
day 1 : electrode construction;
day 4 : stabilisation time, 11-step calibration, 4-step calibration, polarogram;
day 12 : stabilisation time, 4-step calibration;
day 14 : temperature dependency;
day 27 : stabilisation time, 4-step calibration;
day 50 : stabilisation time, 4-step calibration, sensitivity to CO_2.
Because of sample size (n = 10) and the unknown distribution of the data non-parametric statistical techniques are applied. Differences between groups are evaluated applying the randomisation test for independent samples or the Mann-Witney U-test. The latter is used for comparison of the stabilisation times since the exact value is not known if it exceeds the observation time. Differences within groups (according to time or polarisation voltage) are checked applying Friedman two-way analysis of variance. When a difference is indicated this is further analysed using randomisation tests for matched pairs or, regarding stabilisation time, Wilcoxon matched pairs signed ranks..

Stabilisation time
Before performing any experiment the electrode current is recorded for 30 min after switching on V_p (V_p = - 600 mV, 3.75 % O_2, T = 39 °C). In order to evaluate stability the relative variation of the electrode current is calculated in subsequent 3 min time intervals ($g_I = \Delta I / I$). The stabilisation time (t_s) is defined as the first interval in which g_I is smaller than 5 % without being followed by another interval having a g_I exceeding 5 %. A g_I smaller than 5 % is considered as drift, which is evaluated afterwards in the context of short-term and long-term stability. Most of the electrodes show a rather variable stabilisation time (between 1.5 and 30 min, or more). All electrodes of group C stabilise within 30 min; 2 electrodes in group B and 5 electrodes in group A have at least once a stabilisation time exceeding 30 min.

Polarogram slope
Polarograms are obtained by sweeping the external polarisation voltage with a sufficiently low frequency in order to avoid hysteresis phenomena (triangular waveform, - 900 to - 200 mV, 0.0016 Hz, 2.24 mV/s, 7.5 % O_2, 39 °C). The slope of the curve is evaluated by the following parameter

$$g = \frac{\Delta I}{\bar{I}} \frac{1}{\Delta V_p} \qquad [13]$$

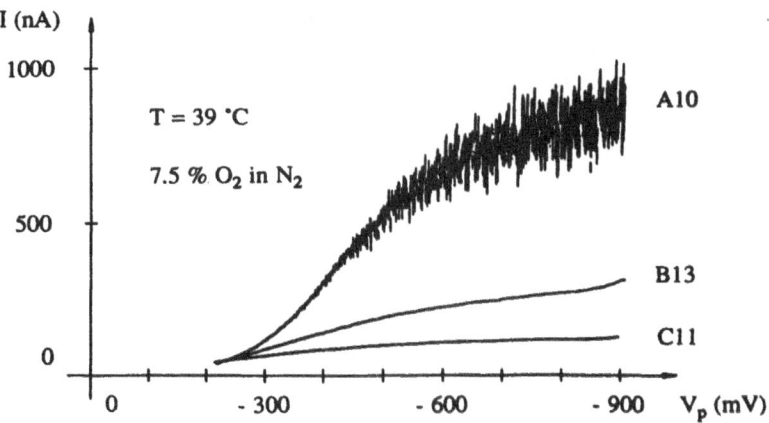

Figure 4 Polarograms of electrodes A11, B13 and C11 (39 °C, 7.5 % O_2). Large fluctuations are observed on the recording of the bare electrode (probably due to turbulences in the fluid).

with ΔV_p representing the difference in V_p, ΔI the difference in current and $\bar{I}$ the mean current ($\Delta V_p = 100$ mV). Values of g are obtained at a $V_p = -600, -650, -700$ and -750 mV respectively. Figure 4 shows typical polarograms of three electrodes, one of each group. Table 1 displays mean values and standard deviations of slopes. Averagely g decreases as V_p increases in the interval studied. This decrease is significant except for the two higher polarisation voltages in group A and C; indeed, some of the electrodes show a higher g at -750 mV in comparison with the former value indicating transition to the steep part of the polarogram. All electrodes show the largest g at -600 mV. g values are significantly higher in group A (p < 0.015). Group C shows the lowest slopes but does not differ significantly from group B.

Table 1 Mean values ($\bar{g}$) and standard deviations (σ_g) of slopes in group A, B and C as a function of polarisation voltage V_p.

V_p	$\bar{g} \pm \sigma_g$ (% per 100 mV)		
(mV)	A	B	C
- 600	26.2 ± 7.7	14.1 ± 4.0	12.3 ± 3.8
- 650	21.0 ± 6.4	11.1 ± 3.5	9.3 ± 3.4
- 700	16.3 ± 5.5	9.0 ± 2.5	7.8 ± 3.0
- 750	15.1 ± 4.5	8.2 ± 2.7	7.6 ± 3.6

Calibration curves : linearity, residual current, sensitivity and stability

In order to evaluate linearity and to determine the sensitivity to oxygen (s_{O_2}) and the residual current (I_0 for $f_{O_2} = 0$), static calibration is performed with 11-step or 4-step procedures ($V_p = -600$ mV, T = 39 °C). pO_2 steps are controlled by computer; a 7 min

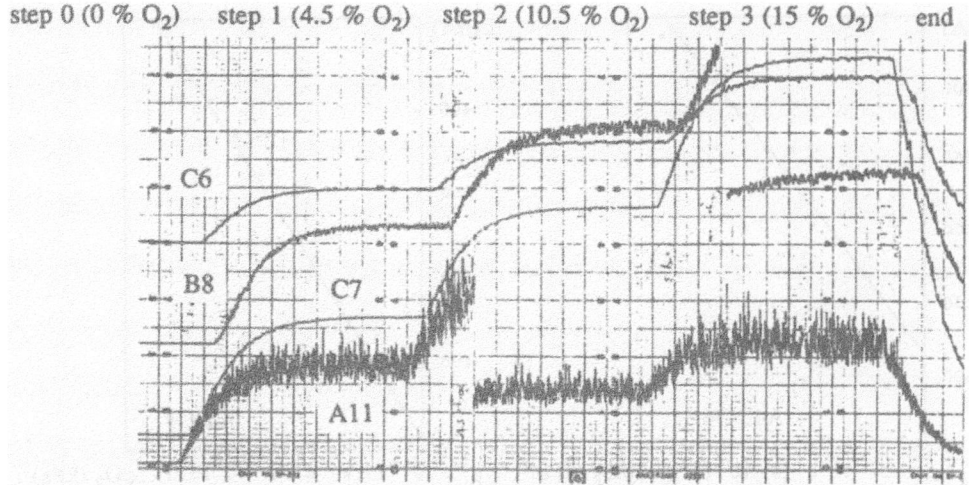

Figure 5 Record of a 4-step calibration of four electrodes (B8, C6, C7, A11; day 4). Scale factors and pen off-set are not the same for each electrode and were changed if necessary.

period appeared to be sufficiently long for reaching a new stable pO_2 level. 11-step procedures are performed on day 4 in order to have an optimal evaluation of linearity. 4-step calibrations are applied for evaluation of long-term and short-term stability. Linear regression analysis is used to determine linearity (r^2, the squared correlation coefficient) and sensitivity (the regression coefficient). The residual current I_0 ($f_{O_2} = 0$) is directly read from the recording. Figure 5 shows recordings of typical calibration curves. With pure nitrogen bubbling through the fluid, the electrode current decreases to reach the residual current. The latter appears to be very stable and a horizontal line is recorded. As pO_2 is increased, electrode current raises to a higher level; fluctuations around a mean value can be observed and the thickest membrane shows the lowest amplitude. These fluctuations are most probably due to turbulences in the fluid.

Table 2 Statistical parameters on r^2 of the five calibrations : average of standard deviations with respect to different calibrations ($\overline{\sigma}_{r^2}$), mean r^2 ($\overline{r^2}$) and standard deviation on mean r^2 of different electrodes ($\sigma_{\overline{r^2}}$).

	A	B	C
$\overline{\sigma}_{r^2}$	0.0615	0.0052	0.0023
$\overline{r^2} \pm \sigma_{\overline{r^2}}$	0.9525 ± 0.0225	0.9966 ± 0.0039	0.9980 ± 0.0013

Linear regression analysis is performed on each calibration curve (Figure 6) (Table 2). The mean r^2 of group A appears significantly lower than in the other groups ($p < 0.0001$). Electrodes of groups B and C behave in a much more linear way. One electrode of group A shows a mean r^2 below 0.99 only once; electrodes of group C perform even better, none of them showing a value below 0.995. The mean standard deviation on r^2 ($\overline{\sigma}_{r_2}$) in group C is smaller than the one of group B. However, statistically group C does not differ significantly from group B ($p < 0.175$).

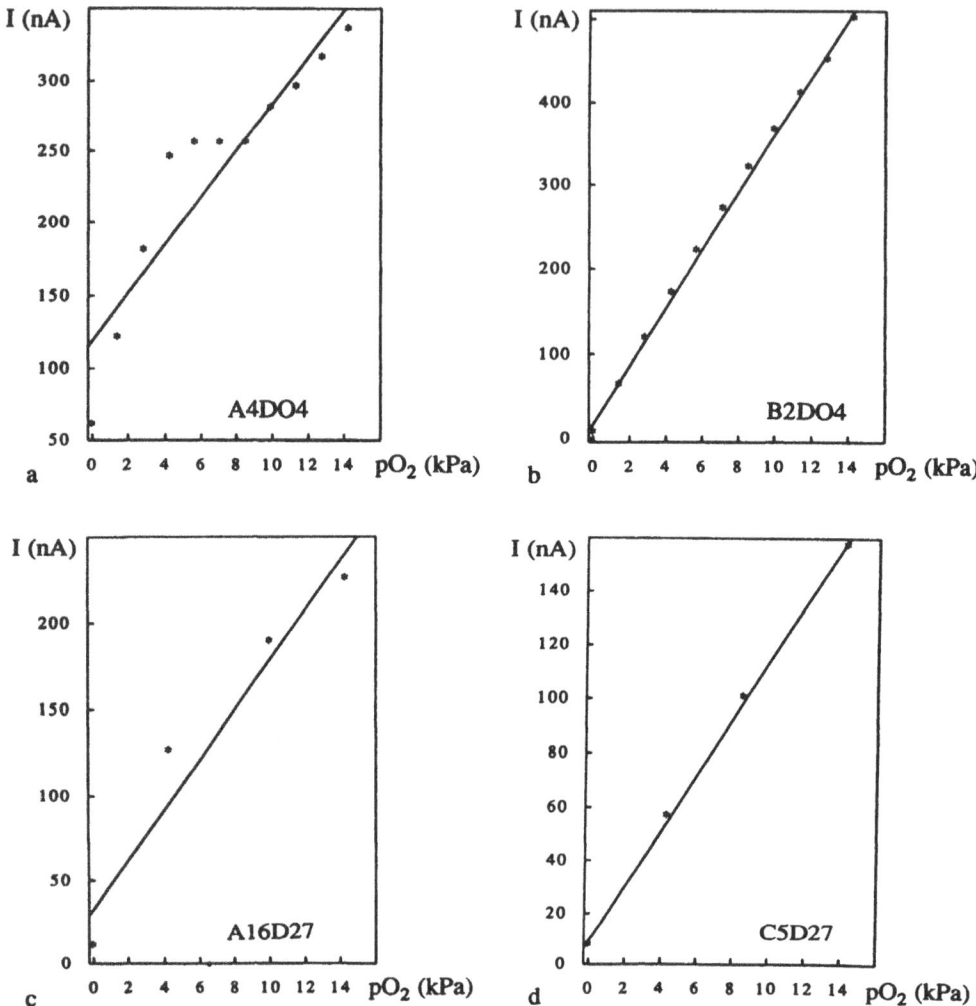

Figure 6 Typical 11-step and 4-step calibration curves of bare (a and c) and membrane covered (b and d) electrodes. Best fitting straight lines are drawn.

During calibration experiments sometimes surprisingly high residual currents are noted. These currents can be caused as a result of leaking paths originating from water vapor condensation on gold connectors and platinum wires by the bubbling effect. It appears that the calibration device insufficiently prevents this phenomenon although rigourous precautions have been taken. Although a considerable amount of calibrations (69 of 150) showed residual currents below 10 nA, we consider the interpretation to be inaccurate as any current might be an overestimation as a result of water vapor condensation.

Table 3 displays statistical parameters on sensitivity to oxygen (s_{O_2}) which is calculated as the regression coefficient. Electrodes of group C had significantly lower

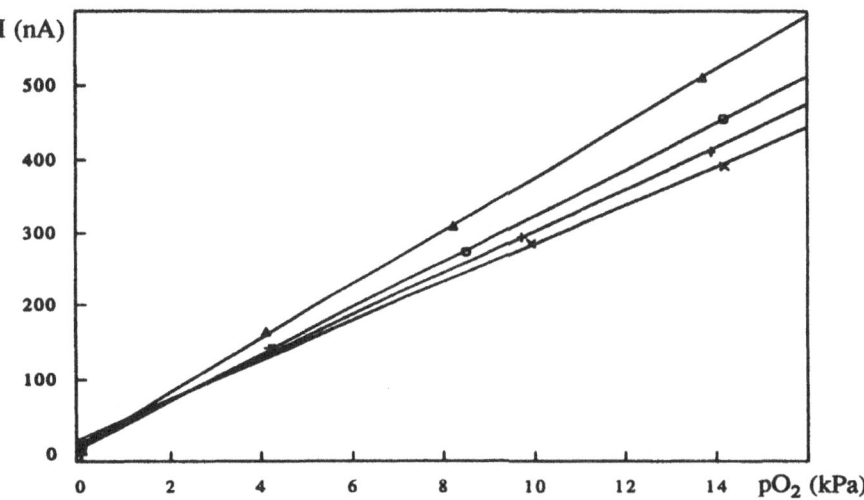

Figure 7 Four calibration curves of electrode B10 obtained at day 4, 12, 27 and 50.

sensitivities compared to group B ($p < 0.00001$) and group A ($p < 0.003$). The mean s_{O_2} of group C (12.97 nA kPa^{-1}) lies very close to halve the value of group B (13.89 nA kPa^{-1}) which can be expected on the basis of membrane thickness in an ideally stirred medium (Equations [5] and [10]). However, the values within a group are considerably spread and some electrodes of different groups show comparable sensitivities. Surprisingly, the overall means of group A and B are almost identical.

Table 3 Statistical parameters on s_{O_2} of the five calibrations : average of standard deviations with respect to different calibrations ($\bar{\sigma}_s$), mean sensitivity ($\bar{s}$) and standard deviation on mean sensitivities of different electrodes ($\sigma_{\bar{s}}$).

	A	B	C
$\bar{\sigma}_s$	10.38	4.43	1.99
$\bar{s} \pm \sigma_{\bar{s}}$	27.22 ± 17.10	27.78 ± 9.64	12.97 ± 2.19

The variation coefficient v_s, calculated for each electrode over four 4-step calibrations (day 4, 12, 27 and 50), is taken as a measure for long term stability. Mean values ± the standard deviations (in %) : 32.68 ± 21.89 for group A, 15.15 ± 6.01 for group B and 16.64 ± 7.70 for group C. Groups B and C do not differ significantly ($p < 0.3$); in group A variation coefficients are higher with a high level of significance ($p < 0.01$). Figure 7 displays the variability of sensitivity for a representative membrane covered electrode. The evolution of sensitivity in relation with time does not show a specific trend.

Short-term stability is evaluated via the relative difference in sensitivity between the 11-step and the 4-step calibration performed on the same day (4). Mean values ± the standard deviations (in %) are 12.44 ± 18.13 for group A, 15.56 ± 16.65 for group B, 3.96 ± 3.22 for group C. Short-term stability is significantly better in group C compared to group

A ($p < 0.002$) and group B ($p < 0.015$). The difference between groups A and B does not appear significant ($p < 0.145$).

Temperature dependency

By altering the temperature of the bath the temperature dependency of s_{O_2} (f_T) can be determined ($V_p = -600$ mV, 7.5 % O_2 and 0 % O_2, 35 °C to 43 °C, 2 °C/min). For a given oxygen fraction an increase of temperature from 35 °C to 43 °C will cause a 3.02 % decrease of pO_2 as a consequence of the increase of water vapor pressure. Hence, in order to determine f_T for constant pO_2 a correction factor (f_{Tv}) of 0.38 % °C^{-1} should be added.

$$f_T = f_{Tv} + \frac{\Delta I}{\Delta T(I^{39 °C} - I_0^{39 °C})} \qquad [14]$$

with ΔI representing the increase in current for a $\Delta T = 8$ °C, $I_0^{39 °C}$ the residual current, $I^{39 °C}$ the electrode current at 39 °C and f_{Tv} the correction factor (0.0038 °C^{-1}). Mean values ± standard deviations (in % per °C) are 0.68 ± 1.17 for group A, 2.99 ± 0.31 for group B, 3.16 ± 0.49 for group C. Group A shows significantly lower values ($p < 0.001$); group B and C do not differ significantly from each other ($p < 0.78$).

Sensitivity to CO_2

The sensitivity to CO_2 is determined by changing the gasmixture from pure N_2 to pure CO_2 ($V_p = -600$ mV, T = 39 °C). s_{CO_2} is then calculated as follows

$$s_{CO_2} = \frac{\Delta I}{P_a - P_v^{39 °C}} \qquad [15]$$

and can also be expressed as a fraction of the sensitivity to oxygen. The latter mean values ± standard deviations (in % of s_{O_2}) are 0.82 ± 0.66 for group A, 0.28 ± 0.15 for group B and 0.32 ± 0.16 for group C. Group A has a significantly higher CO_2-sensitivity compared to group B and C ($p < 0.015$). There is no significant difference between the latter two groups ($p < 0.3$).

ELECTRODE IMPLANTATION AND EFFECT ON TISSUE

In vivo studies on bare electrodes have not been performed because of implantation failure. Because of the poor in vitro results of these electrodes a new implantation is not considered necessary.

Electrode preparation and implantation

Eight electrodes of group B and C are constructed as described previously. A pair of electrodes is fixed in a polymethylmetacrylate frame (6 x 6 x 3 mm) which represents a simplified model of the frame used for photochemical infarction (van Rossem et al., in this issue). Two holes (diameter 0.4 mm) are drilled at a distance of 3 mm from each other. The electrodes are positioned in the holes with the aid of a micromanipulator and an operation microscope is used to evaluate whether the base of the measuring tip is situated at the same level as the bottom of the frame and whether the tip stands perpendicularly to the frame surface. When both conditions are fulfilled the electrodes are fixed with cyanoacrylate glue using a fine needle. A 4-step calibration and a polarogram are recorded.

Two female Dutch rabbits are used for the experiments. Four frames are implanted in the rabbit's skull at different times : at day 1 frame 1 is implanted, at day 22 frame 2, at day 43 frame 3 and at day 52 frame 4. Each time the same procedure is followed, except for the last day as the animal is killed immediately after fixation of the fourth frame. The implantation technique and post-operative care of the animals is described in detail by van Rossem et al. (in this issue). The Ag/AgCl reference electrode is attached to the skin at the inside of the ear. Therefore the epidermal layer is slightly scratched off with a scalpel and a salt bridge (electrode gel) is applied to reduce electrode resistance. A polarising voltage (- 600 mV with respect to the Ag/AgCl electrode) is applied to one electrode of each pair 8, 10, 11, 12, 15, 21, 28 and 50 days following implantation as far as the implantation period allows. pO_2 is recorded for 30 min (except for day 10, 3 h). A polarogram is recorded 10 days after implantation. This schedule corresponds to the protocol that is followed when focal infarctions are induced.

In vivo measurements

It should be stressed that in first instance in vivo measurements are performed in order to check possible morphological changes in cortical tissue as a result of the application of V_p. However, in spite of the limited number of electrodes these measurements can provide interesting and suggestive information. All recordings show a similar pattern with respect to time course of electrode current (i_{O_2}). Stabilisation time varies between 1.5 and 9 min (4.82 ± 1.86 min for group B and 4.93 ± 2.36 min for group C). It is interesting to note that during in vivo stabilisation electrode current decreases, whereas it increases in vitro. Following stabilisation slow variations of the mean current seldomly exceed 10 %; small fast variations are continuously present. Slopes of polarograms also differ from in vitro results. The electrode currents appear to be very constant at two electrodes; two other electrodes show values remaining constant the first days and then suddenly increasing to a higher level. Values between 20 and 85 nA are recorded. The ratio between electrode current and in vitro sensitivity varies between 1.5 and 6 kPa.

Morphologic study

Immediately after fixation of the last frame, two catheters are inserted in the common carotic arteries. 5 ml of nembutal (sodiumbarbiturate 60 mg/ml) is injected intravenously (marginal ear vein) to kill the animal and after cutting the left femoral vein, 250 ml of a 12 % solution of formaldehyde (FA) is flown through each catheter. The animal is then decapitated and the head is kept in FA for 24 hours. After removing the crown of the skull the cerebrum is carefully taken from the cranial vault and kept in FA for two weeks. The small shafts created by electrode implantation are detected using an operation microscope. Four pieces of cortical tissue (approximately 7 x 7 x 2 mm), each of them containing the two shafts of an electrode pair, are cut from the brain and embedded in paraffin. Transversal paraffin sections of 7 μm are cut with a microtome and stained with haemotoxylin-eosin.

Electrode tracks are cut perpendicularly and look like clearly delineated circles in the transversal sections. Sections according to newly implanted electrodes show either hardly any damage or limited bleeding or edema (Figure 8). In utmost superficial sections some meningeal cells pulled down with the electrode are situated around the shaft. Ten days after implantation a small rim of cells surrounds the tracks (Figure 8). Probably these cells are glial cells. However, we don't have evidence to state this possibility since glial cells and lymfocytes can hardly be distinguished in a haematoxylin-eosin staining. Tissue adjacent to the encapsulating sheath appears to be completely normal. Edema is not noticed. Thirty days after implantation the cellular rim is still partially present (Figure 8). However, some regions around the track shows vacuole-like structures, possibly representing glial tissue. In these regions only a few cells are noticed. One track even shows a very limited number of cells. Fifty-one days after implantation no accumulation of cells is noticed around the tracks (Figure 8). Tissue appears to be organised circularly around the track, most probably glial tissue. Only beneath the shaft cells are still accumulated. Up to this postsurgical time period cortical tissue adjacent to the organised circular edge looks perfectly normal. No marked differences can be noticed between sections obtained from the two rabbits. Furthermore, no difference in tissue reaction is noticed regarding polarised and non-polarised electrodes.

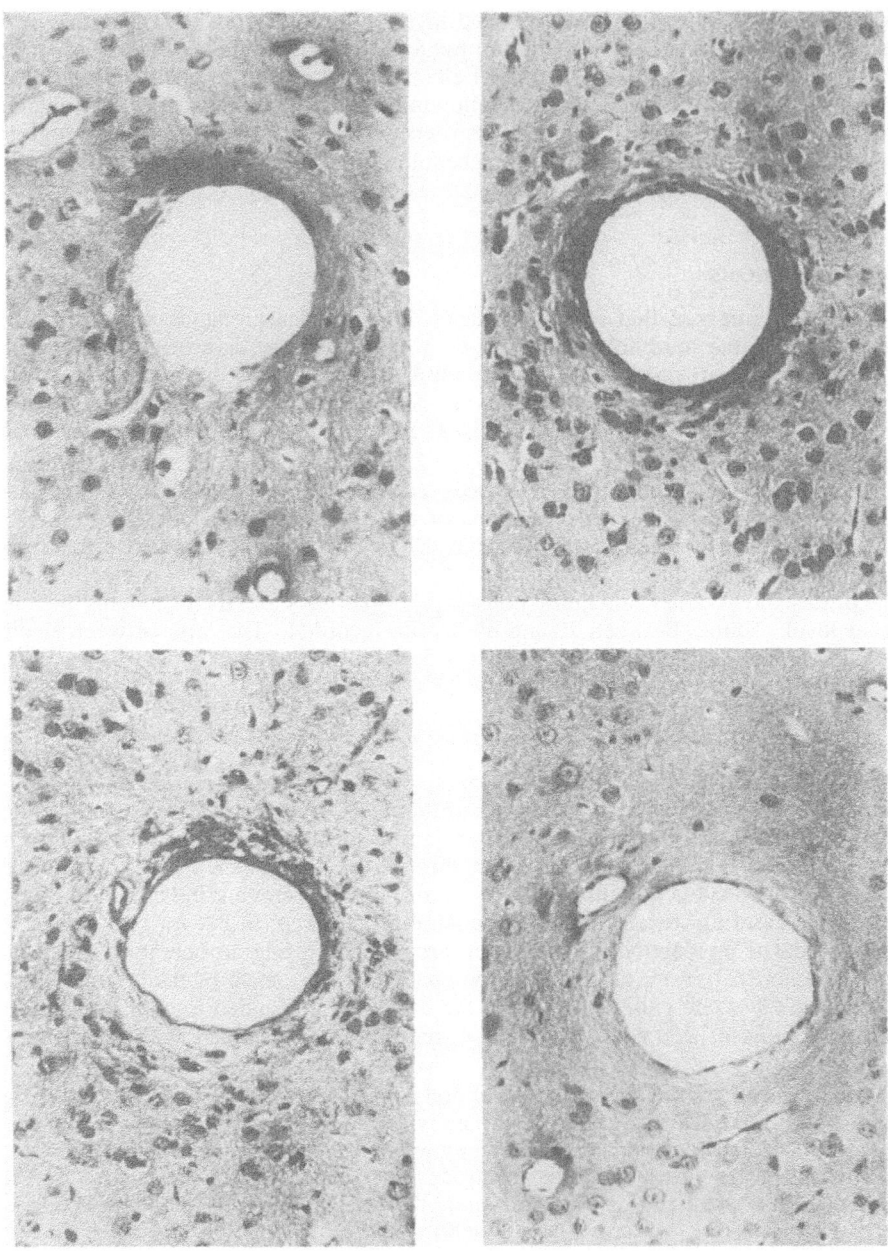

Figure 8 Light microscopic images of transversal sections of the electrode shaft 5 min (upper left), 9 days (upper right), 30 days (lower left) and 51 days (lower right) after implantation.

DISCUSSION AND CONCLUSIONS

As could be expected from theoretical considerations, naked electrodes perform poorly under calibration tests; the application of a homogenous membrane representing an artificial diffusion resistance to oxygen is necessary. Accurate application of the membrane material is a necessity and we are convinced that experience and routine will result in more homogenous membrane coatings. Apart from the measuring tip the insulation of the wire should be completely reliable. Cyanoacrylate glue appears to function well but care has to be taken to avoid cracks as a consequence of accidental bending of the wire.

Although rigourous precautions have been taken, leaking paths between electrode connectors in the calibration chamber, probably caused by water vapor condensation as a result of the bubbling effect, can affect the measurement of electrode current, more specifically the residual current. This is considered a disadvantage of the calibration device which has been developed.

In vitro all membrane covered electrodes stabilise within 30 min (except for two of 10 μm). In vivo stabilisation times are below 10 min. This can be explained by the fact that electrodes being calibrated in vitro are subjected to a number of changes which is not the case for implanted electrodes.

Membrane covered electrodes behave in an almost perfectly linear way during the complete timecourse of the protocol. Thus for routine implantation extensive calibration seems to be a redundant procedure. Two-point calibrations, one in air-saturated Ringer solution and one in oxygen-free Ringer, may determine sensitivity and residual current. Oxygen-free liquid can be obtained by adding glucoseoxidase to a glucose containing solution; moreover, this method shows less risk on leaking paths by water vapor condensation since gas perfusion is not applied.

Although some electrodes from different groups show comparable sensitivities, the mean sensitivity of the 20 μm membrane group is very close to halve the mean sensitivity of the 10 μm membrane group. Assuming ideal stirring and thus homogenous pO_2 in the calibration chamber, this finding agrees with theoretical considerations on the effect of membrane thickness. Nevertheless small current fluctuations are observed in vitro on the 10 μm membrane group suggesting that the stirring is not totally perfect.

Electrodes with a 20 μm membrane show the best linearity, the smallest variation coefficient in sensitivity, the smallest polarogram slope and the best short-term stability as compared with the 10 μm membrane. With regard to temperature dependency, CO_2 sensitivity and long-term stability differences between both membrane groups are very small (no statistically significant difference). Regarding all in vitro tests we may conclude that electrodes covered with a 20 μm membrane are more suitable for in vivo measurements. Nevertheless, sensitivity still varies considerably in this group. Other authors (Beran et al., 1978; Hahn et al., 1981; Baumgärtl and Lübbers, 1983) also have stated limited stability in time as a major disadvantage in oxygen polarography. Electrode currents obtained at different days cannot be compared accurately. Remarkably, in spite of the limited stability in vitro, some of the implanted electrodes showed almost constant current levels over a 50 days period.

Calculated oxygen consumption in vivo appears to vary between 3 and 13 pmol/min. These values respectively correspond to oxygen consumption of 0.001 and 0.004 mm^3 of rat brain cortical tissue (Siesjö, 1978). As the volume of the measuring tip is 0.016 mm^3 one can assume that the electrodes charge their environment to only a small extent. Moreover, the morphological study indicates that viability of cortical neurons adjacent to the electrodes is not jeopardized. Nevertheless, in order to keep oxygen consumption as low as possible - 600 mV is selected as polarising voltage (despite of the larger slope of the polarogram).

The implantation technique developed appears to be quite satisfactory as can be deduced from the morphological study. Electrode shafts appear perfectly circular and implantation causes only limited bleeding and edema. Except at 51 days following implantation the electrode track is surrounded by a small rim of cells. Our observations at 51 days after implantation make us tend to believe that these cells are glial cells producing well organized glial tissue. Our results are comparable with those of Tomida et al. (1989) and Schmidt et al. (1988). The probability of a scarr being formed around the electrode during the first weeks after implantation brings along a number of considerations. As scarr tissue possibly changes diffusion of oxygen to the electrode and contains less or no oxygen consuming cells, measurements could give variable results over time in spite of an identical environment. Therefore, it is necessary to standardize the time period between surgery and induction of the infarction.

ACKNOWLEDGEMENTS

This investigation was made possible through the skilled technical assistance of Decuyper K., Jacqueloot J., and Vandewoude D., and was partially supported by F.G.W.O. Contract 3.0017.88 (Fund for Medical Scientific Research) and by the OZR VUB.

REFERENCES

Baumgärtl, H., and Lübbers, D. W., 1983, Microaxial needle sensor for polarographic measurement of local O_2 pressure in the cellular range of living tissue. Its construction and properties, *in* : "Polarographic Oxygen Sensors", Gnaiger and Forstner, eds., Springer-Verlag, Berlin, 37-65.

Beran, A. V., Shigezawa, G. Y., Whiteside, D. A., Yeung, H. N., and Huxtable, R. F., 1978, In vitro evaluations of monopolar intravascular oxygen sensors, *J. Appl. Physiol. : Respirat. Environ. Exercise Physiol.*, 44 (6) : 969-973.

Clark, L. C., 1956, Monitor and control of blood and tissue oxygen tensions, *Trans. Am. Soc. Artif. Intern. Organs*, 2 : 41-48.

Fennema, M., Wessel, J. N., Faithfull, N. S., and Erdmann, W., 1989, Tissue oxygen tension in the cerebral cortex of the rabbit, *in* : "Oxygen transport to tissue XI", K. Rakusan, G. P. Biro, T. K. Goldstick and Z. Turek, eds., Plenum Press, New York and London, 451-460.

Hahn, A. W., Nichols, M. F., Sharma, A. K., and Hellmuth, E. W., 1981, Glow discharge polymer coated oxygen sensors, *Polymer Science and Technology*, 14 : 85-96.

Hale, J. M., 1983, Factors influencing the stability of polarographic oxygen sensors, *in* : "Polarographic Oxygen Sensors", Gnaiger and Forstner, eds., Springer-Verlag, Berlin, 3-17.

Hitchman, M. L., 1983, Calibration and accuracy of polarographic oxygen sensors, *in* : "Polarographic Oxygen Sensors", Gnaiger and Forstner, eds., Springer-Verlag, Berlin, 18-30.

Proctor, K. G., and Bohlen, H. G., 1979, Tonometer for calibration and evaluation of oxygen microelectrodes, *J. Appl. Physiol. : Respirat. Envir. Exercise Physiol.*, 46 (5) : 1016-1018.

Schmidt, E. M., McIntosh, J. S., and Bak, M. J., 1988, Long-term implants of paralyne-C coated microelectrodes, *Med. & Biol. Eng. & Comput.*, 26 : 96-101.

Shek, J. W., Wen, G. Y., and Wisniewski, H. M., 1986, Atlas of the rabbit brain and spinal cord, Karger, Basel.

Siesjö, B. K., 1978, Brain energy metabolism, John Wiley & Sons, Chichester.

Silver, I. A., 1966, The measurement of oxygen tension in tissue, *in* : "Oxygen measurements in blood and tissues and their significance", J. P. Payne and D. W. Hill, eds., Churchill, London, 135-145.

Tomida, S., Wagner, H. G., Klatzo, I., and Nowak, T. S. Jr., 1989, Effect of acute electrode placement on regional CBF in the gerbil : a comparison of blood flow measured by hydrogen clearance, [^{3}H]nicotine, and [^{14}C]iodoantipyrine techniques, *J. of Cerebral Blood Flow and Metabolism*, 9 : 79-86.

van Rossem, K., Vermariën, H., Jacqueloot, J., and Bourgain, R., 1989, Continuous sensing of oxygen tension in rabbit brain cortical tissue during and after photochemically induced infarction, *Proceedings V Mediterranean Conf. Med. Biol. Eng.*, Patras, 190-191.

Watson, B. D., Dietrich, W. D., Busto, R., Wachtel, M. S., and Ginsberg, M. D., 1985, Induction of reproducible brain infarction by photochemically initiated thrombosis, *Ann. Neurol.*, 17 : 497-504.

PHOTOTHROMBOSIS IN RABBIT BRAIN CORTEX :

FOLLOW UP BY CONTINUOUS pO$_2$ MEASUREMENT

Koen van Rossem, Herman Vermariën, Karin Decuyper
and René Bourgain

Laboratory of Physiology and Physiopathology, University of Brussels
VUB, Laarbeeklaan 103, B-1090 Brussels, Belgium

ABSTRACT

Continuous recording of changes in local pO$_2$ during and after brain infarction in sur-
viving animals which can be followed for months or years, may provide interesting information concerning pathophysiology and treatment of stroke and thrombosis. We performed such measurements before, during and till 4 weeks after photochemical induction
of a cerebrocortical infarction in three rabbits.

Rose bengal - a photosensitive dye which sticks to endothelial cells and gives rise to
endothelial damage and thrombosis when illuminated - was injected intravenously. After
injection, a circular area (diameter 3 mm) of the brain cortex was illuminated using an
optic fiber conducting light from a halogen lamp, whether or not filtered by heat and colour
filters.

In order to enable pO$_2$ measurement in and near the infarct zone, we constructed a
transparent plastic frame in which pO$_2$ electrodes were fixed beneath and 1 mm besides a
shaft permitting mounting of the optic fiber. A black adhesive ring (inner diameter 3 mm)
was attached to the bottom of the frame providing a perfectly demarcated illumination
area. After fixation the electrodes were calibrated and the frame was implanted in the
rabbit's skull.

Ten days later an infarction was induced; pO$_2$ was monitored continuously before,
during and till 4 hours after this induction. Furthermore, pO$_2$ was recorded 24 hours, 48
hours, 5 days, 14 days and 4 weeks after infarction.

Parameters describing the time course of pO$_2$ were determined. In the illuminated
area pO$_2$ decreased after a certain latency time to reach a very low level, probably zero
level, where it remained for at least 24 hours. Gradually, recovery was observed during
the following days, and four weeks after infarction both level and pattern of electrode
current appeared to be normal again. In the border zone pO$_2$ decreased but did not reach
zero level. Recovery was observed earlier than in the illuminated area.

INTRODUCTION

Various animal models are used in pathophysiologic, pathologic and pharmacologic
studies of focal brain ischemia and ischemic stroke (Graham, 1988). Most of them are

based upon occlusion of a major cerebral artery (intra or extra cranial) or induction of cerebral embolism. Variations in collateral circulation and differences in onset and distribution of embolisms are major causes of insufficient reproducibility regarding these models. Recently, Watson et al. (1985) developed a reproducible photochemical rat model which can be applied to induce focal cortical infarctions. In this model brain cortex is illuminated through the skull after intravenous injection of the photosensitive dye rose bengal (absorption peak at 560 nm). Due to photo exitation of the dye, singlet oxygen is generated leading to membrane damage of endothelial cells and subsequent platelet aggregation and thrombosis in all illuminated vessels. By adjusting the dosis of rose bengal, intensity and spectrum of the light, illumination time and finally the illuminated area, one can determine onset velocity and size of the infarction.

Time course of ischemic damage, microvascular alterations and changes in cerebral morphology, cerebral blood flow and metabolism during and after infarction have already been studied applying microscopic and autoradiographic methods (Watson et al., 1985; Dietrich et al., 1986; Grome et al., 1988; Van Reempst et al., 1987; Ginsberg et al., 1989). These techniques imply a number of limitations : they only provide instant information and require killing of the animal so that longitudinal studies cannot be performed. At the present the only study performed longitudinally is one in which sensimotor deficits after neocortical infarction were evaluated (De Ryck et al., 1989).

We have developed a technique which enables continuous recording of local oxygen tension in rabbit brain cortical tissue, in and near the infarct zone, before, during and after photochemically induced photothrombosis. Brain cortex is illuminated through a chronically implanted transparent frame in which oxygen electrodes are fixed. In this way pO_2 can be monitored during both infarction and recovery, providing interesting information about the dynamics of the thrombotic event. Moreover, our technique allows simultaneous recording of other parameters (evoked potentials, tissue perfusion, etc.) since various electrodes may be positioned into the frame. Finally, the rabbits don't have to be killed so that longitudinal studies are enabled. Therefore, this technique may offer a lot of perspectives in stroke and thrombosis research.

MATERIALS AND METHODS

Oxygen sensor configuration and evaluation

Miniature polarographic pO_2 electrodes of the open type were used for oxygen measurement. Construction, calibration, measuring device and a general description of electrode properties are outlined in detail elsewhere (van Rossem et al., in this issue). In short, platinum wire (100 μm diameter) was soldered to a gold connector and insulated with cyanoacrylate except for the measuring tip (1 mm length) which was coated with a homogenuous cellulose acetate membrane (20 μm thickness). The soldered connection was covered with shrinking tube which was sealed with glue. Calibration was performed after fixation of electrodes in the transparent frame by bubbling predetermined gasmixtures (0 % to 15 % O_2) successively through Ringer solution at 39 °C. Electrode sensitivity was determined and linearity of response was checked. Both calibration and in vivo measurements were performed applying a polarising voltage of - 600 mV and using a Ag/AgCl reference electrode, which was fixed to the animals ear (after scratching off the upper layer of the skin) when cortical tissue pO_2 was recorded. Since electrode sensitivity was found to vary considerably over 50 days in vitro (van Rossem et al., in this issue) we did not extrapolate electrode current to quantitative pO_2 values.

Frame construction and electrode fixation

We constructed a polymethylmetacylate (PMMA) frame in which an optic fiber can be mounted and electrodes can be fixed in and near the illumination area (Figure 1). A PMMA ring (3 mm thickness, inner diameter adapted to the optic fiber) was clinged to a circular PMMA bottom plate (1 mm thickness). The border of the resulting cylinder was

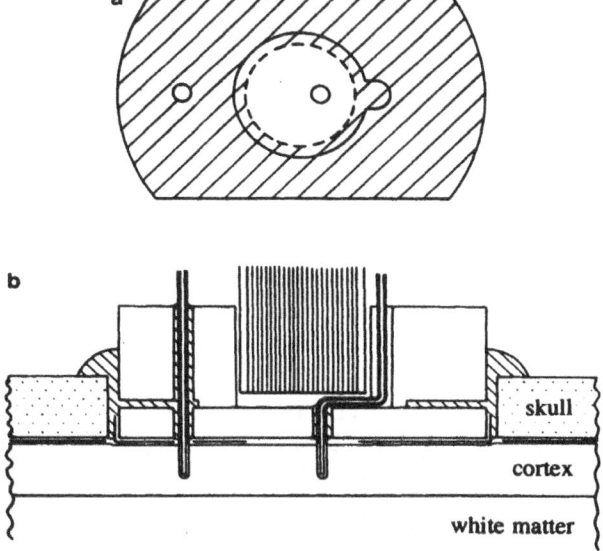

Figure 1 a) Transparent frame, view from above (upper part). Oblique lines indicate the black ring attached to the bottom of the frame. b) Cross-sectional diagram of an implanted frame with fixed pO_2 electrodes and an optic fiber mounted (lower part). Shaded parts indicate glue or dental resin for fixation of the frame.

milled at opposite sides providing two parallel edges at a distance of 6 mm from each other in order to enable positioning. A thin adhesive black ring (inner diameter 3.17 mm) was fixed to the bottom plate providing a perfectly delineated and standardised illumination area. Small holes (0.4 mm diameter) were drilled in and 1 mm near this area (border zone) and electrodes were dropped through these holes. An operation microscope was used to evaluate whether the base of the measuring tip was at the same level of the bottom of the frame and whether the tip was rectangular to the frame. When both conditions were fullfilled the electrodes were fixed with cyanoacylate glue. In order to put the optic fiber as close as possible to the bottom plate, the pO_2 electrode situated in the illumination area was bended and guided through a groove along the cylindrical hole.

Animal preparation

Three female dutch rabbits (3 to 3.5 kg) were used for the experiments. A transparent frame containing two electrodes was implanted in the animals skull. Rabbits were anesthetized by intramuscular injection of 1.5 ml hypnorm (fluanison 10 mg, fentanyl 0.2 mg pro ml). The scalp was shaved and the rabbits head was positioned and fixed into a stereotactic apparatus. After incision and retraction of the scalp the periost was scratched from the skull and a hole was drilled according to the contour of the frame. The center of the hole was positioned 2 mm behind and 2 mm lateral from the bregma (projection area of the fore paw). Dura mater was removed and the frame was implanted with the aid of a micromanipulator (Figure 2). A hollow cylinder was mounted to a micromanipulator and connected with a vacuum pump. The frame was sucked against the cylinder and positioned right above the trepanation hole with the bottom plate parallel to the brain surface. The whole was carefully lowered and the measuring tips were pushed into the brain cortex applying a little bit of overpressure to ensure complete insertion of the tips. Vacuum was then

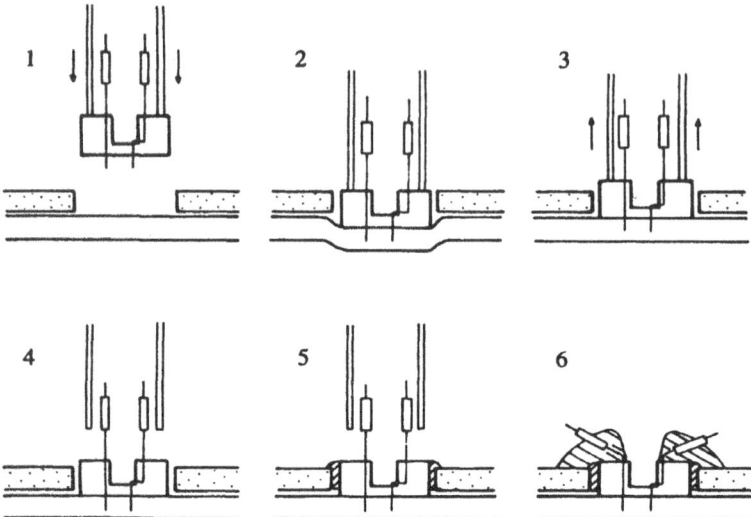

Figure 2 Schematical representation of the implantation of the frame in the rabbit's skull. A vacuum hollow cylinder, mounted to a micromanipulator, is used for insertion of the frame. Cyanoacrylate gel fixes the frame to the skull and the whole is covered by dental resin. Subsequent steps are indicated by numbers (1, 2, 3, 4, 5 and 6).

switched off and the cylinder was slowly elevated until contact with the frame was lost. Dental resin was used to fix the frame to the skull and to cover the wound. After implantation the animals were treated with chloromycetine (chloramphenicol 2 x 100 mg IM per day) during 5 days. Three days later a controle pO_2 monitoring was performed.

Photochemical induction of cerebrocortical infarction

Ten days after implantation of the frame an aliquot of 7.5 mg/ml saline solution of rose bengal, which had been subjected to 0.45 μm filtration, was injected intravenously (0.133 ml/100 g rabbit weight) via the marginal ear vein over a 2 min interval. One minute after finishing injection illumination was started with a Volpi 250 HL fiber optic light source provided with an Osram 250 W halogen lamp. The fiber (3 mm diameter) was mounted into the implanted frame. One animal (rabbit A) was illuminated without using special colour filter setting (light intensity 48 mW over the surface at the end of the fiber, i.e. 600 mW/cm^2). An infrared filter, built into the light source by the manufacturer partially cooled the considerably warm light. Regarding two other rabbits light was additionally filtered respectively using a green wratten filter (Kodak nr 74, spectral width 480 - 630 nm, transmission peak 53 %; rabbit B) and a green dichroic filter (OCLI, spectral width 490 - 600 nm, transmission peak 94 % at 550 nm; rabbit C) light intensity was adjusted to 110 mW/cm^2. Light intensity was checked with a digital power meter (Newport Model 835).

Illumination lasted 20 min. Animals were restrained but not anesthetised during the experiments.

Follow up during and after infarction

pO_2 electrode current was monitored continuously before (30 min), during and until four hours after infarction. Furthermore, measurements lasting 30 min were performed 24 h, 48 h, 5 d, 14 d and 28 d after infarction. Electrode stabilisation time (< 10 min) was included in the measuring period. Two control animals (rabbits 1 and 2) which were also used to study morphologic changes due to electrode implantation (van Rossem et al., in this issue), were followed according to the same measurement procedure without undergoing photochemically induced infarction.

RESULTS

General remarks

During and after infarction animals remained very quite, showing no clear clinical signs pointing at pain or discomfort. No changes in behaviour or movement were observed, although motoric functioning could not be evaluated accurately as a consequence of restricted mobility of the animals due to the limited space in their housing cages. Rabbit A died three weeks after infarction from starvation, due to a trichobézoar (hair ball) present in the stomach. As a consequence controle on day 28 could not be performed.

The green color of the wratten filter used with respect to rabbit B appeared to be modified to green-yellow after illumination. This was probably due to the heat production of the halogen lamp.

Control rabbits

Electrode current (i_{O_2}) showed a similar pattern in both control rabbits. Fluctuations around a mean electrode current ($\bar{i}_{O_2}$) were present in all recordings. After starting measurement $\bar{i}_{O_2}$ decreased during a certain stabilisation time which lasted a few minutes, never exceeding 10 min. Following stabilisation, $\bar{i}_{O_2}$ appeared to be fairly constant. Variations of $\bar{i}_{O_2}$ were very limited and slightly exceeded 10 % only twice. On sight, the pattern of fluctuations could be described in terms of shape, frequency and amplitude. In all records a principal taperly shaped fluctuation appeared, characterised by a frequency (principal frequency f_p) varying from 5 to 10 cycli per min and a variable amplitude (A_p) amounting 10 % to 40 % of $\bar{i}_{O_2}$. Fast minor variations irregularly interfered with the principal fluctuation, which in its turn was occasionally masked by or carried along with limited low-frequent changes. Comparing the two rabbits, slight differences in f_p, A_p, and interfering fluctuations were noted, whereas the pattern appeared to be almost perfectly reproducible over the 50 day time period, regarding each animal separately.

Time course of $\bar{i}_{O_2}$ in control rabbits is represented in table 1. Mean electrode current was approximately constant over the 50 day period, in spite of the overall insufficient long-term stability of our electrodes in vitro.

Table 1 Time course of mean electrode current (nA) over a 50 day time period in control animals.

	days following implantation							
	8	10	11	12	15	21	28	50
rabbit 1	53	42	50	50	50	46	50	50
rabbit 2	27	22	22	24	24	20	22	22

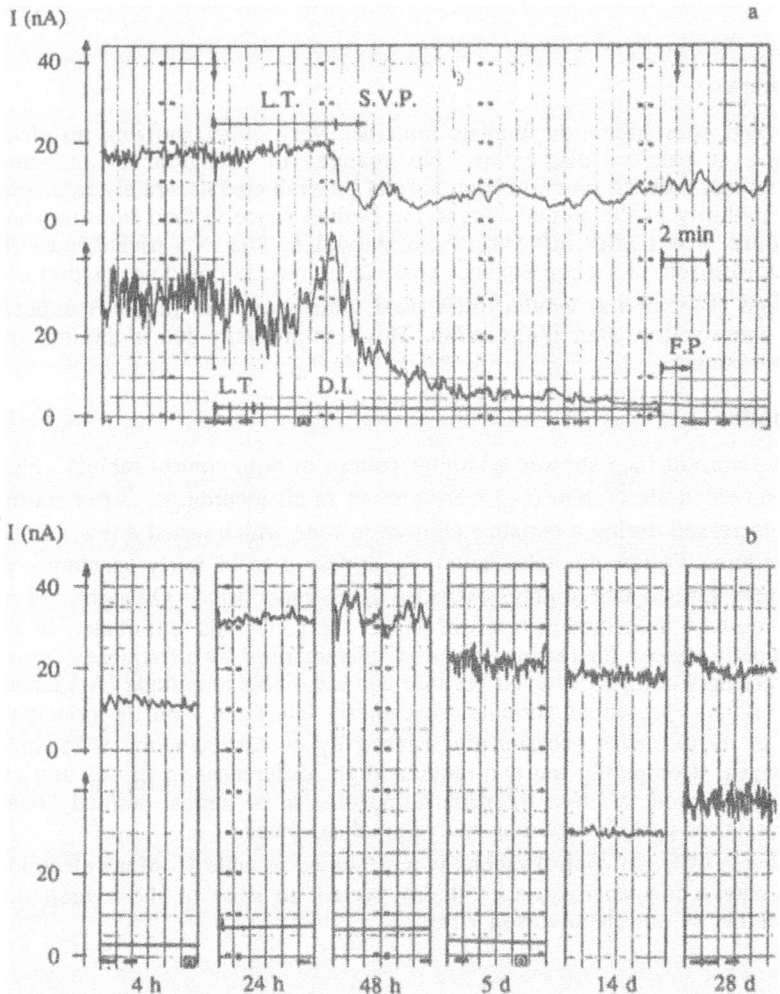

Figure 3 a) Recording of i_{O_2} during infarction of the rabbit cerebral cortex, in (lower curve) and near (upper curve) the illuminated area. Vertical arrows indicate beginning and ending of light exposure. Specific parameters are indicated. L.T. : latency time; D.I. : decline interval; F.P. : flat period; S.V.P. : slow variation period. b) Time course of i_{O_2} in (lower curve) and near (upper curve) the illuminated area (same time scale) : 4 h, 24 h, 48 h, 5 d, 14 d and 28 d after infarction.

Experimental rabbits

Based upon the records, we defined a number of specific parameters in order to describe the timecourse of i_{O_2} during and after infarction. Following onset of light exposure both pattern and $\bar{i}_{O_2}$ remained constant during a certain "latency period". Regarding the illuminated area a "decline interval" was defined starting at the occurance of a 20 % decrease of $\bar{i}_{O_2}$ whether or not accompanied by changes in pattern, and finishing when the "flat period" was reached. During this flat period fluctuations were absent and i_{O_2} remained at a constant minimum level, probably the residual current. In the border zone the latency period was followed by a "slow variation period" characterised by predominant irregular slow variations, on which the principal fluctuation (with a lower A_p) was possibly superposed. The flat period was never reached in this area. Figure 3 shows the record of rabbit B (illuminated with green light) indicating the different period intervals. Results are summarized in table 2. In all animals both flat period and slow variation period were reached within illumination time, and continued after switching off light exposure. Concerning the illuminated area latency time was 30 % shorter in rabbit A than in rabbits B and C. The same proportion was found with respect to the decline interval between rabbits A and B, whereas rabbits A and C showed comparable values.

At a distance of 1 mm from the illuminated area all animals showed comparable latency periods, which always lasted longer than respective latency periods in exposed areas. Slow variation periods showed a decreased $\bar{i}_{O_2}$ in rabbits A and B. In contrast, regarding rabbit C, $\bar{i}_{O_2}$ remained constant in the border zone during and after infarction.

Table 3 describes the time courses of $\bar{i}_{O_2}$ and pattern during the days and weeks following infarction. In general, recovery of both $\bar{i}_{O_2}$ and pattern was noted after a certain period. Recovery went very similarly in rabbits A and B. Concerning these two animals the flat period lasted at least five days while two weeks after infarction $\bar{i}_{O_2}$ appeared to be normalised and principal fluctuations reoccurred. Four weeks after infarction both $\bar{i}_{O_2}$ and pattern were completely normalised in rabbit B. In the border zone slow variations maintained for at least two days in rabbit A and for one day in rabbit B. In both animals a normal pattern was noted from the fifth day on while $\bar{i}_{O_2}$ exceeded original values already 24 h after infarction.

Regarding rabbit C partial recovery occurred faster in both areas. In the border zone $\bar{i}_{O_2}$ did not appear to be decreased at any time, and slow variations had disappeared already 24 h after illumination, being replaced by a normal pattern. In the illuminated area slow variations were observed together with a restoration of $\bar{i}_{O_2}$ after 48 h. Five days after infarction a discrete principal fluctuation was superposed on these slow waves and at two weeks the pattern was back to normal, $\bar{i}_{O_2}$ exceeding its initial value.

Table 2 Time intervals (min) during infarction.

		A	B	C
illum. area	lat. time	0.5	1.5	1.5
	decl. int.	5	16	4
border zone	lat. time	3.5	5	4

Capitals indicate different animals. Rabbit A was illuminated using unfiltered white light. Regarding rabbits B and C filtered green light was applied.

Table 3 Time courses of mean electrode current (nA) and pattern during and after implantation.

time following onset of illum.	A illum. area	A border zone	B illum. area	B border zone	C illum. area	C border zone
0	35 ++++	43 ++++	30 ++++	18 ++++	13 ++++	40 ++++
20 min	3 -	25 ++	3 -	7 +	3 -	40 +
4 h	0 -	11 ++	3 -	10 ++	3 -	40 +
24 h	0 -	65 ++	7 -	33 ++	4 -	40 ++++
48 h	1 -	76 ++	7 -	33 ++++	14 +	40 ++++
5 d	3 -	45 ++++	3 -	22 ++++	22 ++	44 ++++
14 d	32 ++	60 ++++	30 +++	18 ++++	35 ++++	52 ++++
28 d	†	†	35 ++++	21 ++++	35 ++++	52 ++++

Capitals indicate different animals. † : animal A died 3 weeks following infarction as a consequence of a trichobézoar. Rabbit A was illuminated using unfiltered white light. Filtered green light was applied regarding rabbits B and C. Different patterns are represented in an ordinal scale : normal fluctuations (++++); pattern only showing a decreased A_p (+++); principal fluctuation carried along with slow variations (++); slow variations without principal fluctuation (+); flat curve (-).

DISCUSSION

We have modified the light/dye model of Watson et al. (1985) in order to monitor local tissue oxygen tension in rabbit brain cortex before, during and after photochemically induced focal infarction. An implantable transparent frame was constructed enabling positioning of the pO_2 electrodes and mounting of an optic fiber, also providing a perfectly standardised illumination area. Other authors (Watson et al., 1985; Dietrich et al., 1986; Van Reempst et al., 1987; Grome et al., 1988; Ginsberg et al., 1989; De Ryck et al., 1989) illuminate rat cerebral cortex through the skull, facing an unknown scattering of light which might vary between different animals. Moreover, since rabbit skull is considerably thicker, illumination had to be modified for our purpose. Applying each time the same dosis rose bengal we have tried out various illumination intensities and spectra in order to study common events and to determine the most suitable and standardised illumination technique. Indeed, results of Van Reempst et al. (1987) have stressed the importance of illuminating in a standardised way.

A general pattern was observed in the time course of pO_2 during and after infarction. In the illuminated area both pattern of fluctuations and mean electrode current showed major changes, reaching an invariable low value within 20 minutes. Most probably this was residual current, but since we could not fully rely on our determination of residual current (van Rossem et al., in this issue) we could not prove this assumption. After one or more days recovery commenced and electrode current regained or even exceeded initial values four weeks after infarction. In the border zone zero level was certainly never reached and recovery was observed earlier.

Differences observed between the three rabbits were most probably due to different illumination techniques. In the future we will apply filtered light using the green dichroic OCLI filter for two main reasons. Its spectrum was not influenced by long-term illumination in contrast with the Kodak wratten filter. Secondly, unfiltered light contained considerable heat which undoubtly warmed up cerebral cortex. Busto et al. (1987) have shown the critical role of small differences in brain temperature during the ischemic event. As a consequence we prefer to apply cold light which we may obtain approximately using the dichroic green filter (only 0.5 °C increase at the end of the optic fiber). In order to standardize the method, other parameters have to be controlled too. Blood pressure, plasma glucose concentration, and platelet count should be comparable between the experimental animals (Graham, 1988). Body temperature, arterial pO_2 and pCO_2 should be checked.

Although physiologic parameters were unknown in present experimental animals, and in spite of application of different illumination techniques, our findings are in accordance with those of other authors. Watson et al. (1985) found that occlusion of vascular channels progressed after irradiation and was completed within 4 hours. Dietrich et al. (1986) showed a severe hypoperfusion in the illuminated area 30 minutes following irradiation which recovered partially 15 days after infarction. They found a slightly increased CBF in the border zone 30 min after illumination, whereas our results show an equal or decreased pO_2, being consistent with the evolution of secondary edema around the infarct zone (Watson et al., 1985; Grome et al., 1988). Royster et al. (1988) have demonstrated reperfusion and endothelial regeneration in rabbit retinal vessels after photothrombosis. Seven days after occlusion these vessels were completely reopened. The same event might occur in cerebrocortical vessels, although Grome et al. (1988) observed necrosis of the walls of the bloodvessels. This might delay or even impede regeneration so that neovascularisation has to be considered as another possible explanation for recovery of tissue pO_2. Anyway, restoration of both electrode current and pattern indicated recovery of circulation in the infarct zone.

The major advantage of our method is that it enables monitoring of the dynamic ischemic process continuously at any moment and as long as desired. As a consequence it may add a considerable amount of alternative information to the present knowledge concerning focal cortical infarction, obtained using steady state measurements. Yet, a number of improvements can be introduced. Quantification of changes in i_{O_2} pattern can be performed using signal frequency spectral analysis. Integrity and functioning of local blood vessels may be checked by response to CO_2 inhalation. Finally, evoked potentials, tissue perfusion and possibly other parameters may be followed simultaneously, as already performed with respect to global ischemia (Colin et al., 1981; Van Waeyenberge et al., 1985).

ACKNOWLEDGEMENTS

This investigation was made possible through the skilled technical assistance of Andries R., Jacqueloot J., Smets W. and Vereecke F., and was partially supported by F.G.W.O. Contract 3.0017.88 (Fund for Medical Scientific Research) and by the OZR VUB.

REFERENCES

Busto, R., Dietrich, W. D., Globus, M. Y.-T., Valdes, I., Scheinberg, P., and Ginsberg, M. D., 1987, Small differences in intraischemic brain temperature critically determine the extent of ischemic neuronal injury, *J. of Cerebral Blood Flow and Metabolism*, 7 : 729-738.

Colin, F., Manil, J., and Bourgain, R. H., 1981, The effect of acute anoxia on the cortical somatosensory evoked potential in the rabbit, *Neurological Research*, 3(4) : 393-407.

De Ryck, M., Van Reempst, J., Borgers, M., Wauquier, A., and Janssen, P. A. J., 1989, Photochemical stroke model : flunarizine prevents sensorimotor deficits after neocortical infarcts in rats, *Stroke*, 20(10) : 1383-1390.

Dietrich, W. D., Ginsberg, M. D., Busto, R., and Watson, B. D., 1986, Photochemically induced cortical infarction in rat. 1. Time course of hemodynamic consequences. 2. Acute and subacute alterations in local glucose utilization, *J. of Cerebral Blood Flow and Metabolism*, 6 : 184-202.

Ginsberg, M. D., Castella, Y., Dietrich, W. D., Watson, B. D., and Busto, R., 1989, Acute thrombotic infarction suppresses metabolic activation of ipsilateral somatosensory cortex : evidence for functional diaschisis, *J. of Cerebral Blood Flow and Metabolism*, 9 : 329-341.

Graham, D. I., 1988, Focal cerebral infarction, *Journal of Cerebral Blood Flow and Metabolism*, 8 : 769-773.

Grome, J. J., Gojowczyk, G., Hofmann, W., and Graham, D. I., 1988, Quantitation of photochemically induced focal cerebral ischemia in the rat, *J. of Cerebral Blood Flow and Metabolism*, 8 : 89-95.

Royster, A. J., Nanda, S. K., Hatchell, D. L., Tiedeman, J. S., Dutton, J. J., and Hatchell, M. C., Photochemical initiation of thrombosis Fluorescein angiographic, histologic and ultrastructural alterations in the choroid, retinal pigment epithelium and retina, *Arch. Ophtalmol.*, 106 : 1608-1614.

Van Reempst, J., Van Deuren, B., Van de Ven, M., Cornelissen, F., and Borgers, M., 1987, Flunarazine reduces cerebral infarct size after photochemically induced thrombosis in spontaneously hypertensive rats, *Stroke*, 18 (6) : 1113-1119.

van Rossem, K., Vermariën, H., Jacqueloot, J., and Bourgain, R., 1989, Continuous sensing of oxygen tension in rabbit brain cortical tissue during and after photochemically induced infarction, *Proceedings V Mediterranean Conf. Med. Biol. Eng.*, Patras, 190-191.

Van Waeyenberge, M., Vermariën, H., De Backer, H., Manil, J., and Bourgain, R. H., 1985, Discriminant parameters representing cerebral cortical function during anoxic anoxia investigations, *In* : "Advances in Experimental Medicine and Biology", F. Kreuzer, S. M. Cain, Z. Turek and T. K. Goldstick, eds., vol. 191 : Oxygen Transport to Tissue VII, Plenum, New York, 149-161.

Watson, B. D., Dietrich, W. D., Busto, R., Wachtel, M. S., and Ginsberg, M. D., 1985, Induction of reproducible brain infarction by photochemically initiated thrombosis, *Ann. Neurol.*, 17 : 497-504.

INTRAVITREAL AND INTRARETINAL OXYGEN TENSION IN THE RAT EYE

Stephen J. Cringle, Dao-Yi Yu and Valerie A. Alder

Lions Eye Institute and Department of Surgery
University of Western Australia
Nedlands, Western Australia 6009
Australia.

INTRODUCTION

The intravitreal and intraretinal oxygen distribution has been studied in several larger animals[1,2,3] but never before in the rat. In fact the small vitreal volume ($55\mu l$)[4] and large lens has previously dissuaded attempts at intraocular microelectrode work of any description. However, the advent of several rat models of retinal vascular disease has made this animal an attractive model in which to study retinal oxygenation, with a view to a better understanding of human diseases of a vascular nature. We therefore modified our techniques developed for larger animals to allow comparable measurements to be made in the rat. Vitreal oxygen tension distribution has been determined, with particular attention to the effects of retinal arteries and veins. The intraretinal oxygen tension profile has also been categorised. The effects of hyperoxic ventilation were also investigated intravitreally and intraretinally. This control data will be vital to our future studies of the effects of vascular impairment in models of vascular disease.

METHODS

Adult male Sprague-Dawley rats were anaesthetized, intubated, and artificially ventilated with either 20% or 100% oxygen. The animal was held in a stereotaxic frame and the eye sutured to a stabilising eye ring. A polarographic oxygen sensitive microelectrode[1] was inserted into the eye through a small incision at pars plana ciliaris. The microelectrode and the fundus was viewed through an operating microscope in conjunction with a plano-concave contact lens. A system of orthogonal arcs allowed the electrode to be oriented to the desired retinal location and the electrode then stepped forwards or backwards by a piezoelectric translator under computer control. The sytemic blood pressure, heart rate and expired CO_2 were monitored continuously and the arterial blood gases measured at critical stages in the experiment. The systemic parameters, the electrode location and the oxygen current were all monitored by the computer through a data aquisition card, and also recorded on a multi channel chart recorder. For a typical intravitreal oxygen profile measurement the microelectrode was placed on the internal limiting membrane (ILM), adjacent to either an artery, vein or intermediate location (an area with no visible vessels). The electrode was then stepped backwards in 10 μm increments for 200 μm and then in 50 μm steps for another 800 μm with the oxygen tension being measured at each location after a 2 second stabilisation time. This procedure was than reversed and the measurements repeated during the reinsertion. For a typical intraretinal penetration the tip of the microelectrode was positioned at an intermediate location. The electrode was then stepped into the retina in 10 μm increments. The maximum penetration depth was set to ensure that the entire retinal thickness was traversed before the measurements were repeated during the subsequent withdrawal.

RESULTS

A set of typical intravitreal profile measurements are presented in Figure 1. It is evident that considerable oxygen tension gradients exist in the vicinity of retinal arteries. In contrast, the bulk of the vitreous has a very homogenous oxygen distribution and is unaffected by retinal vasculature. Hyperoxic ventilation increases the oxygen tension throughout the vitreous and enhances the gradients seen near the arteries still further. This is in complete agreement with the data from the cat.

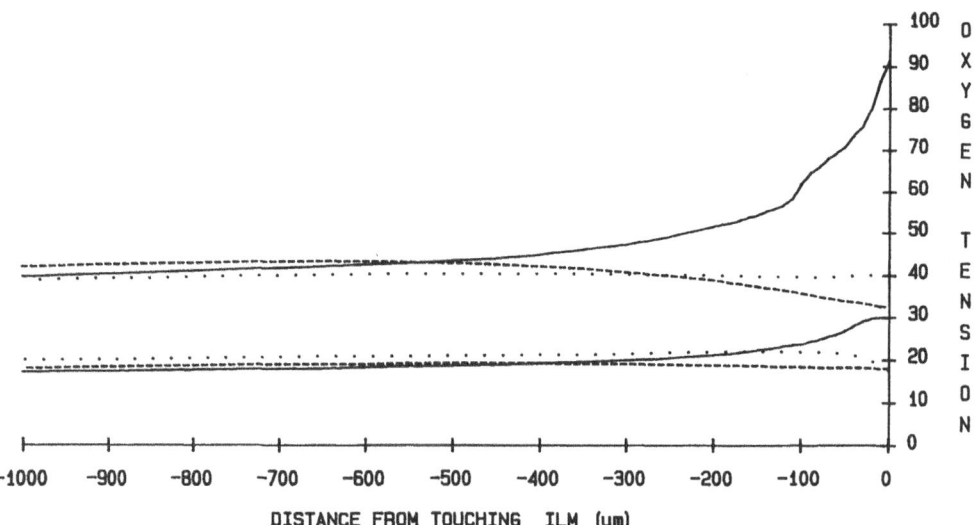

Fig. 1.   A typical set of intravitreal oxygen profiles for air (lower curves) and oxygen (upper curves) breathing. The initial retinal location was either an artery (continuous line), vein (dashed line) or intermediate area (dotted line). The oxygen tension is plotted in terms of distance from the initial touching location at the ILM.

Intraretinal results are presented as oxygen tension as a function of percentage retinal depth. The actual depth is a function of the obliquity of the microelectrode penetration and the true retinal thickness at that location. It is assumed that the peak PO_2 corresponds to the choriocapillaris at 100% retinal depth. This normalisation of retinal thickness allows individual profiles to be averaged. Figure 2 shows the mean ($\pm$SE) oxygen tension profile for 36 penetrations in 5 rats whilst the animals were respired with 20% oxygen. It may be seen that the intraretinal PO_2 decreases with penetration depth to a minimum at approximately 50% retinal thickness, before rising again to a peak at the choriocapillaris in the outer retina. This distribution was similar in all animals and retinal locations used. Figure 3 shows the mean profile ($\pm$SE) obtained when the animal was breathing 100% oxygen (3 animals, 9 penetrations). The 20% oxygen data is replotted here also for comparison. The result of hyperoxic ventilation is to increase oxygen tension throughout the retina. The minimum in the PO_2 profile is no longer distinct

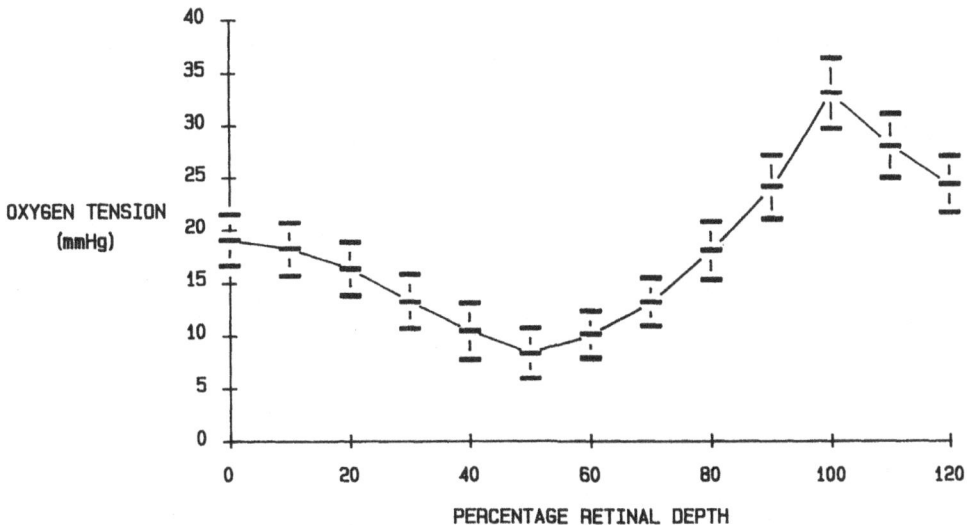

Fig. 2.   Mean oxygen tension ± SE as a function of percentage retinal depth for air breathing rats. 0% retinal depth is the ILM and 100% retinal depth is the choriocapillaris.

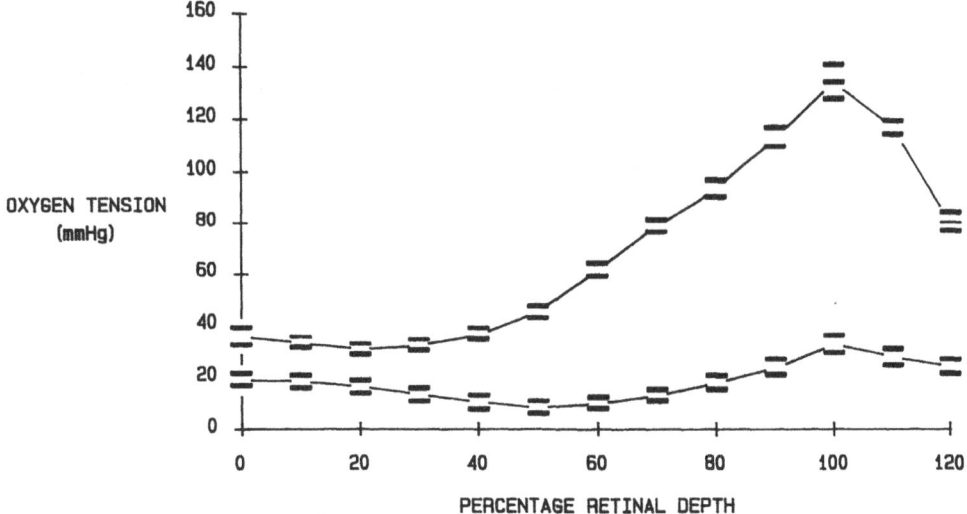

Fig. 3.   Mean oxygen tension ± SE as a function of percentage retinal depth for oxygen breathing rats (upper trace), with that for air breathing replotted on the same scale for comparison (lower trace).

and the curve shows an almost monotonic increase in PO_2 as the choriocapillaris is approached.

DISCUSSION

The intravitreal oxygen distribution in the rat is remarkably similar to that previously shown for the cat[7], as is the response to hyperoxic ventilation. The leakage of oxygen from the retinal arteries is clearly evident. The intraretinal profiles illustrate the contributions of the both the retinal and choroidal circulations to retinal oxygenation. In the air breathing case the minimum in the profile represents the balance point of oxygen diffusion with the deeper retina being supplied by the choriocapillaris and the more proximal retina supported by the retinal circulation. This has been noted before in the cat[1,2] and miniature pig[3]. Linsenmeier also noted the tendency for this minimum to become more pronounced and deeper into the retina when outer retinal oxygen consumption was increased by dark adaptation[5]. The appearance of very low oxygen tensions in his studies indicates that the avascular retina may easily be at risk from hypoxia following an increase in oxygen consumption, and presumably therefore, equally as sensitive to reductions in oxygen supply from either circulation. Our results show that the rat has a similar oxygen distribution to the cat and suggests that intraretinal oxygen profiles in the diseased rat may be sensitive to vascular abnormalities.

With hyperoxic ventilation the majority of the retinal depth is now supplied with oxygen from the choroidal circulation, again a similar finding to that in the cat[6]. The significant increase in oxygen tension throughout the retina and vitreous was not observed in the case of the miniature pig[8], where it seems that autoregulation of the retinal circulation maintains a stable inner retinal and vitreal oxygen tension despite the systemic hyperoxia.

SUMMARY

We have succesfully measured the intravitreal and intraretinal oxygen distribution in normal rats, during both normoxia and hyperoxia. The feasibility of such measurements in the very small rat eye has opened up the opportunity for the use of the many rat models of vascular disease in physiological experiments of retinal oxygen supply and consumption.

REFERENCES

1. V.A. Alder, S.J. Cringle, and I.J. Constable, The retinal oxygen profile in cat. Invest. Ophthalmol. Vis. Sci. 24:30-36 (1983).

2. R.A. Linsenmeier, and C.M. Yancey, Effects of hyperoxia on the oxygen distribution in the intact cat retina. Invest. Ophthalmol. Vis. Sci. 30, 612 (1989).

3. M. Tsacopoulos, R. Baker, and S. Levy, Studies on retinal oxygenation. Adv Exp Med Biol 75:413 (1976).

4. A. Chaudhuri, P.E. Hallett, and J.A. Parker, Aspheric curvatures, refractive indices and chromatic aberration for the rat eye. Vis. Res. 23:1351 (1983).

5. R.A. Linsenmeier, Effects of light and darkness on oxygen consumption and distribution in the cat retina. J Gen Physiol 88:521 (1986).

6. V.A. Alder, and S.J. Cringle, Intraretinal and preretinal PO_2 response to acutely raised intraocular pressure in cats. Am J Physiol. 256 (Heart. Circ. Physiol. 25):H1627 (1989).

7.   V.A. Alder and S.J. Cringle, The effect of the retinal circulation on vitreal oxygen tension. Curr. Eye Res. 4:121 (1985).

8.   C.J. Pournaras, C.E. Riva, M. Tsacopoulos, and K. Strommer, Diffusion of O_2 in the retina of anaesthetized miniature pigs in normoxia and hyperoxia. Exp. Eye Res. 49:347(1989).

BRAIN OXYGENATION STATE : PREPARATION OF ISOLATED PERFUSED RAT BRAIN

AND NEAR-INFRARED SPECTROPHOTOMETRY

M.Inagaki and M.Tamura

Biophysics Division, Research Institute of Applied
Electricity, Hokkaido University
Sapporo   060 Japan

SUMMARY

   The method of blood-free isolated perfused rat brain was described,
of which EEG was within the normal range. The simultaneous measurement
of heme a+a$_3$ and copper in cytochrome oxidase could be performed under
the various conditions such as graded hypoxia and seizure. The
relationship between these two chromophores in the brain was same to
that of isolated mitochondria.

INTRODUCTION

   The in vitro study of the redox behavior of heme a+a$_3$ and copper in
cytochrome oxidase gives the possibility to know the energy state of
living tissue (Hoshi et al. 1989). Normal blood circulated conditions,
in stead, do not allow the measurement of heme a+a$_3$ in visible region,
since hemoglobin masks almost completely this chromophore. Thus, we
have tried to prepare the hemoglobin-free isolated perfused rat brain,
which can be applied to the visible and near-infrared photometry. The
our perfusion model is free from the effect of circuratory system, which is
usually observed with fluorocarbon-substituted rat. Present paper
describes the simultaneous measurement of heme a+a$_3$ and copper of
cytochrome oxidase in the perfused rat brain. The data _in situ_ agrees
with that _in vitro_.

Experimental setup

   As shown tentatively in Fig.1, both carotid arteries were
cannulated and brain was perfused with O_2-saturated fluorocarbon(FC)
suspension under the constant pressure. The oxygen concentration of
inflowing FC-suspension was measured by oxygen electrode. The
absorption changes of heme a+a$_3$ and copper of perfused brain were
measured by quadruple wavelength photometer with two-channel dual-
wavelength mode. The absorbance difference at 602-622 nm and 782-840 nm
were employed for the measurement of heme a+a$_3$ and copper, respectively.
The monochromatic lights obtained by filters with rotating disc (120 Hz)
were illuminated to perfused rat head successively. The transmitted
lights through cranial bone and brain tissue were collected by other
light guide and conducted to detection systems.

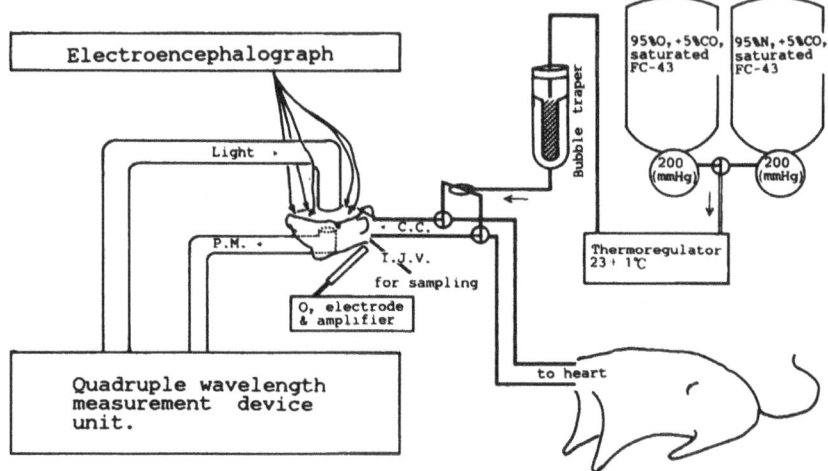

Fig.1 Shematic presentation of the experimental set
       up. C.C; Common carotid artery. I.J.V, internal
       jugular vein.

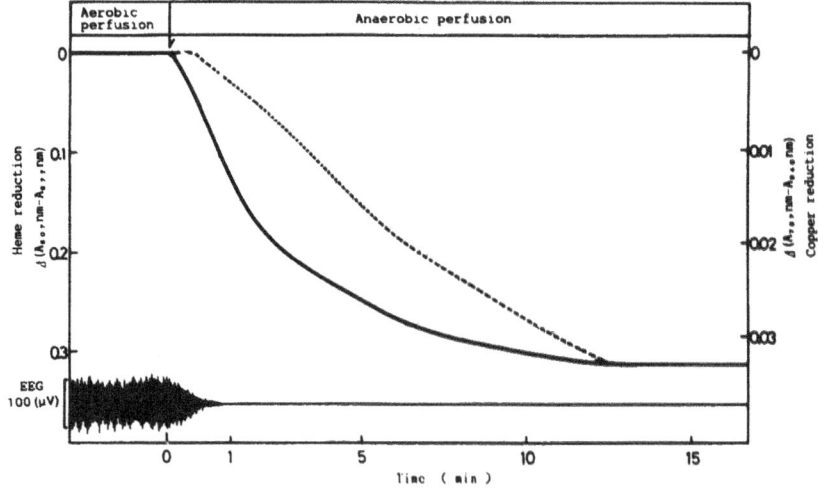

Fig.2 Time cause of the absorption changes of heme
       a+a$_3$ and copper of rat brain induced by the
       transition from aerobic to anoxic state. Heme
       a+a$_3$ was measured at 602–622 nm (——) and copper
       at 787–840 nm (---).

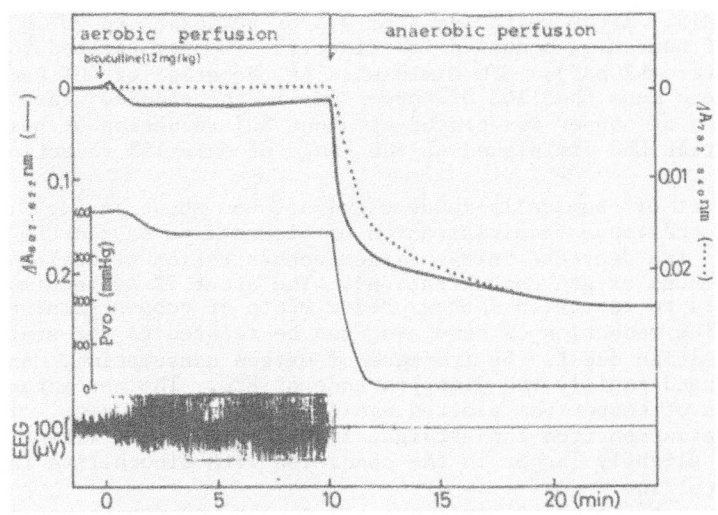

Fig. 3　Time cause of the absorption changes of heme a+a$_3$ and copper of rat brain induced by seizure and anoxic treatment.　Bicuculline (1.2 mg/Kg) was administrated in the aerobic condition.

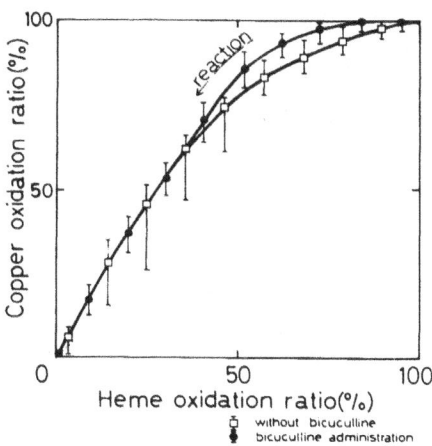

Fig. 4　Relationship between the redox changes of heme a+a$_3$ and copper in the isolated rat brain.

RESULTS

Fig.2 shows the time-course of the absorption changes of heme a+a$_3$ and copper of brain induced by the transition from aerobic state to anoxic state. The reduction of heme a+a$_3$ proceeded significantly faster than copper did. After switching from 95% O$_2$-perfusate to 95% N$_2$, reducation of heme a+a$_3$ occurred immediately. Copper strated to reduction later (~20sec). EEG diminished at the stage of 40% heme a+a$_3$ reduction where less than 10% of copper was in the reduced state. The half reduction of copper was placed at about 80% reduction of heme a+a$_3$. It is noted that EEG diminished at the stage of only 15% reduction of copper.

The effect of chemically-induced seizure was shown in Fig.3. In the aerobic conditions, administration of bicuculline caused the burst of EEG spike and the decrease in the oxygen concentration of outlet from the brain (venous oxygen concentration). The about 7% of heme a+a$_3$ shifted toward to reduction whereas redox state of copper remained unchanged. The reduction of heme a+a$_3$ can be related to the state 3-state 4 transition due to the increase of oxygen consumption. Anaerobic perfusion caused rapidly the disappearence of EEG. The percentage of the reduction of copper was plotted against that of heme a+a$_3$. There is the marked deviation from the straight line (Fig.4). The extent of the deviation is slightly larger in the condition with bicuculline than without bicuculline.

The oxygen affinity of heme a+a$_3$ and copper was determined in the perfused rat head, as shown in Fig.5. The percent changes of redox-states of both chromophores are plotted against oxygen concentration of inflowing perfusate. It is clearly seen in the figure that oxygen affinity of copper is higher than that of heme a+a$_3$, which has been observed in the isolated mitochondria (Hoshi et al. 1989). The oxygen concentrations giving half maximul oxidation of heme a+a$_3$ and copper are 450 mm and 350 mmHg respectively, which are extremely higher values of normal blood circulated conditions.

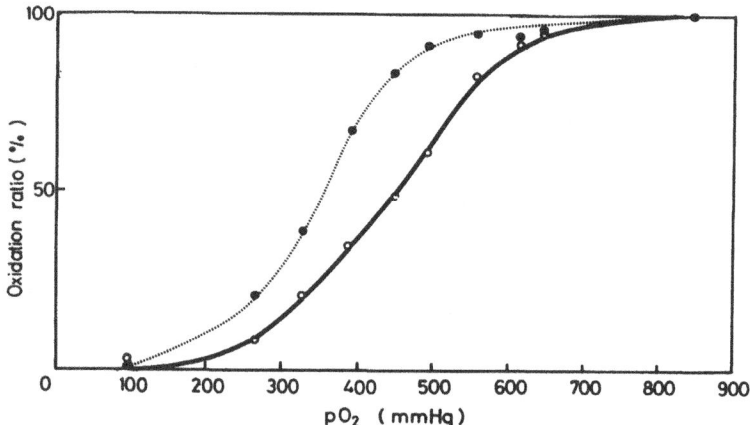

Fig. 5 Relationship of the redox states of heme a+a$_3$ and copper to oxygen concentration of the inflowing perfusate. Oxygen concentration of inflowing perfusate was changed in a step-wise fashion to maintain the steady-state.

DISCUSSION

Our perfusion model of hemoglobin-free isolated rat brain showed the several characteristics of brain function within four hours, such as EEG and reactivity to some drugs. An advantage of perfused brain is that visible and near-infrared spectrophotometry can be applied. The observed change of partial reduction of heme $a+a_3$ not copper by seizure can be explained by the state 3 and 4 transition but not partial hypoxia (Fig.3). The relationship between the redox states of heme $a+a_3$ and copper in the brain is found to be nearly identical to that of isolated mitochondria. The deviation from the straight line (Fig.4) comes from the difference of the oxygen affinity between these two chromophores as observed in vitro (Hoshi et al. 1989) and Fig.5. Thus, the in vitro data can be applied to in situ data.

REFERENCE

Hoshi, Y., Hazeki, O., and Tamura, M., 1989, "The oxygen dependency of the redox state of heme and copper in cytochrome oxidase in vitro." Adv.Exp.Med.Biol. 1989:248

THE SIMULTANEOUS MEASUREMENT OF THE REDOX STATE OF CYTOCHROME

OXIDASE IN HEART AND BRAIN OF RAT IN VIVO BY NIR

Yasuyuki Kakihana, Kinji Ito and Mamoru Tamura

Biophysics Division, Research Institute of Applied
Electricity, Hokkaido University
Sapporo 060, Japan

INTRODUCTION

Cytochrome oxidase(Cyt.ox.), which contains heme and copper, is a
terminal enzyme of respiratory chain and transfers electrons to
molecular oxygen (Erecinska & Wilson 1978). For this reason, monitoring
oxidation-reduction state of Cyt.ox. provides the information about the
pathologic state of tissues caused by impaired oxygen supply. The
Cyt.ox., hemoglobin(Hb) and myoglobin(Mb) have a broad absorption band
in near infrared(NIR) region and the light in this region easily
penetrates biological tissues. Thus NIR spectrophotometry can measure
noninvasively the redox state of Cyt.ox. and oxygenation of Hb and Mb
through skin and bone. In this paper, to obtain the mutual relationship
between the oxygenation state of cardiac and brain tissues, we examined
the redox behavior of copper in Cyt.ox. with anesthetized rat.

METHODS

Animal preparation

The experiments were performed on adult male wister rats, weighting
450-500g, which were anesthetized with pentobarbital (nembutal;50mg/kg).
Tracheostomy was performed and a femoral artery and a femoral vein were
cannulated using poryethylene tubing. The animals were paralyzed with
pancuronium bromide(1.3mg/kg i.v.) and mechanically ventilated with 100%
O_2. Median sternotomy was performed and heart was exposed without
opening pericardium. Tidal volume was adjusted to maintain $PaCO_2$ near
35-40 mmHg. Arterial blood pressure was continuously monitored and the
medicine was infused through a femoral vein.

Optical monitoring

Two sets of the quadruple wavelength spectrophotometer were used to
measure simultaneously the optical absorption changes in the brain and
heart. The light, obtained from Halogen lump, was illuminated a rotating
disc with four interference filter(700, 730, 750, 805 nm) through lens
system. For brain measurement, one light guide (5mm diameter) was fixed
on the center of the head of which skin and muscle were removed. The
light transmitted through the head (cranial bone and brain tissue) was

collected to L shaped light guide (5mm diameter) placed within the
mouth and guided to a photomultiplier. For heart measurement, one
flexible light guide (1mm diameter) was inserted into the left ventricle
through the right carotid artery. The light transmitted through the
left ventricular wall was detected by another light guide placed at the
outside of exposed heart. The oxygen saturation of Hb and Mb, and the
redox state of Cyt.ox. in the heart and brain were obtained by the
analyses of the quadruple wavelength methods(Hazeki & Tamura 1989). A
schematic diagram of the experimental arrangement is shown in Fig.1.

RESULTS

The changes of oxyhemoglobin (oxyHb) and the redox state of
Cytochrome oxidase (Cyt.ox.) in rat brain are shown in Fig.2a.   With
increasing the carbon dioxide concentration of the inspired gas, well
known cerebral vasodilator, oxyHb increased gradually but Cyt.ox.
unchanged. OxyHb returned to the original level after 5min of the

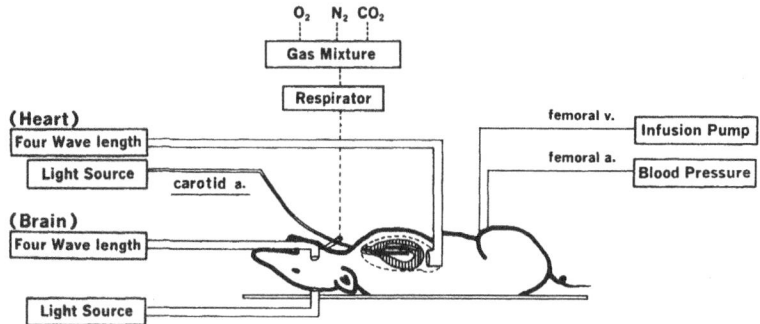

Fig.1   A schematic diagram of the experimental
         arrangemant.

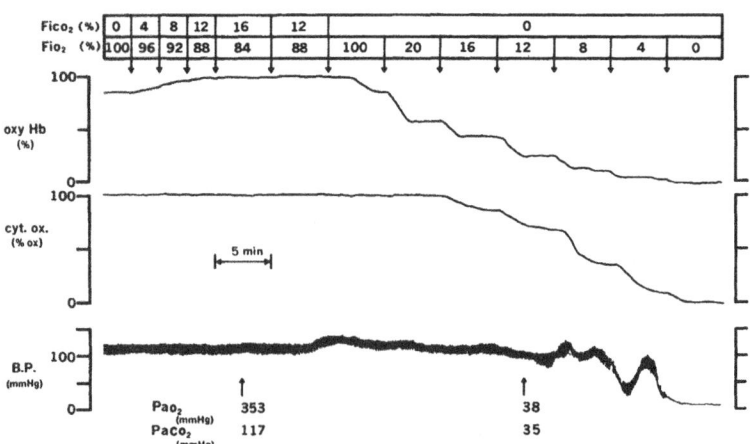

Fig.2a   Changes of oxyHb and the redox state of Cyt.ox.
          in rat brain during hyperoxia, hypercapnia and
          hypoxia. The oxygen and carbon dioxide
          concentration of inspired gas are
          changed as shown in abscissa.

cessation of the carbon dioxide. When the hypoxia induced by decreasing the oxygen concentration of the inspired gas from 100% to 0%, oxyHb decreased step-wisely but not Cyt.ox. until the oxygen concentration of the inspired gas reached below 16%. The plot of the percent oxidation of Cyt.ox. against the percent oxygenation of Hb in brain is shown in Fig.2b. The plot showed an upper deviation from the straight line. The reduction of Cyt.ox. begins to start at 55% oxyHb.

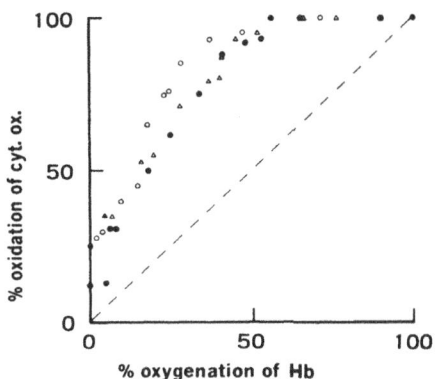

Fig.2b   Relationship between the redox state of Cytochrome oxidase and the oxygenation state of hemoglobin in rat brain. The experimental conditions are as described in Fig.2a. Different symbols indicate the results from different rats.

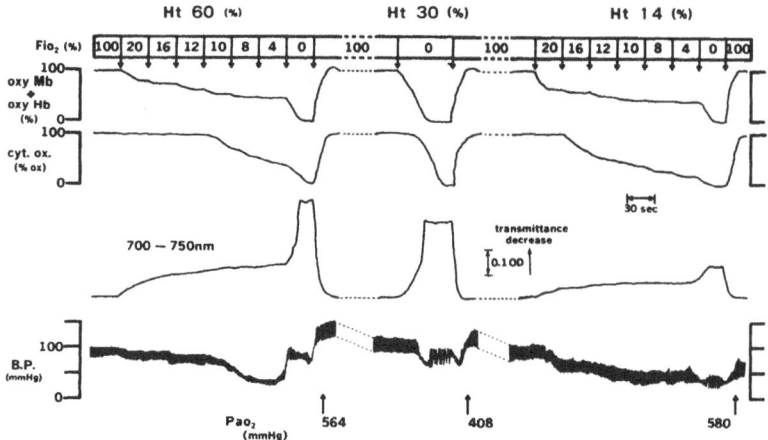

Fig.3   Changes of oxyHb+oxyMb and the redox state of Cyt.ox. in Fluosol-43 exchanged rat heart during hypoxia under various hematocrit conditions. The oxygen concentration of the inspired gas is changed as shown in abscissa. OD means the optical density.

Fig.3 shows the changes of oxyHb+oxyMb and the redox state of Cyt.ox. in rat heart under the various hematocrit conditions. In this experiment, the blood of rats were exchanged by the transfusion with Fluosol-43 and aerobic-anaerobic conditions were obtained by lowering the oxygen concentration of the inspired gas from 100% to 0%. The decrease of hematocrit caused the decrease of the total optical density(OD) by transition. However, the relationship between the redox state of Cyt.ox. and the oxygenation of Hb+Mb showed a upper deviation from the straight line and almost same with various hematocrit level (data not shown).

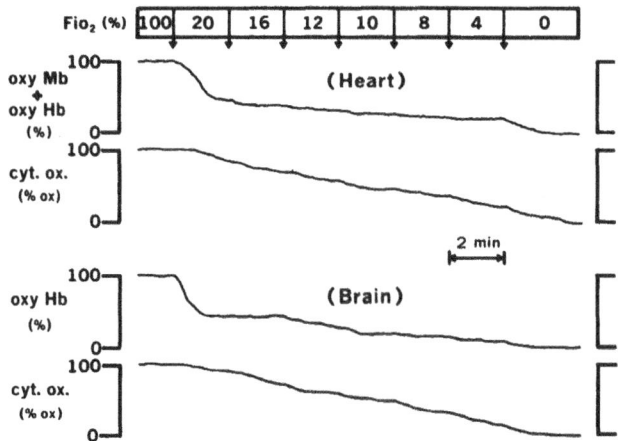

Fig.4a   Changes of oxyHb(+oxyMb) and the redox state of Cyt.ox. in rat brain and heart during hypoxia. The oxygen concentration of the inspired gas is changed as shown in abscissa.

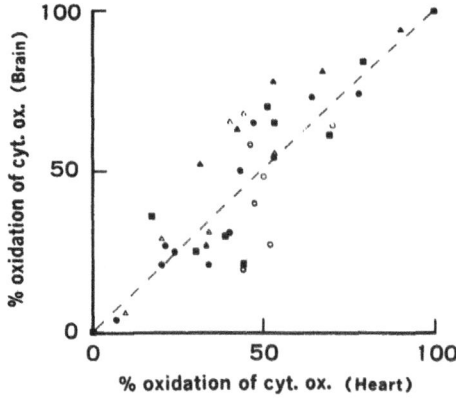

Fig.4b   Relationship between the redox state of Cytochrome oxidase in brain and heart. The experimental conditions are described in Fig.4a. Different symbols indicate the results from different rats.

Fig.4a shows the changes of oxyHb(+oxyMb) and the redox state of Cyt.ox. during aerobic-anaerobic transition. The relationship between Cyt.ox. and oxyHb in brain was similar to Fig.2b and that in heart was similar to Fig.3. The plots of the percentage reduction of Cyt.ox. in brain against in heart showed a linear relationship (Fig.4b).

DISCUSSION

The quadruple wavelength analysis was first developed to monitor the oxygenation of Hb and the redox state of Cyt.ox. in brain. We assumed that the principle of this method would be applicable to the heart, because absorption spectra of Mb was very similar to that of Hb in near infrared region. The relationship of the redox state of Cyt.ox. against the oxygenation of Hb+Mb did not change when Hb concentration was decreased by exchanging blood with Fluosol-43. This suggests that our heart measurement picks up mainly myoglobin in cardiac tissue rather than blood. Present simultaneous optical measurements demonstrated that the oxidation states of the copper in Cyt.ox. of both cardiac and brain tissues behaved almost similarly during progressive hypoxia, even though heart and brain would have probably the significant difference of the oxygen supply and consumption and capacity of energy. And oxygenation state of brain completely depended on the oxygenation state of cardiac tissue which determined the cardiac output. In any case, the simultaneous optical measurement of different organs will provide further quantitative information about the circulatory system in vivo.

SUMMARY

The mutual relationship of the redox state of cytochrome oxidase in both cardiac and brain tissues were examined by the simultaneous optical measurement. The oxygenation states of cardiac and brain tissues behaved in a same manner during progressive hypoxia. We concluded that the oxygenation state of brain completely depends on the oxygenation state of cardiac tissue which determines the cardiac output.

REFERENCES

Erecinska, M., Wilson, D.F. (1978) "Cytochrome C oxidase ; A synopsis."
    Arch.Biochem.Biophys. 188:1

Hazeki, O., and Tamura, M., (1989) "Near infrared quadruple wavelength
    spectrophotometry of rat head" Adv.Exp.Med.Biol. 248:63

PICOSECOND TIME OF FLIGHT MEASUREMENT OF LIVING TISSUE:

TIME RESOLVED BEER-LAMBERT LAW

Yasutomo Nomura and Mamoru Tamura

Biophysics Division, Research Institute of Applied
Electricity, Hokkaido University
Sapporo 060 Japan

INTRODUCTION

In the turbid systems such as living tissues, the optical
pathlength differs significantly from geometrical pathlength due to the
effect of multiscattering. With living tissues, optical absorbers, for
example, hemoglobin and myoglobin coexist together with the scattering
materials. And, there are several arguments concerning the effect of
such absorber on the distribution of optical pathlength. Previously, we
showed that, using picosecond photometry, light intensity along the non-
linear path taken by photons through scattering media is exponentially
attenuated (Nomura, et al. 1989). Present paper expands this
conclusion into living tissues of rat _in situ_ in the range from visible
to near-infrared region.

MATERIALS AND METHOD

Pulsed laser lights in the red and near-infrared regions were
incidented to the scattered biological models and anesthetized rats. The
transmitted lights were time-resolved by the streak camera (Fig.1). With
animal experiments, the light guide with 2mm diameter was employed for
illuminating and collecting the photons.
The scattering medium used was a suspension of defatted dried milk
solution with varying the concentration of hemoglobin.
Fluorocarbon substitution of rat was performed according to the
method previously reported (Hazeki and Tamura 1989).

RESULTS

Time-resolved Beer-Lambert Law

According to Fig. 2 of time of flight analysis, we define I_0 as the
intensity of light as a function of time taken by photons transmitted
through the sample without absorber. I is the intensity in the presence
of absorber. Assuming that the scattering property of photons does not
change by absorption, the following equations can be derived at a
certain time, t, after the light pulse.

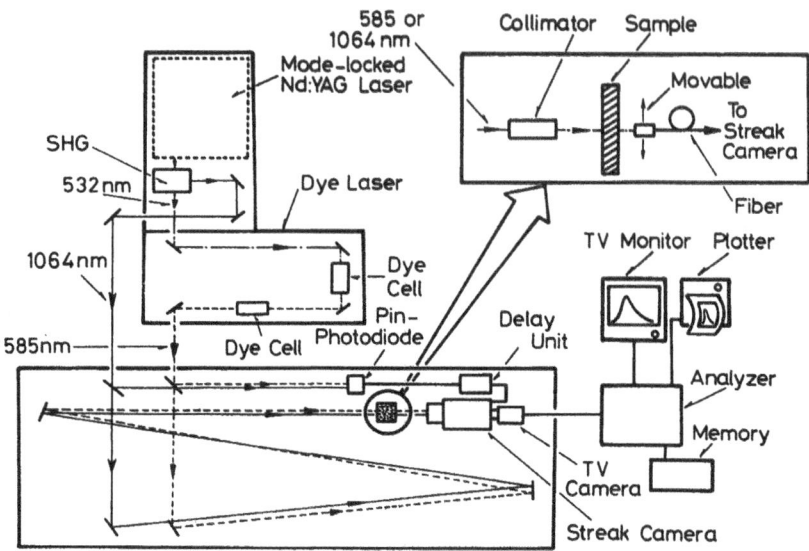

Fig.1   Block diagram of picosecond time of flight measurement
        at visible and near-infrared lights.

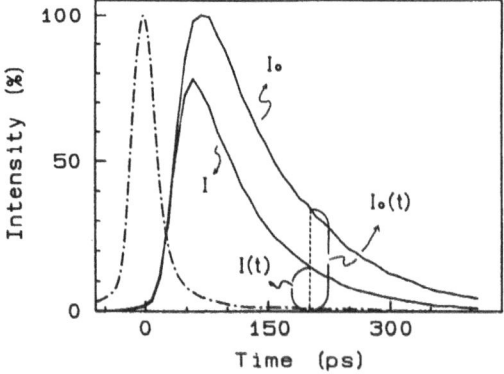

Fig.2   Analysis of the time of flight incidented to the
        scattered media.   I,I_0; Thansmitted light intensity
        with and without absorber.   I(t), I_0(t);   Intensity at
        time t of I and I_0.   Incident light pulse is shown as
        (-----).

$$A(t) = \log I_0(t)/I(t) = ECVt \qquad (1)$$

A(t) is the absorbance at time t, and E, C, V are molar extinction coefficient of absorber, concentration of absorber and velocity of light in water (0.23 mm / psec).

Rat thigh muscle at visible light

Fig. 3 shows the changes in the profile of time of flight of 585 nm light transmitted through the rat thigh muscle, where the hematocrit of rat was changed by fluorocarbon transfusion. Decrease in the hematocrit value caused the increase in the transmitted light intensity. The half-width of the transmitted light at 16% hematocrit was about 50 psec which was slightly broader than that of incident light (40 psec).

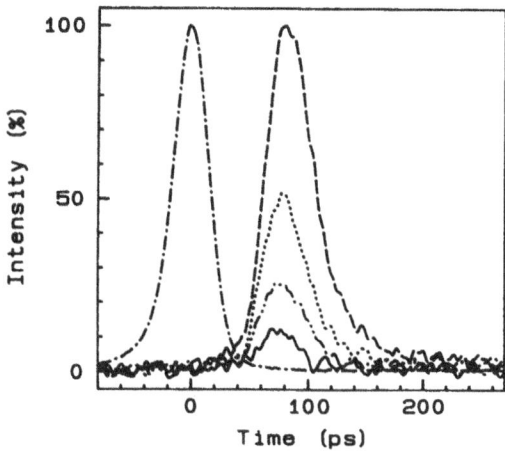

Fig.3   Profile of the light intensity against time. The pulsed light (·−··−·−) was illuminated an rat thight muscle through light guide. The transmitted lights at various hematocrit are shown where solid line is normal (45% hematocrit). Hematocrit; 16% (-----), 22%(····), 37% (-··-··-)

The plot according to Eqn(1) is shown in Fig. 4, where the curve of 16% hematocrit is taken as I_0, and therefore absorbance is given as the difference. The plot gives straight line with three different hematocrit values. Since Eqn(1) stands in Fig. 4, we can calculate the mean concentration of hemoglobin (mean hematocrit) in the thigh muscle using the value of E. The mean hematocrit in the muscle is 0.25%.

Near-infrared region

Eqn(1) is also tested at 1064 nm (Fig. 5) in the model system. The incidented light (30 psec) diffuses significantly by scattering in our model system (3 cm thickness, defatted milk (130g/l)). The addition of hemoglobin in the solution accompanied the decrease in the light intensity and shift of the peak position toward the earlier after the pulse. The inset shows the plot of Eqn(1), which gives the straight line. Thus, Eqn(1) can be used at 1064 nm.

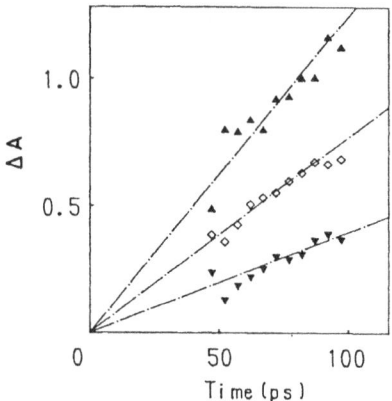

Fig.4 Plot of the absorbance difference against time, where
       the curve of 16% is taken as I_0 in Eqn(1). Hematocrit,
       45% (▲), 37% (◊), 22% (▼).

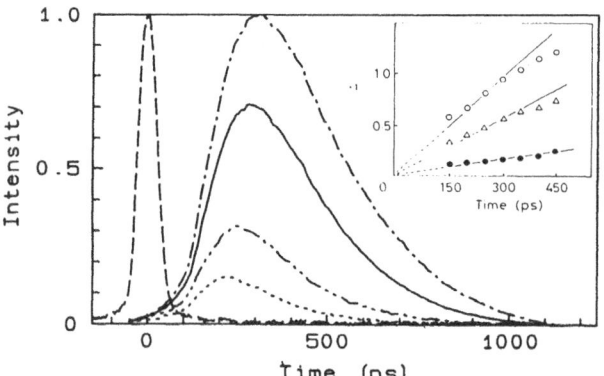

Fig.5 Time of flight at 1064 nm. Defatted milk solution
       without hemoglobin (-----). Inset, plot of Eqn(1).
       0.2mM hemoglobin (●), 0.4mM (△), 0.8mM (○).

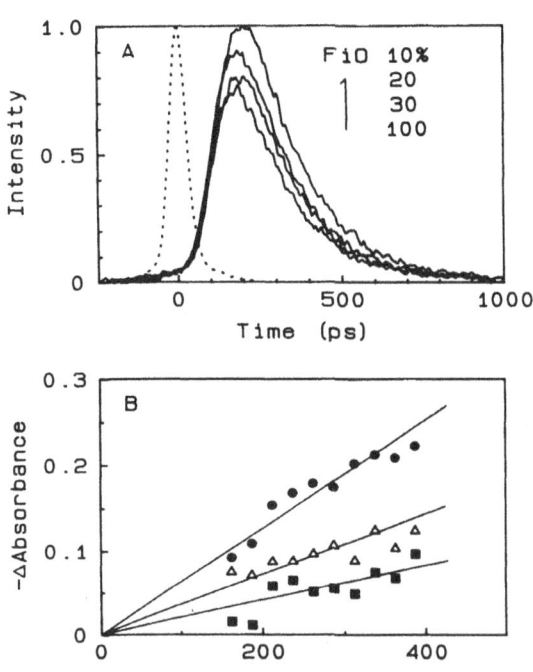

Fig.6 Time of flight measurement at 1064 nm of rat brain.
(A), Changes in profile of light intensity against time
with changing FiO_2 in anesthtized rat. (B), Plot of
the absorbance difference against time according to
Eqn(1). The curve of 100% O_2 is taken as I_0 and
therefore the absorbance difference is given as
minimus. FiO_2m 30% (■), 20% (△) and 10% (●).

Fig. 6 shows the temporal profile of transmitted light of anesthetized rat head with changing the oxygen concentration of inspired gas (FiO$_2$). In A, decreasing FiO$_2$ accompanied the increase in the intensity of transmitted light. The absorbance difference referred to 100% FiO$_2$ is shown in Fig. 6-B according to Eqn(1). The absorbance difference was found to be proportional to time.

The quantitation of the absorbance in the turbid system like living tissue has encountered the serious problem since multisccattering phenomenon alters the optical pathlength. Thus, it has been accepted that Beer-Lambert law that stands for clear solution can not be simply applied to the biological tissue. As shown previously, (Nomura, et al. 1989), we demonstrated that time-resolved Beer-Lambert law stands in the scattered media, and concluded that the distribution of pathlength remains unchanged with changing the absorption. Present paper confirms this for living tissue as well as 1064 nm of near-infrared region. Thus our approach for quantitation of absorption using picosecond photometry differs from those of Delpy et al (Delpy, 1988) and Chance et al (Chance, 1988).

In summary, Beer-Lambert law, which requires the independence of the optical pathlength against absorption, can be used in the living tissue.

REFERENCE

Chance, B., Leigh, J.C., Miyake, H., Smith,D., Nioka, S., Greenfield, R., Finander, M., Kaufman, K., Lery, M., Yong,M., Cohn, P., Yoshika, H., and Boretsky, R., 1988, "Comparison of time-resolved and unresolved measurement of deoxy hemoglobin in brain. Proc.Natl.Acad.Sci. U.S.A 85:4971-4975

Delpy, D.T., Cope, M., van der Zee, P., Arridge, S., Wray, S., and Wyatt, J. 1988, "Estimation of optical pathlength through tissue from direct time of flight measurement2 Physics in Med. and Biol. 33: 1433-1442

Hazeki, O., and Tamura, M., 1988, "Quantitative analysis of hemoglobin oxygenation state of rat brain in situ by near-infrared spectrophotometry." J.Appl.Physiol. 64:796-802

Nomura, Y., Hazeki, O., and Tamura, M., 1989, "Exponential attemuation of light along non-linear path through the biological model. Adv.Exp.Med.Biol. 248:77-80

CEREBRAL OXYGENATION STATE IN CHEMICALLY-INDUCED SEIZURES IN THE

RAT--STUDY BY NEAR INFRARED SPECTROPHOTOMETRY

Yoko Hoshi and Mamoru Tamura

Biophysics Division, Research Institute of Applied
Electricity, Hokkaido University
Sapporo 060 Japan

INTRODUCTION

It is well known that cerebral blood flow (CBF) is markedly
enhanced during seizure. However, whether the increase in blood flow
results in the sufficient oxygen delivery to meet the increased demand
is still controversial. The cerebral oxygen concentrations during
seisure have been monitored by using oxygen electrode or
spectrophotometry which measured the redox states of cytochrome
oxidase(Cyt.ox.) or pyridine nucleotide in mitochondria. In previous
paper, Hazeki and Tamura reported the quantitative analysis of
hemoglobin(Hb) oxygenation state and the redox state of Cyt.ox. in the
rat brain by using near infrared quadruple wavelength
spectrophotometry(Hazeki et al. 1988). In this study, we will report
the Hb oxygenation state and the redox state of Cyt.ox. in the rat
brain during and after seizure.

METHODS

20 Wister rats(male, 250–300g) were anesthetized with urethan
(ethylcarbamate, 8mg/Kg, i.p.). They were tracheotomized and femoral
arteries and vein were cannulated. They were paralyzed with panchronium
bromide(2μg/Kg, i.v.) and were mechanically ventilated with 21%O_2 –79%
N_2 gas, the rate beeing initially adjusted to give an arterial carbon
dioxide tension(PaCo$_2$) of 37–42mmHg.
Light from the halogen lump was illuminated on the head through a
light guide and light emerging from the palate was guided to
photodetection system. Changes in cerebral blood volume(CBV), cerebral
venous Hb oxygenation state and the redox state of Cyt.ox. in the brain
were estimated by quadruple wavelength analysis(wavelength; 700, 730,
750, and 805 nm).
Electroencephalogram(EEG) and arterial blood pressure(BP) were
monitored simultaneously. EEGs were recorded from the bilateral
occipital bis electrodes placed on the skull.
Convulsions were induced by administration of Pentylenetetrazol
(PTZ; 50mg/Kg,i.v.).

RESULTS AND DISCUSSION

Fig.1 shows the changes in cerebral Hb oxygenation states, the redox state of Cyt.ox., CBV in the brain and BP caused by PTZ administration. Fig.2 shows the changes in EEG monitored simultaneously. PTZ administration caused transient reduction of Cyt.ox. in the brain, slightly later, desynchronization appeared on EEG, and then BP began to increase concomitantly with the increase of CBV. When BP reached the maximum, burst of spikes appeared on EEG. Oxygenated Hb(oxy-Hb) increased and Cyt.ox. was reoxidized in this stage. Hyperoxidation of Cyt.ox. was not observed, which had been observed by previous workers(Kreisman et al. 1981)(Hempel et al. 1980). In postictal phase, although BP and oxy-Hb remained higher and more than preseizure state, Cyt.ox. was in the certain reduced state.

Changes in the hemoglobin oxygenation state and the radox state of cytochrome oxidase in the brain during PTZ seizure

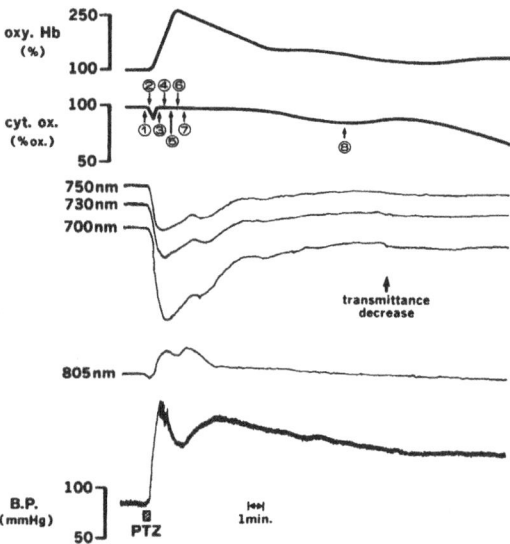

Fig. 1.   PTZ was administrated at the point shown by PTZ. In percent oxygenation state of Hb and percent oxidation of Cyt.ox. are shown in upper traces, where changes caused by a normoxic-anoxic transition were taken as 100%.

Fig.3 shows the effect of successive administration of PTZ on the cerebral oxygenation state and BP. The first administration caused the similar changes to those shown in Fig.1. In postictal phase, when Cyt.ox. was in the certain reduced state, the second administration caused the transient reduction of Cyt.ox. further and then BP began to increase concomitantly with the increase of CBV and reoxidation of Cyt.ox.. Changes in EEGs caused by both administrations were similar to those in Fig.2.

Change in EEG during PTZ seizure

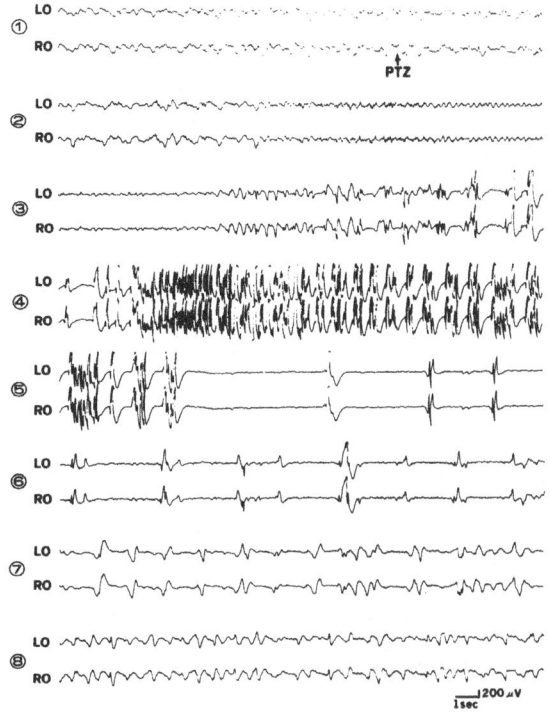

Fig. 2. EEGs were recorded in the stage shown in Fig 1.
1-8 were recorded at the points marked 1-8 in Fig 1.

Effect of succcessive PTZ administration on the cerebral hemoglobin oxygenation state, the redox state of cytochrome oxidase in the brain and blood pressure

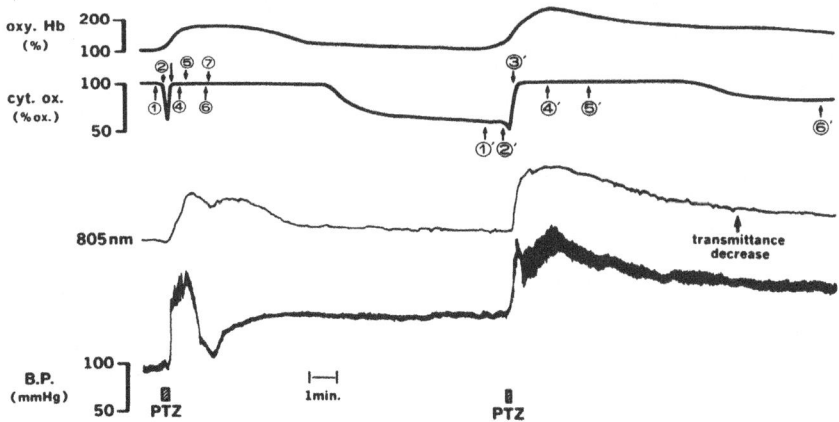

Fig. 3. The same dose of PTZ (50mg/Kg) was administrated
successively.

Effect of successive PTZ administration on EEG

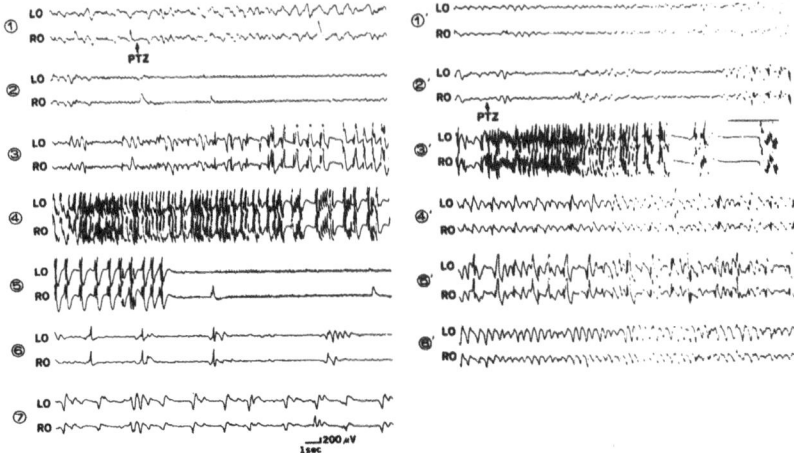

Fig. 4. EEGs were recorded in the stage shown in Fig 3.
Left EEGs were recorded during the first seizure.
1-7 were recorded at the points marked 1-7 in Fig 3.
Right EEGs were recorded during the second seizure.
1-6 were recorded at the points marked 1-6 in Fig 5.

Relationship between the cerebral oxygenation state and EEG during PTZ seizure

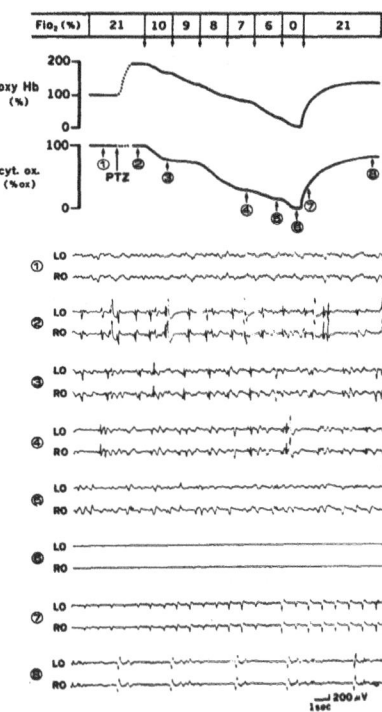

Fig. 5. An overdose of PTZ (200mg/Kg) was administrated. FiO_2
was gradually decreased. EEGs were recorded at the
points marked 1-8 in the trace of Cyt.ox..

Desynchronization, which means activation of the cerebral activity, always appeared on EEG before BP began to increase. These results suggest that the increased metabolic demand in the brain may cause transient tissue hypoxia which triggers the increase of CBF.

It is accepted that the increase of CBF during seizure is due to an increased cerebral perfusion pressure, a decreased cerebrovascular resistance and the insult of cerebral autoregulation(Magnaes et al. 1974)(Dymond et al. 1976). An increased oxy-Hb and no hyperoxidation of Cyt.ox. during seizure show the excessive blood supply through A-V shunt. Therefore, the later reduction of Cyt.ox. observed in postictal phase may be due to continuing A-V shunt.

Fig.5 shows the relationship between the cerebral oxygenation state and EEG during seizure. Administration of an overdose of PTZ(200mg/Kg,iv) caused prolonged epileptic activities on EEG which continued for 15-20 minutes. A rat was administrated by an overdose of PTZ and inspired oxygen concentrations(FiO_2) were gradually decreased while epileptic activities were continuing. Swithing FiO_2 from 21% to 10% caused about 20% reduction of Cyt.ox. concomitantly with the decrease of BP. When FiO_2 was reduced to 6%, Cyt.ox. was reduced almost maximally. In a rat without PTZ administration ,about 50% of Cyt.ox. was reduced(data not shown). This result suggests that even under mild hypoxic conditions, seizure can cause severe brain tissue hypoxia because of relatively less oxygen supply to the increased metabolic demand and also the decreased cardiac function.

Spikes did not disappear until about 85% of Cyt.ox. was reduced. EEG became isoelectric slightly later than that Cyt.ox. was maximally reduced. When a rat was reoxygenated after 1.5 minutes anoxic conditions, spikes appeared on EEG before Cyt.ox. was reoxidized completely, while in a rat without PTZ administration, EEG remained isoelectric at this stage(data not shown).

These results reflect the fact that under ill conditions such as hypoxia,threshold of neuron activity decreases and hypersynchronization of alive neurons occures. Thus, even under severe hypoxic conditions at least a part of neurons remain to be able to fire.

SUMMARY

The hemoglobin oxygenation state and the redox state of Cyt.ox. in the rat brain during and after seizure induced by PTZ were measured by using near-infrared spectrophotometry. PTZ administration caused transient reduction of Cyt.ox. in the brain, which might be a trigger for the increase of CBF during seizure. In postictal phase, although BP remained high, Cyt.ox. was in the certain reduced state, which might be due to A-V shunt. Hypoxic loading during seizure caused more reduction of Cyt.ox. than under non-epileptic conditions, which meant that seizure even under mild hypoxic conditions could cause severe hypoxic brain damage.

REFERENCES

B.Magnaes, B., and Nornes,H., 1974, "Circulatory and Respiratory Changes in spontaneous epileplic seizures in man." Europ.Neurol. 12:104

Dymond, A.M., and Crandall, P.H., "Oxygen availability and blood flow in the temporal lobes during spontaneous epileptic seizures in man." Brain Research, 102:191

F.G.Hempel, F.G., Kariman, K., and Saltzman, H.A., 1980 "Redox transitions in mitochondria of cat cortex with seizures and hemorragic hypotension." Am.J.Physiol. 238:H249

Hazeki, O., and Tamura, M., 1988, "Near infrared quadruple wavelength spectrophotometry of the rat head." Adv.Exp.Med.Biol. 248:63

Kreisman, N.R., J.C.Lamanna, J.C., Rosenthal, M., and Thomas, J., 1981, "Oxidative metablic responses with recurrent seizures in rat cerebral cortex : Role of systemic factors." Brain Research, 218:175

EXPERIMENTALLY MEASURED OPTICAL PATHLENGTHS FOR THE ADULT HEAD, CALF AND

FOREARM AND THE HEAD OF THE NEWBORN INFANT AS A FUNCTION OF INTER OPTODE

SPACING

P. van der Zee,[*] M. Cope,[*] S.R. Arridge,[*] M. Essenpreis,[*] L.A. Potter,[#] A.D. Edwards,[#]
J.S. Wyatt,[#] D.C. McCormick,[#] S.C. Roth,[#] E.O.R. Reynolds,[#] D.T. Delpy.[*]

Departments of Medical Physics[*] and Paediatrics[#], University College London
London WC1.

INTRODUCTION

The technique of near infrared spectroscopy (NIRS) is being increasingly used in clinical measurements of blood and tissue oxygenation (Wyatt et al 1986, Edwards et al 1990), and there are now several commercially produced instruments undergoing clinical testing. Whilst early NIR measurements were solely of a qualitative nature (Brazy et al 1985,1986, Fox et al 1985, Ferrari et al 1986), techniques have now been developed to quantitate some parameters such as cerebral blood flow (Edwards et al 1988a), cerebral blood volume (Wyatt et al 1990a) and muscle oxygen consumption (Cheatle et al 1990).

All these quantitative techniques require a knowledge of the pathlength travelled by the light as it traverses the tissue. Due to the effects of scattering of light by the tissue, this distance is greater than the geometrical spacing between the optodes by an amount which depends upon the optical properties of the tissue and the measurement geometry (Delpy et al 1989). We have previously shown that for measurements in the near infra red region, it is possible with reasonable accuracy to apply a modified Beer Law to the measured changes in attenuation (Cope et al 1988). The modification involves the addition of a simple constant differential pathlength factor (DPF) to the inter optode spacing. The DPF can be determined from time of flight measurement of an ultrashort optical pulse through the tissues (Delpy et al 1988). Monte Carlo modelling and experimental measurements have shown that this method of deriving the DPF is valid both for slab and spherical tissue geometries (Delpy et al 1988, van der Zee 1990). For spherical geometries, the inter optode spacing is the length of the chord between the optodes. Values for the DPF have been derived for the transilluminated rat head (Delpy et al 1988), adult wrist (Edwards et al 1988b) and post mortem pre term infant head (Wyatt et al 1990b). Several alternative methods have been suggested for determining the DPF. These include derivation from the absorption of light by tissue water (Wray et al 1988, Cope et al 1989), measurement of the log slope of a transmitted ultra short light pulse (Chance et al 1988, Wilson et al 1989) or by measurement of the phase shift of amplitude modulated light (Lakowicz et al 1990). However, none of these alternative techniques has yet been validated by experimental measurement.

When NIRS measurements are made on large objects, it is often not possible to transilluminate the tissues and measurements must be made in partial transmission or reflection mode. In these cases, data are needed for the value of the DPF as a function of optode spacing. We report here values of the DPF derived from time of flight measurements in reflection mode from the adult head, calf and forearm and post mortem on the head from the newborn infant.

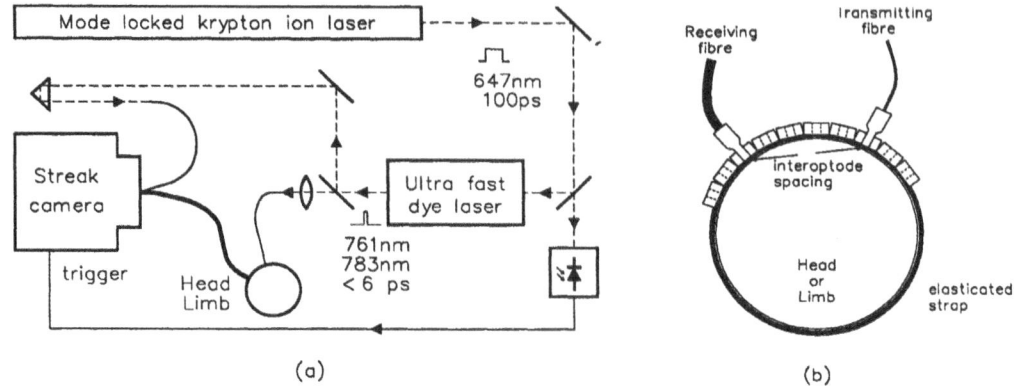

Figure 1. (a) Experimental system for time of flight measurement. (b) Method for attachment of the optical fibres.

EXPERIMENTAL METHOD

Equipment

The measurement system is shown schematically in Figure 1a. Ultra short light pulses (<6pS duration, 76 MHz repetition rate, wavelength 761nm or 783nm) were generated by a synchronously pumped dye laser (Coherent 701-3) pumped by a mode locked Krypton laser (Coherent KR3000). These pulses were coupled into a single 125 µm diameter, 60 cm long low dispersion optical fibre (Corning SDF). This permitted flexible positioning of the input light onto the tissue surface. Light emerging from the tissue was collected in an optical fibre bundle of the same low dispersion fibre. This fibre bundle consists of 100 fibres each exactly one metre long, arranged in a 1.9 mm diameter circle at the proximal end, but aligned in a single row at the distal end. This single row of fibres formed the input slit of the synchroscan streak camera (Hamamatsu Photonics C1587) which was used to detect the emerging light. A single additional fibre was used to couple a reference light pulse into the streak camera. This pulse was used to time the photons emerging from the tissue.

Subjects

Measurements of the DPF for the adult forearm were made on a group of 11 subjects, six male, five female aged 22 - 54 years (median 30). For the calf, measurements were made on 10 subjects, five male, five female aged 22 - 54 years (median 31) and for the head studies, 10 subjects, six of whom were male and four female, aged 22 - 54 years (median 26). All subjects were in good health. The measurements were made at a laser wavelength of 761nm.

The heads of ten infants were studied after death. They all had been admitted to the Neonatal Unit at University College Hospital. Five were male and five female. Their birthweights varied from 494 to 3220 (median 1720) grams. The principle clinical details are given in Table 1. Death occurred at a postnatal age of 0.75 to 384 (median 145) hours, and they were studied 12 to 55 (median 36) hours after death. These measurements were made at a wavelength of 783nm, except for infant 10 where a wavelength of 761nm was employed.

Measurement Procedure

To couple the transmitting and receiving fibres to the tissues, a flexible elasticated strap was placed around the object (see Figure 1b). Attached to the strap was an array of one centimetre wide black plastic blocks. Holes in these enabled the fibres to be held against the tissue at intervals of

Table I. Principle clinical details of the infants studied. Ultrasound images were obtained in life and confirmed on the day of study.

Infant No.	Sex	Gestation (weeks)	Age at death (hours)	Time at study, post mortem (hours)	Diagnosis	Ultrasound appearance	Biparietal diameter (cm)	Birth weight (g)
1	F	37	32	48	Diaphragmatic hernia	Normal	9.1	3220
2	F	38	0.75	25	Cerebral birth trauma	Normal	8.9	2850
3	M	28	19	32	Pulmonary interstitial emphysema	Bilateral periventricular echos	6.8	1260
4	F	26	30	37	Pulmonary interstitial emphysema	Haemorrhagic parenchymal infarction	6.7	1140
5	M	24	14	48	Pulmonary interstitial emphysema	Normal	5.3	600
6	M	35	384	12	Truncus arteriosus	Normal	9.0	2530
7	F	33	14	12	Pulmonary hypoplasia	Normal	6.6	1299
8	M	24	8	12	Extreme prematurity	Normal	5.0	494
9	M	34	16	32	Polycystic kidneys	Normal	7.7	1420
10	F	27	2.5	55	Rhesus isoimmunisation, hyaline membrane disease	Normal	7.6	1020

approximately one centimetre. During measurements, the receiving fibre bundle was usually held fixed in one block while the more flexible transmitting fibre was moved. At the end of the measurements, the fibres were removed and the positions of the holes in the blocks marked on the skin surface using a felt tip pen. The elasticated strap was then removed and the inter optode spacing measured using callipers. The accuracy of the measurement, allowing for possible skin movement was ±2mm.

Measurements from the forearm were made at a point mid way between the wrist and elbow, the arm being supported in a relaxed position. The optical fibres were positioned over the muscles on the medial aspect of the forearm away from any palpable bone. Measurements from the calf were made on the rear surface of the muscle at the broadest part of the leg. The subjects were seated with the leg muscles relaxed. Measurements from the adult head were carried out on the upper part of the forehead just below the hair line. The receiving fibre was positioned over the temple, and the transmitting fibre moved outward along the forehead. For the studies on the head of the newborn infant, the elasticated strap was not used, the fibres being held in movable clamps. The receiving fibre was positioned over the coronal suture midway between the anterior fontanelle and the external auditory meatus. The transmitting fibre was positioned successively at approximately one centimetre intervals along the medial portion of the coronal suture.

As a final study to assess the repeatability of the DPF measurement, a series of five separate measurements were made on the forearm of one adult subject at an inter optode spacing of 3 cm. The elasticated band was left in position on the arm, but the transmitting and receiving fibres were removed and then replaced between each measurement.

RESULTS

Measurement Repeatability

The DPF calculated from the repeated measurements from the adult forearm showed a standard deviation (S.D.) σ_r of 0.07. This is much smaller than the S.D. that arose from errors in the measurement of d, the inter optode spacing: $\sigma_d = 2$ mm. The vertical error bars in the subsequent data for the DPF were determined as follows:

$$S.D._{total} = \sqrt{\frac{DPF^2}{d^2} \sigma_d^2 + \sigma_r^2}$$

Adult Forearm

The results are shown in Figure 2. It can be seen that the DPF was almost constant beyond an inter optode spacing of 2.5 cm. The mean value of 3.59 ± 0.32 was the same for both male and female subjects

Adult Calf

Figure 3 shows that the DPF of the adult calf was almost constant beyond an inter optode spacing of 2.5 cm, but that there was a considerable difference in the DPF between males and females. The mean value for all the subjects was 4.65 ± 0.73, the values for the males was 3.98 ± 0.46 and for females 5.14 ± 0.43.

Adult Head

The DPF of the adult head was almost constant beyond an inter optode spacing of 2.5 cm as shown in Figure 4 and no difference between males and females was observed. The mean value of the DPF was 5.93 ± 0.42, which is considerably higher than the value observed in either the forearm or the calf.

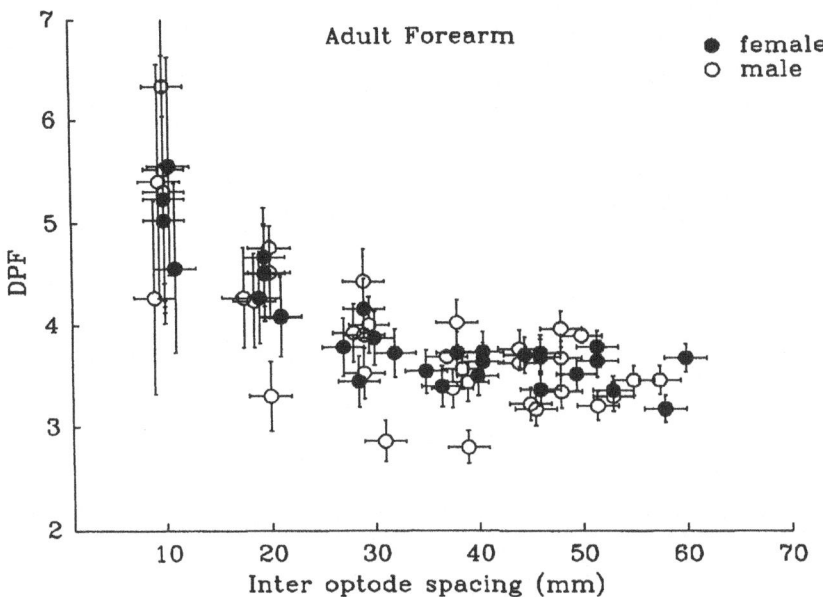

Figure 2. Experimentally derived values for the DPF in the adult forearm. The error bars indicate one standard deviation.

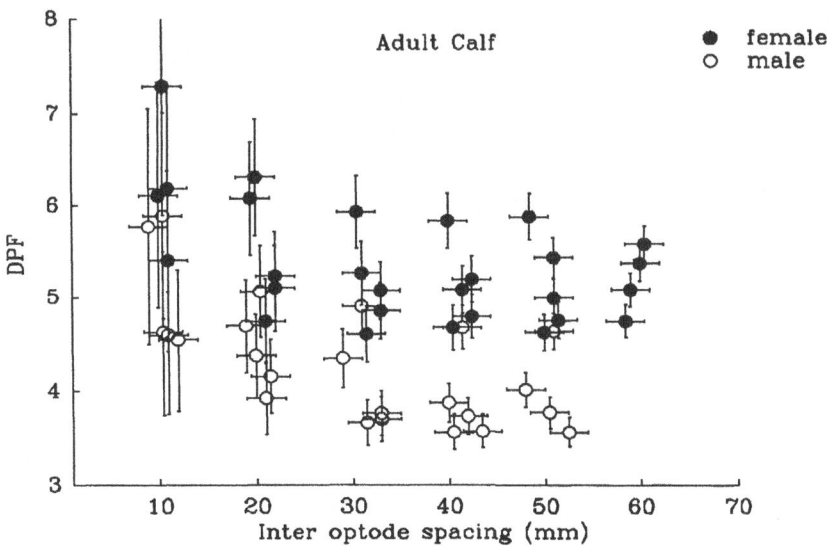

Figure 3. Experimentally derived values for the DPF in the adult calf. The error bars indicate one standard deviation.

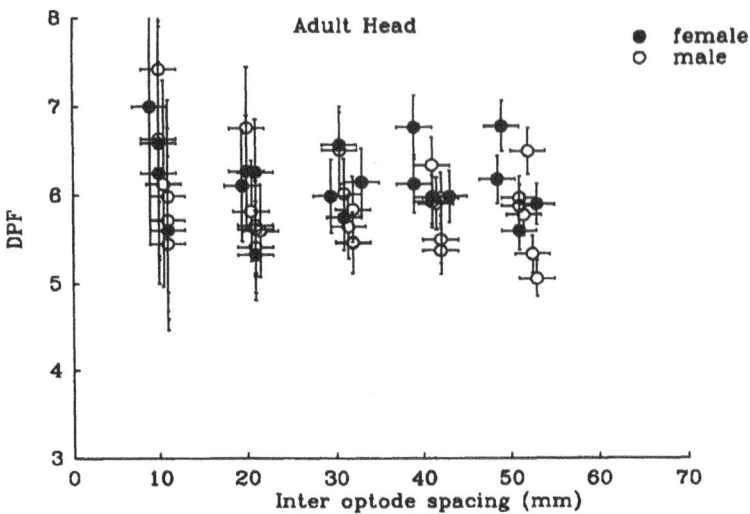

Figure 4. Experimentally derived values for the DPF in the adult head. The error bars indicate one standard deviation.

Baby Head

The results are shown in Figure 5, the data being displayed for two gestational age ranges, 24 - 30 and 33 - 38 weeks gestation. In common with the previous data, the DPF was almost constant beyond an inter optode spacing of 2.5 cm. The mean value of the DPF for all the infants was 3.85 ± 0.57. There was a statistically insignificant tendency for the DPF to increase with gestation.

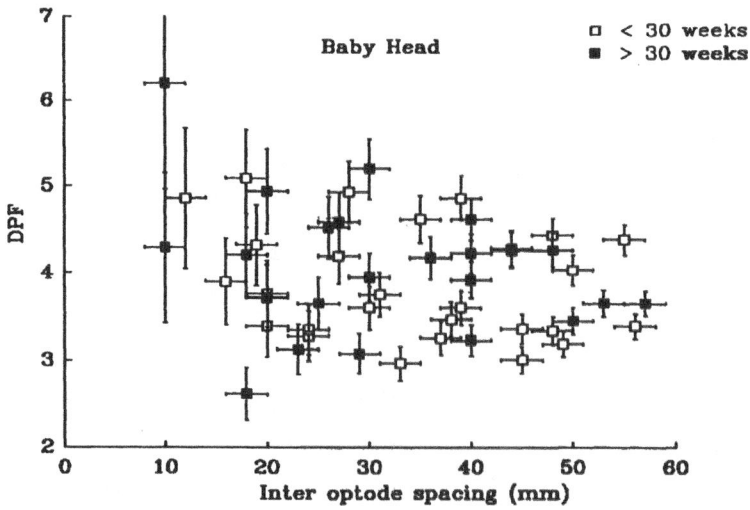

Figure 5. Experimentally derived values for the DPF in the baby head. The error bars indicate one standard deviation.

DISCUSSION

The results obtained in these studies reveal features which require discussion both in terms of the variations in tissue type and geometry, and the differences between observed and theoretically predicted behaviour.

The first general point of importance was that in all tissues the DPF fell with increasing inter optode spacing, the value becoming almost constant beyond 2.5cm. The absolute magnitude of the final DPF was dependent on the tissue being examined. These results are at variance with predictions based upon Monte Carlo modelling of light transport in a homogeneous sphere (van der Zee 1990). The model predicted that the DPF should increase with inter optode spacing in the manner shown in Figure 6. There are three possible explanations for this discrepancy. The first is that the model is incorrect in its treatment of light transport in tissue. This is considered unlikely since for slab geometries, the results obtained from the model have been verified against both experimentally measured light distributions and theoretical calculations. (Delpy et al 1988). Also, model predictions for cylindrical geometries are in qualitative agreement with experimental data (Arridge et al 1990). The second explanation is that the increase in DPF at shorter fibre spacings is a result of the inhomogeneity of real tissue and in particular the presence of superficial layers of skin, fat and bone. Although this may account for some of the observed differences, it is not thought to be the major cause of the discrepancy. If it were, the effects should be most marked in the DPF measurements on inhomogeneous tissues such as the adult and neonatal head, and less so on the more homogeneous tissues of the arm and leg. The experimental data tend if anything to show the reverse. The third explanation is that the difference is caused by the limited acceptance angle of the receiving fibre/streak camera in the measurement system (approximately 7° half angle). The modelled results were obtained for an imaginary receiving fibre, capable of accepting light emerging at any angle. Thus, in the model, light which is scattered obliquely as it first enters the tissue may be detected by the receiving fibre with a high relative intensity (since its pathlength is short). These detected photons

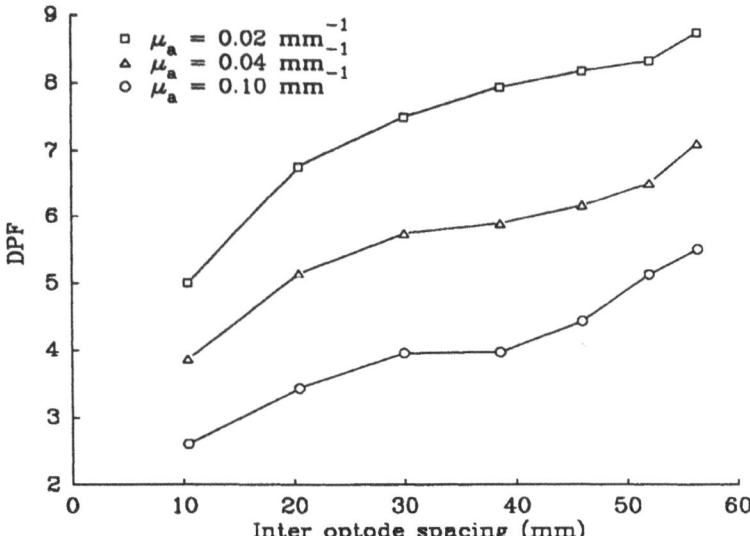

Figure 6. Monte Carlo predictions for the change in the DPF with inter optode spacing on a sphere of six centimetre diameter. The data are shown for three different absorption coefficients, the scattering coefficient being 5 mm^{-1}.

will tend to weight the calculated DPF towards lower values. In the measurement system these photons would not be detected. To check this supposition, the Monte Carlo model will be modified to take into account the acceptance angle of the receiver and the previous calculations repeated.

The second general feature of the measured DPF's is that the absolute value differed with tissue type. This is in general attributable to the differences in the absorption coefficient (μ_a) and reduced scattering coefficient $\mu_s(1-g)$. The predictions of the Monte Carlo model show that the DPF increases with scattering and decreases with absorption, with the absorption dependence being the dominant factor (see Figure 6). A higher absorption will tend to attenuate photons which have travelled a greater distance thus shortening the DPF. There are few data available on the absorption and scattering coefficients for the various tissues at 760 - 800nm. Values reported at 630nm for the brain lie in the range μ_a = 0.02 - 0.18 mm^{-1} and $\mu_s(1-g)$ = 0.64 - 5.7 mm^{-1} (Patterson et al 1987, Sterenborg et al 1989, Karagiannes et al 1989, van der Zee - unpublished observations). For muscle, the reported values are μ_a = 0.15 - 0.17 mm^{-1} and $\mu_s(1-g)$ = 0.44 - 0.7 mm^{-1} (Wilson et al 1986). Although there are no data on the absorption and scattering coefficients for the brain of the preterm infant, it is known that the attenuation is lower than for adult brain (Svaasand et al 1983). This is thought to be due to the lesser degree of myelination in the immature brain, myelin being an important contributor to scattering.

The DPF's from the arm were lower than for any other tissue (3.59 ± 0.32), (with no difference between males and females), probably because the highly absorbing muscle at this point on the arm is close to the surface with, in these subjects, only a thin layer of surface fat. The corresponding data for the calf of men is slightly higher (3.98 ± 0.49), and for females it is much higher (5.14 ± 0.43). This finding presumably reflects differences in fat-muscle ratio, and possibly differences in the thickness of the superficial fat layers. The size of this difference is rather surprising since all these subjects were of normal weight, and the majority took regular exercise. If the difference arose for the reasons given above, then it is likely that the DPF will be significantly different for sedentary subjects or those with specific pathologies.

The adult head gave the largest values for DPF (5.93 ± 0.42). This was not unexpected, since the relatively high albedo of adult brain tissue would lead to a longer pathlength   The relatively low value for the standard deviation was surprising. The adult head is an extremely heterogeneous structure containing many differing tissue types and convoluted tissue boundaries which vary from subject to subject. These could be expected to give rise to scattering which might be subject dependent. However this was not the case, the S.D. being only 7.1% of the mean DPF, and of this, almost 6% can be explained by the inaccuracy in the measurement of the inter optode spacing.

Results for the head of the newborn infant gave a mean value for the DPF of 3.85 ± 0.57. Although this is lightly less than the value of 4.39 ± 0.28 previously obtained for the transilluminated infant head (Wyatt et al 1990b), the difference does not reach statistical significance. The value is lower than that for the adult head, and presumably this reflects the lower scattering coefficient for the newborn infant brain. This explanation would be consistent with the small difference in the DPF that appears to exist between those infants <30 weeks gestation and those >30 weeks. The fact that these measurements were made post mortem should not have caused a large error in the DPF. Previous studies in the rat brain have shown that the DPF changed by less than 4% before and after death. In the only other reported study of pathlength in infants, using the as yet unvalidated phase shift technique (Benaron et al 1990), values for the DPF in the range 3.96 - 6.13 were obtained at inter optode spacings of 1.8 - 3.0 cm.

SUMMARY

The Differential Pathlength Factor (DPF) has been measured for several different tissues. The results showed that the DPF varied with the type of tissue studied, and in the case of the adult calf with sex. However, the DPF for all tissues studied was constant once the inter optode spacing exceeded 2.5 cm.

Thus, measurements can be made by NIR spectroscopy at a range of inter optode spacings, and a single DPF used in the calculation of chromophore concentration. The results also showed that the major source of error in the DPF lay in the measurement of the inter optode spacing. To improve accuracy, two options are possible. Firstly, some means of continuous measurement of inter optode spacing could be incorporated in the NIR instrumentation. The better alternative would be an instrument incorporating a method of directly measuring the optical pathlength at each wavelength. This could be done either by time of flight measurement, or if it can be validated, by phase shift measurement.

ACKNOWLEDGMENTS

This work was supported by the Medical Research Council, the Science and Engineering Research Council, the Wellcome Trust, the Wolfson Foundation and Hamamatsu Photonics KK.

REFERENCES

Arridge, S.R., van der Zee, P., Cope, M., Delpy, D.T. 1990. New results for the development of infrared absorption imaging. Proc. SPIE, 1245, (in press)

Benaron, D.A., Gwiazdowski, S., Kurth, C.D., Steven, J., Chance, B., Delivoria-Papadopoulos, M. 1990. Cerebral changes with growth in infants by haemoglobin phase shift spectroscopy. Pediatric Research, XXX, 38A, 216

Brazy, J.E., Lewis, D.V., Mitnick, M.H., Jöbsis, F.F., 1985, Noninvasive monitoring of cerebral oxygenation in preterm infants: preliminary observations. Pediatrics, 75, 217-225

Brazy, J.E., Lewis, D.V., 1986, Changes in cerebral blood volume and cytochrome aa3 during hypertensive peaks in preterm infants. Pediatrics, 108, 983-987

Chance, B., Leigh, J.S., Miyake, J., Smith, D.S., Nioka, S., Greenfield, R., Finander, M., Kaufman, K., Levy, W.E., Young, M., Cohen, P., Yoshioka, H., Borestky. R. 1988. Comparison of time resolved and unresolved measurements of deoxy haemoglobin in the brain. Proc. Nat. Acad. Sci., 85, 4971-4975

Cheatle, T.R., Potter, L.A., Cope, M., Delpy, D.T., Coleridge-Smith, P.D., Scurr, J.H. 1990. Near infrared spectroscopy - A new technique for metabolic assessment in peripheral vascular disease. Br. J. Surg. (in press)

Cope, M., Delpy, D.T., Reynolds, E.O.R., Wray, S., Wyatt, J., van der Zee, P. 1988. Methods of quantitating cerebral near infrared spectroscopy data. Adv. Exp. Med. & Biol., 222, 183-189

Cope, M., Delpy, D.T., Wray, S., Wyatt, J.S., Reynolds, E.O.R. 1989. A CCD spectrometer to quantitate the concentration of chromophores in living tissue utilising the absorption peak of water at 975 nm. Adv. in Exp. Med. & Biol., 247, 33-40

Delpy, D.T., Cope, M., van der Zee, P., Arridge, S.R., Wray, S., Wyatt, J.S. 1988. Estimation of optical pathlength through tissue from direct time of flight measurement. Phys. Med. & Biol., 33, 12, 1433-1442

Delpy, D.T., Arridge, S.R., Cope, M., Edwards, D., Reynolds, E.O.R., Richardson, C.E., Wray, S., Wyatt, J.S., van der Zee, P. 1989. Quantitation of pathlength in optical spectroscopy. Adv. in Exp. Med. & Biol., 247, 41-46

Edwards, A.D., Wyatt, J.S., Richardson, C.E., Delpy, D.T., Cope, M., Reynolds, E.O.R. 1988a. Cotside measurement of cerebral blood flow in ill newborn infants by near infrared spectroscopy. Lancet, ii, 770-771

Edwards, A.D., Reynolds, E.O.R., Richardson, C.E., Wyatt, J.S. 1988b. Estimation of blood flow in man using near infrared spectroscopy (NIRS). J. Physiol., 410, 50P

Edwards, A.D., Wyatt, J.S., Richardson, C.E., Potter, A., Cope, M., Delpy, D.T., Reynolds, E.O.R. 1990. Effects of indomethacin on cerebral haemodynamics and oxygen delivery investigated by near infrared spectroscopy in very preterm infants. Lancet. 335, 8704, 1491-1495

Ferrari, M., De Marchis., Giannini., Nicola, A., Agostino, R., Nodari, S., Bucci, G., 1986, Cerebral blood volume and haemoglobin oxygen saturation monitoring in neonatal brain by near infrared spectroscopy. Adv. Exp. Med. Biol., 200, 203-212

Fox, E., Jobsis, F.F., Mitnick, M.H. 1985. Monitoring cerebral oxygen sufficiency in anaesthesia and surgery. Adv. Exp. Med. & Biol., 191, 849-854

Lakowicz, J.R., Berndt, K. 1990. Frequency domain measurements of photon migration in tissue. Chem. Phys. Lett. 166, 3, 246-252

Karagiannes, J.L., Zhang, Z., Grossweiner, B., Grossweiner, L.I. 1989. Applications of the 1-D diffusion approximation to the optics of tissues and tissue phantoms. Applied Optics, 28, 2311-2317

Patterson, M.S., Wilson, B.C., Feather, J.W., Burns, D.M., Pushka, W. (1987) The measurement of dihematoporphyrin ether concentration in tissue by reflectance spectrophotometry. Photochemistry and Photobiology, vol 46, 337-343

Sterenborg, H.J.C.M., van Gemert, M.J.C., Kamphorst, W., Wolbers, J.G. (1989) The spectral dependence of the optical properties of human brain. Lasers in Medical Science, vol 4, 221-227

Svaasand, L.O., Ellingson, R. (1983) Optical Properties of human brain. Photochemistry and Photobiology. Vol 38, 293-299

van der Zee, P., Arridge, S.R., Cope, M., Delpy, D.T., 1990. The effect of optode positioning on optical pathlength in near infrared spectroscopy of brain. Adv. Exp. Med. & Biol. (in press)

Wilson, B.C., Patterson, M.S. 1986. The physics of photodynamic therapy. Phys. Med. Biol., 31, 4, 327-360

Wilson, B.C., Park, Y., Hafetz, Y., Patterson, M., Madsen, S., Jacques, S. 1989. The potential of time resolved reflectance measurements for the non invasive determination of tissue optical properties. Proc. S.P.I.E., 1064, (in press)

Wray, S., Cope, M., Delpy, D.T., Wyatt, J.S., Reynolds, E.O.R. 1988. Characterisation of the near infrared absorption spectra of cytochrome aa_3 and haemoglobin for the non invasive monitoring of cerebral oxygenation. Biochim. Biophys. Acta, 933, 184-192

Wyatt, J.S., Cope, M., Delpy, D.T., Wray, S., Reynolds, E.O.R. 1986. Quantification of cerebral oxygenation and haemodynamics in sick newborn infants by near infrared spectrophotometry. Lancet, 2, 1063-1066

Wyatt, J.S., Cope, M., Delpy, D.T., Richardson, C.E., Edwards, A.D., Wray, S.C., Reynolds, E.O.R. 1990a. Quantitation of cerebral blood volume in newborn infants by near infrared spectroscopy. J. Appl. Physiol. 68, 3, 1086-1091

Wyatt, J.S., Cope, M., Delpy, D.T., van der Zee, P., Arridge, S.R., Edwards, A.D., Reynolds, E.O.R. 1990b. Measurement of optical pathlength for cerebral near infrared spectroscopy in newborn infants. Dev. Neuroscience. 12, 140-144

NEAR-INFRARED IMAGING *in vivo* (I): IMAGE RESTORATION TECHNIQUE APPLICABLE TO THE NIR PROJECTION IMAGES

Ryuichiro Araki and Ichiro Nashimoto

Department of Hygiene and Environmental Physiology
Saitama Medical School
38 Morohongo, Moroyama, Iruma-gun, Saitama 350-04,Japan

INTRODUCTION

Near-infrared (NIR) imaging *in vivo* has potential capability to measure spatial distribution of tissue oxygen insufficiency noninvasively. Attempts to apply this technique to the living tissues have, however, been hampered by strong light scattering due to optical turbidity of the tissues. There seem the following possible approaches to enhance spatial resolution of NIR imaging:

1) picosecond time-resolved spectroscopy (Chance et al., 1988)
2) coherent detection imaging (Toida et al., 1989)
3) computer-aided digital image restoration (Rosenfeld and Kak, 1976)

The first and second approaches are selective detection of unscattered light transmitted through the tissue. However, intensity of the unscattered light component is so weak that it becomes difficult to obtain sufficient quality of images. On the other hand, the third technique can utilize also scattered component of light transmitted through the tissue.

Digital image restoration, in other words, image refocusing, has been used to refocus blurred or degraded images in the field of medical imaging as well as remote sensing or other practical fields. We have performed refocusing of NIR projection images of human forearm and obtained enhanced resolution.

MATERIALS AND METHODS

NIR imaging system. Light from a 150 W xenon lamp was introduced into an optical system which consisted of a pinhole and a lens to obtain parallel beam light-flux, then the beam was guided onto a volunteer's forearm. The projection images were measured with a Pertier-cooled integration CCD video camera (Model TM-840NP equipped with a custom-made Pertier device, PULNiX, U.S.A.) via NIR interference filters. Integration period of the CCD was controlled with a controller unit (Model VF-500, PULNiX, U.S.A.). The integrated image was then digitized (512 x 512 pixels) using a 8-bit video digitizer (Model CT-9800B, Cybertek Co., Japan). Sixteen times of image accumulation was performed to reduce quantized

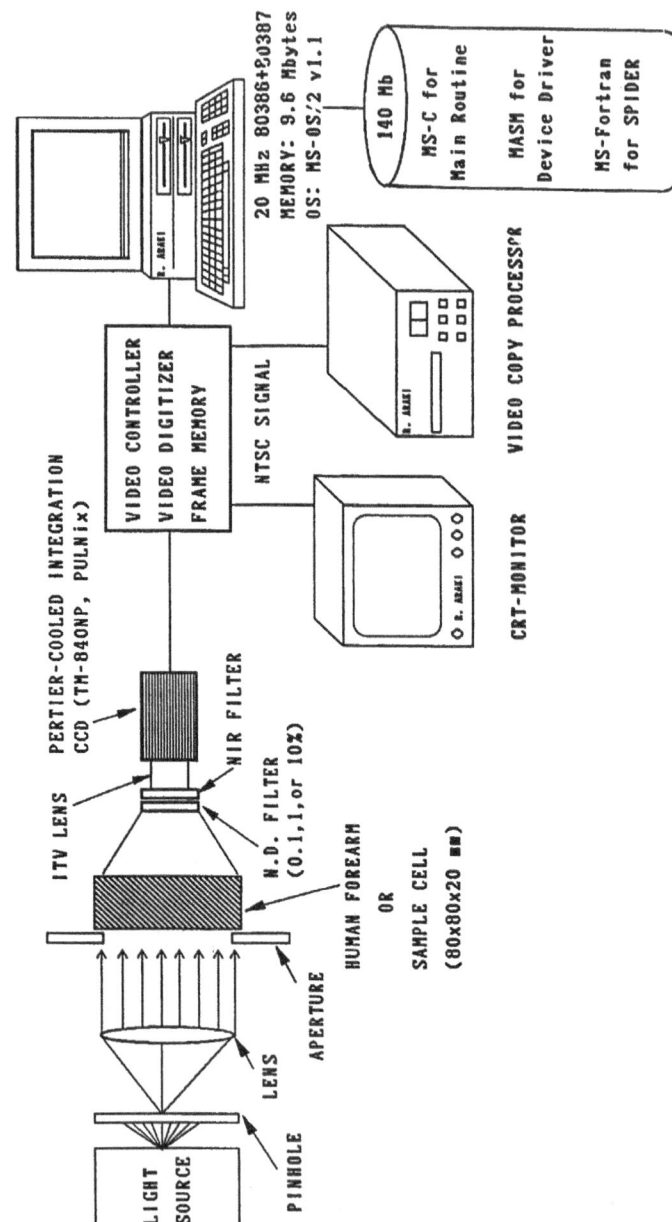

Fig. 1 Block diagram of the experimental setup. Details are described in the text.

156

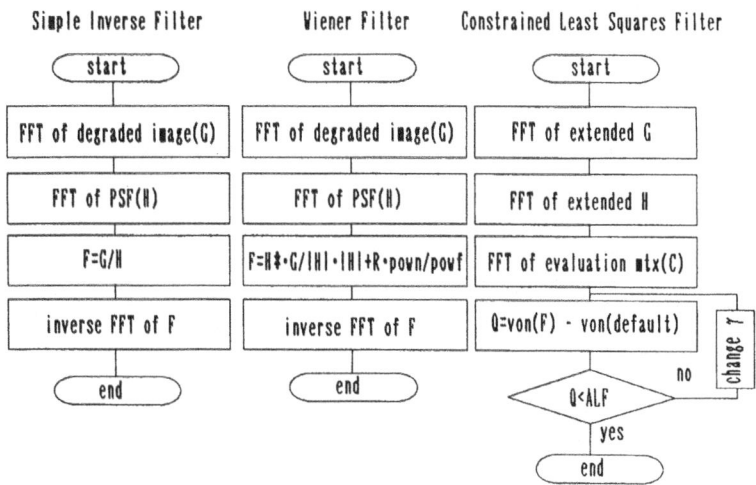

Simple Inverse Filter | Wiener Filter | Constrained Least Squares Filter

Simple Inverse Filter
- start
- FFT of degraded image(G)
- FFT of PSF(H)
- F=G/H
- inverse FFT of F
- end

Wiener Filter
- start
- FFT of degraded image(G)
- FFT of PSF(H)
- F=H‡·G/|H|·|H|+R·povn/povf
- inverse FFT of F
- end

Constrained Least Squares Filter
- start
- FFT of extended G
- FFT of extended H
- FFT of evaluation mtx(C)
- Q=von(F) - von(default)
- Q<ALF
- change ?
- no / yes
- end

Flow chart of image restoration algorithms. G and F are degraded and original images, H is point spread function, R is factor for Wiener filtering, povn and povf are power spectra of noise and that of original image, von(F) and von(default) are variance in noise of original image and that of default value,and ALF is threshold of variance in noise in the restored image.

1 4 m i n	2 3 m i n	9 1 d a y s

Comparison of computation time of three algorithms for images which consist of 512 x 512 pixels executed with a 20 MHz 80386+80387-based computer (NEC PC-98RL). The computation time for constrained least squares filter was estimated from the result for 32 x 32 pixels test data.

Fig. 2 Comparison of refocusing algorithms.

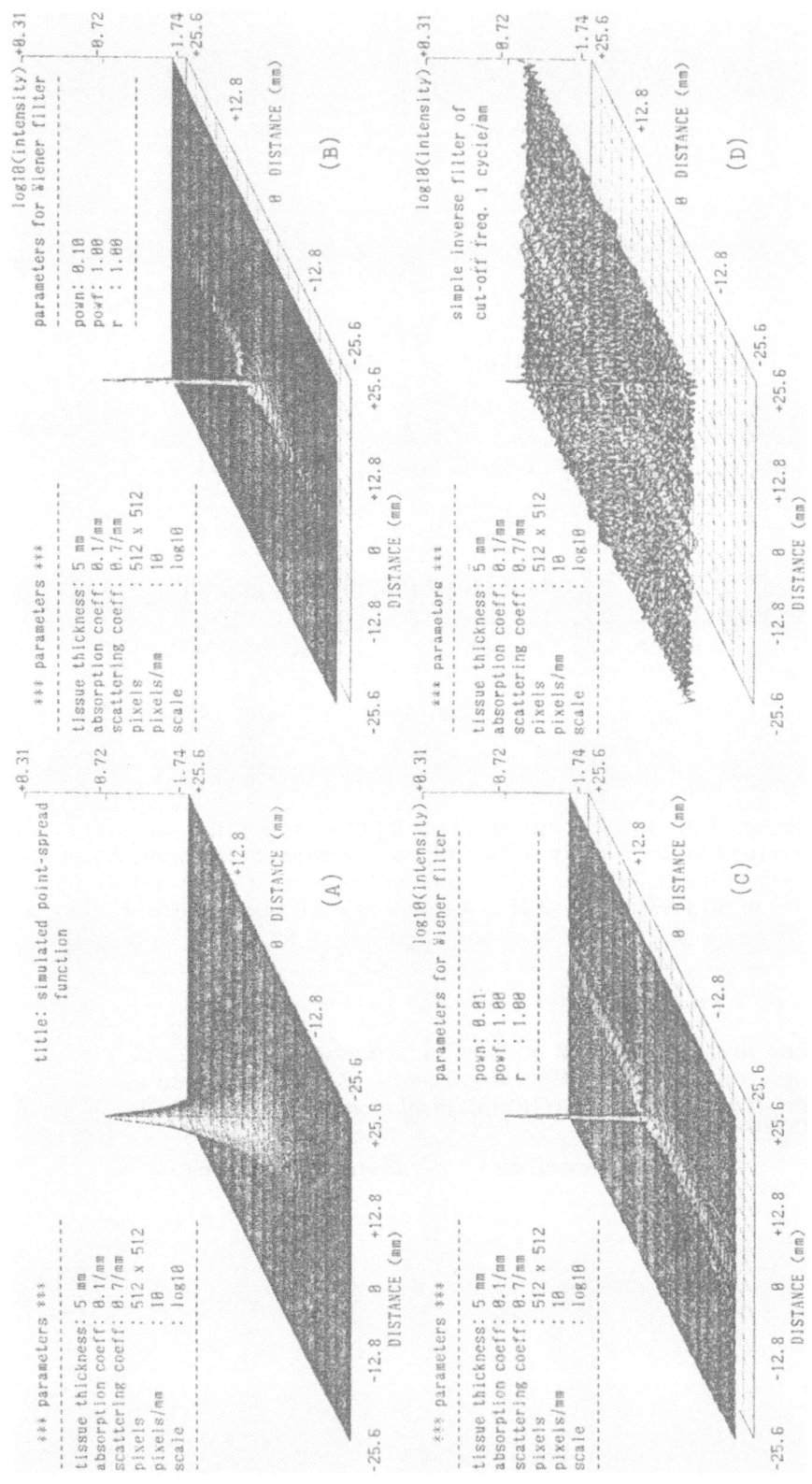

Fig. 3 Simulated 2-dimensional PSF of light through the tissue and results of refocusing computation. Part A, simulated PSF; Part B and C, restoration by Wiener filter; Part D, restoration by simple inverse filter.

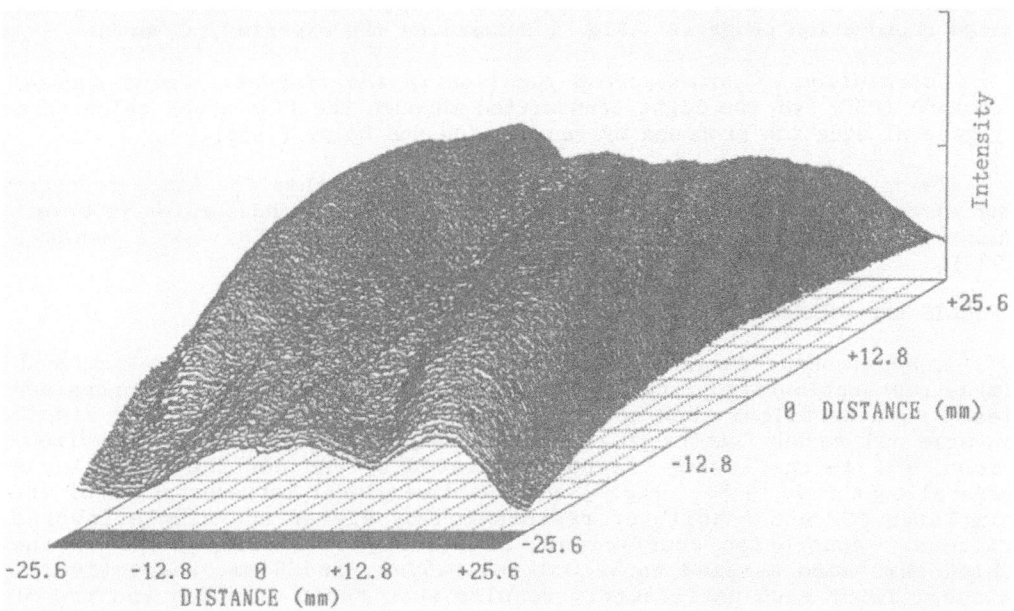

Fig. 4 3-dimensional representation of a NIR projection image of human forearm at 750 nm. The z-axis represents relative intensity of transmitted light.

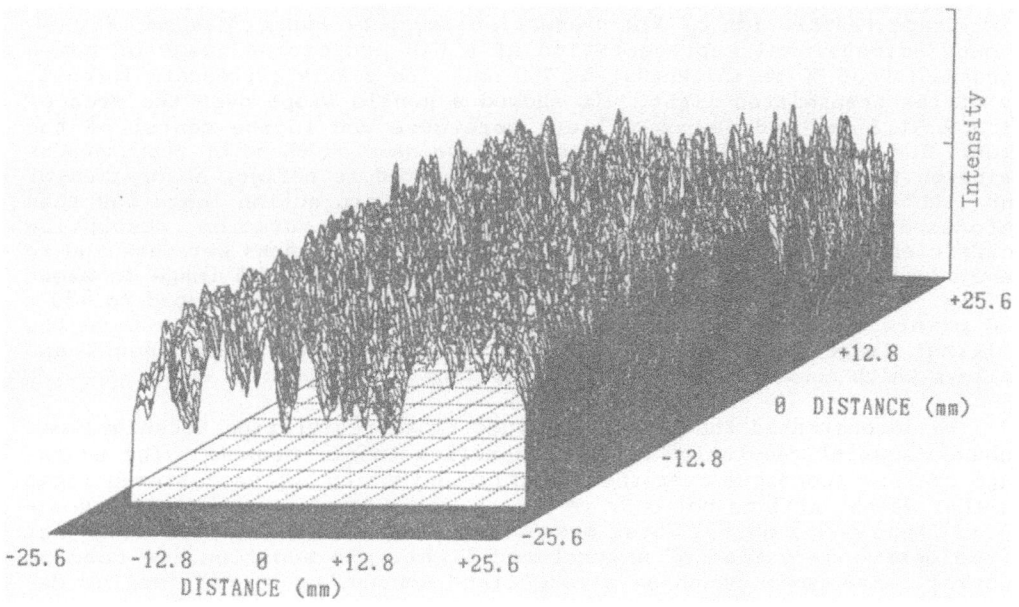

Fig. 5 3-dimensional representation of a NIR projection image of human forearm refocused by Wiener filter.

errors due to A/D conversion. The digitized images were then processed with an 80386-based computer equipped with a 80387 math-coprocessor (Model PC-9801RL with MS-OS/2 version 1.1, NEC, Japan) using preprocessing and image restoration programs. Fig. 1 summarizes the experimental setup.

Calculation of point-spread function in the tissues. Point-spread function (PSF) of the light transmitted through the tissue was calculated by general equation proposed by van der Zee and Delpy (1988).

Image restoration algorithms. Several algorithms for image refocusing have been already developed and improved. We tested simple inverse, constrained least squares, and Wiener filters (SPIDER user's manual, 1982).

RESULTS AND DISCUSSION

Comparison of image restoration algorithms In Fig. 2, we compared image restoration algorithms and their execution time. Since constrained least squares filter took too long computation time, we applied simple inverse and Wiener filters to refocusing computation of simulated 2-dimensional PSF in the living tissue calculated by van der Zee and Delpy's general equation (1988). Fig. 3 shows 3-dimensional representation of the simulated PSF and results of restoration by Wiener and simple inverse filters. Absorption coefficient, scattering coefficient, and tissue thickness were assumed to be 0.1/mm, 0.7/mm, and 5 mm, respectively. Wiener filter gave satisfactory results when ratio of power spectra of noise and that of the original image were assumed to be 0.1:1 or 0.01:1, whereas simple inverse filter caused remarkable deterioration of signal-to-noise ratio in the background area. Thus Wiener filter was considered to be the most appropriate algorithm for image restoration of NIR projection images in the living tissues.

Image restoration of NIR projection image of human forearm Fig. 4 shows 3-dimensional representation of a NIR projection image of human forearm (about 50 mm thickness) at 750 nm. The z-axis represents intensity of the transmitted light. It showed a gentle slope over the area of 51.2 x 51.2 mm, and narrow valleys were observed in the center of the plot. The positions of the valleys were the same as those of subcutaneous veins on the detector side, and thus considered to reflect absorption of the light by the subcutaneous veins. The NIR projection image was then refocused using Wiener filter. For refocusing computation, absorption coefficient, scattering coefficient, and tissue thickness were assumed to be 0.1/mm, 0.7/mm, and 5 mm, respectively. The refocused image is shown in Fig. 5. To perform image restoration, image size was reduced to 480 x 480 pixels to apply faster computation algorithm. In contrast to the original image (Fig. 4), the refocused image showed sharp peaks and valleys which corresponded to relatively fine blood vessels.

We demonstrated that computer-aided image restoration technique can enhance spatial resolution of NIR projection images *in vivo*. The advantage of this technique over the selective detection of unscattered light is that we can utilize not only very weak unscattered light but also scattered light components. Thus, measurement time can be shorter than that of selective detection of unscattered light. In addition, we need no special instruments such as strong light source, sensitive imaging devices, and picosecond time-resolved spectroscopic instruments. However, this technique also has the following limitations. 1) Since living tissues are optically heterogeneous, the PSF for image restoration has to be adaptively varied as absorption coefficient, scattering coefficient, and tissue thickness change. Such computation cannot be achieved by existing image restoration techniques, and thus we have to develop new computation algorithm. 2) Quality of the restored image depends on S/N ratio of the original image including quantized errors and accuracy of the PSF used for refocusing computation. Therefore, both more detailed and precise information on PSF of the light transmitted through living tissues as well as more sensitive imaging device and higher resolution of A/D conversion are required to fulfill the requirements for practical use of this technique.

SUMMARY

To enhance spatial resolution of NIR projection images *in vivo*, we performed refocusing of NIR projection images of human forearm of about 50 mm thickness. A volunteer's forearm was illuminated by parallel light beam-flux, and then projection images at 750 nm was measured with a Pertier-cooled CCD video camera and digitized. For refocusing computation, PSF of the light transmitted through the tissue was calculated by general equation proposed by van der Zee and Delpy (1988). Simple inverse, constrained least squares, and Wiener filters were tested as refocusing algorithms. Wiener filter gave the best result in terms of image quality and computation time. By applying Wiener filter to the image refocusing of NIR projection images of human forearm, we obtained enhanced spatial resolution.

REFERENCES

Chance, B. et al., 1988, Comparison of time resolved and unresolved measurements of deoxyhemoglobin in brain, *Proc. Natl. Acad. Sci. USA*, 85:4971.

Rosenfeld, A. and Kak, A.C., 1976, Digital picture processing, Academic Press, Inc., New York.

SPIDER (Subroutine Package for Image Data Enhancement and Recognition) user's manual, 1982, the Ministry of International Trade and Industry of Japan.

Toida, M. et al., 1989, Optical heterodyne technique for achieving excellent image detection in highly scattering media such as biological substances and tissues, *in*: OSA Annual Meeting, 1989 Tech Digest Series, Vol. 18, Paper FI6, Opt. Soc. Am. Washington DC 1989, 233.

Van der Zee, P. and Delpy, D.T., 1988, Computed point spread functions for light in tissue using a measured volume scattering function, *Adv. Exp. Med. Biol.*, 222:191-197.

MUSCLE OXYGENATION BY FAST NEAR INFRARED SPECTROPHOTOMETRY (NIRS) IN ISCHEMIC FOREARM

Roberto Alberto De Blasi*, Enrico Quaglia*, Alessandro Gasparetto*, Marco Ferrari^

*Department of Anesthesiology and Critical Care, I University of Rome "La Sapienza", Policlinico Umberto I, 00161 Rome, Italy
^Department of Biomedical Sciences and Technology, and Biometrics, University of L'Aquila, 67100 L'Aquila, Italy and Cell Biology Laboratory, Istituto Superiore Sanita',Viale Regina Elena 299, 00161 Rome, Italy

INTRODUCTION

Most of the knowledge about cell respiration is based on studies of isolated enzymes and mitochondria. Although essential to identify the biochemical mechanisms and sites of control, the metabolic function of an intact cell depends on complex interactions that should be studied "in vivo" to take into account the system properties. The measurement of tissue oxygen saturation has been widely investigated by many techniques. "In vivo" investigations have been limited by the invasivity of these methods that prevent to know accurately the oxygen dynamic gradients occurring around and into the cell.

Skeletal muscle oxygenation has been extensively studied because this tissue has the greatest range of energy turnover compared to oxygen demand. Furthermore muscle omogeneous anatomy offers the opportunity for evaluating easier the tissue oxygen gradients.

Near Infrared Spectroscopy (NIRS) has been developed to obtain a non-invasive monitoring of changes in tissue oxygenation on intact organs like brain and muscle (Jobsis, 1977; Cope and Delpy, 1988; Ferrari et al., 1989). NIRS research has been focused to distinguish changes in absorption due to variations of relative concentration of the most abundant absorbers: i.e. oxygenated and deoxygenated hemoglobin (Hb) and myoglobin (Mb) and copper ions of cytochrome a-a3.

An absolute quantitation of the absorbers could be

obtained with the knowledge of the optical pathlength of photons in the examinated tissue (Chance et al., 1988; Delpy et al., 1988). Muscle algorithms are based on the assumption that myoglobin does not modify its oxygenation in most of the explored circumstances (Hampson et. al. 1987) so that only net changes in absorption results from oxy-deoxyHb. Although the high Mb affinity for oxygen should lead to the prevalence of the oxygenated form in resting condition and in submaximal muscular work (Gayesky et al., 1987), the same could not be true in oxygen limited conditions.

The interest of clinicians and physiologists is the knowledge of the "in vivo" oxygen-limited cellular function, i.e. the oxygen tension below which oxygen consumption (VO2) is limited by the availability of oxygen (disoxia) (Connett et al., 1990). Recent studies (Gayesky et al., 1988) gave a clear answer to most of the questions concerning this matter performing invasive spectroscopic measurements and a sophysticated statistical analysis on frozen skeletal muscle samples.

Fast scanning spectrophotometry and derivative spectroscopy, extensively applied in agricolture, have been used recently to measure non invasively the brain Hb saturation on dogs (Ferrari et al., 1989) and human volunteers during hypoxic hypoxia (Hanley et al, 1990). In comparison to multiwavelenght photometers this instrumentation offers a global view of the spectral tissue characteristic in a wide wavelenght range and allows for the investigation of absorption peaks changes during the measurement.

The aim of the present study was to evaluate the human forearm muscle oxygenation by near IR continuous fast scanning spectroscopy in oxygen limited conditions. The quality of spectra and the fast collection time can allow for obtaining a clear picture of the Hb/Mb oxygenation changes in muscle.

METHODS

Measurements were obtained on 15 adult volunteers (mean age 30.6 ± 4.7) using a microprocessor controlled scanning spectrophotometer (mod. 6250 NIRSystem, Silver Spring, MD USA). A 680-1050 nm spectrum with a high S/N ratio can be performed in 30 sec averaging 50 consecutive scans. Scans were recorded every 60 sec for the duration of the experiments. Fiber optics (200 cm length and 0.5 cm active diameter) were applied on the forearm at the distance of 2.5 cm by a black rubber support so that a stable fibers geometry was achieved. All measurements were performed on the proximal right forearm brachio-radial muscle. A pneumatic cuff was previously loosely wrapped around the arm. In some experiments a pulse oxymeter (Omheda, Biox 3700) was inserted on the tip of the first finger of the contralateral arm. Before each experiment, the two fibers were positioned to transilluminate two etched nickel screens. This spectrum provided a 3 optical density reference against which all subsequent spectra were compared.

The following protocols were performed : a) hypoxic hypoxia, b) forearm ischemia, and c) venous occlusion . In the first protocol subjects were breathing an hypoxic mixture with an inspired oxygen concentration (FiO2) graded from 21%

to 9%. FiO2 was analysed by an oxygen monitor (OM 15, Sensor Medics, Anaheim, Ca). All subjects underwent to venous cannulation and blood samples were drawned after that a stable FiO2 and stable spectra were recorded. Blood Hb saturation was analyzed by a Co-oxymeter (Radiometer, Copenhagen). In the second protocol a 5 min stabilization period, during which all spectra were superimposable, was followed by a 10 min arterial occlusion obtained by cuff inflation at a pressure of 260 mmHg. After the cuff release, a 5 min recovery phase followed. In the third protocol, after the 5 min stabilization period, a venous occlusion was obtained by inflating the cuff at 60 mmHg. The pressure value has been choosen to not interfere with the arterial blood inflow. The recovery phase was similar to that of the second protocol.

The NIRS signals are processed to provide a quantitative information on changes in oxygenated and deoxygenated Hb/Mb relative to baseline condition. At this purpose the derivative method, well known in "in vitro" biochemistry (O'Haver, 1979), was utilized with the aim to minimize scattering and non specific absorption such as geometrical and volume changes during the measurement. Derivative spectra maintain the same relationship between concentration and wavelength absorption as it occurs for the original spectra but, because constant offsets of optical density are removed, the log (1/Td) varies linearly with concentration (Williams and Norris, 1987).

RESULTS

In normoxia and resting conditions the absorption peak of deoxyHb/Mb is clearly observed at 757-760 nm, while another absorption band due to the O-H harmonic of water is clearly shown around 968 nm (Wray et al.,1988) (fig.1). In hypoxic hypoxia the increase of the 757-760 nm absorption peak reflects the Hb/Mb oxygen desaturation. The spectral shift, provoked by severe hypoxia, was due to Hb volume increase in the optical field.

In the first minutes of arterial occlusion a 757-760 nm peak raise was shown accompained by spectral shifts due to Hb volume increase (upper panel fig. 2). The wavelength corresponding to the maximum peak did not change during the occlusion and recovery phase. Lower panel of fig. 2 shows the same spectral effects using first derivative. The difference between the maximum of Hb/Mb absorption (760 nm) and the Hb/Mb isosbestic point (800 nm) (Δ O.D.) was measured on 10 adult volunteers (fig.3). No further significant Δ O.D. changes were shown after the 4th minute up to the end of 10 min occlusion. The corresponding derivative spectra showed a minimum peak around 773 nm. The amplitude of this peak (fig. 4) changed within the first four min of occlusion with the same time course of fig. 3. A high correlation between Δ O.D. changes and first derivative maximum changes was found suggesting that 800 nm is an appropriate correction for a relative measurement of Hb/Mb oxygen saturation.

As shown in fig. 2 the recovery phase was characterized by a sudden decrease of 760 nm absorption peak below the control value. OxyHb/Mb increase was accompanied by a shift of the whole spectrum due to the increase of Hb volume occurring during the reflow phase. The return to baseline

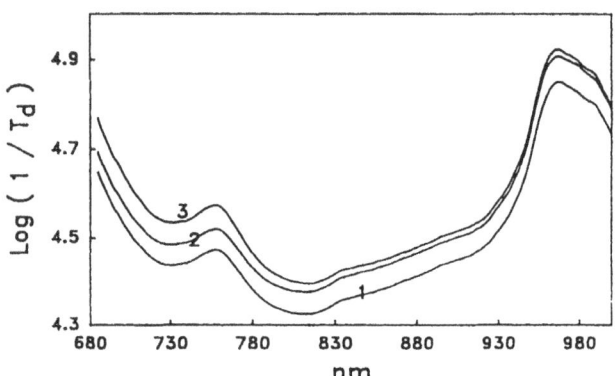

Fig. 1. Human forearm near IR spectra. Spectrum 1: FiO2 21%, SvO2 91 %; spectrum 2: FiO2 13%, SvO2 86.9%; spectrum 3: FiO2 9%, SvO2 48.9%. During hypoxic hypoxia the increase of absorption peak at 757-760 nm reflected the raise of deoxyHb/Mb in the optical field. Spectrum shift up was attributable to hemoglobin volume increase.

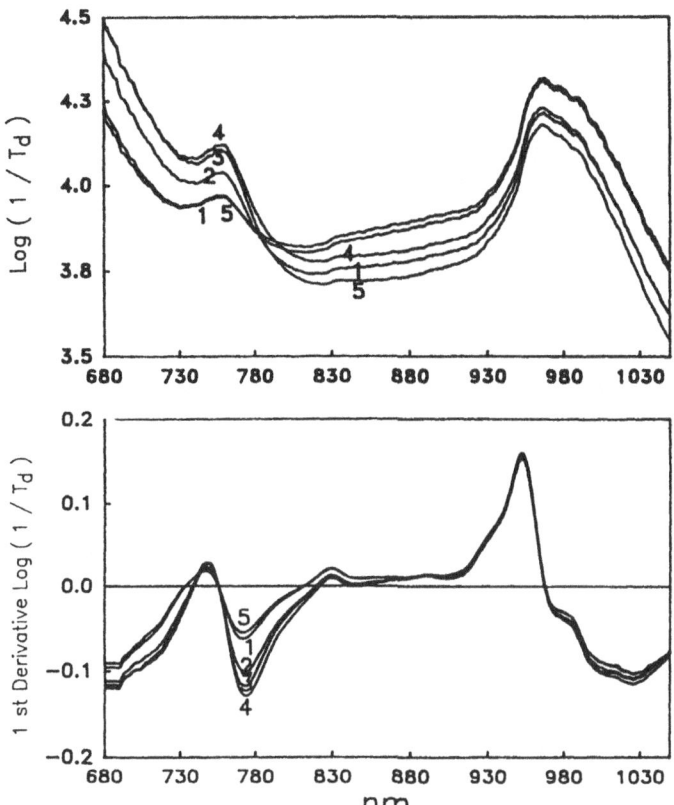

Fig. 2. Near infrared spectra in human forearm ischemia: control condition (1), 2 min ischemia (2), 4 min ischemia (3), 10 sec recovery (4), and 5 min recovery (5). In the first minutes of occlusion a 757-760 nm peak raise was observed accompained by an initial moderate spectral shift due to Hb volume increase. A complete return to baseline values was obtained after 5 min. Lower panel shows corresponding first derivative spectra.

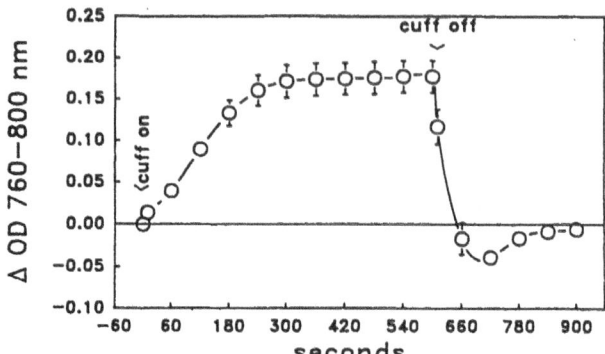

Fig. 3. Time course of muscle oxygenation during ischemia induced by application of a tourniquet. Oxygenation was measured at 760 nm and corrected for hemoglobin volume changes measured at the isosbestic wavelength 800 nm. Note that after the 4th minute of occlusion a plateau was obtained until cuff release.

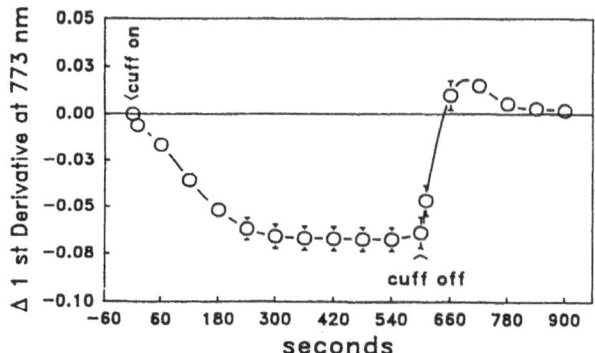

Fig. 4. Time course of response to ischemia induced by application of a tourniquet measured at 773 nm on first derivative spectra. The time course is similar to that of fig.3

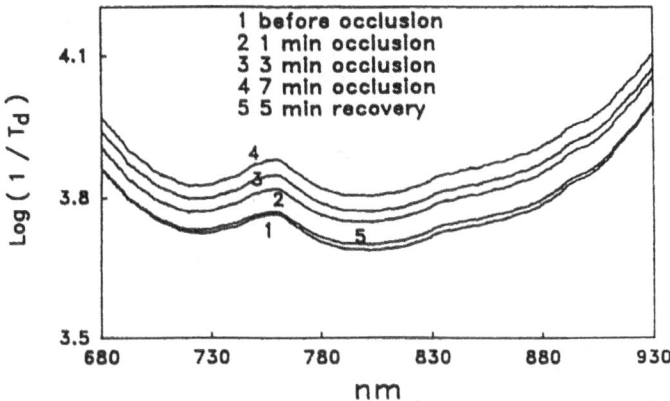

Fig. 5. Near infrared spectra during forearm venous occlusion induced by a 60 mmHg cuff inflation. 1: baseline; 2: 1 min; 3: 3 min; 4: 7 min; 5: recovery 5 min. A continuous raise of spectra was observed up to the end of occlusion due to hemoglobin volume increase; Hb/Mb saturation was almost unchanged.

conditions was obtained in about 5 min. The derivative spectra confirm the Hb oxygenation increase occuring after the cuff release mainly due to the rapid arterial blood inflow.

During venous occlusion of the forearm (fig. 5), a continuous raise of spectra was observed up to the end of occlusion. No consistent oxygenation changes could be observed also by derivative spectroscopy.

DISCUSSION

The brain has been the first organ to be studied by NIRS (Jobsis, 1977; Wray et al. 1988; Ferrari et al. 1983). More recently studies indicate the clinical relevance of NIRS measurements on muscle. In particular NIRS could be applied to evaluate muscle performances during exercise (Chance et al. 1990), to identify patients with mitochondrial myopathies (Sobolewski et al., 1990) and to verify the farmacological treatment (Fisher et al., 1990).

In this study we provided detailed spectra of human muscle during different challenges. We observed an initial moderate spectral shift soon after the inflation that can be interpreted as some blood squeezed into the veins of the optical field by the inflated cuff. During ischemia a plateau was reached in 3-4 min. A rapid plateau phase was observed also, using the same protocol, by Hampson (1988) who interpreted it as a total or near-total Hb and Mb deoxygenation. Other authors observed a biphasic desaturation (Chance et al., 1988). We confirm the findings of the first author providing first accurate spectral evidences of muscle oxygenation changes. Because the short time from the start of the occlusion it is not likely that a complete Hb desaturation occurred. In fact a mean venous pO2 of 24mmHg only was measured after 30 min of tourniquet ischemia (Wilgis, 1971). In a separate study we measured an arterial pO2 of 40 mmHg after 5 min occlusion and a corresponding venous pO2 of 24 mmHg. We found that the arterial pO2 values approchead to the venous pO2 prolonging the occlusion time.

A possible explanation of the plateau phase can be found taking into account the target of NIRS investigation. Many experimental data lead to the conclusion that NIRS measures a balance between the Hb/Mb saturation content of the different districts of circulation and tissue oxygen consumption. Large vessels represent only a small fraction of the total absorption, the large part of it being due to the Hb/Mb oxygen saturation in capillary and venous circulation related to the cellular oxygen consumption. A support to this interpretation is also given by our study. Using multiple regression analysis we found no correlation between SvO2 values, measured in the venous blood, and the corresponding spectra performed during different levels of hypoxic hypoxia. The conclusion is that spectral changes in muscle tissue are not related to venous Hb saturation contrarly at what was recently demonstrated in the dog brain during hypoxic-hypoxia (Ferrari et al., 1989). The fact that after 4 min of ischemia no more spectral changes are observed could be related to a lack of oxygen consumption from hypoxic cells rather than to total Hb desaturation in the blood. The oxygen gradient from capillary to mitochondria normally present gives rise to the totally oxygenated form of myoglobin. This condition is present also in near-maximum exercise when there is no oxygen limitation (Gayewski et al., 1987). During the arrest of circulation, the oxygen delivery to tissue is drastically reduced so that an anoxic state is rapidly reached at mitochondrial level and no further Hb desaturation occurs.

The degree of Hb oxygen saturation could also be an expression of the mitochondrial respiratory activity as suggested by Sobolewski (1990). This recent work shows that performing a forearm ischemia on patients with mitochondrial myopathy a significant lower Hb-desaturation was obtained respect to normal patients and subjects with other defined myopathies.

During venous outflow restriction an increase of blood volume sudden occurred, because of the persisting arterial blood inflow, until steady state was reached. Interestingly no significant variation of the Hb/Mb oxygen saturation was observed during venous outflow restriction as measured by absorption and derivative spectra. On the contrary for the same protocol Hampson found an increase of reduced Hb/Mb content. A possible explanation of our data is that below the diastolic pressure value although the vein outflow is obstructed the arterial inflow is sufficient to provide an adequate O2 support.

In conclusion recent clinical data (Chance, 1990; Fisher, 1990) indicate that NIRS can measure muscle oxygenation changes in many physiological and clinical circumstances. In addition quantitation of these measurements could be reached using time of flight (Chance et al., 1988). Unfortunately previous studies gave few spectroscopic details. We were able to perform accurate spectra during different oxygen limited conditions and utilize derivative spectroscopy to identify oxygenation changes.

AKNOWLEDGMENTS

This research has been funded in part by C.N.R. contribution N.89.04165.04

SUMMARY

Fast scanning near infrared spectroscopy (680-1050nm) was utilized to evaluate human forearm muscle oxygenation in 15 adults volunteers. Spectra were recorded in hypoxic hypoxia, ischemia and venous outflow restriction. Derivative spectra were performed with the aim to obtain a quantitative information of Hb/Mb oxygen saturation free from volume and scattering changes. The absorption spectra O.D. demonstrate an increase of deoxy-Hb/Mb in hypoxic condition with a moderate volume changes. In ischemia a rapid Hb/Mb desaturation occurred until a plateau was reached at 4th min. The cuff release was followed by hyperemia with Hb volume raise and oxy-Hb/Mb increase above the control. Spectral data support the hypothesis that derivative NIRS can be used to identify muscle oxygenation changes.

REFERENCES

Chance, B., Nioka, S., Kent, J., McCully, K., Fountain, M., Greenfeld, R., and Holtom, G., 1988, Time-resolved spectroscopy of hemoglobin and myoglobin in resting and ischemic muscle. Anal. Bioch. 174: 698.
Chance, B., Reddy, K.S., Budinger, T., Kakihara, H., 1990, Effect of O2 debit on myoglobin deoxygenation in exercising limb. FASEB J., 4 : 1598.
Connett, R.J., Honig, C.R., Gayeski, T.E. and Brooks, G.A., 1990 Defining hypoxia : a system view of VO2, glycolysis, energetics,and intracellular PO2. J. Appl. Physiol. 68: 833.
Cope, M., Delpy, D.T., 1988, System for long-term measurement of cerebral blood and tissue oxygenation on newborn infants by near infra-red transillumination. Med. & Biol. Eng. & Comput. 26: 298.
Delpy, D.T., Cope, M., Van der Zee, P., Arrige, S., Wray, S., and Wyatt, J., 1988, Estimation of optical pathlength through tissue from direct time of flight measurement. Phys.Med.Biol. 33: 12: 1433.
Ferrari, M., Giannini, I., Carpi, A., and Fasella, P.,1983, Non-invasive near infrared spectroscopy of brain in fluorocarbon exchange-transfused rats. Physiol. Chem. Phys. Med. NMR. 5: 107.
Ferrari, M., Wilson, D.A., Hanley, D.F., Hartman, J.F., Traystman, R.J., Rogers, M.C., 1989, Non invasive determination of cerebral venous hemoglobin saturation in the dog by derivative near infrared spectroscopy. Am. J. Physiol. 256: H149.

Fisher, M., Guyot, A., Sobolewski, E., Chance, B., Peterson, P.L., 1990, The evaluation of treatment strategies in mitochondrial myopathy (MM) with near infrared reflectance spectroscopy (NIRS). Neurology, 40 suppl.1: 648S.

Gayeski, T.E.J., Connett, R.J., and Honig, C.R., 1987, Minimum intracellular PO2 for maximal cytochrome turnover in red muscle in situ. Am.J.Physiol. 252: H906.

Gayeski, T.E.J., Federspiel, W.J., and Honig, C.R.,1988, A graphical analysis of the influence of red cell transit time, carrier-free layer thickness and intracellular PO2 on blood-tissue O2 transport. Adv. Exp. Med.Biol. 222: 25.

Hampson, N.B., Jobsis-Vander Vliet, F.F., and Piantadosi, C., 1987 Skeletal muscle oxygen availability during respiratory acid-base disturbances in cats. Respiration Physiol. 70: 143.

Hampson, N.B., and Piantadosi, C.A., 1988, Near infrared monitoring of human skeletal muscle oxygenation during forearm ischemia. J. Appl. Physiol. 64: 2449.

Hanley, D.F., Cross, K., Norris, K., Ferrari, M., Wilson, D., Traystman, R.J., Borel, C., and Diringer, M., Infrared prediction of human cerebral venous hemoglobin saturation, 1990,Neurology. 40 suppl.1: 229S.

Jobsis-Vander Vliet, F.F., 1977, Noninvasive infrared monitoring of cerebral and myocardial oxygen sufficiency and circulatory parameters. Science. 198: 833.

O'Haver, T.C., 1979, Potential clinical application of derivative and wavelength modulation spectrometry. Clin.Chem. 25: 1548.

Sobolewski, E., Guyot, A., Fisher, M., Chance, B., Peterson, P.L., 1990, Near infrared reflectance spectroscopy (NIRS) of Mitochondrial Myopathy (MM). Neurology vol.40 suppl.1: 645S.

Wilgis, E.F.S., 1971, Observations on the effects of tourniquet ischemia. Journal of Bone and Joint Surgery,53-A : 7.

Williams, P., and Norris, K., 1987 Near-infrared technology in the agricultural and food industries. St. Paul, MN: Am. Assoc.of Cereal Chemists.

Wray, S., Cope, M., Delpy, D.T., Wyatt, J.S., Reynolds, E.O.R., 1988, Characterization of the near infrared absorption spectra of cytochrome a, a3 and hemoglobin for the non-invasive monitoring of cerebral oxygenation. Biochem. Biophis. Acta 933: 184.

NEAR-INFRARED IMAGING *in vivo* (II): 2-DIMENSIONAL VISUALIZATION OF TISSUE

OXYGENATION STATE

Ryuichiro Araki and Ichiro Nashimoto

Department of Hygiene and Environmental Physiology
Saitama Medical School
38 Morohongo, Moroyama, Iruma-gun, Saitama 350-04, Japan

INTRODUCTION

Oxygenation-deoxygenation state of Hb in living tissues has been measured by means of near-infrared (NIR) spectrophotometry. Several authors have examined quantitative analysis of Hb oxygenation in brain, skeletal muscle, and other organs by using 2- or 3-wavelength NIR spectrophotometry (Hazeki and Tamura, 1988; Wray *et al.*, 1988) and multicomponent analysis of NIR spectra (Araki and Nashimoto, 1989). Measurement of spatial distribution of this parameter by NIR spectrophotometry, however, has not been established yet. We have measured 2-wavelength NIR projection images of the human forearm and obtained an image showing 2-dimensional distribution of Hb oxygenation state *in vivo*.

MATERIALS AND METHODS

Measurement of NIR projection images. NIR projection images of a test phantom and a volunteer's forearm were measured with a 150 W xenon lamp as a light source and a Pertier-cooled CCD video camera via NIR interference filters centered at 700 and 800 nm (Araki and Nashimoto, this issue). Optical attenuance values of well-oxygenated and deoxygenated human RBC suspension (10% hematocrit) at these wavelengths were measured and a set of two simultaneous equations was derived from these optical attenuance values. Then equations for selective quantification of oxy-Hb, deoxy-Hb, and Hb oxygen saturation (SO_2) were determined as follows:

$$[oxy\text{-}Hb] = 0.56 \times 0 \text{ at } 800 \text{ nm} - 0.95 \times 0 \text{ at } 700 \text{ nm} \qquad 1)$$
$$[deoxy\text{-}Hb] = 1.00 \times 0 \text{ at } 700 \text{ nm} - 0.24 \times 0 \text{ at } 800 \text{ nm} \qquad 2)$$
$$SO_2 = [oxy\text{-}Hb]/([oxy\text{-}Hb] + [deoxy\text{-}Hb]) \times 100 \qquad 3)$$

where [oxy-Hb] and [deoxy-Hb] are relative quantity of oxy-Hb and that of deoxy-Hb.

Test phantom: A test phantom which simulates optical properties of living tissues was used to validate the above mentioned equations. Well-oxygenated and deoxygenated RBC suspension of 50% hematocrit was packed in hematocrit tubes of 1 mm diameter to simulate artery and vein. These hematocrit tubes were soaked in RBC suspension of 3% hematocrit containing Al_2O_3 powder (3-10 microns) of about 2.0 O.D. in an acrylic cell (80 x 80

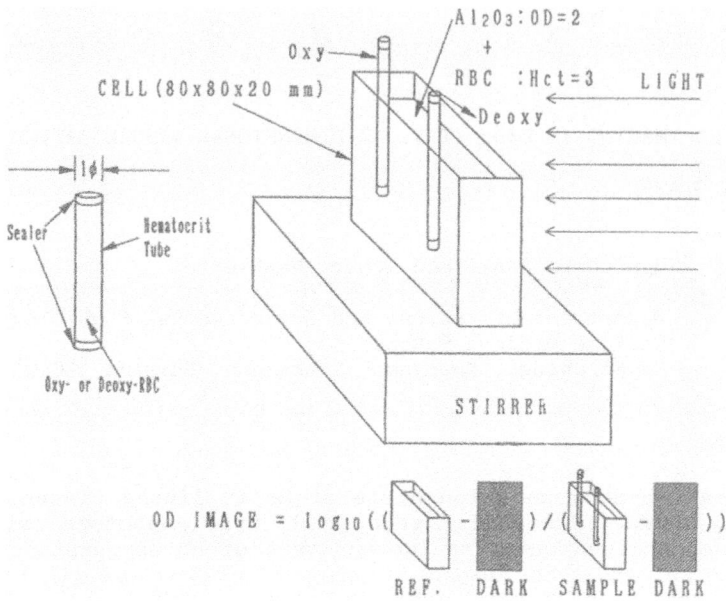

Fig. 1 Block diagram of the test phantom and procedure to obtain O.D. images.

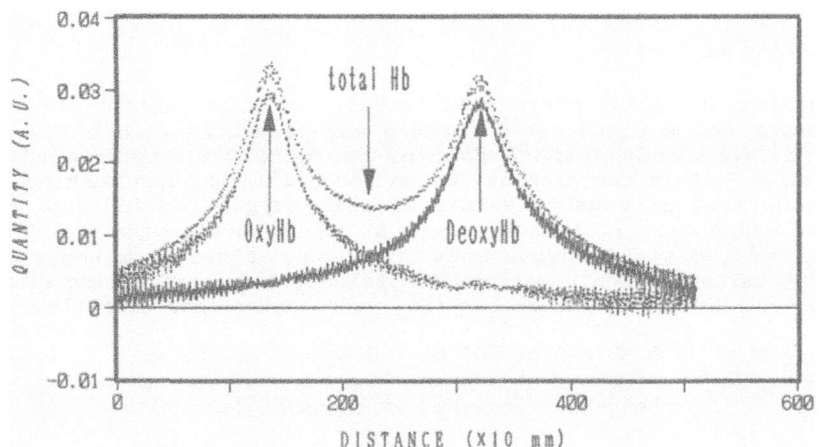

Fig. 2 Cross sections of oxy-, deoxy-, and total-Hb distribution in the test phantom. The left and right hematocrit tubes contained well-oxygenated and deoxygenated RBC suspension.

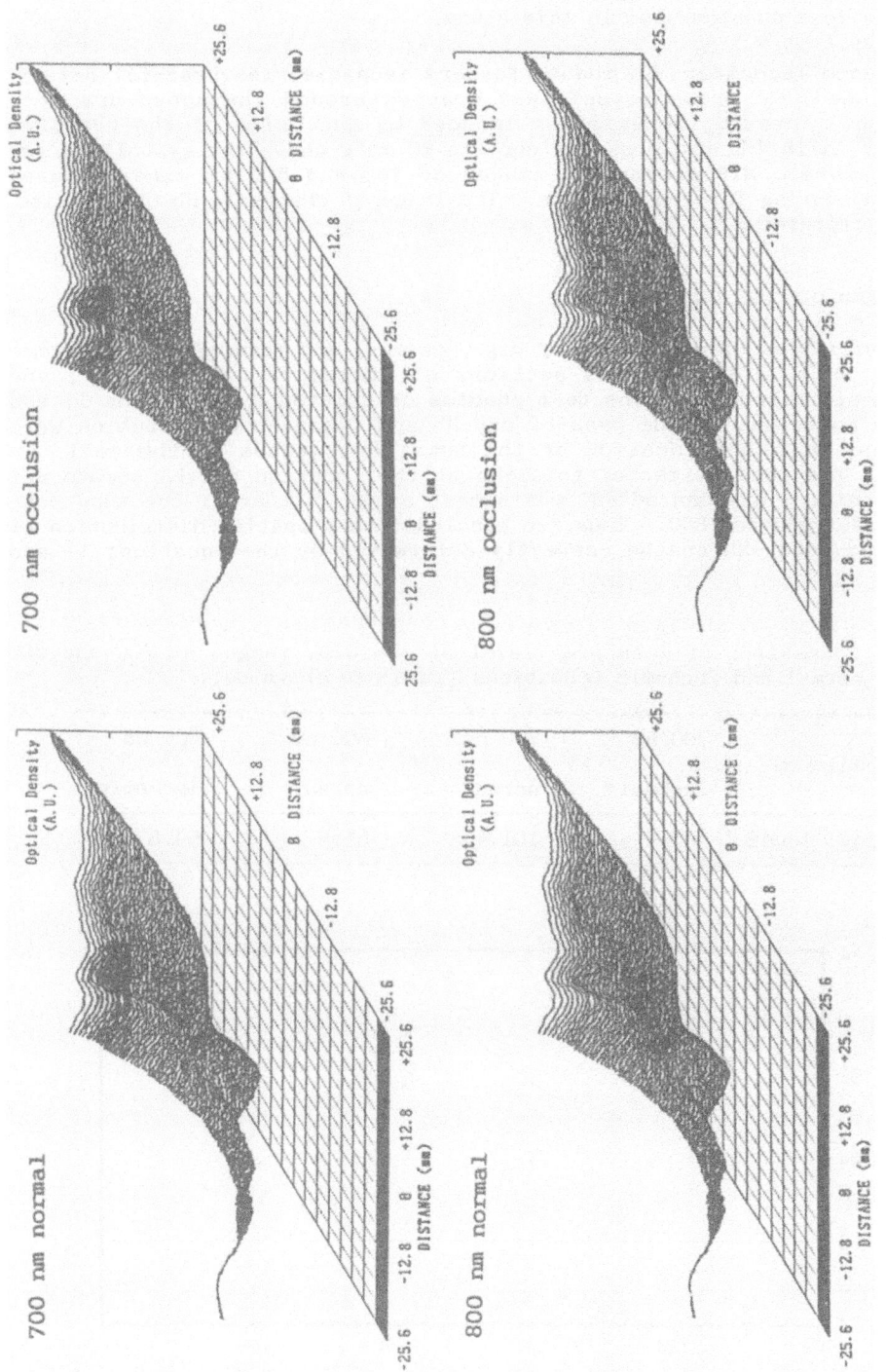

Fig. 3 Three-dimensional representation of O.D. images of the human fore-arm at 700 and 800 nm under normal and ischemic conditions.

x 20 mm). The RBC suspension in the cell was gently stirred with a magnetic stirrer to prevent precipitation of RBC and Al_2O_3 powder. Fig. 1 shows the test phantom used in this study.

Forearm ischemia: To induce forearm ischemia, the brachial artery was occluded. A pressure cuff was wrapped around the upper arm of a volunteer. Forearm ischemia was induced by occlusion of the brachial artery by inflating the pressure cuff to 50 mmHg above the systolic arterial pressure. NIR projection images at 700 and 800 nm were measured before and during forearm ischemia. The Image of change in Hb oxygenation was then calculated.

RESULTS AND DISCUSSION

Quantitative visualization of oxy-, deoxy-, and total-Hbs in the test phantom. Fig. 2 shows cross sections of images of oxy-, deoxy-, and total-Hb distribution in the test phantom calculated by equations 1) and 2). The positions of the peak of oxy-Hb and deoxy-Hb distribution were consistent with the location of the hematocrit tubes in the cell. In addition, the peak height of total-Hb at the location of the hematocrit tube containing oxygenated RBC was almost equal to that of the tube containing deoxygenated RBC. Thus, we concluded that spatial distribution of oxy-Hb and deoxy-Hb can be correctly determined by the equations 1) and 2).

Table 1. Comparison of mean gray level of the O.D. images at 700 and 800 nm under normal and ischemic conditions (in order of value).

condition	700 nm ischemia	>	700 nm normal	>	800 nm normal	>	800 nm ischemia
gray level	131.4		101.4		61.5		54.6

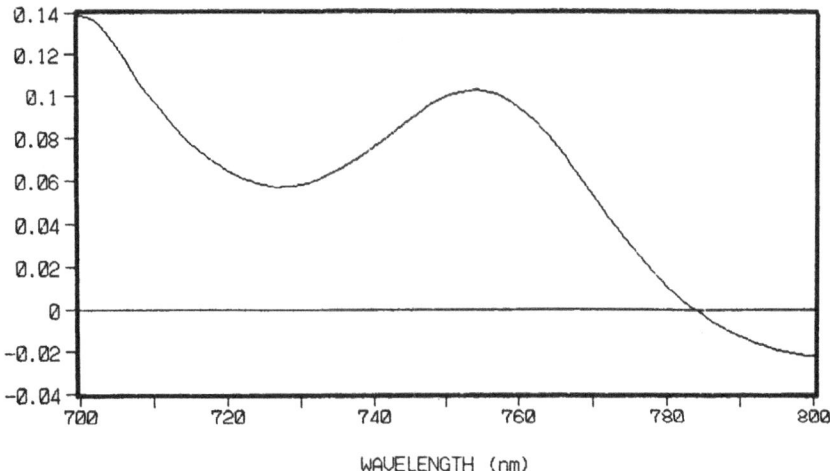

WAVELENGTH (nm)

Fig. 4 Ischemia-induced change in the reflectance NIR spectrum of the human forearm.

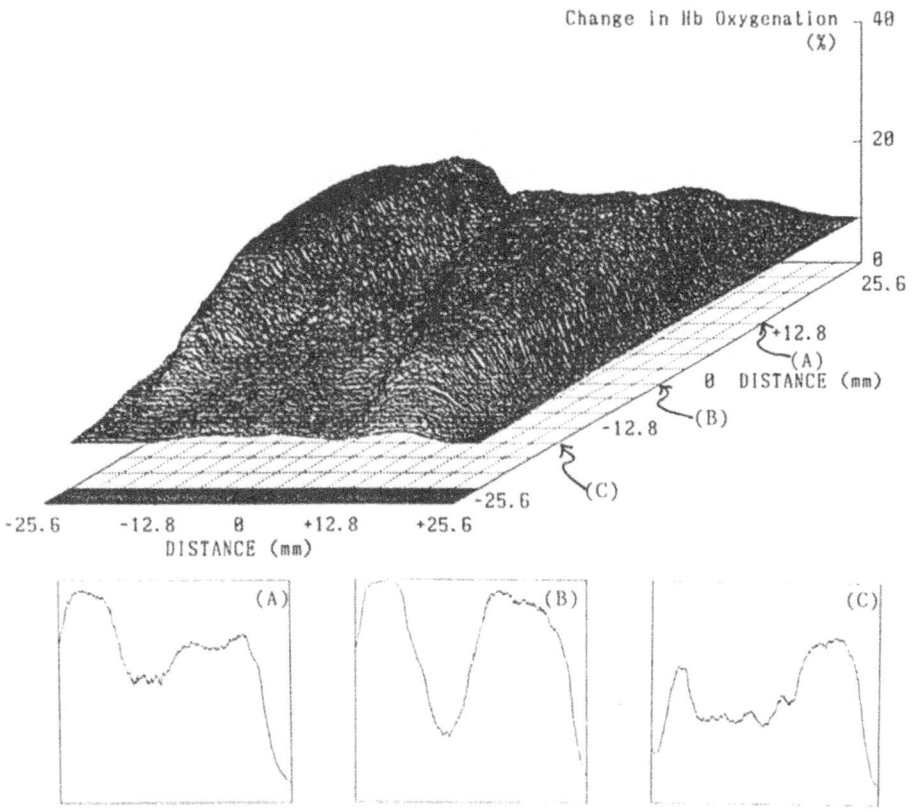

Fig. 5 3-dimensional representation and its cross sections of ischemia-induced change in SO_2 in the human forearm (normal minus ischemic conditions).

NIR projection images of the human forearm under normal and ischemic conditions. Fig. 3 shows 3-dimensional representation of O.D. images of the human forearm at 700 and 800 nm under normal and ischemic conditions. The z-axis was represented as relative optical density because absolute optical density could not be determined due to shading of incident light intensity. Table 1 shows comparison of mean gray level of the O.D. images (in order of value). Occlusion of the brachial artery caused increase in O.D. at 700 nm and decrease in O.D. at 800 nm. The change at 700 nm was about 4-fold larger than the change at 800 nm. Fig. 4 shows ischemia-induced change in the reflectance NIR spectrum of the forearm. The NIR spectra were measured with a scanning NIR spectrophotometer reported previously (Araki and Nashimoto, 1989). The isosbestic point of this difference spectrum was around 785 nm, and shorter than that of oxygenation-deoxygenation of Hb at 805 nm. This shift can be explained by slight decrease in blood volume caused by the forearm occlusion (Hampson and Piantadosi, 1988). This spectral change is consistent with the results shown in Fig. 3 and Table 1, and explained by deoxygenation of Hb and slight decrease in blood volume.

2-dimensional distribution of Hb oxygenation state in the human forearm. By applying the equation 3), we estimated 2-dimensional distribution of Hb oxygenation state in the human forearm under normal and ischemic conditions. Fig. 5 shows 3-dimensional representation and its cross-sections of an image showing spatial distribution of change in Hb oxygenation in the forearm (normal condition minus ischemic condition). The 3-dimensional plot showed a gentle slope, and narrow valleys were observed in the center of the plot. Since these valleys are attributable to the subcutaneous veins, the smaller changes in Hb oxygenation as compared with changes in the other area seem reasonable. However, the change in Hb oxygenation on the edges of the image seemed to be underestimated because of incomplete compensation of incident light shading.

We have demonstrated that 2-dimensional visualization of Hb oxygenation in living tissues up to 50 mm thickness is feasible by using our current experimental setup, although several uncertain errors seem to remain in our results. These errors may arise from light shading due to uneven incident light flux as well as insufficient image quality in terms of signal-to-noise ratio, and insufficient accuracy of the imaging device and A/D conversion. To overcome these problems, x-y scanning of collimated and well stabilized incident light, longer integration time, compensation of linearity of the imaging device, and higher resolution of A/D conversion are required.

SUMMARY

By measuring NIR projection images at 700 and 800 nm, we examined 2-dimensional visualization of Hb oxygenation state in the human forearm of about 50 mm thickness. Equations for selective quantification of oxy-Hb, deoxy-Hb, and SO_2 were derived from *in vitro* experiment, and validity of the equations were checked by using a test phantom which simulates optical properties of living tissues. Forearm occlusion caused increase in 2-dimensional O.D. image at 700 nm, while slight decrease was observed in the image at 800 nm. These results were consistent with changes in NIR spectra caused by the forearm occlusion and explained by Hb deoxygenation and slight decrease in blood volume. Although several uncertain errors seemed to remain in our results, we obtained an essentially reasonable image which demonstrates 2-dimensional changes in Hb oxygenation in the human forearm caused by forearm ischemia.

REFERENCES

Hazeki, O., and Tamura, M., 1988, Quantitative analysis of hemoglobin oxygenation state of rat brain *in situ* by near-infrared spectrophotometry, *J. Appl. Physiol.*, 64:796-802.

Wray, S., Cope, M., Delpy, D.T., Wyatt, J.S., and Reynolds, E.O.R., 1988, Characterization of the near infrared absorption spectra of cytochrome aa3 and haemoglobin for the non-invasive monitoring of cerebral oxygenation, *Biochim. Biophys. Acta*, 933:184-192.

Araki, R., and Nashimoto, I., 1989, Multicomponent analysis of near-infrared spectra of anesthetized rat head: (II) Quantitative multivariate analysis of hemoglobin and cytochrome oxidase by non-negative least squares method, *Adv. Exp. Med. Biol.*, 248:11-20.

Hampson, N.B., and Piantadosi, C.A., 1988, Near infrared monitor human skeletal muscle oxygenation during forearm ischemia, *J. Appl. ol.*, 64:2449-2457.

MONITORING OF THE OXYGEN PRESSURE IN THE BLOOD OF LIVE ANIMALS

USING THE OXYGEN DEPENDENT QUENCHING OF PHOSPHORESCENCE

Marek Pawlowski and David F. Wilson

Department of Biochemistry and Biophysics, Medical School, University of
Pennsylvania, Philadelphia, PA 19104

INTRODUCTION

The oxygen dependent quenching of phosphorescence has proven a powerful method
for measuring oxygen pressure in biological samples (see for examples Vanderkooi et al,
1987; Wilson et al, 1988, Rumsey et al, 1988; Robiolio et al, 1989). The method has a rapid
response time (msec) and can accurately measure oxygen pressure throughout the
physiologically important range (760 Torr down to 10^{-2} Torr). One of it's most promising
applications is to measure the oxygen pressure in the tissue of living animals (Wilson et al,
this book). Addition of the phosphorescent probes to the blood, for example, allows
measurement of the phosphorescence by non invasive optical techniques, thereby making
possible direct calculation of the oxygen pressure in the observed area as the changes occur.
In order to most effectively utilize this new tool it is necessary to develop both an instrument
capable of making the measurements and a thorough understanding of the behavior of the
phosphorescence in this environment. One important aspect of that environment is the
presence of a continuum of oxygen pressures in the field of observation, ranging from the
arteriols to the veinous drainage vessels. Thus the phosphorescence decay curve is expected
to be a summation of many decay curves of unequal initial values.

In this communication we present a new phosphorimeter (based on that of Green et al,
1988) specifically designed for measurement *in vivo*. We have used that phosporimeter to
measure the phosphorescence decay as observed in the blood of the cortex of an anesthetized
piglet, and have designed software for deconvolution of the decay curve to identify regions
of différent oxygen pressures.

METHODS OF APPROACH

Instrumentation: The phosphorimeter has been designed to use light guides both for
the illuminating light and for collecting the tissue phosphorescence. A schematic diagram of
the instrument is given in Figure 1. A flashlamp (EG&G FX249 with a PS450 power
supply) was selected which had a light flash of with a 5 usec decay constant. The flash was
mounted in a metal can with the 1.5 mm arc close to the glass face of the can. This allowed
efficient collection of the light by a collecting lens which imaged the arc on a 9 mm
diameter light guide. There was a 0.8 cm space provided for optical filters between the
collecting lens and the light guide. This light guide was one branch of a bifrucated light
guide which was randomly mixed (salt and pepper) with the light guide which returned the

Oxygen Transport to Tissue XIII, Edited by T.K. Goldstick *et al.*
Plenum Press, New York, 1992

phosphorescence to the photomultiplier. A lens system was placed on the common end of the light guide and focused the excitation light on an approximately 1 mm spot at about 6 mm from the lens. Phosphorescence from the sample was collected by the lens and carried to the photomultiplier by the second half of the bifrucated light guide. At the housing of the photomultiplier, the light was columnated with a lens and passed through a optical filters. These filters were designed to eliminate the excitation light and pass the phosphorescence (such as a long pass cutoff filter with a wavelength at half transmission of 630 nm) before reaching the Hammamatsu R928 photomultiplier.

The electronics and data processing system was similar to that previously described (see Green et al, 1988). A 1 MHz, 12 bit A/D board (MetraByte Corp. DAS-50) was chosen to make use of its 12 bit range while the 1 MHz digitizing rate allowed resolution of phosphorescence decay with lifetimes of down to a few microseconds. The timing sequence of the flash lamp and data acquisition was controlled using a 5 channel counter-timer board with a 1 MHz onboard clock. The microcomputer used was a ZEOS international 80386 microcomputer.

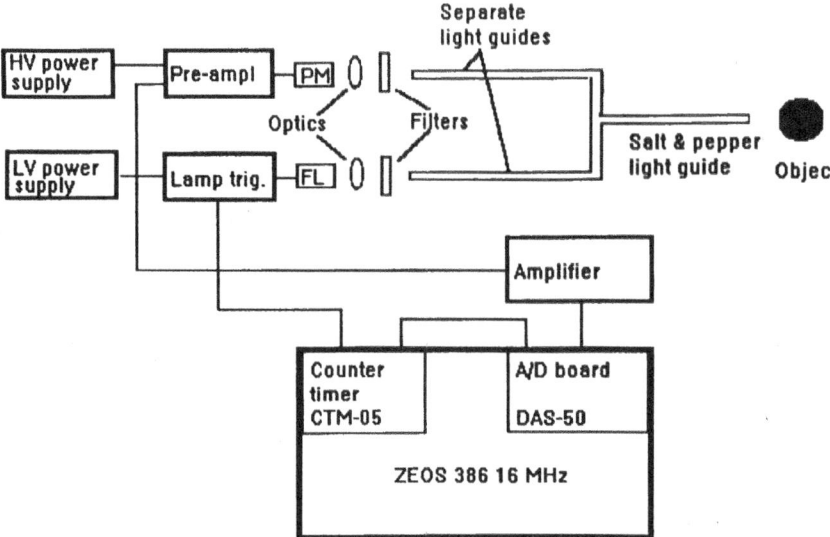

Figure 1. A schematic diagram of the phosphorimeter for tissue phosphorescence measurements.

Software: Data acquisition was controlled by a program written in C language which allowed the operator either manual or automated control of several parameters; timing of the flash, delay between the flash and the beginning of data acquisition, baseline determination, and the rate of collection of digital values (A/D board digitizing rate). Parameters which were set by the operator included the flash intensity, high voltage on the photomultiplier, the gain on the amplifier for the photomultiplier output, the number of data points taken, the number of flashes to be averaged and the type of analysis used (single exponential fit or deconvolution). Setup of the operating parameters was aided by an "oscilloscope" operating mode in which the photomultiplier output from each discharge of the flashlamp was digitized and displayed on the microcomputer monitor. In this mode the effect of changes in the operating parameters could be immediately observed.

Data analysis could be by either fit to a single exponential or by deconvolution to give the number of exponentials and their initial intensity values. The single exponential is determined with a chi-square algorithm which fits the data to a straight line (Bevington, 1969). In the case of multiexponential curves, we have developed a linear prediction (LP) method which requires no *a priori* knowledge of the number of exponentials present. This method assumes that the data is a superposition of a finite number of exponentials with the data points taken at regular intervals. The decay is linearized and solved by a linear least squares procedure using a Hauseholder decomposition (QRD) (Tang et al, 1985) or a singular value decomposition (SVD) (Barkhuijsen et al, 1985) routine. Both QRD and SVD allow determination of the rank of the data matrix (i.e. number of independent equations) and to reject noise by excluding the small eigenvalues usually associated with noise. The ability of the procedure to resolve the number of exponentials is limited only by the noise in the data. Thus the signal to noise ratio is of critical importance in evaluating the data for more than one exponential.

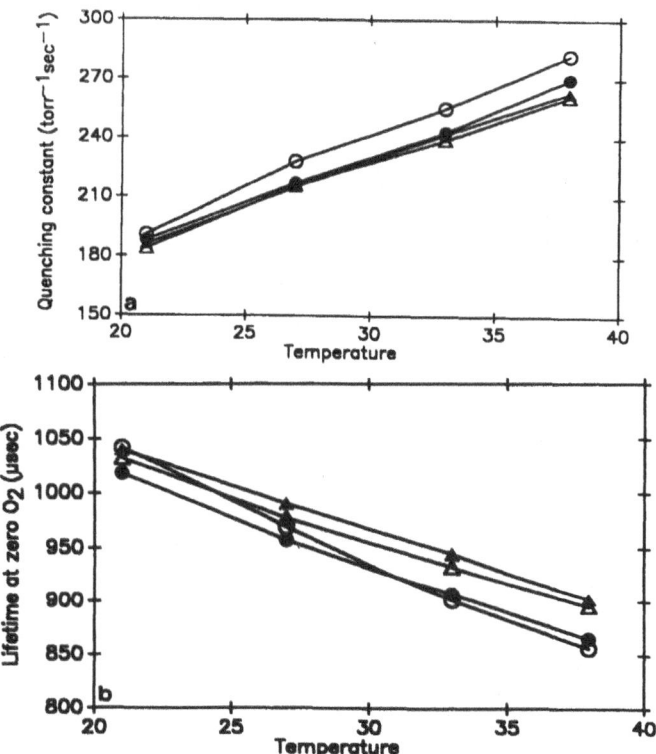

Figure 2a,b. The effect of oxygen on the phosphorescence of Pd-coproporphyrin. The Pd-coproporphyrin was dissolved in a medium containing 150 mM NaCl, 20 mM HEPES, 1% bovine serum albumin, 10 mM glucose and 1 mM EDTA with the pH adjusted to 6.4 (O), 6.8 (●), 7.2 (△) and 7.75 (▲) with Tris base. The probe concentration was 2 uM. Duplicate samples were incubated in open Erlenmeyer flasks while other samples were treated with glucose oxidase and catalase and sealed in vials. The glucose oxidase consumed the oxygen present in the medium to provide an anaerobic medium for measurement of T^0. The samples were placed in a Dubnoff shaking metabolic water bath and equilibrated at the lowest of the indicated temperatures. The phosphorescence lifetimes in each sample was measured, the temperature raised to the next temperature and the measurements repeated. After the measurements at 38^0 the temperature was progressively lowered again. Each point is the average of the measurements made during the increasing and decreasing phases of the temperature cycle (there were no systematic differences between the values).

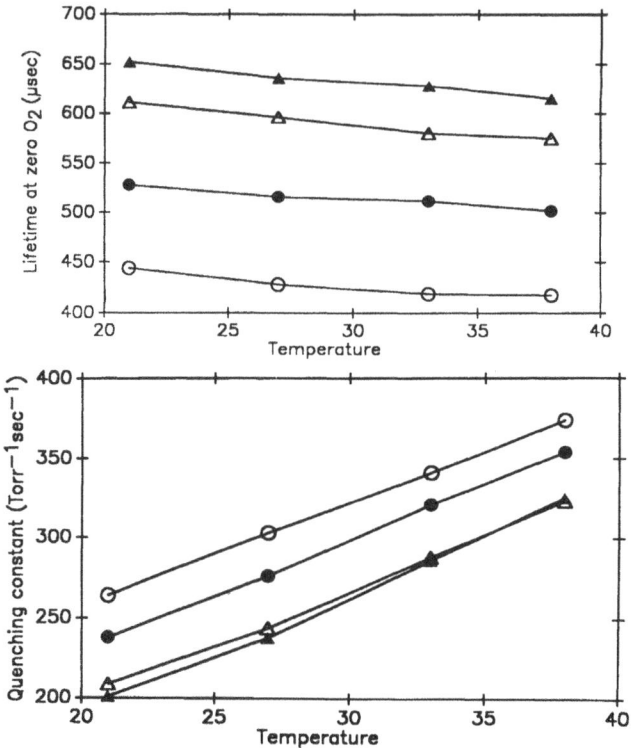

Figure 3a,b. The oxygen dependence of the phosphorescence of Pd-tetra (4-carboxyphenyl) porphyrin. The probe was dissolved and analyzed as described in the Legend of Figure 2.

RESULTS

Calibration of the oxygen probes; dependence on temperature and pH.

The Paladium-porphyrin complexes are currently the best (most readily used) oxygen probes but they have generally been used at room temperature. The probe which has been used for most of the previous studies, Pd-coproporphyrin, is too expensive for extensive experimentation in animals because of the relatively large amounts of required (several mg) per experiment. The oxygen dependence of phosphorescence is described by the Stern-Volmer equation:

$$T^0/T = 1 + k_Q * T^0 * PO_2 \qquad (1)$$

Where T^0 and T are the phosphorescence lifetimes in the absence of oxygen and at an oxygen pressure PO_2. The quenching constant (k_Q) is a second order rate constant related to the frequency of collisions between oxygen and the excited triplet state of the porphyrin and the probability of energy transfer when collision occurs. Thus the oxygen dependence of phosphorescence is described with two constants, k_Q and T^0. Figure 2a presents the measured values of k_Q in 150 mM NaCl media at four different pH values, 6.4, 6.8, 7.2 and 7.75. There is no significant effect of pH on either k_Q or T^0 for this probe over the pH range from 6.8 to 7.75, but from pH 6.4 to pH 6.8 there appears to be a small increase in k_Q. The temperature dependence in k_Q is approximately an increase of 8% per degree. The value of T^0 decreases approximately 7% per degree.

In animal experiments we typically use Pd-tetra (4-carboxyphenyl) porphyrin, which is much more reasonably priced while maintaining good solubility and lifetime characteristics. The pH and temperature dependence of the quenching constant and lifetime at zero oxygen are given in Figure 3. This probe has a pH dependence of k_Q which is negligible between pH 7.2 and 7.8 but changes approximately 3% per 0.1 pH unit between pH 6.2 and 7.2. The value of T^O is more dependent on pH than is that of Pd-coproporphyrin, decreasing approximately 4% per 0.1 pH unit from pH 6.2 to 7.2 but less from 7.2 to 7.8 (about 1% per 0.1 pH unit). On the other hand the value of the quenching constant increases only about 3.7% per degree and the value of T^O decreases less than 0.2% per degree. Most importantly the individual measurements are highly reproducible and independent of most other experimental conditions.

<u>Measurements of the decay of phosphorescence in the cortex of anesthetized animals</u>

The Pd-tetra (4-carboxyphenyl) porphyrin was injected into the blood of an anesthetized newborn piglet and the surface of the cortex observed through a window in the skull (see Wilson et al, this book, for details). The end of the light guide was brought near the surface of the window and observations of the cortex made through the window and superfusate of artificial cerebrospinal fluid. In this configuration the data collected by averaging 10 flashes indicated most of the phosphorescence came from blood with a nearly uniform oxygen pressure. When the lifetime was calculated using a single exponential fit, the instrument and software are capable of obtaining 2-3 measurements per second, suitable for following even the most rapid transients *in vivo*. Figure 4 shows an example of data taken during a brief (2 minute) period of interrupted respiration. The curve resolved into two major components with 2/3 of the signal having lifetimes of about 120 usec and the remaining 1/3 lifetimes of about 304 usec. In the data given in Figure 4, the signal to noise was too great to further resolve the component exponentials of the phosphorescence decay.

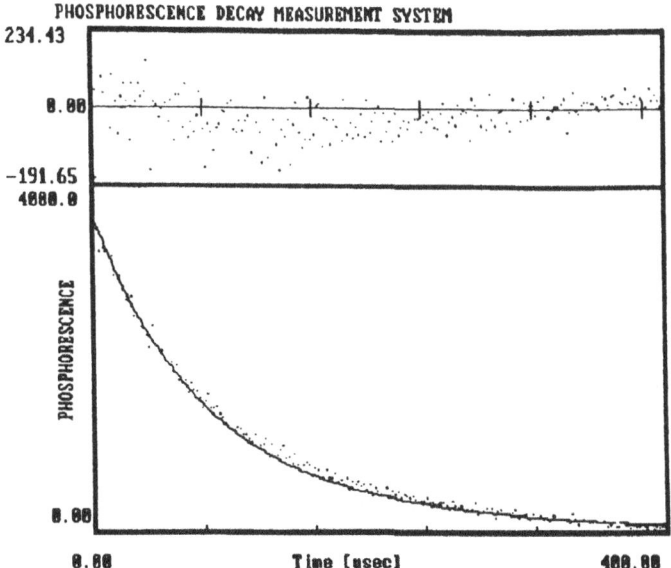

Figure 4. The phosphorescence decay of Pd-tetra (4-carboxyphenyl) porphyrin in the blood of the cortex of a newborn piglet. The probe was injected (25 mg/kg) into the femoral vein as a 1:1 complex with bovine serum albumin. The averaged data from ten lamp flashes and fit by deconvolution (solid line) gave two exponents with lifetimes of 124 usec and 304 usec and initial values of 2596 and 1298 units respectively. The correlation coefficient for the fit is 0.9976. The upper graph is the deviation of the measured intensity values from solid line.

DISCUSSION

The oxygen dependent quenching of phosphorescence is well suited to measurements of the oxygen pressure in tissue. The change in phosphorescence lifetime is large, the amount depending somewhat on the probe of choice, and the lifetime can be measured with high accuracy. The calibration data for the two probes show that Pd-coproporphyrin has a essentially no dependence on pH in the physiological range. This is a useful property when pH changes are expected during the experimental protocol. The quenching constant for Pd-tetra (4-carboxyphenyl) porphyrin, while greater than that of the coproporphyrin, is also pH independent on the alkaline side of pH about 7.0. In the latter case the value of T^O is pH dependent and this must be taken into account. In the Stern-Volmer relationship (equation 1), the value of T^O is important when the oxygen pressure is low and T^O/T is close to one. When this ratio is much greater than 1, equation 1 reduces to:

$$1/T = k_Q * PO_2 \qquad\qquad (2)$$

Thus, for relatively high oxygen concentrations the effect of T^O on the calculations is minimal. For the probes discussed in this paper, the value of T^O is important to calculations of oxygen pressure in the range found in the veinous blood and below. Even in the case of the Pd-tetra (4-carboxyphenyl) porphyrin this effect is becomes important only at pH values more acidic than 7.2. The temperature dependence of both probes is too small to be significant for experiments *in vivo* where temperature variations are small. It should be emphasized that the characteristics of the probes are those of the probe-albumin complex. Serum albumin binds the metaloporphyrins with high affinity and is present in a large molar excess of the probe concentration in the blood. The albumin acts to make the probes soluble in aqueous media, to hold it in a well defined matrix with somewhat restricted access to oxygen and to supress any possible interactions between the probes and other tissue components.

Measurements of the phosphorescence emission from probe in the blood of piglets indicate most of the phosphorescence arises from probe in an environment with an oxygen pressure near that of the veinous blood. This is consistent with the vascular system of the cortex in which the drainage vessels run over the surface of the tissue. This maximizes the exposure of veinous blood both to the excitation light and for collection of the emitted phosphorescence. The deconvolution routines, although still in their early development, show that it will be possible to observe and quantitate regions of relative hypoxia interspersed in normoxic tissue. With the data analysis system we have developed, the ability to resolve multiple exponentials is primarily determined by the signal to noise. This parameter will be substantially improved as the instrumentation is optimized.

SUMMARY

Oxygen dependent quenching of phosphorescence is a powerful new method for measuring oxygen pressure in biological systems (see Vanderkooi et al, J. Biol. Chem. 262 (1987) 5476; Wilson et al, J. Biol. Chem. 263 (1988) 2712). This technology has now been extended to include measurements of the phosphorescent of oxygen probes dissolved in the blood as a complex with albumin. In this communication, we report on a new microcomputer controlled phosphorimeter in which the tissue is illuminated by a flash lamp and the phosphorescence observed through flexible light guides designed to make measurements in regions down to approximately 1 mm in diameter. Measurements of the phosphorescence lifetimes of these probes in regions of tissue necessarily report a distribution of oxygen pressures due to the presence of blood in veins and arteriols as well as that present in the capillaries. Thus the phosphorescence decay is the sum of a continuum of exponentials with different decay constants and initial intensities. The complete phosphorescence decay curve is digitized using a 12 bit, 1 MHz A/D board and deconvoluted using numerical methods to yield a distribution of exponentials contributing to the total decay curve.

REFERENCES

Barkhuijsen, H., de Beer, J., Bovee, W.M.M.J., and van Ormondt, D. (1985) Retrieval of frequencies, amplitudes, damping factors, and phases from time-domain signals using a linear least-squares procedure, J. Magn. Reson. 61: 465-481.

Bevington, P.R. (1969) Data reduction and error analysis for physical sciences (McGraw-Hill, New York) Chapters 1-4.

Green, T.J., Wilson, D.F., Vanderkooi. J.M., and DeFeo, S.P. (1988) Phosphorimeters for analysis of decay profiles and real time monitoring of exponential decay and oxygen concentrations, Analy. Biochem. 174:73-79.

Robiolio, M., Rumsey, W.L., and Wilson, D.F. (1989) Oxygen diffusion and mitochondrial respiration in neuroblastoma cells, Amer. J. Physiol. 256: C1207-C1213.

Rumsey, W.L., Vanderkooi, J.M., and Wilson, D.F. (1988) Imaging of phosphorescence: a novel method for measuring oxygen distribution in perfused tissue, Science 241: 1649-1651.

Tang, J., Lin, C.P., Bowman, M.K., and Norris, J.R. (1985) An alternative to Fourier Transform spectral analysis with improved resolution, J. Magn. Reson. 62: 167-171.

Vanderkooi, J.M., Maniara, G., Green, T.J., and Wilson, D.F. (1987) An optical method for measurement of dioxygen concentration based on quenching of phosphorescence, J. Biol. Chem. 262: 5476-5482.

Wilson, D.F., Rumsey, W.L., Green, T.J., and Vanderkooi, J.M. (1988) The oxygen dependence of mitochondrial oxidative phosphorylation measured by a new optical method for measuring oxygen concentration, J. Biol. Chem. 263: 2712-2718.

NUCLEAR MAGNETIC RESONANCE SPECTROSCOPY AND THE

STUDY OF TISSUE OXYGEN METABOLISM: A REVIEW

Robert Vink

*Department of Chemistry and Biochemistry, James Cook
University of North Queensland, Townsville, Queensland, 4811
Australia*

ABSTRACT

Nuclear magnetic resonance (NMR) techniques are being increasingly utilised as an *in vivo* method to monitor tissue oxygen concentration in various organs. In muscle and heart, proton NMR spectroscopy of myoglobin has been used to calculate local oxygen tension through the oxygen sensitivity of the histidine group signal intensity. Similarly, spin lattice relaxation times of perfluorocarbon emulsions are oxygen sensitive, and this property has been taken advantage of to produce oxygen maps of brain by fluorine NMR imaging. Phosphorus NMR spectroscopy has also been extensively used to monitor bioenergetic state, which under some conditions, is directly related to tissue oxygen tension. This review will focus on these NMR techniques for oxygen determination, and will critically assess their utility for further studies.

INTRODUCTION

It is universally accepted that characterisation of tissue oxygen metabolism is essential to understanding organ biochemistry. To this end, much effort has been directed toward developing techniques that allow monitoring of tissue oxygen levels. These methods have included arterial-venous differences in haemoglobin saturation, the measurement of metabolite ratios that are sensitive to oxygen tension (eg., lactate/pyruvate), electrode methods, and optical methods such as phosphorescence detection and near infrared absorption spectrophotometry (1-4). The optical methods have the advantage of being relatively noninvasive, and as such may be regarded as *in vivo* techniques. Nonetheless, they have their limitations including injection of potentially toxic phosphorescent compounds, and difficulties in resolving overlapping infrared absorption spectra of relevant proteins. Consequently, the search for a clinically useful noninvasive monitor of tissue oxygen tension is ongoing.

One technique that has received increasing attention as an oxygen probe is nuclear magnetic resonance (NMR). Nuclear magnetic resonance is a totally noninvasive, nondestructive technique receiving increasing application to the study of organ biochemistry (5-6). Most *in vivo* studies have centred around the phosphorus nuclei which permit the continuous monitoring of high energy phosphates such as phosphocreatine (PCr), adenosine triphosphate (ATP), and inorganic phosphate (Pi), in addition to determinations of pH and free magnesium concentration. More recently, proton NMR spectroscopy has been used to characterise the role of lactate and amino acids in pathologic situations (6). Both phosphorus and proton NMR studies have been applied in an effort to determine tissue oxygen concentration. In addition to NMR spectroscopy, NMR imaging has been used in an effort to map tissue oxygen concentration. This review will focus on some of the more recently developed NMR techniques currently being used in the characterisation of tissue oxygen metabolism. General principles of NMR and it's application to the study of biological systems will not be discussed here, and readers are referred to some excellent reviews elsewhere (7-8).

MYOGLOBIN

Myoglobin is a well characterised muscle protein which serves both to store and facilitate the movement of oxygen within muscle tissue. Under oxygen replete conditions, the proton NMR signals from oxymyoglobin are obscured *in vivo* by a very intense water signal. However, when myoglobin converts to it's deoxy state, the heme iron of myoglobin becomes paramagnetic and shifts these proton resonances away from water to a position clearly visible in the proton NMR spectrum. The more myoglobin that is present in the deoxy state, the more intense the shifted signal. The intensity of the deoxymyoglobin can therefore be used as a marker of local tissue oxygenation. This property of myoglobin, first reported by La Mar and colleagues (9), has been exploited recently by Jue and Anderson (10) in isolated rat hearts, and by Wang *et al.* (11) in human skeletal muscle. Spectra with adequate signal to noise were obtained in time blocks ranging from 1 to 4 minutes, at frequencies of either 77 MHz or 300 MHz. In both tissues, under normal oxygenated conditions, there was no observable signal from deoxymyoglobin. Decrease in oxygen tension resulted in the development of a clearly distinguishable deoxymyoglobin peak. In the skeletal muscle experiments, Wang *et al.* (11) demonstrate that application of a pressure cuff to prevent blood flow results in an increase in deoxymyoglobin up until a maximum amplitude was achieved at 6 min. At this time, total myoglobin concentration in human skeletal muscle was calculated to be 340 μM, which the authors claim is consistent with previous reports. The authors also reported that their calculated initial resting oxygen pressure was 40 Torr with a resting respiration rate of 80 μM/min.

The sensitivity of the technique has been estimated to be sufficient to detect oxygen levels in the range of 10^{-7} to 10^{-5} M, corresponding to 10% to 90% O_2 saturation. While the technique has neither the sensitivity of the phosphorescence technique, or the time resolution of near infrared spectrophotometry, it has the advantages of being localised, unambiguous, and safe for clinical use. Furthermore, there does not seem to be any contributions from a deoxyhaemoglobin signal, which some would consider an advantage. In contrast, Wang *et al.* (11) believe that a deoxyhaemoglobin signal would provide information concerning oxygen transport from blood to muscle. There are, however, disadvantages to the technique. Firstly, it is obviously restricted to muscle. Secondly, there are some technical limitations. In addition to a requirement for excellent water suppression, the quantitation of NMR signals has been historically extremely difficult, particularly in *in vivo* surface coil experiments. Thus, quantitation of deoxymyoglobin is no trivial matter. In this respect, contributions from the extracellular myoglobin has to be established, particularly if the technique is to be used in pathologic situations.

PROTON METABOLITE RATIOS

The determination of oxygen tension from lactate/pyruvate ratios has been extensively applied in a number of tissues. However, whereas the classical technique depends upon tissue extraction principles, proton NMR provides an avenue whereby this ratio can be continuously and noninvasively monitored in a localised volume of tissue. The difficulty in determining this ratio by NMR, however, stems from the fact that many metabolite peaks in a proton spectrum overlap, thus making identification and quantification difficult. This difficulty has been largely overcome by the development of procedures that select out individual metabolites such that they no longer have any interference from neighbouring metabolites. These editing, or selective excitation schemes have been used with great success in monitoring lactate and certain amino acids (12). However, to date, no selective excitation sheme for pyruvate has been devised. Nonetheless, Jue *et al*. (13) have proposed a method whereby pyruvate can be monitored in a proton spectrum. These authors demonstrate that pyruvate in a proton NMR spectrum has contributions from a limited number of metabolites. These overlapping metabolites can be filtered out of a proton spectrum by selectively exciting these resonances and subtracting them from the composite spectrum. Using a perfused liver system, the feasibility of such a technique was demonstrated. The advantages of such a method lie in the general properties of NMR: the technique is noninvasive, provides kinetic information about the changes in lactate/pyruvate ratio, there are no quantitation difficulties when using ratios, and it is potentially highly localised. The disadvantages include that efficient water suppression is critical, and the editing scheme employed to edit out non-pyruvate metabolites is time consuming and technically tedious. Nonetheless, the technique is worthy of further pursuit.

PROTON IMAGING

A number of reports have suggested that the relaxation times of proton NMR images are sensitive to oxygen concentration and are therefore a reflection of local oxygen tension (14-15). However, a number of factors are known to affect proton relaxation times. In a controlled study, Grant *et al*. (16) have demonstrated that proton relaxation times of cerebrospinal fluid exposed to air for different periods of time undergoes changes in relaxation time. However, the changes in relaxation time were associated not only with differing oxygen concentration, but also with changing CO_2 concentration and pH. Clearly, more studies are required to establish which physiologic variables are responsible for altered proton relaxation times, and their relative contributions, before the technique shows promise as an oxygen probe.

PERFLUOROCARBON EMULSIONS

Originally developed as oxygen carrying blood substitutes,

perfluorocarbon (PFC) emulsions have been gaining increased attention as a method by which to determine tissue oxygen tension. These compounds are rich in [19]fluorine atoms, an isotope virtually absent from biological tissues. [19]F is, however, readily amenable to NMR studies with an 83% sensitivity relative to proton. Therefore, provided sufficient ^{19}F signal can be detected after PFC is injected, NMR imaging of fluorine is a distinct possibility. Of particular interest is the fact that spin-lattice relaxation rates (T_1) of these perfluorocarbon emulsions are affected by the paramagnetic properties of oxygen such that the relationship between $1/T_1$ and oxygen tension is linear (17-20). Thus, T_1 relaxation times may be determined from the ^{19}F images and vascular oxygen tension calculated. Eidelberg et al. (18-19) have recently applied this technique using the PFC perfluorotributylamine to obtain ^{19}F NMR images of cat brain, and with a calibration curve for this PFC, calculated an oxygen tension map of the image. These authors found that mean oxygen tension ranged from 82 mm Hg in the vein of Galen to 186 mm Hg in the carotid and rete. Mean vascular oxygen tension in the cerebral cortex was 154 mm Hg.

The authors also point out several difficulties with the technique. The low ^{19}F abundance after PFC injection necessitates long imaging times and large intravenous infusions. Together with chemical shift artifacts that need to be suppressed, the experimental time is quite long. Furthermore, being restricted to the vasculature, the PFC calculated oxygen tension tends to overestimate tissue oxygen tension. Nonetheless, the cortical values derived by ^{19}F NMR are consistent with values obtained by invasive techniques. In a model of focal ischemia, the authors have demontrated that the technique does hold promise as a tool for localising regions of anoxia and for obtaining qualitative measurements (19).

PHOSPHORUS NMR

Phosphorus NMR spectroscopy has long been recognised as a noninvasive, nondestructive approach to cell bioenergetics (21). The relationship between cell bioenergetic state and oxygen supply has since been pioneered by Chance and colleagues (22-24) who have applied phosphorus NMR to the study of various organs in an attempt to monitor localised oxygen tension. These authors have shown in detail that since phosphorus NMR can monitor most of the high energy phosphates in tissue, the relationship between the phosphorus NMR spectrum and oxygen tension can be described with the use of the equation for ATP synthesis by oxidative phosphorylation.

$$3ADP + 1/2O_2 + 3P_i + NADH + H^+ \leftrightarrow 3ATP + NAD^+ + H_2O$$

Using Michaelis-Menten kinetics, this relationship can be described as

$$\frac{V}{V_{max}} = \frac{1}{1 + \dfrac{K_1}{ADP} + \dfrac{K_2}{O_2} + \dfrac{K_3}{P_i} + \dfrac{K_4}{NADH} + \dfrac{K_5}{H^+}}$$

In most instances, $P_i \gg K_3$, $NADH \gg K_4$, and $H^+ \gg K_5$. Therefore, these terms can be eliminated and the equation simplified to:

$$\frac{V}{V_{max}} = \frac{1}{1 + \dfrac{K_1}{ADP} + \dfrac{K_2}{O_2}}$$

This simplified equation predicts that as oxygen tension decreases, ADP concentration will increase, a prediction that has been extensively verified. The difficulty, however, is that ADP concentration *in vivo* is on the order of 20 μM which is well below the limit of NMR detection. Chance and colleagues (22) then introduce the creatine kinase equation to calculate ADP concentration.

$$PCr + ADP + H^+ \leftrightarrow ATP + Cr$$

where the equilibrium can be described as

$$K_{obs} = \frac{[ATP][Cr]}{[PCr][ADP]}$$

and $K_{obs} = K_{CK} [H^+]$. Substituting the creatine kinase equilibrium for ADP, we obtain

$$\frac{V}{V_{max}} = \frac{1}{1 + \dfrac{K_1 \, K_{obs}}{ATP} \dfrac{PCr}{Cr} + \dfrac{K_2}{O_2}}$$

This equation predicts that provided ATP and K_{obs} are constant, then PCr/Cr will decrease with a decrease in oxygen tension. Furthermore, Chance and colleagues (23) substitute Pi for Cr on the basis that coupled hydrolysis of PCr and ATP will give equimolar amounts of Cr and Pi. Thus, PCr/Pi obtained by phosphorus NMR may provide information on oxygen metabolism in tissue.

There are several assumptions implicit in using PCr/Pi ratios as an indicator of tissue oxygen tension. Firstly, the tissue must be in a steady state with ATP remaining constant. This is easily confirmed from the NMR spectra. Secondly, the Pi pool must equal the Cr pool. While this may be true for some tissues (eg. muscle), this is not the case in others (eg. brain). Nonetheless, the changes in PCr/Cr through PCr hydrolysis will equal PCr/Pi, therefore substitution of Pi for Cr in this respect seems appropriate. Finally, K_{CK} must remain constant. This would seem a reasonable assumption since only changes in pH and free magnesium can potentially affect the equilibrium constant, and pH is already incorporated into the K_{obs} term. However, recent studies have shown that free magnesium can change significantly (25), with resultant affects on the creatine kinase equilibrium constant (26). Despite this apparent drawback, the use of PCr/Pi ratios as a marker for tissue oxygen status has proven remarkably successful. In one of the early descriptions of it's utility as a bioenergetic marker, Chance and colleagues (22) demonstrated in mitochondria that PCr/Pi ratio and cytosolic phosphorylation potential could be used as indicators of ATP synthesis or respiration. Later studies by these authors and others (5,6,23,24) demonstrated similar results in a variety of tissues, including brain. Finally, when taking free magnesium into account, Vink *et al.* (26) have confirmed that PCr/Pi ratios are indeed a reflection of cytosolic phosphorylation potential. One can therefore conclude that PCr/Pi ratios are a reflection of tissue oxygen metabolism.

While phosphorus NMR offers many advantages, there are a number of disadvantages. Firstly, the phosphorus nuclei have a very low sensitivity relative to proton and spectra can only be obtained with a time resolution of minutes. Secondly, the derived equations can only be related to oxygen tension if the tissue is in a steady state. Nonetheless, the technique has proven very useful in studies of tissue oxygen metabolism, and has even been proposed as a sensor of oxygen concentration in brain (27).

CONCLUSION

This review has critically examined a number of recently developed NMR techniques for the determination of tissue oxygen tension. While a number of the techniques appear promising for future studies, at the present time the techniques have serious limitations that prevent routine application to quantitation in *in vivo* systems. Nonetheless, as a noninvasive, qualitative technique, NMR is a technique that offers distinct advantages. With the field of NMR expanding as rapidly as it has in the past decade, there is no doubt that the technology will yet make significant contributions to the study of oxygen transport to tissues.

References

1. R. Rakusan, Oxygen in the Heart Muscle, Thomas, Springfield (1971).
2. M. Tamura, N. Oshino, B. Chance, and A.I. Silver, Arch. Biochem. Biophys. 191:8 (1978).
3. R.S.C. Cobbold, in Transducers for Biomedical Measurements: Principles and Applications, pp 380-398, Wiley, New York (1974).
4. B. Chance, S. Nioka, J. Kent, K. McCully, M. Fountain, R. Greenfeld, and G. Holtom, Anal. Biochem. 174:698 (1988).
5. G.K. Radda and D.J. Taylor, Int. Rev. Exp. Path. 27:1 (1985).
6. S.M. Cohen (ed), Physiological NMR Spectroscopy: From Isolated Cells to Man, Ann. N.Y. Acad. Sci. 508 (1987).
7. D.G. Gadian, Nuclear Magnetic Resonance and Its Applications to Living Systems, Clarendon Press, Oxford (1982).
8. T.L. James and A.R. Margulis (eds), Biomedical Magnetic Resonance, Radiology Research and Education Foundation, San Francisco (1984).
9. D.J. Livingston, G.N. LaMar, and W.D. Brown, Science 220:71 (1983).
10. T. Jue, and S. Anderson, Magn. Reson. Med. 13:524 (1990).
11. Z. Wang, E.A. Noyszewski, and J.S. Leigh, Magn. Reson. Med. 14:562 (1990).
12. H.P. Hetherington, M.J. Avison, and R.G. Shulman, Proc. Natl. Acad. Sci. U.S.A. 82:3115 (1985).
13. T. Jue, Y. Chung, and R.G. Shulman, Abstr. Soc. Magn. Res. Med. 1:276 (1987).
14. S.F. Akber, Eur. J. Radiol. 9:56 (1989).
15. S. Ogawa, T.-M. Lee, A.S. Nayak, and P. Glynn, Magn. Reson. Med. 14:68 (1990).
16. R. Grant, B. Condon, S. Moyns, J. Patterson, and G. Teasdale, Magn. Reson. Med. 6:397 (1988).
17. J. Taylor, and C. Deutsch, Biophys. J. 53:227 (1988).
18. D. Eidelberg, G. Johnson, D. Barnes, P.S. Tofts, D. Delpy, D. Plummer, and W.I. McDonald, Magn. Reson. Med. 6:344 (1988).
19. D. Eidelberg, G. Johnson, P.S. Tofts, J. Dobbin, H.A. Crockard, and D. Plummer, J. Cereb. Blood Flow. Metab. 8:276 (1988).

20. J.E. Fishman, P.M. Joseph, M.J. Carvlin, M. Saadi-Elmandjra, B. Mukherji, and H.A. Sloviter, Invest. Radiol. 24:65 (1989).
21. B. Chance, S. Eleff, and J.S. Leigh, Proc. Natl. Acad. Sci. U.S.A. 74:7430 (1980).
22. L. Gyulai, Z. Roth, J.S. Leigh, and B. Chance, J. Biol. Chem. 260:3947 (1985).
23. B. Chance, J.S. Leigh, J. Kent, and K. McCully, Fed. Proc. 45:2915-2920 (1986).
24. B. Chance, J.S. Leigh, S. Nioka, T. Sinwell, D. Younkin, and D.S. Smith, Ann. N.Y. Acad. Sci. 508:309 (1987).
25. R. Vink, T.K. McIntosh, P. Demediuk, M.W. Weiner, and A.I. Faden, J. Biol. Chem. 263:757 (1988).
26. R. Vink, A.I. Faden, and T.K. McIntosh, J. Neurotrauma 5:315 (1988).
27. B. Chance, J.S. Leigh, and S. Nioka, Abst. Soc. Magn. Reson. Med. 5:1368 (1986).

ON-LINE OXYGEN UPTAKE MEASUREMENT (V̇O₂): A COMPUTER FEED-BACK CONTROLLED REBREATHING CIRCUIT FOR LONG TERM OXYGEN UPTAKE REGISTRATION

A.P.K. Verkaaik*, W. Erdmann, G. van Dijk** and B. Westerkamp**

*Dept. of Anesthesiology, Erasmus University Rotterdam, The Netherlands and
**Research Laboratories, The Netherlands

INTRODUCTION

One of the most wanted basic parameters in all study areas concerning oxygen transport to tissue is on-line registration of total oxygen consumption, a so far generally not available parameter that gives insight into the general aspect of any existing or occurring impairment of oxygen delivery to the tissue. With adequate oxygen transport capacity of the blood and normal cardiac function reduction of oxygen consump-

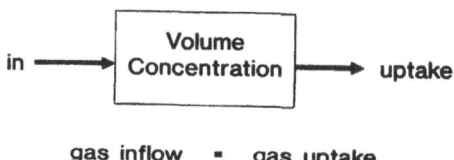

gas inflow ▪ gas uptake

Fig. 1
General principle of closed circuit measuring devices.

tion as a sign of tissue oxygenation impairment is only valuable if disturbance of all functions influencing oxygen uptake in the lungs are excluded: inspiratory oxygen concentration and lung function parameters. The first onset to develop a suitable system in the described direction has been made as early as 1915 by the physiologist Jackson, but further development to a system to fulfil described demands never occurred. A closed rebreathing spirometer system has been developed: once the experimental animal (or patient) is connected to the system inspiratory oxygen concentration is computer feed-back controlled to preset values, CO_2 is on-line measured in the connection piece to the evaluated subject and further on fully absorbed from respiration gas. Lung function is continuously controlled by on-line monitoring of volume, flow and respiratory pressure.

Oxygen Transport to Tissue XIII, Edited by T.K. Goldstick *et al.*
Plenum Press, New York, 1992

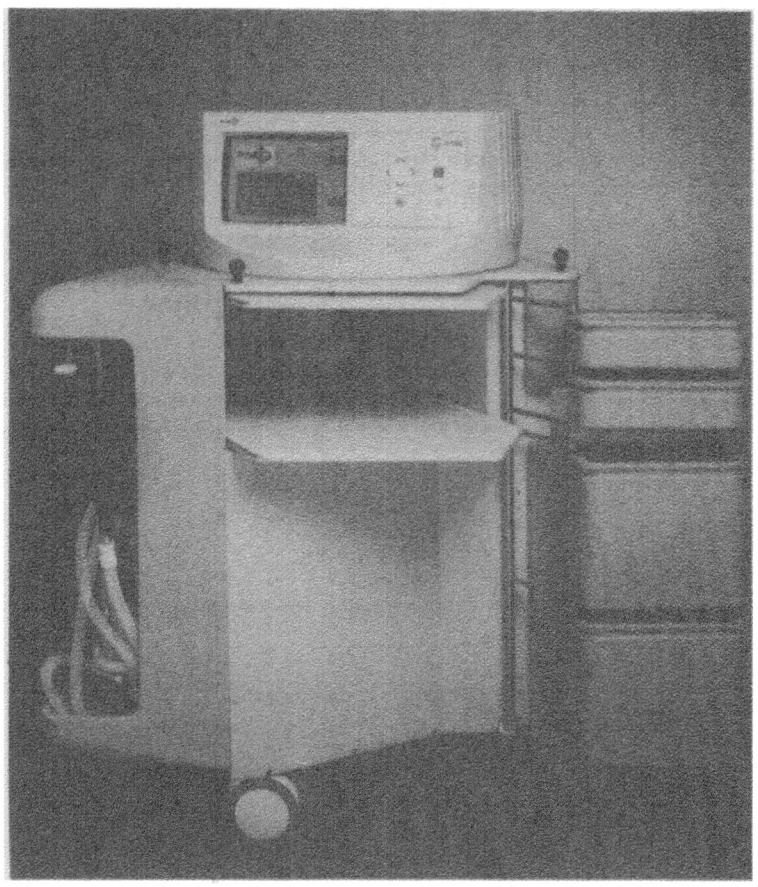

Fig. 2

System design: triangle form with connection to patient respectively experimental animal in the nose. The service-display screen and keyboard is turnable to all three sides. The membrane chambers and the electronic hard-ware are built into the closed bottom part of the apparatus.

METHODOLOGY

The basis of measurements via a closed rebreathing system is the principle that fresh gas delivered to the system equals gas uptake by the subjects. Furthermore, to guarantee exact measurement of O_2-consumption volume consistency is a prerequisit. If measurements are performed with two different gases, one gas can be regulated via a preset concentration while the other is regulated via a preset system volume (Fig. 2).

A system has been designed combining feed back controlled system volume consistency (N_2 or N_2O) and oxygen concentration feed back control to preset values.

System design: the system is a totally closed circuit in which the respiration gas is rotated unidirectional with 70 l/min (no rebreathing of expired air). The gas passes through a membrane chamber included into the circuit and any displacement of the metal membrane (fixed freely moveable in a rubber rolling seal) which separates the internal patient side circuit from the external part of the chamber, is capacitively monitored (volume and flow) while respiratory pressure is on-line monitored by a pressure transducer (static and dynamic compliance). Oxygen is measured paramagnetically in the inspiratory part of the circuit and exactly adjusted to preset values via feed-back control of an electronic frequency valve connected to pressurized oxygen, the oxygen inflow ($\dot{V}O_2$) is registered. An infrared spectrometer measures CO_2 (integrated with the flow: CO_2-production) via sampling from the connection tube. Via changes of air pressure in the external part of the membrane chamber the membranes can be moved if arteficial ventilation is wanted, according to a predrawn ventilation curve on the computer screen (computer control of two inspiratory valves and one expiratory electronic valve) (Fig. 3).

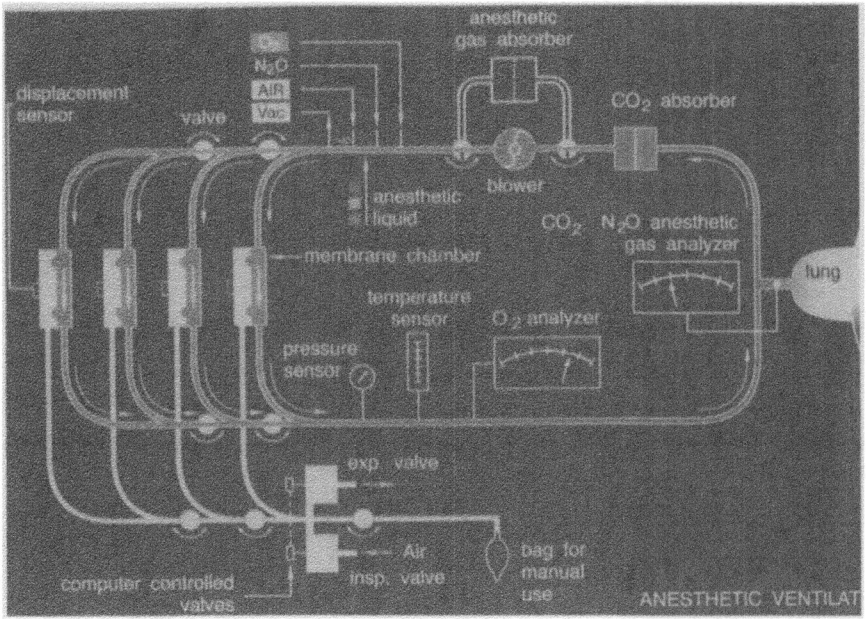

Fig. 3

Schematic description of the Rotterdam Physio Flex for application in physiology, lung physiology, anesthesia and intensive care. At the end of each expiration whether spontaneous or artificial, the metal membranes in the flow through membrane chambers are brought into the original position by nitrogen respectively nitrous oxide inflow, a feed back controlled electronic oxygen valve regulates inspiratory oxygen concentration to the desired preset value. The number of membrane chambers included into the circuit is related to the tidal volume, thus compressible circulating volume is kept as small as possible. An infrared spectrometer measures respiratory gases. The system volume is rotated by a blower with 70 l/min (> maximal inspiratory flow) to avoid possibility of rebreathing.

During spontaneous respiration in volunteers breathing normal air (20% O_2, 75% N_2 and 5% H_2O as the closed system is fully water saturated), N_2 inflow exactly determines leakage due to non-adequate closure of the subject's mouth piece or inadequate cooperation of the subject. Known leakage is taking into account and exact oxygen uptake measurement is the result. Furthermore, the subject (or experimental animal) is breathing normal oxygen concentration. This has been the great drawback of conventional closed rebreathing respirometers where measurements are performed with varying and unphysiological high oxygen concentrations.

The purpose of this study was to compare measured oxygen consumption to so far known and generally accepted values due to calculations based on body weight and body surface area.

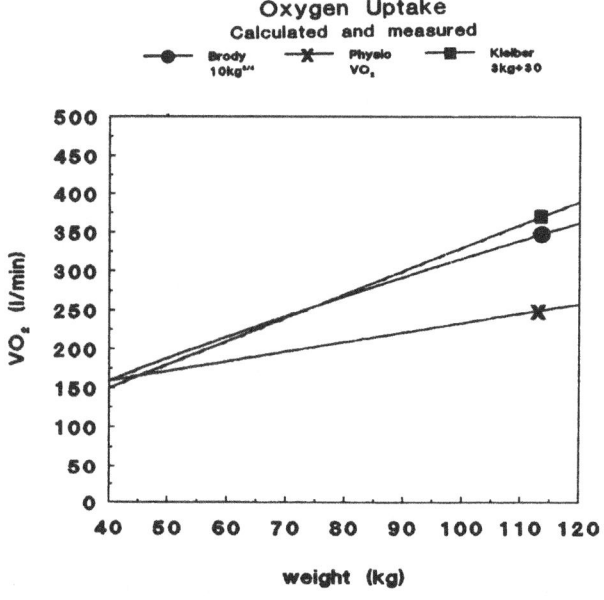

Fig. 4

Calculated (Brody 1945, Kleiber 1947) oxygen consumption and oxygen consumption measured with the Rotterdam Physioflex[R]

RESULTS

Oxygen consumption of 100 normal subjects (40-120 kg body weight) in rest has been determined using the Physio Flex (physiological flexible closed circuit system). A linear relationship is found. Compared to Brody (1945, oxygen consumption = 10 x $kg^{3/4}$ ml O_2/min) and Kleiber (1947, oxygen consumption = 3 x kg + 30 ml O_2/min), the measured oxygen consumption is slightly lower with increasing differences in the upper weight group. E.g., the measured values are 30% lower than so far assumed in patients with 120 kg (Fig. 4).

Oxygen consumption related to length reveals a linear degression with increasing length. However, comparison could only be made between 120 and 160 cm of length, in the larger subjects body weight was always higher and comparison was therefore invalid (Fig. 5).

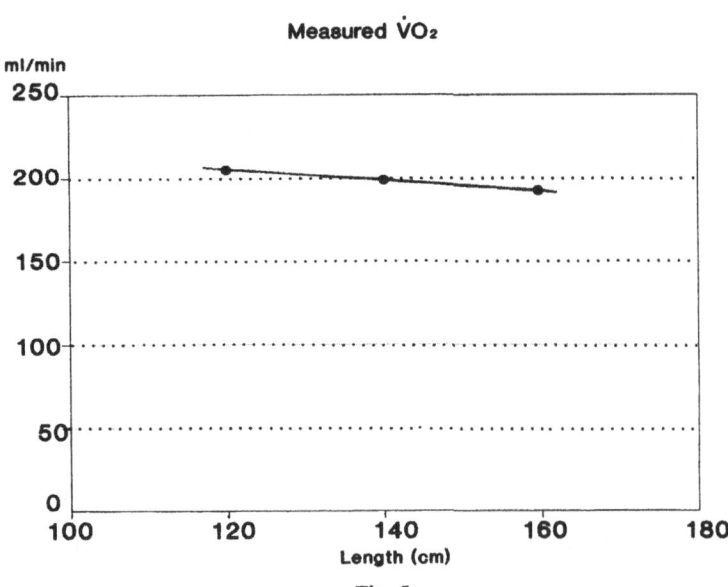

Fig. 5

Oxygen consumption (Physio Flex) with increasing subject length.

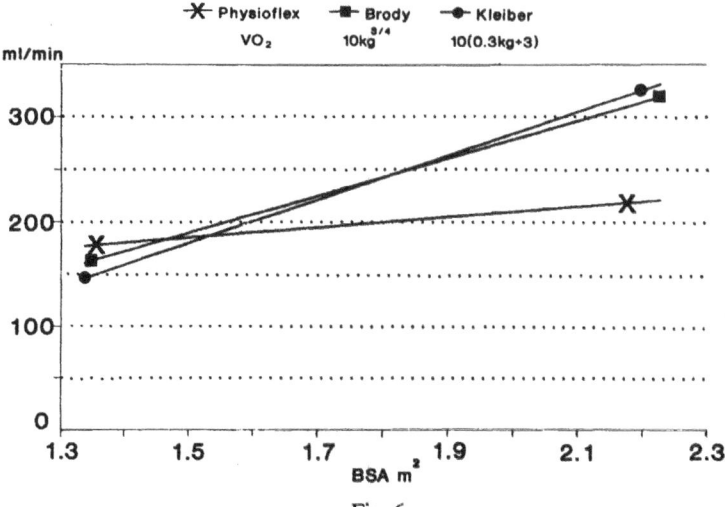

Fig. 6

Calculated and measured oxygen consumption with changing body surface area ($0.007184 \times \text{weight}^{0.425} \times \text{length}^{0.725}$).

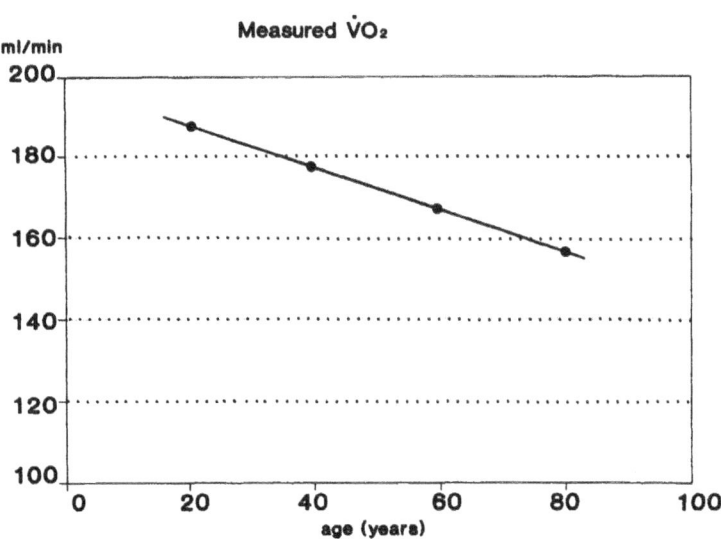

Fig. 7

Measured oxygen consumption (same weight) with increasing age.

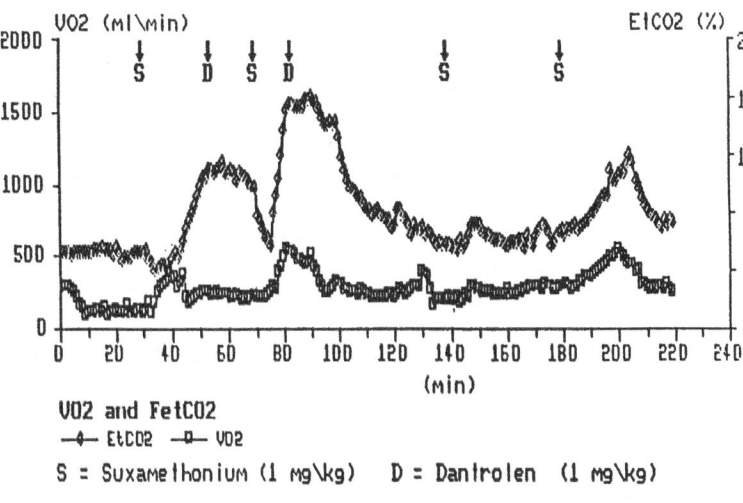

S = Suxamethonium (1 mg\kg)   D = Dantrolen (1 mg\kg)

Fig. 8

Suxamethonium (1 mg/kg) induced hyperthermia in a Pietrain pig (basic anesthesia: O_2, N_2O, halothane). Lower curve: early increase of oxygen consumption before endexpiratory CO_2 increase (upper curve), and temperature increase (not noted). Dantrolene[R] 1 mg/kg stops the development of a fully developed hyperthermic crisis but the dose is not high enough to prevent a reaction to a second Suxamethonium injection. A second cummulating Dantrolene injection of 1 mg/kg has a protective effect for the following hour (see third Suxamethonium injection).

In the following, oxygen consumption was correlated to body surface area in m^2 (0.007184 x kg body weight$^{(0.425)}$ x body length cm$^{(0.725)}$).

According to the classical calculation, oxygen consumption increases linearly with surface area, the measured values, however, only show a very small increase with increasing body surface area, with the result that subjects with a body surface area of $1.3m^2$ consume 134 ml O_2/min per m^2, while subjects with $1.8m^2$ body surface area have only a consumption of 111 ml O_2/min per m^2 (Fig. 6).

Another interesting result is reduction of oxygen consumption with increasing age, whereby the weight in the four groups (± 20, ± 40, ± 60, ± 80) did not differ significantly. Oxygen consumption reduced from 187.0 ml O_2/min in 20 years old to 156 ml O_2/min in 80 years old patients; thus a 17% reduction with linear relationship to age is seen (Fig. 7).

The described system offers a broad range of physiological parameters needed for further interpretation of measurement results in other areas of the oxygen transport chain and becomes a valuable asset to the experimental laboratory setup for oxygen transport studies. Example: a starting hyperthermic crisis in a pig (pietrain strain) is preceded by hugh increase of oxygen consumption and CO_2-production while respiratory parameters remain unchanged excluding alveolar ventilation. Or: a pig develops hypoxia although high FiO_2. Compliance drops, ventilation drops and diffuse alveolar atelectasis is diagnozed, inversed I/E ratio is introduced and oxygen uptake in the lung normalizes.

Looking to the example of hyperthermia on-line oxygen consumption measurement reveals to be a most important parameter not only experimentally but also clinically (Fig. 8). Increase of oxygen consumption allerts in a very early stage that metabolisme is excessively increasing, before increase of expiratory CO_2 is seen and far before temperature increase. Early treatment with Dantrolen[R] can be started cutting through the vicious circle of this life endangering complication (mortality > 50%).

SUMMARY

A totally closed feed back controlled anesthesia- and ventilation circuit has been developed feasible to be applied for artificial ventilation with tidal volumes as low as 5 ml at a rate of up to 60/min and for spontaneous ventilation with on-line measurement of physiological lung parameters (pressure, volume, flow). Oxygen inflow is regulated via actual-set value comparison, oxygen inflow is measured and recorded on-line (= oxygen uptake by the connected subject).

On-line oxygen uptake (consumption) measurement furnishes a valuable, so far not available parameter to monitor changes in the oxygen transport chain to the tissue and to register physiological oxygen consumption values and derangement of metabolisme.

First results show that total body oxygen consumption of man in rest is lower than so far expected in the high weight and body surface area groups and that oxygen consumption is decreasing with length and age. Metabolic derangements such as in developing hyperthermia crisis are noticed in a very early stage when therapy is still possible before severe damage has occurred.

REFERENCES

Brody S. Bionergetics and Growth. Reinhold, New York, 1945.
Jackson D.E. A new method for the production of general analgesia and anesthesia with a description of the apparatus used. J Lab & Clin Med 1: 1-12, 1915.
Kleiber M. Body size and metabolic rate. Physiol Rev 27: 511, 1947.

OXYGEN DELIVERY (DEL O_2) DEPENDENT AND INDEPENDENT OXYGEN CONSUMPTION ($\dot{V}O_2$)

A. Verkaaik*, H. van der Zee** and W. Erdmann*

*Dept. of Anesthesiology, Erasmus University, Rotterdam, The Netherlands
**Dept. of Anesthesiology, Albany Medical College, NY, U.S.A.

INTRODUCTION

Oxygen uptake ($\dot{V}O_2$) of man is directly determined by the metabolic rate and not by the oxygen flux to the tissues (oxygen delivery - $\dot{D}O_2$ as the product of cardiac output and arterial oxygen content). Under physiologic conditions a broad range of metabolic rates (Cain, 1977) is adequately autoregulated to tissue oxygen demand via changes in tissue perfusion and augmentation of the extraction ratio with proportional changes of oxygen consumption. Below a critical $\dot{D}O_2$ level of approximately 300 ml/min/m² (Shibutani et al., 1983) autoregulative and extraction capacity are fully exploited and tissue oxygen debt develops with the consequence that oxygen consumption becomes delivery dependent (Fig. 1).

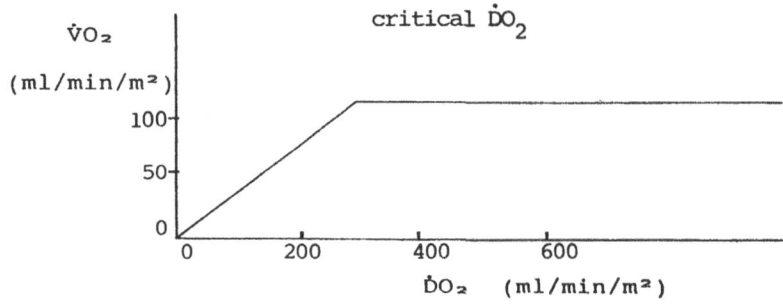

Fig. 1

Relationship between $\dot{V}O_2$ and $\dot{D}O_2$ under physiologic conditions.

In critically ill patients adequate oxygen transport to tissue is often severely disturbed (Vermeij et al., 1990). Unfortunately monitoring equipment was not yet available to detect impairment of tissue oxygenation in an early stage of development and immediately investigate the underlying problem to start restoring therapy before irreversible damage has occurred. In animal experiments it has been indicated that measurement of oxygen uptake could be a most valuable parameter (Cain, 1977; Cain, 1984; V.d. Zee and Verkaaik, 1990), but lack of suitable monitoring equipment for on-line registration of oxygen uptake ($\dot{V}O_2$) made it impossible to study the interrelationship of oxygen uptake and oxygen delivery to the tissue in a large group of animals and patients with different underlying diseases and problems (e.g. sepsis, multitrauma, major surgery). The aim of the presented study was to get more insight into the $\dot{V}O_2$-$\dot{D}O_2$ relationship under the various pathophysiological conditions of daily clinical patient management. The studies were performed with a newly developed system (Verkaaik and Erdmann, 1990) for on-line measurement of $\dot{V}O_2$ under experimental as well as under clinical conditions.

METHODS

Oxygen consumption is continuously monitored after connection of the animal (respectively patient) to a closed-loop feed back-controlled totally closed rebreathing circuit (Erdmann et al., 1989; Van der Zee, and Verkaaik, 1990; Verkaaik et al., 1991).

The so-called "Rotterdam Physio Flex" (Verkaaik, Grogono, and Erdmann, 1990) displays on-line oxygen consumption and the decisive pulmonary function parameters (volume, pressure, flow, and capnogram) and inspiratory oxygen concentration, which makes exclusion of pulmonary uptake problems as well as inspiratory oxygen concentration changes possible (Fig. 2).

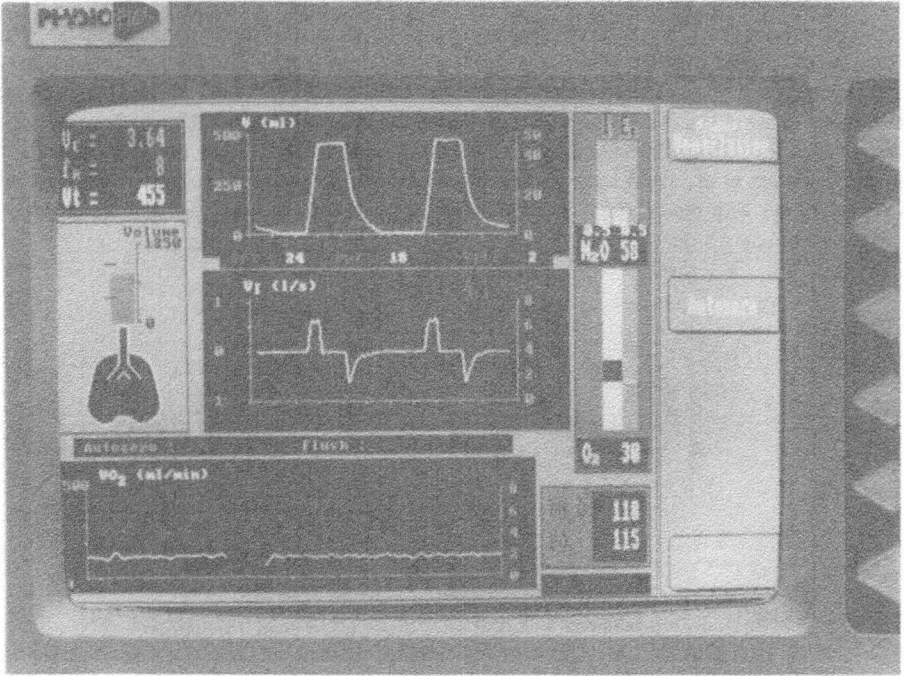

Fig. 2

On-line screen-display of the "Rotterdam Physio Flex[(R)]. The upper trace shows instantaneous volume and pressure, the middle trace inspiratory-expiratory flow and capnogram (on-line respiratory CO_2-concentration) and the lower trace trend recording (1 h display) of oxygen uptake and endexpiratory CO_2-concentration respectively instead CO_2-production. Inspiratory oxygen is displayed in bar form and numbers on the right side of the screen together with anesthetic concentrations. All preset and measured values are (averaged over a minute) stored in the computer memory with a 75 h registration capacity and can be transfered to computer discs for long time storage and data management.

Animal experiments were performed in healthy pigs and oxygen delivery to tissue was manipulated in two ways:

1)   decrease in cardiac output and mean arterial pressure by sodium nipride;
2)   introduction of a septic shock by intravenous administration of endotoxin.

The aim of study was to investigate the $\dot{D}O_2$-$\dot{V}O_2$ relationship under life endangering pathophysiologic conditions and further the validity of on-line oxygen uptake measurement as a diagnostic parameter for early detection of developing impairment of tissue oxygenation supply.

RESULTS

Oxygen uptake and oxygen delivery was measured in Yorkshire pigs ranging from 25 to 30 kg body weight with an average oxygen uptake of 194 ml/min (± 7 ml/kg/min) and an oxygen delivery of 680 ml/min (± 25 ml/kg/min). Mean arterial pressure was decreased by controlled infusion of sodium nipride (NaP) with concomitant decrease of cardiac output to 69 (I), 65 (II) and 50% (III, Fig. 3) of the original value. As a consequence, $\dot{D}O_2$ dropped to 470 ml/min (17.2 ml/kg/min) (I), 445 ml/min (16.3 ml/kg/min) (II) and 350 ml/min (12.8 ml/kg/min) (III, Fig. 3). As long as oxygen delivery remained above 16 ml/kg/min, no change in total oxygen consumption was observed ($\dot{D}O_2$ independent $\dot{V}O_2$), but $\dot{V}O_2$ accutely became $\dot{D}O_2$ dependent when $\dot{D}O_2$ dropped further in lower cardiac output states. Stop of sodium nipride infusion at 60 min led to an immediate increase of mean arterial pressure and a recovery of cardiac output to 62% of the control value with concomitant increase of $\dot{D}O_2$ to 16.1 ml/kg/min in 20 min (IV, Fig. 3) resulting in $\dot{V}O_2$ values above the control (repay of oxygen debt), increasing to an excess $\dot{V}O_2$ of 13.5% above control (220 ml/min or 8.1 ml/kg/min) at maximum (V, Fig. 3). The repay procedure lasted for 80 min until control values were reached without further intervention. During the repay period $\dot{D}O_2$ never surpassed the control value in any of the six experiments, all of them showing the same characteristic changes as described above.

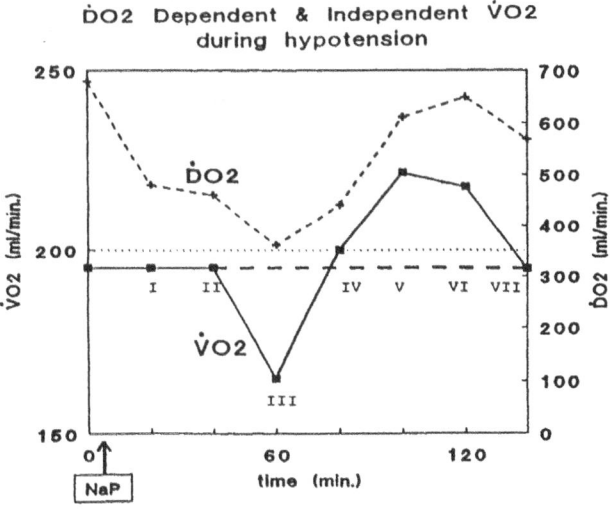

Fig. 3

$\dot{V}O_2$ - $\dot{D}O_2$ relationship in the pig undergoing artificial hemodynamic shock (sodium nipride): when $\dot{D}O_2$ (and cardiac output) is half its control value, $\dot{V}O_2$ becomes $\dot{D}O_2$ dependent.

In another series of six pig experiments $\dot{V}O_2$ and $\dot{D}O_2$ were followed after induction of a septic shock syndrome by endotoxin infusion. In the first 60 min $\dot{V}O_2$ remains 15-20% above the original value although a significant drop of cardiac output (vasodilatation) and concommitantly $\dot{D}O_2$ is seen. A typical hyperdynamic phase is following for the next 60 min with forse increase of cardiac output, oxygen delivery and, simultaneously following, oxygen consumption. Central venous oxygen saturation and oxygen extraction ratio remain practically unchanged during these first two hours of septic shock syndrome. The following period is marked by a gradual drop of cardiac output and tissue oxygen delivery. $\dot{V}O_2$ decreases to the control level, remains there for a period while oxygen extraction ratio increases with a drop of central venous oxygen saturation compensatorily. Oxygen consumption can be thus maintained on the control value for a period of 3 h, whereby cardiac output and oxygen saturation stabilize on 75% of the control (stabilized hypodynamic phase). At 5 h (300 min) a sudden drop of cardiac output (to 70% of the original) and $\dot{D}O_2$ (to 60% of the control) occurs with a $\dot{D}O_2$ dependent decrease of $\dot{V}O_2$ to 80% of the control, 20 min later at minute 340 the animal dies after cardiac fibrillation. Noticeable seems to be the close dependency of oxygen uptake on oxygen delivery starting with the development of the hypodynamic phase.

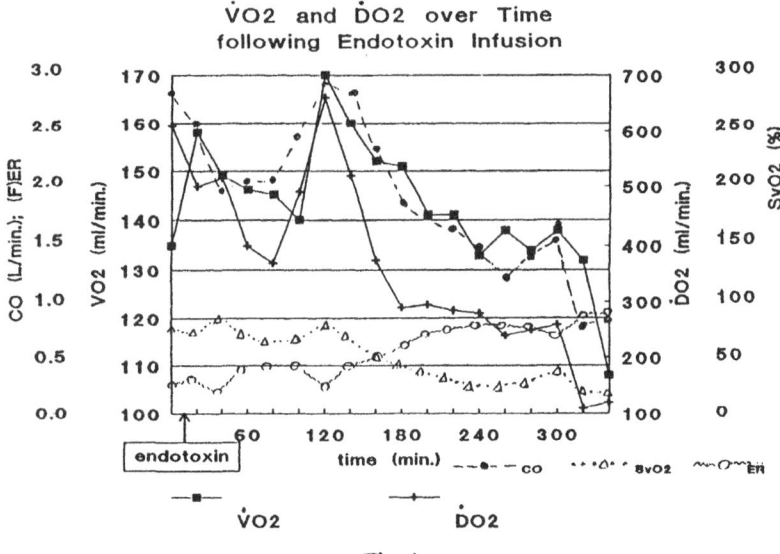

Fig. 4

Simulated septic shock (endotoxin shock), note the VO_2 changes following DO_2 changes. Cardiac output (CO), fraction of extraction ratio ([F]ER) and venous oxygen saturation (SvO_2) are displayed as well.

DISCUSSION

The phenomenon of $\dot{V}O_2$ dependency on $\dot{D}O_2$ has been broadly discussed in the literature and pro and contra, whether and when this pathophysiological process begins, still remains not clearly answered (Cain, 1977, 1984; Shibutani 1983; Vermeij et al., 1990). The broad range of compensatory mechanisms to keep $\dot{V}O_2$ and adequate oxygen supply independent from tissue oxygen delivery (oxygen flux) remains a fact under physiological conditions (Fig. 1). Under pathophysiological conditions, however, physiological autoregulative response may be severely disturbed. To get more insight into the basic problems leading to derangement is of outmost importance for the practicing intensive care physician and his clinical decisions for the right therapeutical approach. The tissue oxygen supply depriving problems on the IC mainly consist of two major groups of patients, those that underwent severe hemodynamic shock and those that suffer from severe septic shock.

Hypodynamic hemodynamic shock was introduced by NaP (Fig. 3). A distinct moment can be recognized when $\dot{V}O_2$ cannot be further maintained by increase of extraction and becomes fully dependent on $\dot{D}O_2$ with further decrease of this latter parameter. The development op $\dot{V}O_2$ dependency on $\dot{D}O_2$ during hemodynamic shock thus, follows the physiological response mechanism described in Fig. 1.

However, the $\dot{V}O_2$ in the post-hypodynamic phase (recovery phase from hemodynamic shock) is higher than control values and remains so for a prolonged time in an attempt to restore O_2 deficit (debt). It even appears that repayment of O_2 debt in the recovery period is larger than the O_2 deficit incurred during the hemodynamically induced period of tissue oxygen supply impairment. Looking to just the parameter $\dot{V}O_2$ continuously controlled, it can be said that this parameter adds to the overall patient assessment with respect to whether tissue oxygenation supply impairment with drop of oxygen uptake occurs and to what degree.

More complicated is the interrelationship of oxygen uptake and tissue oxygen delivery in septic shock (Fig. 4). The typical hemodynamic response is biphasic, a primary hyperdynamic period with increase of cardiac output and a secondary hypodynamic phase mostly ending in death. $\dot{V}O_2$ dependency on $\dot{D}O_2$ occurs during both periods clearly separated from each other by a third interim phase (Fig. 5).

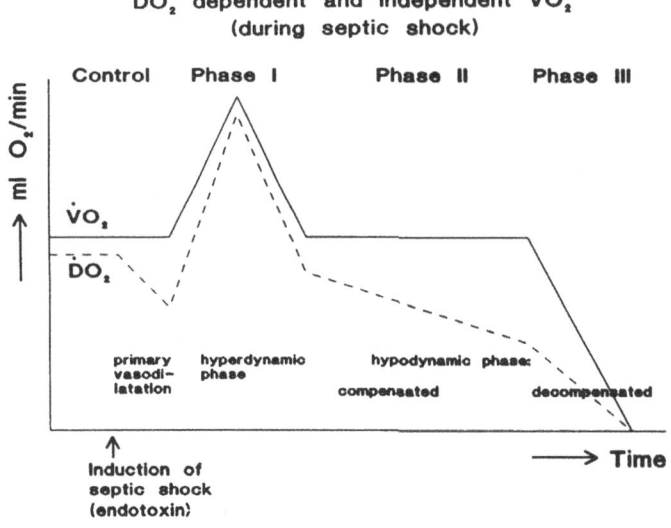

Fig. 5

$\dot{D}O_2$ dependent (phase I and III) and independent (phase II) $\dot{V}O_2$ during septic shock.
$\dot{V}O_2$ during induced septic shock shows three characteristic phases:
1a. $\dot{D}O_2$ dependent increase of $\dot{V}O_2$
2a. $\dot{D}O_2$ independent stabilization of $\dot{V}O_2$ to control values
3a. $\dot{D}O_2$ dependent decrease of $\dot{V}O_2$

1. During the hyperdynamic phase $\dot{V}O_2$ increases simultaneously to $\dot{D}O_2$ without any response in extraction ratio ($\dot{D}O_2$ dependent $\dot{V}O_2$ in hyperdynamic phase).
2. An interim phase follows in which hyperdynamia gradually develops with gradual drop of $\dot{D}O_2$, but $\dot{V}O_2$ returns to control values unaffected by the $\dot{D}O_2$ changes. $\dot{V}O_2$ remains stable while extraction ratio is increased with a drop of central venous oxygen saturation.
3. With a drop of cardiac output below 75% and $\dot{D}O_2$ below 70% of the original, $\dot{V}O_2$ again becomes $\dot{D}O_2$ dependent in the terminal hypodynamic phase. In this phase maximally increased O_2-extraction and a further decreasing $\dot{D}O_2$ brings the $\dot{V}O_2$ on the brink of becoming $\dot{D}O_2$ dependent. Furthermore, it cannot be excluded that at this low level of $\dot{D}O_2$ some organs may be already in oxygen deficit (e.g. intestestines due to vasoconstriction) while other organs may be still hyperperfused (e.g. vasodilation in muscles). This maldistribution in blood flow may be due to direct impairment of autonomic myogenic control during endotoxemia.

CONCLUSION

The on-line registration of $\dot{V}O_2$, nevertheless, seems to be an outmost valuable parameter to indicate when hemodynamically or septically induced impairment of oxygen delivery ($\dot{D}O_2$) has reached a degree where tissue oxygen supply is impaired, namely at that moment where $\dot{V}O_2$ falls below control and becomes $\dot{D}O_2$ dependent: in hemodynamic shock this occurs monophasic when $\dot{D}O_2$ drops below a critical level.
In septic shock this occurs threephasic: after a $\dot{D}O_2$ dependent
phase of $\dot{V}O_2$-increase in the hyperdynamic phase and a $\dot{D}O_2$ independent $\dot{V}O_2$-stable phase during moderate hypodynamia follows a third phase of $\dot{D}O_2$ dependent $\dot{V}O_2$ during severe hypodynamia.

REFERENCES

Cain S.M., 1977, Oxygen delivery and uptake in dogs during anemic and hypoxic anemia, J Appl Physiol, 42: 228-234.

Cain S.M., 1984, Supply dependency of oxygen uptake in ARDS: myth or reality? Am J Med Sc, 288: 119-124.

Erdmann W., Veeger A.I., Verkaaik A.P.K., 1989, Narkosebeatmungsgeräte: Gegenwart und Zukunft, in: Narkosebeatmung, Low Flow, Minimal Flow, Closed Circuit. J.P. Jantzen and P.P. Kleemann, eds., Schattauer, Stuttgart.

Shibutani K., Komatsu T., Kubal K. et al., 1983, Critical level of oxygen delivery in anesthetized man. Crit Care Med 11: 640-643.

Verkaaik A.P.K., Erdmann W., 1990, Respiratory diagnostic possibilities during closed circuit anesthesia. Acta Anesth Belg, 41: issue 3 in press.

Verkaaik A.P.K., Grogono A.W., Erdmann W., 1990, Informative imaging of multiple parameters for the practice of the anesthesiologist. Acta Anaesth Belg, 41: issue 3, in press.

Verkaaik A.P.K., Erdmann W., Van Dijk G., Westerkamp B., 1991, On-line oxygen uptake measurement (VO_2): a computer feed-back controlled rebreathing circuit for long term oxygen uptake registration. Oxygen Transport to Tissue XIII, Plenum Press, New York and London.

Vermeij C.G., Feenstra B.W.A., Bruining H.A., august 1990, Mathematic coupling of oxygen delivery and oxygen uptake in postoperative and septic patients. Chest, in press.

Van der Zee H., Verkaaik A.P.K., 1990, Cardiovascular implementation of respiratory parameters. Acta Anesth Belg, 41: issue 3 in press.

FREE RADICALS

MICROTOPOGRAPHIC ANALYSIS OF OXIDATIVE STRESS IN ORGAN MICROCIRCULATORY UNITS

Makoto Suematsu and Masaharu Tsuchiya

Department of Internal Medicine, School of Medicine
Keio University
Tokyo 160, Japan

INTRODUCTION

In eukaryotic cells, molecular oxygen functions primarily as the terminal acceptor of electrons during mitochondrial electron transport. Most of oxygen consumed in this process is reduced to water by cytochrome oxidase. Only a small percentage of the total O_2 consumption appears as partially reduced species such as O_2^- and H_2O_2. It is likely, however, that under pathological situations, these active oxidants may be highly generated and can exert, in turn, their cytotoxic properties. It is technically difficult to obtain spatial and temporal information of intravital oxidative stress. Since most observations which suggested participation of active oxidants or lipid peroxides in tissue injury are based upon preventive effects of antioxidants , establishment of the microscopic oxidant-visualizing system has been awaited to further clarify microtopographic relationship between oxyradical generation and tissue breakdown. This report is therefore aimed to summarize the methodology for oxyradical visualization *in vivo*.

CURRENT APPROACH FOR OXYRADICAL VISUALIZATION

Chemical Probes

Because active oxygen metabolites are highly reactive and short-lived compounds that react with the nearest molecule, oxidative stress can be categorized regarding the spatial relationship between oxidant-generating sites and their target as follows; (i) phagocyte-mediated oxidative stress, in which extracellularly released oxyradicals exert their cytotoxicity on the target cell. In this case, contact interaction between phagocytes (effector) and target cells or molecules such as bacteria, malignant cells, vascular endothelium and parenchymal cells. (ii) intracellular oxidative stress, where cytosolic and/or mitochondrial radical-generating enzymes mediate oxidant formation and the subsequent cell death.

Oxyradical-sensitive chemical probes may be useful tools for analysis of the spatial and temporal distribution of oxyradical generation *in vitro* and *in vivo*. DiGregorio et al. (1987) applied nitroblue tetrazolium (NBT), a chemical probe which is oxidized by O_2^- to form insoluble blue deposits, formazan, at the sites of O_2^- generation. To determine O_2^- release from individual cells, they quantified the turbidity of formazan in PMA-stimulated single alveolar macrophages by means of microscopic videodensitometry. Another probe to visualize oxidant activities is dichlorofluorescin (DCFH) diacetate. Since Cathcart et al. (1983) reported its usefulness to determine picomolar levels of hydroperoxides, the DCFH study has been applied to investigate oxyradical release from individual cells such as neutrophils (Bass, et al., 1983), renal epithelial cell lines (Scott, et al., 1988) or

Table 1 Chemiluminescence probes for assessment of oxidant generation

probe	specificity	references
luminol	H_2O_2-MPO-Cl^- system-derived oxidants (ClO^-)	Allen, et al., 1976 DeChatelet, et al., 1982 Brestel, et al., 1985 Edwards, et al., 1987 Suematsu, et al., 1988
lucigenin	O_2^- and/or H_2O_2	Greenlee, et al., 1962
cypridina luciferin analogue (CLA, MCLA)	O_2^- and/or $^1O_2$	Nakano, et al., 1986

hepatocytes (Gores, et al., 1989). DCFH diacetate is a stable nonfluorescent compound that can diffuse into cells, is hydrolyzed to DCFH and is thereby trapped within the cells. DCFH is known to be oxidized by hydroperoxides to yield dichlorofluorescein (DCF), a fluorochrome molecule. This compound may be therefore a powerful tool to assess the dynamics of intracellular oxidant-generating system such as cytochrome P-450 or xanthine oxidase.

Besides the use of NBT or DCFH diacetate, the third strategy for oxyradical visualization is the chemiluminescence method. Compared with the fluorescence methods, this technique has several advantageous properties for evaluating *in vivo* oxidative stress (Suematsu, et al., 1990a): (i) the absence of photobleaching effect; (ii) no need for excitation illumination, resulting in attenuation of cellular damages; (iii) availability of different types of chemilumigenic probes which makes it possible to elucidate what kinds of active oxidants are involved in different experimental models (Table 1). Since, however, the chemilumigenic response induced by oxyradical production is generally too weak to be observed as a visible image, even in the presence of chemiluminescence sensitizers, a special device for the low-level chemiluminescence visualization and its analytical evaluation is required.

Microscopic Oxidant-Visualizing System

Applying the aforementioned chemical probes, we have studied the dynamics of oxidative stress in various experimental models. Since the final goal of our study is to demonstrate the spatial and temporal correlation between oxyradical generation and cell-to-cell interactions ensued by cytotoxic changes or tissue breakdown, development of intravital microscopy (Chambers and Zweifach, 1955) may provide an observation system suitable for *in vivo* monitoring and imaging of oxidative stress. We first started establishment of the methodology to visualize chemiluminescence signal coming through a microscope. The conventional objective of the chemiluminescence study was to count the oxidant-associated scintillating photonic activity by means of a photomultiplier. By the development of a two-dimensional photon-counting system equipped with a computor-assisted digital image processor (Tsuchiya, et al., 1985), recent work with this system was aimed at the collection of quantitative information about the spatial distribution of chemiluminescence within a single cell. Figure 1 showed the experimental setup to visualize luminol-dependent photoemission released from the rat mesenteric microvascular beds treated with proinflammatory agents such as endotoxin, leukotriene B4 or platelet-activating factor (PAF). When an optical image from a microscope is focused on the bialkali photocathode in the front portion of the camera head, it emits photoelectrons in proportion to the input intensity of the microscopic image. These photoelectrons are amplified in a 2-stage microchannel plate by a million-fold and transmitted to a phosphor screen. The scintillating photonic images on the screen are visualized by a low-lag vidicon, and then introduced into an image analyzer for digital processing. To acquire the best quantum efficiency for ultraweak chemiluminescence imaging, the peak wave length of the probe-associated

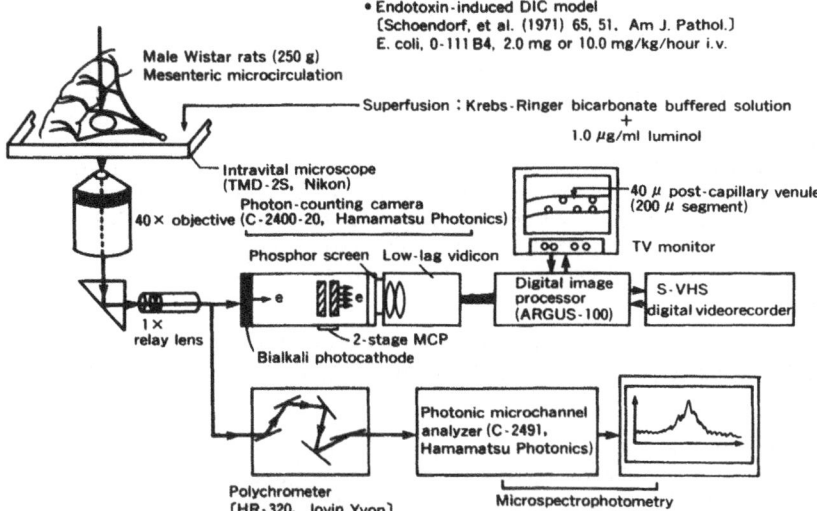

Figure 1. Experimental setup for *in vivo* visualization of luminol-dependent photoemission in the rat mesenteric microvascular beds. All system should be kept in a light-excluding chamber during the experiment. Using a polychrometer assisted by an ultrasensitive photonic microchannel analyzer (C-2491), the spectral profile of the photoemission can be obtained (See Reference 22.).

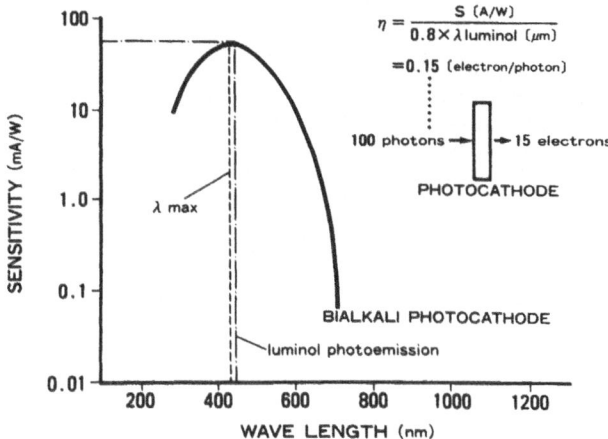

Figure 2. Spectral profile of the sensitivity of a photon-counting camera, C-2400-08. Quantum efficiency (η) was calculated to be 0.15 (electron/photon), namely, 100 photons which strike the same point of the bialkali photocathode yield 15 photoelectrons, if the input photonic image is derived from luminol (the peak wave length is 420 nm.). S; camera sensitivity.

response should be matched with the sensitivity profile of the visualizing system. As shown in Figure 2, the maximum quantum efficiency in our camera head was obtained at 410-420 nm, which just falls on the peak wavelength of luminol-dependent photoemission. To visualize the photochemical signal such as the DCF fluorescence, a silicon intensified target camera is applicable. This camera is sensitive enough to acquire the suitable fluorographs with lower level of illumination for excitation, which is advantageous for reducing the photobleaching effect. Whenever we try to apply the fluorescence method to assess biological phenomena using isolated viable cells or *in vivo* system, it is desirable to check spectral profile of the fluorescence observed through a microscope. Photonic microchannel analyzer incorporating a polychrometer may be a powerful tool for this purpose (Sarukura, et al., 1988). We applied the DCF method combined with the microspectrophotometry to analyze temporal and spatial distribution of the CCl4-induced or hypoxia-induced oxidative stress in hepatic microcirculatory units using the isolated perfused rat liver as described below (Suematsu, et al., 1989c, 1990b).

GRANULOCYTE-MEDIATED OXIDATIVE STRESS IN MICROCIRCULATION - APPLICATION OF CHEMILUMINESCENCE

Using the photon counting imaging system, we have investigated the mechanism of luminol-dependent photoemission derived from single neutrophils. When neutrophils were stimulated by opsonized zymosan, the luminol-dependent photonic burst clearly correlated with the distribution of the phagocytizing cells in the microscopic field and each chemilumigenic site was observed in a restricted region close to the cells (Suematsu, et al., 1987). To further elucidate the role of azulophilic degranulation and the extracellular release of myeloperoxidase (MPO), luminol-dependent chemiluminescence from the cells stimulated by PMA and calcium ionophore A 23187 was also studied. Neutrophils stimulated by PMA alone produced only a negligible level of photonic activities in the presence of luminol. The addition of calcium ionophore, however, induced explosive changes of photonic burst which was closely correlated with the cellular distribution. Of interest is that PMA and calcium ionophore-induced photoemission seemed to splash into extracellular space, while zymosan-induced activities concentrated on the site of phagocytosis (Suematsu, et al., 1989b.). Since these photonic activities were clearly eliminated by catalase, sodium azide (Suematsu, et al., 1988) or replacement with chloride-free medium (unpublished observation), a major active oxidant responsible for luminol-dependent photoemission was proposed to be the H_2O_2-MPO-Cl$^-$ system-derived species such as hypochlorous anion (ClO$^-$).

Figure 3 illustrates *in vivo* endothelium-granulocyte interaction and the visualization of luminol-dependent photoemission in the PAF-treated rat mesenteric microvascular beds (Suematsu, et al., 1989a), showing that each " hot spot " closely corresponds to the site of granulocyte adherence in post-capillary venules. It seems difficult to draw firm conclusions concerning the direct participation of MPO-mediated oxidants like ClO$^-$ in endothelial cytotoxicity. However, so far as we observed, the compartmentalized interface between endothelium and sticking granulocyte may be the most critical domain of oxidative stress. Recently, much attention has been paid to the concurrent action of MPO-derived oxidants and proteinases on *in vivo* tissue breakdown. As reported by Weiss (1989), powerful oxidants such as ClO$^-$ are by nature short-lived and nonspecific, and proteinases are held in check by either their own latency or a high effective antiproteinase screen, namely, α1-proteinase inhibitor, secretory leukoproteinase inhibitor and α2-macroglobulin. Acting alone, either set of weapons would allow the sticking granulocytes to exert only highly localized effects. But if overproduction of ClO$^-$ and proteinases are simply combined, the cell can break through all the intrinsic barriers that have been disgined to protect host tissues from injury. By oxidatively inactivating aforementioned proteinase inhibitors, sticking granulocyte-derived ClO$^-$ may create an microenvironment at the compartmentalized region close to endothelial surface, where elastase, collagenase, gelatinase (Weiss, et al., 1985, Peppin, et al.,1986) or plasminogen activator inhibitor-1 (PAI-1) are able to exert destructive effects more efficiently and with greater specificity than could even enormous doses of oxidants alone.

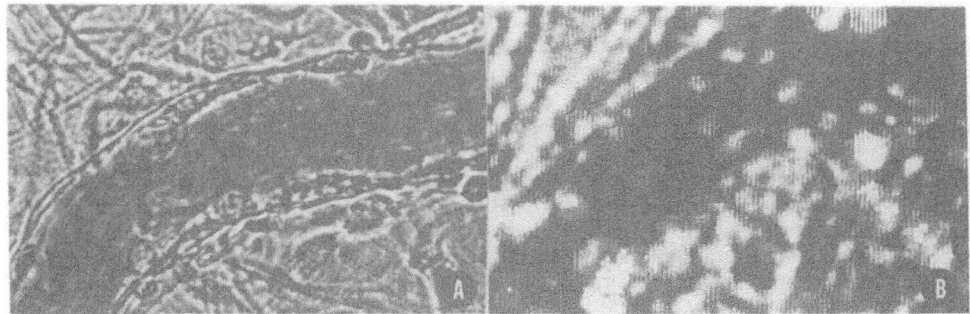

Figure 3. *In vivo* luminol-dependent photoemission during endothelium-granulocyte interaction in the rat mesenteric post-capillary venule superfusedwith 100 nM PAF-acether (Suematsu, et al. with permission.). A : a rare microvascular image, B: the photonic frame superimposed upon the microvascular frame. Exposure time to obtain a suitable photonic image was 180 sec.

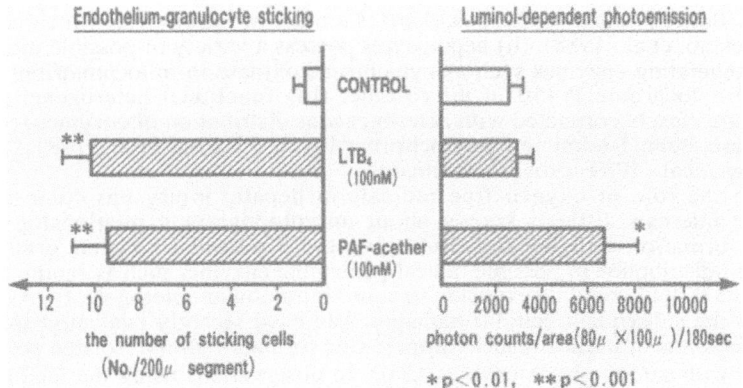

Figure 4. Dissociation between sticking and oxidative activation in leukotriene B4 (LTB4)-treated microvascular beds (Suematsu, et al., 1990a).The photonic intensity was calculated in the region (80 x 100 μ) containing a straight, unbranched segment of a post-capillary venule (40μ).

By contrast, leukotriene B4 did not cause any significant photonic burst at concentrations which are high enough to induce endothelium-granulocyte interaction (Suematsu, et al., 1990a). Our results may imply that secretagogue activation of H2O2-MPO-Cl⁻ system in sticking granulocytes is not apparent in leukotriene B4-induced leukotactic activation *in vivo*. From viewpoints of the MPO-mediated enhancement of proteolysis, leukotriene B4 seems to be a more "biological" mediator of granulocyte diapedesis than PAF (Figure 4). Since the *in vivo* photonic burst reflect overflowing oxidant activities which remain to be eliminated by radical-scavenging system, it is difficult to conclude whether lack of photonic burst in leukotriene B4 -treated microvasculature directly indicates the absence of oxidant generation in the area observed. Our results, however, suggest that the leukotriene B4-induced leukotactic change may not be followed by granulocyte-mediated oxidative stress. As one might expect, the concept that granulocyte diapedesis can occur independently of oxidant production is supported by the fact that NADPH-oxidase-deficient granulocytes in chronic granulomatous disease are accumulated in pus at the sites of infection (Curnutte, et al., 1974). Taken together with the present findings, spatial and temporal information of granulocyte-mediated oxidative stress directly obtained by photonic intensifier microscopy support the significance of MPO-mediated active oxygen metabolites in tissue breakdown and may provide further insight into the oxidative events which would take place during endothelium-granulocyte interaction in various pathological states.

HYPOXIA AND OXIDATIVE STRESS IN HEPATIC MICRO-CIRCULATORY UNITS - APPLICATION OF DCF FLUOROGRAPHY

Several anatomical and biochemical chalacteristics should be considered to analyze the oxidative changes in hepatic microcirculatory units; (i) As established by Rappaport et al. (1954), hepatocytes, a parenchymal component of the liver, are organized in three dimensional structure known as liver acinus, which forms the smallest microcirculatory unit (Figure 5). Hepatocytes surrounding the terminal portal venule (TPV), which receives blood with the highest concentration of incoming solutes including oxygen, are called zone 1 hepatocytes. By contrast, hepatocytes surrounding the terminal hepatic venule (THV), which are perfused with blood that has already suffered metabolic modification derived from the uptake and secretion of solutes by preceding hepatocytes, are called zone 3 hepatocytes. Between these two zones, there is a mass of hepatocytes which is denoted zone 2 (Gumucio, et al., 1988). (ii) hepatocytes possess a variety of possible intracellular oxyradical-generating enzymes such as cytochrome oxidase in mitochondrion, xanthine oxidase and cytochrome P-450 in microsome. (iii) functional heterogeneity among hepatocytes are closely correlated with heterogeneous distribution of enzymes (eg. lactate dehydrogenase, zone 1-dominant ; cytochrome P-450 (Baron, et al., 1981,) xanthine oxidase (Chen, et al., 1989) ; zone 3-dominant).

Although the role of oxygen free radicals in hepatic injury has come to attract considerable interests, little is known about microtopographic relationship between oxyradical formation and hepatocellular injury. Sinusoidal oxygen gradient and heterogeneous distribution of possible radical-generating enzymes such as xanthine oxidase or cytochrome P-450 would be expected to cause intralobular heterogeneity of oxidative changes and the subsequent cellular damages. We have recently evaluated spatial and temporal alterations of intracellular hydroperoxide formation in the isolated perfused rat liver treated with carbontetrachloride (CCl4). In observations using the DCFH-loaded isolated perfused liver, CCl4 markedly activated the DCF fluorescence in the downstream sinusoidal region surrounding THVs, namely, hepatic zone 3 (Figure 6). This change was prevented by either SKF-525A, a potent cytochrome P-450 inhibitor, or retrograge CCl4 perfusion, where the sinusoidal O2 gradient was reversed (Suematsu, et al., 1989c) . These data suggests that oxidative vulnerability of zone 3 hepatocytes to CCl4 may be interpreted not only by predominant localization of cytochrome P-450 but also by hypoxic microenvironvent (PO2 is approximately 50 torr in the outlet perfusate, which is equivalent to the physiological O2 tension in the hepatic venous blood.) which is favorable to proceed CCl4-induced microsomal lipid peroxidation (Burk et al., 1984).

By contrast, in the experimental model of low-flow hypoxia where the rate of portal perfusion was reduced to 25% (0.75 ml/min/g liver), the DCF fluorescence was activated, forming patchy fluorogenic sites in the upstream sinusoid as early as 10 min after starting

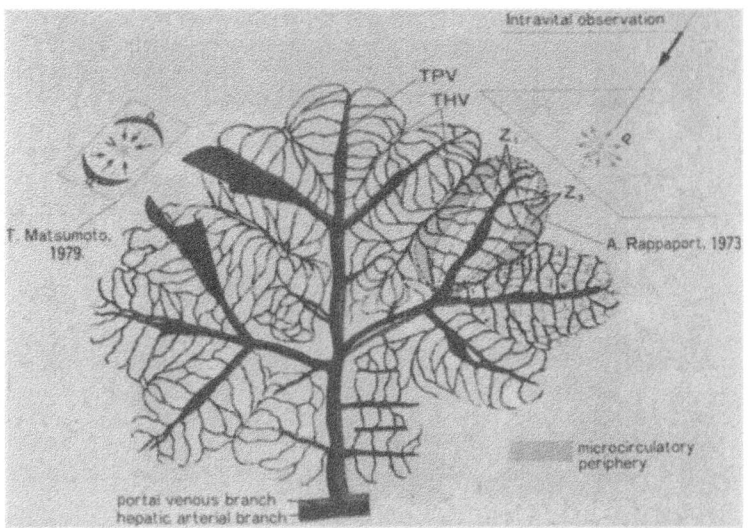

Figure 5. Concepts of hepatic microcirculatory units and the approach for intravital observation. In the rat liver, direct observation of terminal portal venules (TPV) is anatomically difficult. Most larger vessels which are observed from the surface hepatic microvasculature are found to be terminal hepatic venules (THV). Namely, according to the concept of Rappaport, et al. (1973, the right), most hepatocytes surrounding the superficial microangioarchitecture belong to hepatic zone 2 or zone 3 (Z3). Matsumoto, et al. (1979) established another concept, in which the portal venous tree is devided into conducting and parenchymal portions regarding its branching pattern and functional role, by performing stereoscopic angiography and reconstruction of histological serial sections. The beggining of the parenchymal portion (TPVs) conforms to a strict scheme of branching, which terminates to form a spanning septum-like inflow-front (the left).

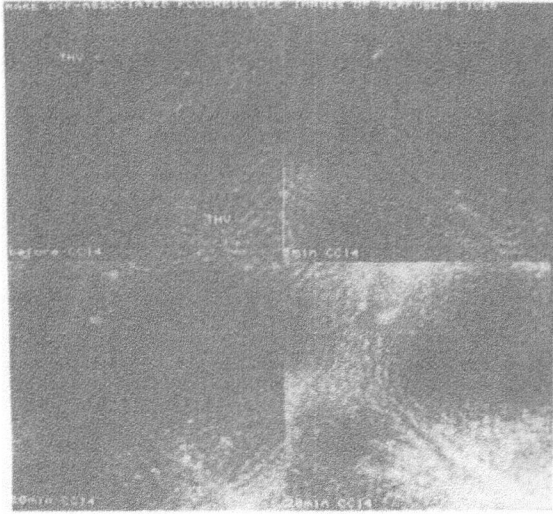

Figure 6. Spatial and temporal alterations of the intralobular DCF fluorescence in the CCl4-perfused isolated rat liver. The DCF activation occurs predominantly in the region surrounding THVs, namely, hepatic zone 3. (the left upper; control, the right upper; 5min, the left lower; 10min and the right lower; 20 min after starting the transportal CCl4 perfusion at a concentration of 1.0 mM.

217

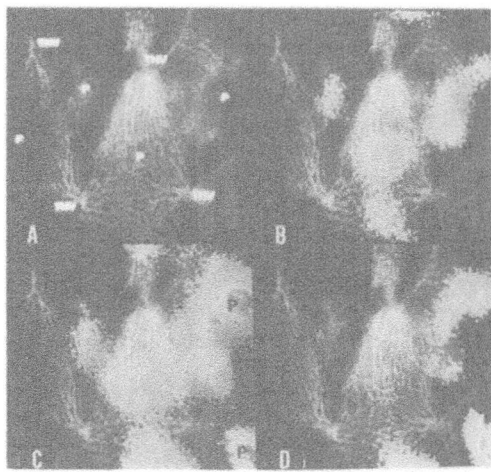

Figure 7. The intralobular DCF activation during 25% low-flow hypoxia in the isolated perfused rat liver. The left upper shows the hepatic microvascular image which was obtained by the FITC-albumin injection after experiments. After digitally composing the fluorographs which indicate the net DCF increase during low-flow procedure, the FITC microvascular frame was superimposed on each of the DCF fluorographs (A; before hypoxia, B; 10 min hypoxia, C; 30 min hypoxia, D; 40 min hypoxia.). In several periportal areas (P), DCF-activated regions showed a ring-shaped (TPV-sparing) pattern. These findings suggest the midzonal hepatocytes apart from THVs may be the most vulnerable to hypoxia-induced oxidative stress.

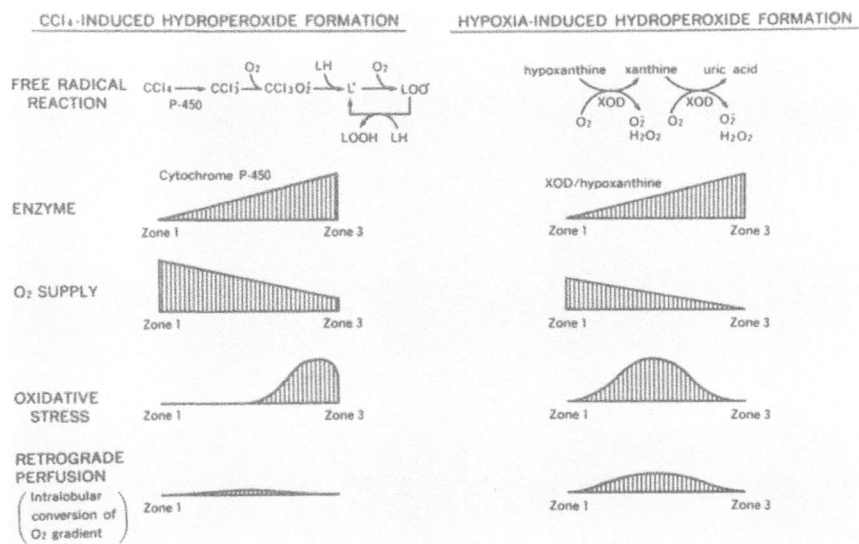

Figure 8. Possible role of intralobular heterogeneity of oxidant-generating enzymes and sinusoidal oxygen gradient in oxidative hepatocellular injury. P-450; cytochrome P-450, XOD; xanthine oxidase.

the low-flow procedure, and was then spread out towards zone 3 (Suematsu, et al., 1990b). In several periportal areas, DCF-activated regions showed a ring-shaped (TPV-sparing) pattern (Figure 7). Taking account of the superficial microangioarchitecture in the liver (Figure 5, see the direction of microscopic observation.), these finding suggest that the intermediate zone (zone2) may be the most vulnerable to hypoxia-induced oxidative stress. The hypoxia-induced DCF activation was prevented by SOD or allopurinol, implicating the involvement of xanthine oxidase-mediated O_2^- in intracellular hydroperoxide formation. Since the retrograde low-flow perfusion also induced midzonal DCF activation and global ischemia caused a rapid depletion of the DCF fluorescence (unpublished data), incompletely oxygenated regions may play more critical role in hypoxia-induced oxidative stress rather than highly anoxic regions as discussed by Marroto et al. (1988). The present findings illustrate the pathological relevance of heterogeneity in tissue oxygenation and distribution of radical-generating enzymes to destine the microtopography of oxidative tissue injury (Figure 8). Furthermore, it should be noted that low-flow hypoxia causes oxidative changes in the marginally oxygenated upstream sinusoids, rather than in the anoxic zone 3, where hypoxia-induced hepatocellular necrosis is predominantly observed. This suggests that spatial distribution of tissue destruction may not always correspond to the oxyradical-generating sites. Although it is still difficult to trace both oxidative changes and cellular viability simultaneously in the same hepatic microcirculatory unit, development and refinement of the present technique will provide a useful method to clarify the microtopografic correlation between oxidative stress and tissue breakdown in organ microcirculatory units.

SUMMARY

Current approaches for visualization of oxidative stress in organ microcirculatory units were summarized. Recent development of digital imaging photonic microscopy has made it possible to analyze spatial and temporal alterations of oxyradical generation during tissue injury. Luminol-dependent photonic imagery revealed granulocyte-mediated oxidative stress during microvascular damages, suggesting that the interface between venular endothelium and sticking granulocytes may be the most critical site of oxidative stress. Fluorographic analysis assisted by dichlorofluorescin (DCFH) diacetate is a powerful tool to visualize intracellular hydroperoxide formation. This method demonstrated intralobular heterogeneity of oxidative stress in the isolated perfused hepatic microcirculatory units exposed to either CCl4 or low-flow hypoxia. CCl4 caused the activation of dichlorofluorescein (DCF) predominantly in perivenular areas, while the 25% low-flow perfusion induced periportal or midzonal DCF activation. Refinement of the present technique will provide further insight into the microtopographic correlation between oxidative stress and tissue breakdown in microcirculation.

ACKNOWLEDGMENTS

We thank Prof. H. Wayland and Prof. E.V. Cilento for their suggestions. This work was supported by a Grant-in-Aid for Scientific Research from the Ministry of Education, Science and Culture of Japan and by a grant from Keio University School of Medicine. A preliminary report of portions of this work was presented at the 37th annual meeting of the Microcirculatory Society for Satellite Symposia,' New Optical Techniques for the Study of the Microcirculation', Washington, DC, March 1990.

REFERENCES

1. Allen, R.C., L.D.Loose, Phagocytic activation of a luminol-dependent chemiluminescence in rabbit alveolar and peritoneal macrophages, *Biochem. Biophys. Res. Commun.*, 69: 245 (1976).

2. Baron, J., J.A.Redick, F.P. Guengerich, An immunohistochemical study on the localization and distributions of phenobarbital- and 3-methyl cholanthrene-inducible cytochrome P-450 within the livers of untreated rats, *J. Biol. Chem.*, 256: 5931 (1981).

3. Bass, D.A., J.W.Parce, L.R.DeChatelet, P.Szedjda, M.C.Seeds, M.Thomas, Flow cytometric studies of oxidative product formation by neutrophils: a graded response to membrane stimulation, *J. Immunol.*, 130: 1910 (1983).

4. Brestel, E.P., Co-oxidation of luminol by hypochlorite and hydrogen peroxide - implications for neutrophil chemiluminescence, *Biochem. Biophys. Res. Commun.*, 126, 482: (1985) .

5. Burk, R.F., J.M. Lane, K.Patel, Relationship of oxygen and glutathione in protection against carbon tetrachloride-induced hepatic microsomal lipid peroxidation and covalent binding in the rat, *J. Clin. Invest.*, 74: 1996 (1984).

6. Cathcart, R., E. Schwiers, B.N. Ames,Detection of picomole levels of hydroperoxides using a fluorescent dichlorofluorescein assay, *Anal. Biochem.*, 134: 111 (1983).

7. Chambers, R and B.W. Zweifach, Topography and function of the mesenteric circulation, *Am. J. Anat.*, 75: 175 (1944).

8. Chen, L., G.J.Davis, L.Lumeng, Zonal distribution of xanthine oxidase in rat liver , *Clin. Res.*, 39: 537 (1989).

9. Curnutte, J.T., D.M.Whitten, B.M.Bavior, Defective superoxide production by granulocytes from patients with chronic granulomatous disease, *N.Engl. J. Med.*, 290: 593 (1974).

10. DeChatelet, L.R., G.D.Long, P.S. Shirley, D.A. Bass, M.J. Thomas, F.W. Henderson and M.S. Cohen, Mechanism of the luminol-dependent chemiluminescence of human neutrophils, *J. Immunol.*, 129: 1589 (1982) .

11. DiGregorio, K.A., E.V Cilento and R.C. Lantz, Measurement of superoxide release from single pulmonary alveolar macrophages.*Am. J. Physiol.*, 252: C677 (1987) .

12. Edwards, S.W., Luminol- and lucigenin-dependent chemiluminescence of neutrophils: role of degranulation,*J. Clin. Lab. Immunol.*, 22: 35 (1987) .

13. Gores, G. J., C.E. Flarsheim, T.L. Dawson, A.L. Nieminen, B. Herman, J.J. Lemasters, Swelling, reductive stress, and cell death during chemical hypoxia in hepatocytes, *Am. J. Physiol.*, 257: C347 (1989) .

14. Greenlee, L., I. Fridovich, P.Handler, Chemiluminescence induced by operation of iron-flavoproteins, *Biochemistry*, 1: 779 (1962).

15. Gumucio, J.J. and J. Chianale, *In The Liver : Biology and Pathobiology* (eds. Arias, W.B. et al.) pp931, Raven Press, New York, (1988) .

16. Marroto, M.E., R.G.Thurman, J.J. Lemasters, Early midzonal cell death during low-flow hypoxia in the isolated, perfused rat liver: protection by allopurinol, *Hepatology*, 8:585 (1988).

17. Matsumoto, T., A study on the normal structure of the human liver, with special reference to its angioarchitecture, *Acta Hepatologica Japonica* (In Japanese), 20: 223 (1976).

18. Nakano, M., K.Sugioka, Y.Ushijima, T. Goto, Chemiluminescence probe with *Cypridina* luciferin analog, 2-methyl-6-phenyl-3,7,- dihydroimidazo (1,2-a) pyrazin-3-one, for estimating the ability of human granulocytes to generate O_2-, *Anal. Biochem.*, 159: 363 (1986) .

19. Peppin, G.J., Weiss, S.J., Activation of the endogenous metalloproteinase, gelatinase, by triggered human neutrophils, *Proc. Natl. Acad. Sci. USA*, 83: 4322 (1986).

20. Rappaport, A.M., Borowy, Z.J., Lougheed, W.M., and Lotto, W.N., Subdivision of hexagonal liver lobules into a structual and functional unit; role in hepatic physiology and pathology, *Anat. Rec.* 119: 11 (1954).

21. Rappaport, A.M., The microcirculatory hepatic unit, *Microvasc. Res.*, 6: 212 (1973).

22. Sarukura, N., M. Watanabe, A. Endoh, S. Watanabe, Single-shot measurement of subpicosecond KrF pulse width by three-photon fluorescence of the XeF visible transition, Opt. Lett., 13: 996 (1988).

23. Scott, J.A., C.J. Homcy, B.A.Khaw, C.A Rabito, Quantitation of intracellular oxidation in a renal epithelial cell line, *Free Rad. Biol. Med.*, 4: 79 (1988).

24. Suematsu, M., C. Oshio, S. Miura, M. Tsuchiya, Real-time visualization of oxyradical burst from single neutrophil by using ultrasensitive video intensifier microscopy, *Biochem. Biophys. Res. Commun.* 149: 1106 (1987).

25. Suematsu, M., C. Oshio, S. Miura, M. Suzuki, S. Houzawa, M. Tsuchiya, Luminol-dependent photoemission from single neutrophil stimulated by phorbol ester and calcium ionophore - role of degranulation and myeloperoxidase, *Biochem. Biophys. Res . Commun.*, 155: 106 (1988).

26. Suematsu, M., I. Kurose, H. Asako, S. Miura and M. Tsuchiya, *In vivo* visualization of oxyradical-dependent photoemission during endothelium-granulocyte interaction in microvascular beds treated with platelet-activating factor, *J. Biochem.*, 106: 355 (1989a).

27. Suematsu, M., S. Houzawa, S. Miura, H. Nagata, T. Kitahora, T. Morishita, C. Oshio, M. Tsuchiya, Effects of serine protease inhibitors on oxyradical burst from phagocytizing neutrophils - Analysis by chemiluminescence counting and its microscopic imaging, *J. Biolum. Chemilum.*, 4: 531 (1989b).

28. Suematsu, M., S. Kato, T. Yanagisawa, H. Asako, A. Yokoyama, H. Ishii and M. Tsuchiya, Intralobular heterogeneity of CCl4-induced oxidative stress in perfused rat liver visualized by digital imaging fluorescence microscopy, *Hepatology*, 10: 612 (1989c).

29. Suematsu, M. and M. Tsuchiya, Platelet-activating factor and granulocyte-mediated oxidative stress - Strategy for *in vivo* oxyradical visualization, *Lipids,* (1990a, in press).

30. Suematsu, M., S. Kato, H. Suzuki, T. Yanagisawa, H. Ishii and M. Tsuchiya, Zonal heterogeneity of oxidative stress and its correlation with lobular perfusion during low-flow hypoxia in rat perfused liver, *Hepatology*, (1990b, in press).

31. Tsuchiya, Y., E. Inuzuka, T. Kurono, M. Hosoda, Photon-counting image acquisition system and its applications, *J. Imag. Tech.*, 11: 215 (1985).

32. Weiss, S.J., G. Peppin, X. Ortiz, C. Ragsdale, S.T. Test, Oxidative autoactivation of latent collagenase by human neutrophils, *Science*, 227: 747 (1985).

33. Weiss, S.J., Tissue destruction by neutrophils, *N. Engl. J. Med.*, 320: 365 (1989).

CYTOCHROME P-450 UNDER CONDITIONS OF OXIDATIVE STRESS:
ROLE OF ANTIOXIDANT RECYCLING IN THE PROTECTION MECHANISMS

E. Serbinova, S. Ivanova, A. Kirova, S. Kitanova,
L. Packer, and V. Kagan

Institute of Physiology, Bulgarian Academy of Sciences
Sofia, 1113 Bulgaria

University of California, Berkeley CA 94720, USA

Institute of Medical and Biomedical Problems
Moscow, USSR

Free radicals initiate lipid peroxidation of microsomal membranes by rapidly forming lipid hydroperoxides (LOOH) from endogenous polyunsaturated fatty acid residues of phospholipids. Lipid hydroperoxides may in turn propagate lipid peroxidation via the cytochrome P-450-dependent mechanism. Indeed cytochrome P-450 catalyzes the oxidative cleavage of lipid hydroperoxides to a pool of alkoxyl- (LO) and peroxyl- (LOO) radicals which induce an additional formation of lipid hydroperoxides. Propagation and termination of lipid peroxidation in the cell are accompanied by a concomitant accumulation of numerous lipid peroxidation products possessing different effects on biomembranes and membrane bound enzymes (Halliwell and Gutteridge, 1985; Kagan, 1988). However, these products originate from hydroperoxides, the primary molecular lipid peroxidation products, which can be reduced to corresponding hydroxy-compounds by peroxidases (Christophersen, 1969; Ursini et al., 1985) thus preventing the formation of various scission products.Lipid peroxidation is considered to be an efficient triggering mechanism of the disassembly of microsomal membranes and cytochrome P-450. An inverse correlation exists between the steady-state concentrations of lipid hydroperoxides and cytochrome P-450 content in liver endoplasmic reticulum membranes (Kagan et al, 1974). This makes it reasonable to believe that lipid peroxidation activation triggers the chain of reactions leading in the end to disappearance of cytochrome P-450.Can it be concluded that lipid hydroperoxides themselves are responsible for cytochrome P-450 degradation? The first shade of doubt appears when one compares kinetic curves of cytochrome P-450 destruction in liver microsomal membranes with the curves of lipid hydroperoxides and secondary lipid peroxidation products (TBA-reacting carbonyl compounds), (Fig.1). In this *in vitro* system the time-course of lipid hydroperoxides passes through a maximum, whereas cytochrome P-450 declines following monotonous kinetics as does the accumulation of TBA-reactive substances. Thus it seems unlikely that primary lipid peroxidation products, hydroperoxides, are directely involved in cytochrome P-450 disassembly.

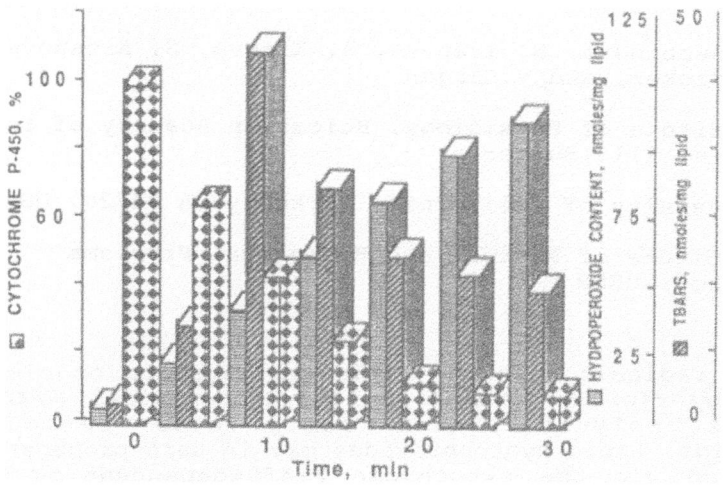

Fig. 1 Accumulation of lipid hydroperoxides, TBARS and degradation of cytochrome P-450 in rat liver microsomes incubated with Fe-ADP (20 µM/20 µl) + NADPH 500µM

This question will remain open until the effect of hydroperoxides only, (but not of the mixture of different lipid peroxidation products including hydroperoxides as is in the case of lipid peroxidation induction by Fe^{2+} + NADPH) is clarified. Accumulation of lipid hydroperoxides as the only (or at least predominant) lipid peroxidation products can be achieved by the use of reticulocyte lipoxygenase (Schewe et al, 1981; Lankin et al, 1985). Typical UV-spectra of rat liver microsomes incubated in the presence of reticulocyte lipoxygenase are shown in Fig.2. Addition of increasing amounts of the enzyme led to the increased absorption at 234 nm, characteristic of lipid hydroperoxides with conjugated dienes (in the reference compartment lipoxygenase and microsomes were present in the different cuvettes), whereas no peaks in the region of 270-280 nm, characteristic of secondary lipid peroxidation products appeared. Under the conditions used, addition of 70μl of lipoxygenase resulted in conversion of approximately 15% membrane phospholipids into phospholipid hydroperoxides. However, this caused neither significant destruction of cytochrome P-450 nor its conversion into cytochrome P-420 (Fig.3). Thus it can be concluded that phospholipid hydroperoxides accumulating among the other lipid peroxidation products in the course of lipid peroxidation are not responsible for cytochrome P-450 destruction.

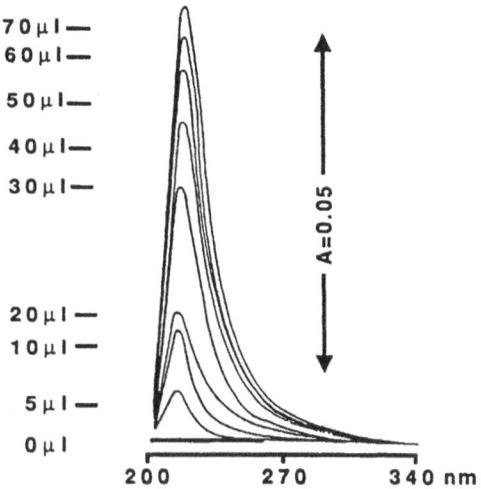

Fig. 2 UV spectra of rat liver microsomal suspension after addition of different amounts of reticulocyte 15-lipoxygenase (0.1 mg protein/ml)

It is known that hydroperoxides of free fatty acids can support oxidative hydroxylation reactions (Weiss and Estabrook, 1986, Weiss et al, 1986). However the polarity of fatty acid hydroperoxides, resulting in their poor incorporation into membranes (preferential partition into aqueous phase), makes them inefficient in stimulation of cytochrome P-450-dependent reactions.Prevention of hydroperoxide scission by free radical scavengers or antioxidant enzymes not only prevents the excessive accumulation of membrane-perturbing compounds, but can facilitate oxidative metabolism.

We suggest that under physiological conditions lipid peroxidation does not destruct cytochrome P-450 until secondary peroxidation products are accumulated in sufficient amounts. Control of lipid peroxidation due to GSH-dependent reactions is one of the mechanisms to prevent scission of hydroperoxides. The other possibility is the interaction of lipid peroxyl-radicals (LOO) with antioxidant molecules, like vitamin E (TOC-OH):

$$LOO\cdot + TOC\text{-}OH \xrightarrow{\hspace{3cm}} LOOH + TOC\cdot.$$
free radical reductase

Unfortunately this reaction results in an exhaustion of membrane vitamin E pools. It was suggested that special enzyme system(s), free radical reductase(s), may operate to recycle antioxidants from their phenoxyl radicals. However, no direct experimental evidence supporting this hypothesis was obtained until recently (Bast and Haenen, 1988, McCay, 1985).

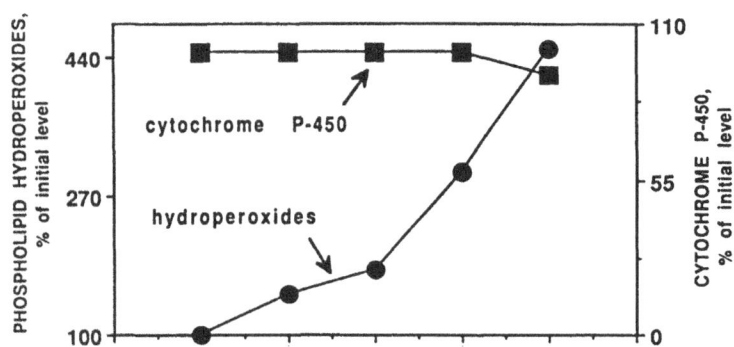

Fig. 3 Effects of phospholipid hydroperoxides generated by reticulo-cyte 15-lipoxygenase on cytochrome P-450 in rat liver microsomes.

We showed (Packer et al, 1989) that in a system containing soy-bean lipoxygenase and linolenic acid and microsomes in phosphate buffer (pH 7.4) alpha-tocopherol was easily oxidized to generate radicals. It might be suggested that generation of tocopheroxyl radicals (TOC-O·) was due to interaction of alpha-tocopherol (TOC-OH) with lipoxygenase-derived lipid peroxide radicals (LOO·):

$$Toc\text{-}OH + LOO\cdot \longrightarrow Toc\text{-}O\cdot + LOOH \qquad (1)$$

Steady-state concentrations of chromanoxyl radicals are determined not only by the rate of their generation but also by the rates of their decay in reactions:

$$Toc\text{-}O\cdot + Toc\text{-}O\cdot \longrightarrow Products \qquad (2)$$
$$Toc\text{-}O\cdot + LOO\cdot \longrightarrow Products \qquad (3)$$
$$Toc\text{-}O\cdot + AH \longrightarrow Toc\text{-}OH + A\cdot \qquad (4)$$

In contrast to reactions (2) and (3) where chromanoxyl radicals are irreversibly lost interaction of chromanoxyl radicals with reductants [reaction (4)] regenerates antioxidant molecules, which can be repeatedly used for formation of chromanoxyl radicals.

Under the conditions used, alpha-tocopherol gave characteristic pentameric ESR spectra of chromanoxyl radicals with g-values of the components 2.0122, 2.0092, 2.0061, 2.0028 and 1.9993. In the presence of ascorbate ESR signal of ascorbyl radical was observed first, which decayed and was progressively substituted by the appearing and growing signal of the tocopheroxyl radical. After reaching the maximal magnitude, the signals of tocopheroxyl radical followed typical decay kinetics (Fig. 4B).

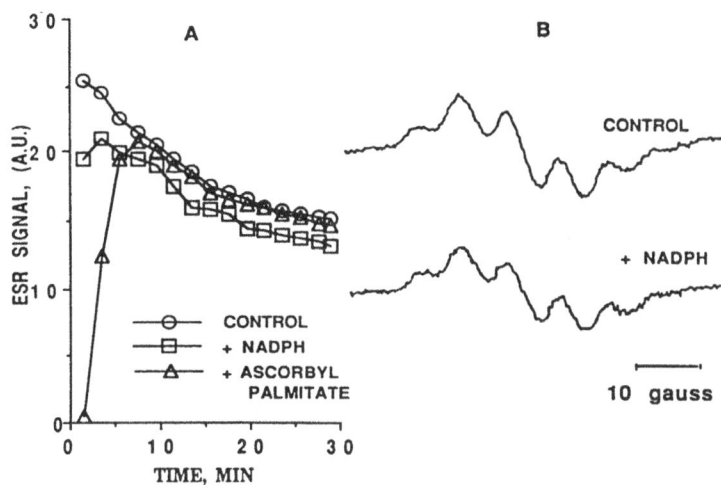

Fig. 4 ESR spectra (B) and time-course of chromanoxyl radicals (A) generated from alpha-tocopherol by lipoxygenase+ linolenic acid oxidation system in the presence of rat liver microsomes and their recycling by NADPH and ascorbate. Conditions: microsomes 27 mg protein/ml, linolenic acid 14 mM, lipoxygenase 90 U/µl, alpha-tocopherol 8 mM, NADPH 5 mM, ascorbyl palmitate 2.5 mM.

The addition of another reductant, NADPH, to the incubation medium caused a pronounced decrease in the magnitude of ESR signals of tocopheroxyl radical (Fig. 4A). The signals reappeared, but their magnitudes increased in the course of subsequent incubation and coincided with the decay curves obtained in the absence of NADPH (Fig. 4B).

This transient NADPH-dependent disappearance of the ESR signal of phenoxyl radicals was found to occur not only with alpha-tocopherol but also with other phenoxyl radicals such as alpha-tocopherol homologues, butylated hydroxy-toluene and its homologues, (Kagan et al, 1990, 1990a)

It is well known that microsomal cytochrome P-450 supported reactions are able to generate radical intermediates from their substrates in the course of monooxygenase reactions

(Utsumi et al., 1990; Epe and Metzler, 1985). These radical
intermediates can be both suicidal for cytochrome P-450 and
damaging for other neighboring macromolecules (Rahimtula,
1983). Our data gives an evidence for the presence of a "self-
defense" mechanism in the monooxygenase system, which is
capable of reducing radical species to nondangerous stable
molecules.

However, under harsh oxidative conditions, this free
radical reductase protective mechanism is probably not
efficient enough to maintain steady-state vitamin E
concentration in membranes.

We investigated the effects of different kinds of
oxidative stress on cytochrome P-450. Oxidative stress was
induced either by long-term exposure of rats to hypokinesia (45
days) or by short-term exposure to iron loading, exhausttive
physical exersise, hyperoxia or a combination of these
factors.The data in Table 1 shows that intramuscular injection
of iron (Ferrum Hausmann, single dose, 500 mg iron/kg b.w.) to
rats results in the development of oxidative stress, viz. a
decrease in vitamin E content and increase in the amount of
lipid peroxidation products. It is documented that accumulation
of lipid peroxidation products in microsomal membranes results
in destruction of cytochrome P-450 (Kitada et al, 1989; Iba et
Mannering, 1987),but in this study we found only a slight
decrease in cytochrome P-450 content in iron-loaded rats (11%
decrease of the control, $p < 0.005$). This small effect of iron
loading on cytochrome P-450 may be due to insufficient amounts
of lipid peroxidation products in the membrane to induce
destruction of cytochrome P-450. Combined treatments of rats by
iron loading coupled with physical stress (swimming) and/or
hyperoxia leads to more pronounced effects. In these cases
vitamin E depletion was even more pronounced and lipid
peroxidation products accumulation greater. Parallel to this,
the decrease of cytochrome P-450 was sharper (Table 1).

Long-term hypokinesia to a pronounced stimulation
of endogenous lipid peroxidation, decrease of vitamin E level,
loss of cytochrome P-450 content, and decrease of cytochrome
P-450 supported enzymic activities (Table 2, Fig. 5). These
changes were even greater in the readaptation period. We
suggest that the level of antioxidant protection achieved

Table I. CONTENT OF VITAMIN E, ENDOGENOUS LIPID PEROXIDATION PRODUCTS
(CONJUGATED DIENES) AND CYTOCHROME P-450 IN LIVER MICROSOMES
FROM RATS EXPOSED TO IRON, PHYSICAL AND OXYGEN LOADING

TREATMENT	VITAMIN E nmoles/mg protein	CONJUGATED DIENES O.D./mg protein	CYTOCHROME P-450 nmoles/mg protein
CONTROL	0.73±0.14	0.38±0.04	0.90 ± 0.02
+IRON	0.44±0.06	0.56±0.04	0.83 ± 0.03
+IRON+HYPEROXIA	0.05±0.01	0.78±0.06	0.44 ± 0.05
+IRON+SWIMMING	0.05±0.03	0.63±0.09	0.59 ± 0.03
+IRON+SWIMMING +HYPEROXIA	0.11±0.02	0.69±0.05	0.61 ± 0.02

There were 8 animals in each treatment group. For all groups $p < 0.05$.

Table 2 CONTENT OF CYTOCHROME P-450, CYTOCHROME b5 AND
ENDOGENOUS LIPID PEROXIDATION FLUORESCENCE PRODUCTS IN
RAT LIVER MICROSOMES FROM RATS EXPOSED TO HYPOKINESIA
AND SUBSEQUENT READAPTATION

| | CYTOCHROME P-450 | CYTOCHROME b5 | FLUORESCENCE PRODUCTS |
	nmoles/mg protein		Ifl.,(A.U.)
CONTROL	0.97±0.15	0.80±0.10	1.26±0.20
CONTROL + ANTIOXIDANTS	1.50±0.30	0.94±0.15	1.24±0.20
HYPOKINESIA	0.60±0.10	0.44±0.10	2.40±0.30
HYPOKINESIA + ANTIOXIDANTS	1.37±0.40	0.64±0.20	1.70±0.20
READAPTATION	0.64±0.15	0.70±0.10	4.10±0.40
READAPTATION + ANTIOXIDANTS	1.10±0.30	0.80±0.15	1.40±0.30

There were 6 animals in each group. For all group $p<0.05$.

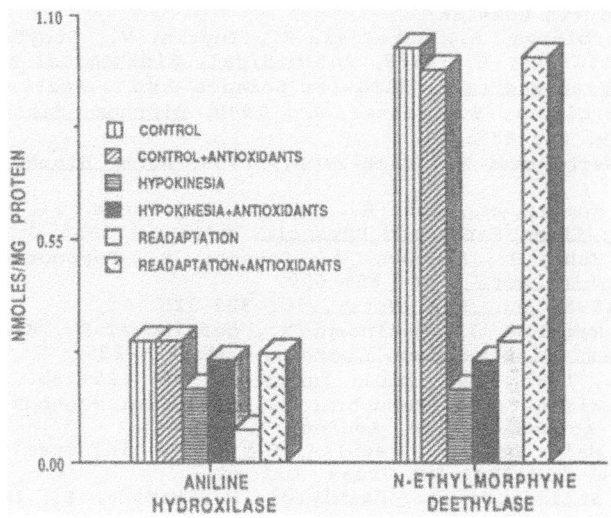

Fig. 5 Changes in monooxygenase activity in liver microsomes
from rats exposed to hypokinesia and subsequent readaptation

during long-term hypokinesia is not sufficient to withstand the activation of oxidative metabolism under conditions of readaptation.

Deficiency of endogenous antioxidants and stimulation of endogenous lipid peroxidation can be prevented by treatment of animals with exogenous antioxidants. Earler we observed that the combination of alpha-tocopherol and the water-soluble radical scavenger, carnosine, gives a synergistic effect against *in vitro* induced lipid peroxidation (Kagan et al, 1989). The results presented in Table 2 and Fig. 7 show that in animals which were treated with a combination of alpha-tocopheryl-succinate + carnosine the effects of hypokinesia were less pronounced.

Thus we conclude that the replenishment of antioxidant pools due to their administration is crucial for antioxidant protection against oxidative stress in vivo.

REFERENCES

1. Bast, A., Haenen, G., 1988, Biochem. Biophys. Acta, 574: 537-541.
2. Christophersen, B., 1974, Biochim. Biophys. Acta, 176: 463-471.
3. Epe, B., and Metzler, M., 1985, Chem. Biol. Interact., 56: 351-361.
4. Halliwell, B. , and Gutteridge, J., 1985, Free radicals in Biology and Medicine, Clarendon Press, Oxford .
5. Iba, M., Mannering, G., 1987, Biochem. Pharmacol., 36: 1447-55.
6. Kagan, V., 1988, "Lipid peroxidation in biomembranes", CRC Press, Florida.
7. Kagan, V., Kotelevtsev, S., Koslov, Yu., 1974, Proc. Acad. Sci. USSR, 217: 213-216 (in Russian).
8. Kagan, V., Serbinova, E., Bakalova, R., Tuyrin, V., Stoytchev, Ts., Erin, A., Prilipko, L., 1989, in: Medical, Biochemical and Chemical Aspects of Free Radicals, Elsevier Science Publ., Amsterdam.
9. Kagan, V., Serbinova, E., Packer, L., 1990, Biochem. Biophys. Res. Commun., 169: 851-857.
10. Kagan, V., Serbinova, E., Packer, L., 1990, Arch. Biochem. Biophys., 280: 33-39.
11. Kitada, M., Komori, M., Ohi, H., Imaoka, S., Funae, Y., Kamataki, T., Res. Commun. Chem. Path. and Physiol., 63(2): 175-88.
12. Lankin, V., Kuhn, H., Hiebsch, C., Schemet, R., Rapoport, S., 1985, Biomed. Biochim. Acta , 44: 655-659.
13. McCay, P., 1985, Ann. Rev. Nutri., 5: 323-325.
14. Packer, L., Maquire, J., Mehlhorn, R., Serbinova, E., Kagan, V., 1989, Biochem. Biophys. Res. Commun., 159: 229-235.
15 Rahimtula, A., 1983, Chem. Biol. Interact., 45: 125-135.
16. Schewe, T., Wiesner, R., Rapoport, S., 1981, in: Methods in Enzymology, Academic Press, London, 71:430-445.
17. Ursini, F., Maiorino, M., Gregolin, C., 1985, in: Free radicals in liver injury, 217-220, IRL Press, Oxford.
18. Utsumi, H., Shiimakura, A., Kashiwagi, M., Hamada, A., 1989, J.Biochem., 105: 239-244.
19. Weiss, R., Arnold, J., Estabrook, R., 1986, Arch. Biochem. Biophys, 252: 334-338.
20. Weiss, R., and Estabrook, R., 1986, Arch. Biochem. Biophys., 251: 336--347.

ISCHEMIA-REPERFUSION INJURY AND FREE RADICAL INVOLVEMENT

IN GASTRIC MUCOSAL DISORDERS

T. Yoshikawa, Y. Naito, S. Ueda, H. Ichikawa,
S. Takahashi, M. Yasuda, and M. Kondo

First Department of Medicine
Kyoto Prefectural University of Medicine
Kyoto 602, Japan

INTRODUCTION

Oxygen-derived free radicals have been implicated as possible mediators in the development of tissue injury induced by ischemia and reperfusion.[1,2] Furthermore, lipid peroxidation mediated by free radicals is believed to be one of the important causes of cell membrane destruction and cell damage.[3] The ischemia itself causes tissue damage and eventual death, but further injuries can occur while oxygen reintroduced to the tissue. Much evidence suggests that free radical and active oxygens including superoxide, hydrogen peroxide, hydroxyl radical, and singlet oxygen, contribute to the tissue injury. At least five possible sources are under investigation for the production of the active oxygen species: 1)the hypoxanthine-xanthine oxidase system ; 2)the activated polymorphonuclear leukocytes; 3)the disrupted mitochondrial electron transport system; 4)the metabolism of arachidonate via the lipoxygenase pathway, and 5) vascular endothelial cell.

The object of this investigation was to study the chronological changes in gastric mucosal injury and TBA-reactive substances in the gastric mucosa after ischemia or ischemia-reperfusion, and also to evaluate the antioxidative effect of several radical scavengers. In addition, to clarify the sources of oxygen radicals, the effect of the treatment with allopurinol or with anti-neutrophil serum on acute gastric mucosal injury was investigated.

ISCHEMIA-REPERFUSION INJURY MODEL IN THE RAT STOMACH

Male Sprague-Dawley rats weighing 180-220 g were used for ischemia-reperfusion. The animals were not fed 18 h prior to the experiments, but allowed free access to water. Ischemia in gastric mucosa was created under intraperitoneal pentobarbital anesthesia by applying small clamps to the celiac artery, and reoxygenation was produced by removal of the clamps.[4] By the clamping of celiac artery, the gastric

Oxygen Transport to Tissue XIII, Edited by T.K. Goldstick *et al.*
Plenum Press, New York, 1992

mucosal blood flow decreased to 10% of that measured before clamping, that was so called low flow state but not complete ischemia, and recovered to the normal range by the subsequent reperfusion. The total area of erosions (TAE), a morphological index of gastric injury, did not increase after 30 min, but significantly increased after 60 min ischemia, and 30 min ischemia with 30 min reperfusion ; however, the increase in TAE in the latter was significantly much higher than that in the former (Fig. 1). Between the cases of 90 min ischemia and 30 min ischemia with 60 min reperfusion, there is little difference in TAE. Therefore, these results are consistent with the view that the injury produced by 30 min reperfusion of a tissue subjected to 30 min ischemia is more severe than produced by ischemia per se.

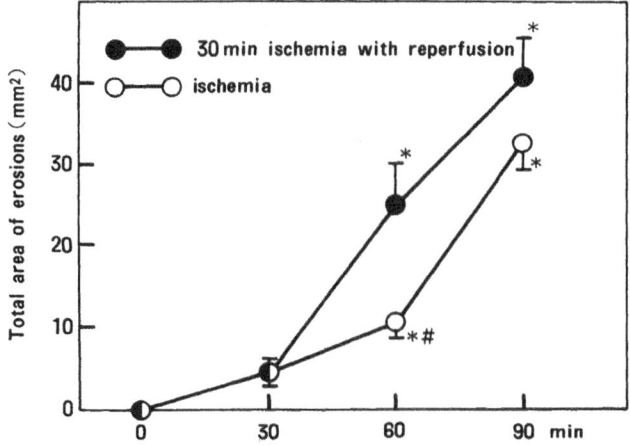

FIGURE 1 Changes in total area of gastric erosions after ischemia or ischemia-reperfusion. Each value indicates the mean ± SE of 5-12 rats. *p<0.001 for difference to the values of rats before clamping of celiac artery. #p<0.05 for difference to the values of rats 30 min after reperfusion following 30 min of ischemia.

ISCHEMIA-REPERFUSION INJURY AND LIPID PEROXIDATION

Important target molecules of biological damage caused by active oxygens are lipids, nucleic acids, enzymes, and proteins. In particular, unsaturated fatty acids located in the lipophilic section of cell membranes are prone to attack by active oxygens which produces lipid peroxides through a chain reaction of lipid peroxidation. Peroxidation is initiated by the attack of any chemical species that has sufficient reactivity to abstract a hydrogen atom from a methylene carbon in the unsaturated fatty acid, including hydroxyl radicals and iron-oxygen complexes. As shown in Fig. 2, the resulting carbon-centered radical (L·) is to undergo molecular rearrangement, followed by reaction with oxygen to give a peroxyl radical (LOO'). Peroxyl radicals can combine with each other to cause singlet oxygen formation, or they can attack membrane proteins, but they are also capable of abstracting hydrogen from adjacent fatty acid side chains in a membrane and so propagating the chain reaction of lipid peroxidation. Hence, a single initiation event can result in conversion of hundreds of fatty acid side chains into lipid hydroperoxides. Chain-braking antioxidants within the membrane interrupt the chain reaction by providing an easily-donatable hydrogen for abstraction by peroxyl radicals. The most important (but not the only) chain-braking antioxidant in human membrane is vitamin E.

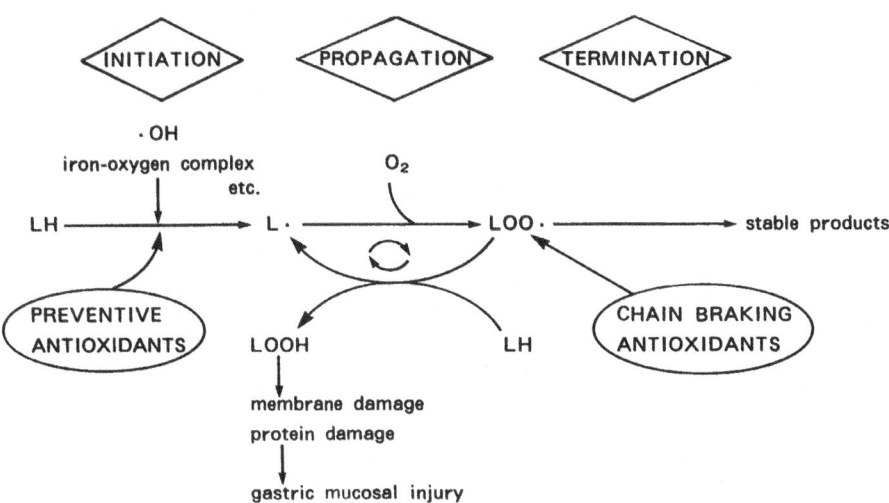

FIGURE 2 An outline mechanism of lipid peroxidation mediated by free radical chain reaction.

Lipid hydroperoxides produced by this process are thought to be main factor in the damage to biological membranes caused by active oxygens. The occurrence of lipid peroxidation in biological membranes causes impairment of membrane functioning, decreased fluidity, inactivation of membrane-bound receptors and enzymes, and non-specific permeability to ions such as Ca^{2+}. These pathological changes may attack the gastric mucosa with acid and pepsin to produce erosions or ulcers. To investigate the role of lipid peroxidation in the pathogenesis of gastric mucosal injury induced by ischemia-reperfusion, thiobarbituric acid (TBA)-reactive substances and α-tocopherol in the gastric mucosa, an index of lipid peroxidation, were measured by the method of Ohkawa et al.[5] and by HPLC method of Abe,[6] respectively. TBA-reactive substances in the gastric mucosa increased scarcely for 30 min ischemia but increased remarkably and significantly after following reperfusion.[4] α-Tocopherol in the gastric mucosa decreased slightly after ischemia, but more decreased significantly during reperfusion (Fig.3).[7] α-Tocopherol/cholesterol ratio also decreased 30 min and 60 min after reperfusion. These results suggests that lipid peroxidation or lipid peroxides may play an important role in the formation of gastric mucosal injury induced by ischemia-reperfusion. Vitamin E, which reacts with lipid peroxyradical and terminates free radical-mediated chain reaction, is speculated to be consumed in serum and gastric mucosa to prevent the development of tissue damage. Conversely, the decrease of the vitamin E suggests that the implication of free radicals in gastric mucosal injury induced by ischemia-reperfusion.

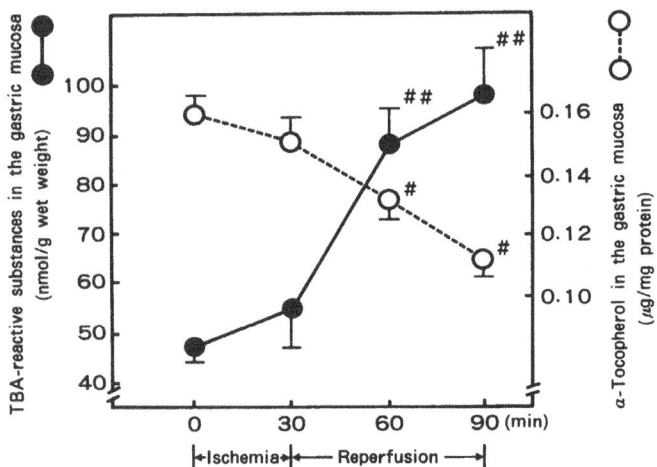

FIGURE 3 Changes in thiobarbituric acid(TBA)-reactive substances and α-tocopherol in the gastric mucosa after ischemia-reperfusion in rats. Each value indicates the mean ± SE of 5-14 rats. #p<0.05 and ##p<0.001 for difference to the values of rats before clamping of the celiac artery.

INHIBITION BY SCAVENGERS AGAINST ISCHEMIA-REPERFUSION INJURY

We then considered the possibility that active oxygens may induce the lipid peroxidation-mediated gastric mucosal injury with ischemia-reperfusion. To test this possibility, human SOD (50,000 U/kg, Nippon Kayaku Co., Ltd., Tokyo), a scavenger of superoxide, and/or catalase (90,000 U/kg, Sigma, St.Louis, MO), a scavenger of hydrogen peroxide, were injected subcutaneously 1 h before ischemia, and 10,000 U/kg of SOD was intravenously injected just before reperfusion, because the active oxygen cannot be measured directly in vivo. As shown in Fig.4, the increase in the total area of the erosions was significantly inhibited by pretreatment with SOD, catalase, and SOD+catalase, and the increase in TBA-reactive substances also significantly inhibited by pretreatment with SOD plus catalase.[8] In addition, SOD and catalase inhibited the aggravation of these changes without recovering the reduced gastric mucosal blood flow or affecting the gastric mucosal protective factors including prostagrandins and mucous secretion. These results indicate that the effectiveness of these enzymes possibly appears through the catalyzation of active oxygens, especially superoxide radicals, and not through the effects on gastric mucosal microcirculation.

In addition, we have reported that Ebselen which shows glutathione peroxidase-like activity, can protect against the gastric mucosal injury induced by ischemia-reperfusion, and also inhibits the lipid peroxidation in the gastric mucosa (Fig.5).[9] The results indicate that scavenging hydrogen peroxides and lipid peroxides is important for protecting the gastric mucosa from ischemia-reperfusion injury.

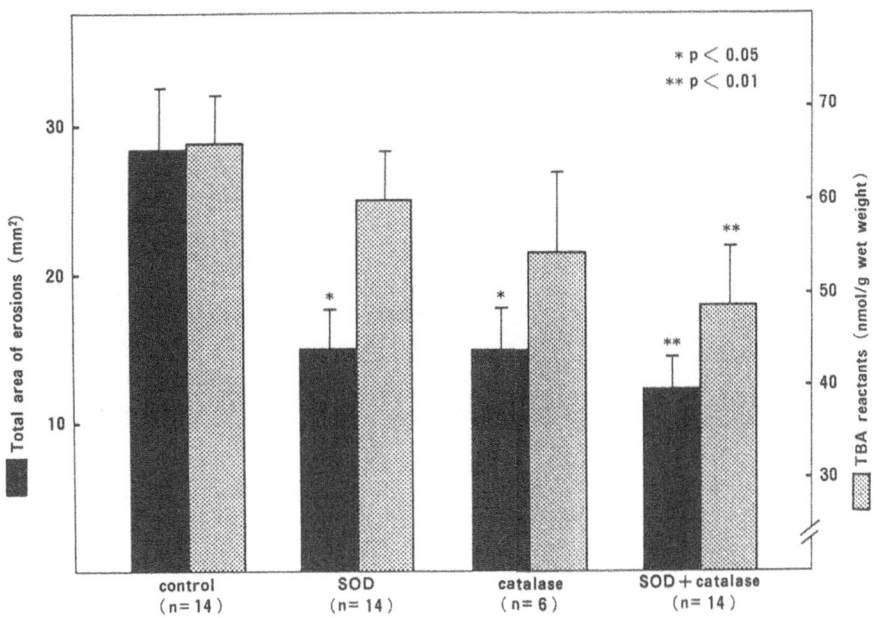

FIGURE 4 Effect of SOD and/or catalase on the total area of erosions and on the increase in TBA-reactive substances in the gastric mucosa after ischemia-reperfusion. SOD at a dose of 50,000 U/kg was injected subcutaneously 60 min before the claming and 10,000 U/kg was injected intravenously 1 min before the reperfusion. Catalase at a dose of 90,000 U/kg was injected subcutaneously 60 min before the clamping. Each value indicates the mean $\pm$ SE of 6-14 rats. *p<0.05 and **p<0.01 for difference to the values of rats treated with physiological saline.

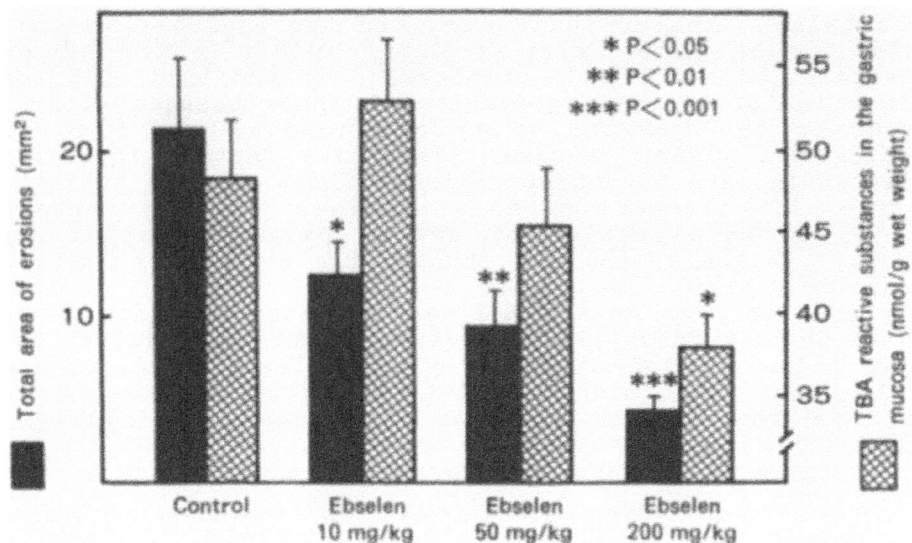

FIGURE 5 Effect of Ebselen on the total area of erosions and on the increase in TBA-reactive substances in the gastric mucosa after ischemia-reperfusion.   Ebselen dissolved in carboxymethyl cellulose(CMC) solution was treated 60 min before the clamping.   Each value indicates the mean ± SE of rats. *p<0.05, **p<0.01 and ***p<0.001 for difference to the values of rats treated with CMC solution.

SOURCES OF OXYGEN RADICALS

In ischemic intestinal tissue, the source of oxygen free radicals is thought to be the action of xanthine oxidase on hypoxanthine.[10]   While hypoxanthine accumulates during ischemia as a result of adenosine 5'-triphosphate degradation, molecular oxygen is provided by reperfusion, which could theoretically result in a burst of superoxide production. In a gastric injury model induced by hemorrhagic shock, Itoh and Guth[11] reported that gastric injuries were significantly reduced by treatment with allopurinol, a competitive inhibitor of xanthine oxidase. To determine whether xanthine oxidase plays a role in producing the gastric mucosal injury associated with ischemia-reperfusion in our rat model, we have utilized allopurinol which was administered orally (50 mg/kg) to rats 48, 24 h prior to the experiment.   Xanthine oxidase activities in the gastric mucosa was 21.6 ± 2.2 U/g wet weight in the control rats, however, the activity fell to undetectable level by the treatment with allopurinol. As shown in Fig.6, treatment with allopurinol significantly attenuated the gastric mucosal injury induced by ischemia-reperfusion, and significantly inhibited the increase in TBA-reactive substances in the gastric mucosa, indicating that xanthine oxidase in a principle source of active oxygens produced during ischemia-reperfusion in the stomach.   However, this xanthine oxidase hypothesis has many unknown elements which must be resolved in the future, including the time required to convert xanthine oxidase, the mechanism of the conversion, and the enzyme distribution within tissue.

Another potential source of oxygen radicals in ischemia-reperfusion is the activated polymorphonuclear leukocyte(PMN). PMNs can generate superoxide anion when exposed to appropriate particles or specific soluble inflammatory mediators. Recent studies have demonstrated the involvement of PMN in the pathogenesis of gastric and intestinal mucosal injuries. Grisham et al.[12] have proposed that ischemia and reperfusion results in xanthine oxidase-generated, superoxide-dependent accumulation of PMNs in the intestinal mucosa, where PMN-derived oxidants or exacerbate injury, or both. From the observation that the reperfusion-induced accumulation of myeloperoxidase (an index of PMN infiltration) in intestinal mucosa is significantly attenuated by pretreatment with either allopurinol, superoxide dismutase, they suggest that xanthine oxidase-derived oxygen radicals play an important role in modulating the leukocyte adherence and extravasation initiated by reperfusion.

In the stomach, Smith et al.[13] have reported that depletion of neutrophils via neutrophil antiserum resulted in a dramatic reduction in the area of gross lesions as well as a reduction in red cell flux into the lumen of the stomach after ischemia-reperfusion. We also produced PMN depletion

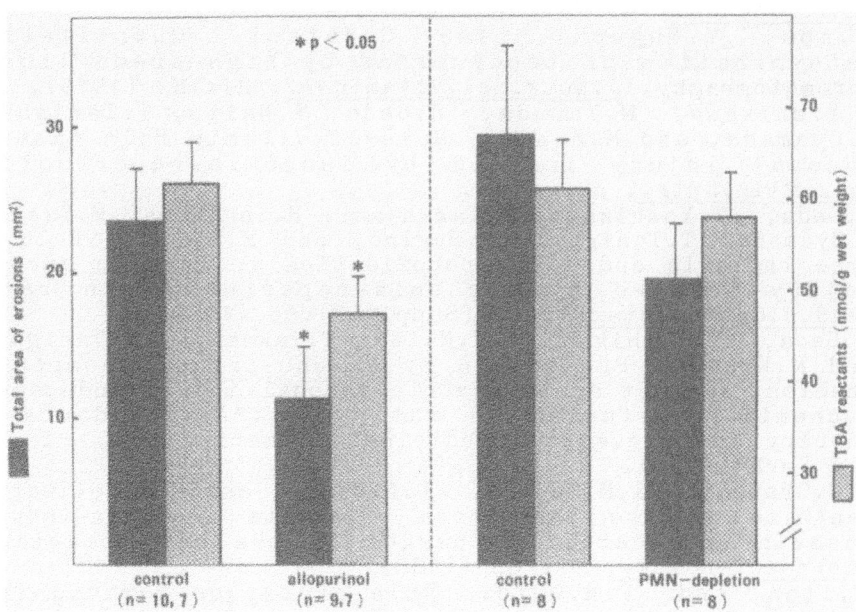

FIGURE 6 Effect of allopurinol or anti-polymorphonuclear leukocyte(PMN) antibody on the total area of erosions and on the increase in TBA-reactive substances in the gastric mucosa after ischemia-reperfusion. Allopurinol at a dose of 50 mg/kg was orally administered for 2 days before the experiments. PMN-depletion rats were produced by administration with anti-rat PMN antibody 18 hr before the clamping. Each value indicates the mean ± SE of 7-10 rats. *p<0.05 for difference to the values of control rats.

rats by administration of the anti-rat PMN serum from the immunized rabbits. However, the depletion of PMN counts did not show significant inhibition against the aggravation of gastric mucosal lesions or the increase in TBA-reactive substances in the gastric mucosa induced by ischemia-reperfusion (Fig. 6). As compared with the hypoxanthine-xanthine oxidase system, PMNs seem to play a relatively small part in the formation of gastric mucosal injury induced by ischemia-reperfusion.

REFERENCES

1. D.N.Granger, G.Rutili, and J.M.McCord. Superoxide radicals in feline intestinal ischemia. Gastroenterology, 81:22 (1981).
2. J.M.McCord. Oxygen-derived free radicals in postischemic tissue injury. N.Engl.J.Med., 312:159 (1985).
3. E.Niki. Antioxidants in relation to lipid peroxidation. Chem.Phys.Lipids, 44:227 (1987).
4. T.Yoshikawa, S.Ueda, Y.Naito, S.Takahashi, H.Oyamada, Y.Morita, T.Yoneta, and M.Kondo. Role of oxygen-derived free radicals in gastric mucosal injury induced by ischemia or ischemia-reperfusion in rats. Free Rad.Res.Comms., 7:3 (1989).
5. H.Ohkawa, N.Ohishi, and K.Yagi. Assay for lipid peroxide for animal tissues by thiobarbituric acid reaction. Anal.Biochem., 95: 351 (1979).
6. K.Abe, Y.Yuguchi, and G.Katsui. Quantitative determination of tocopherols by high-speed liquid chromatography. J.Nutr.Sci.Vitaminol., 21:183 (1975).
7. T.Yoshikawa, M.Yasuda, S.Ueda, Y.Naito, T.Tanigawa, H.Oyamada, and M.Kondo. Role of vitamin E in gastric mucosal injury induced by ischemia-reperfusion. Am.J.Clin.Nutr., in press.
8. S.Ueda, T.Yoshikawa, S.Takahashi, H.Ichikawa, M.Yasuda, H.Oyamada, T.Tanigawa, S.Sugino, and M.Kondo. Role of free radicals and lipid peroxidation in gastric mucosal injury induced by ischemia-reperfusion in rats. Scad.J.Gastroenterol., 24(Suppl.62):55 (1989).
9. S.Ueda, T.Yoshikawa, Y.Naito, T.Takemura, T.Tanigawa, and M.Kondo. Protection by seleno-organic compound, Ebselen, against acute gastric mucosal injury induced by ischemia-reperfusion in rats, in :"Antioxidants in therapy and preventive medicine" I.Emerit, ed., Plenum, New York (1990).
10. D.N.Granger, J.M.McCord, D.A.Parks, and M.E.Hollwarth. Xanthine oxidase inhibitors attenuate ischemia-induced vascular permeability changes in the cat intestine. Gastroenterology, 90: 80 (1986).
11. M.Itoh and P.H.Guth. Role of oxygen-derived free radicals in hemorrhagic shock-induced gastric lesion in rats. Gastroenterology, 88:1162 (1985).
12. M.B.Grisham, L.A.Hernandez, and D.N.Granger. Xanthine oxidase and neutrophil infiltration in intestinal ischemia. Am.J.Physiol., 251:G567 (1986).
13. S.M.Smith, L.Holm-Rutili, M.A.Perry, M.B.Grisham, K-E.Arfors, D.N.Granger, and P.R.Kvietys. Role of neutrophils in hemorrhagic shock-induced gastric mucosal injury in the rat. Gastroenterology, 93:466 (1987).

INFLUENCE OF ANTI-INFLAMMATORY DRUGS AND FREE RADICAL SCAVENGERS ON INTESTINAL ISCHEMIA INDUCED OXIDATIVE TISSUE DAMAGE

A.J. Augustin, R.K. Goldstein, J. Milz and J. Lutz

Physiologisches Institut der Universitaet

Roentgenring 9, D-8700 Wuerzburg, F.R.G.

INTRODUCTION

The model of intestinal ischemia, effected by an occlusion of the superior mesenteric artery, seems to be well suited to assess the contribution of oxygen free radicals in this ischemic disease and, by using different free radical scavengers, the sources of these oxidative metabolites as well. Thus, because there is substantial body of evidence that neutrophils are one of the major sources of free radicals, a treatment of intestinal ischemia by antibiotics can be helpful (Lutz and Augustin, 1989). These antibiotics manage a diminution of chemoattractants for neutrophils, leading to an only slight activation of these cells. There is still a controverse discussion about the tissue damaging activity of myeloperoxidase in vivo by generation of the oxidant HOCl. Many clinical data indicate only a slight toxicity, which is in contradiction to the high reactivity of OCl^- (for review see Weiss, 1989). In spite of that, other findings show an oxidative damage done by hypohalous acids if myeloperoxidase can first bind to its target. This damage is believed to be a hypochlorous acid damage because of the high concentration of Cl^- compared to other halides in body fluids. Moreover, there seems to be a tissue dependence in chloramine toxicity concerning lung and vessel beds above all. Chloramine is the active metabolite of OCl^-, formed by a reaction of OCl^- with endogenous, amine-containing moieties.

Regardless the above presented facts, myeloperoxidase activity itself is a well suited parameter to assess neutrophil immigration into tissues, because of the good correlation between MPO-activity and the number of neutrophils. This is important as far as there are further pathways of production of oxidative metabolites in neutrophils like the superoxide anion producing membrane bound oxidase and the arachidonic acid pathway, activated by phospholipase A2 which anew can be effected by other mediators of inflammation like plasmine or complement C5a (Fig.7). By the reason of a possible contribution of the phospholipase pathway a treatment with drugs, which act as phospholipase inhibitors, should be taken into account.

Oxygen Transport to Tissue XIII, Edited by T.K. Goldstick *et al.*
Plenum Press, New York, 1992

For a long period xanthine oxidase was believed to be the primary source of oxygen free radicals (Granger et al.,1986) which was supported by the protection of tissue by allopurinol and by the fact, that the intestine is one of the richest sources of this enzyme (Della Corte & Stirpe, 1972). Now the xanthine oxidase theory which propagated XO as the major source of oxidative damage is left since Parks and coworkers (1988) could show that D-O conversion of the enzyme needs about 1h for 20% conversion, while Roy and McCord (1983) demonstrated a complete D-O conversion within 1 minute. These data show that it is preferable to use a model of complete ischemia to observe oxidative tissue damage rather than the hypotension model, designed because of unconfirmed data of conversion rapidity (Granger et al., 1986). The action of allopurinol as a hypochlorous acid (Grootveld et al., 1987) and as a hydroxyl radical scavenger (Moorhouse et al., 1987) explains the beneficial effects of this drug. Nevertheless it should be taken into account that xanthine oxidase produces certain amounts of superoxide anion radical. The quantity of the production could be shown to be dependent on ischemia durance. This radical species is primarily harmless, but a precursor of tissue damaging radicals. Beside that xanthine oxidase can liberate iron from ferritine (Biemond et al., 1986) which is superoxide-independent and requires no oxygen. These findings are supported by data of us which show an oxidative damage, already in the ischemic period (Lutz and Augustin, 1989).

Recent results about hydroxyl radical producing iron dependent reactions (Haber Weiss, Fenton raction), former believed as "non reactions" (Del Meastro, 1979), show a contribution of free iron not only in propagation but in initiation as well (Aruoma et al., 1989). The action of the iron binding enzymes transferrin and lactoferrin is dependent on iron loading and pH. At low pH, which is found in ischemic tissues, iron is easily mobilized from both enzymes, taking place in the above mentioned reactions. If the proteins are saturated with iron, a loss of protective effects is remarkable also. Thus, lactoferrin and transferrin can protect against OH·-generation and promote the generation as well (Aruoma and Halliwell, 1987).

The aim of this study was the evaluation of the different free radical sources in our model of intestinal ischemia by determination of lipid peroxides and myeloperoxidase activity in intestinal tissue under the influence of different drugs. Special attention is focused on the two different periods of ischemic disease by experiments, done with different reperfusion times.

MATERIAL AND METHODS

Experiments were performed in male Wistar rats by reversibly occluding the superior mesenteric artery under ether anesthesia. This was done by pulling the vessel against a thread, which was lead via a tube through the abdominal wall and fixed until removal. The rats were given an analgesic (tramadol, 2.5 mg/kg b.wt.) immediately after the initial operation and again when the occlusion was reopened under ether anesthesia. The occlusion period lasted 90 min, the reperfusion time (until sacrificing the animals) 2.5 hours or 30, 60, 120 and 240 minutes. After the end of experiments the tissue content of myeloperoxidase and lipid peroxides was determined.

Determination of lipid peroxides: The lipid peroxide level of intestinal tissue was determined by a modification (Lutz et al., 1990) of the method of Ohkawa et al. (1979). About 1 g of tissue was homogenized in 1.15% KCl (9ml) for 30 sec with an Ultra Turrax ^{R} blender (Janke & Kunkel, Staufen, FRG). The assay mixture consisted of 0.1 ml of the sample, 2 ml of 0.9% NaCl, 0.2 ml of sodiumduodecylsulfat (%), 3 ml of TBA (thiobarbituric acid reagent, containing equal parts of 0.8% aqueous thiobarbituric acid and acetic acid) and was heated for 75 min at 95 C. Thereafter the mixture was cooled in tap water and 5 ml butanol for fluorimetric determinations (Uvasol^R, Merk AG, Darmstadt, FRG) were added. The colouring matter was brought to the organic layer by shaking the mixture vigourosly. After centrifugation at 3000 rpm for 15 min the samples were assayed fluorimetrically at 515 nm excitation and 553 nm emmission. Amounts of 0,1,2,3 nmol of tetramethoxypropane served as external standard and were assayed in the above described way.

Determination of myeloperoxidase: The tissue level of myeloperoxidase (MPO) was determined by a method of Krawisz et al. (1984), first described by Bradley in 1982 for skin tissues. The specimen (200-500 mg) were homogenized for 30 sec (three times) in 3 ml hexadecyltrimethyammoniumbromide (HTBA) solution (0.5% HTBA in 50 nM phosphate buffer, pH 6.0) with the Ultra Turrax blender. HTBA acts releasing MPO from the primary granules of the neutrophils. The homogenate was sonicated for 10 sec, freeze-thawed three times and centrifugated at 40000 g for 15 min at 4° C. This procedure supplies a stabil pellet. The supernatant was assayed for myeloperoxidase activity spectrophotometrically: 2.9 ml of 50 mM phosphate buffer, pH 6.0, containing 0.167 mg/ml O-dianisidine hydrochloride and 0.0005% H_2O_2 combined with 0.1 ml of the supernatant of the tissue samples, prepared as described above.

The change in absorbance at 460 nm was measured with a Shimadzu UV-190 spectrophotometer. MPO data are expressed as values x controls (Grisham et al., 1986).

Survival of animals: Three groups (n.th., pr.tr., CP) were compared concerning survival after 90 min of mesenteric artery occlusion. The abbreviations of drugs are explained below; details about pr.tr. which means an antibiotic pretreatment are reported elsewhere (Lutz and Augustin, 1989). Estimation of survival was done by a test of Kaplan and Meier (1958). We also compared the mean survival time statistically (Lutz et al., 1985).

Statistical analysis: The values are reported as means ± SE. Student's t test and analysis of variance were used to determine wether data groups were statistically different. $P<0.05$ was considered to be significant.

The following drugs were given in comparison to an untreated group (n.th.). If not stated otherwise they were given at the end of the occlusion period: superoxide dismutase (SOD), 20 mg/kg bwt. = 60.000 U/kg, given together with CAT, half the dose i.p. and i.v.; catalase (CAT), 20 mg/kg =680.000 U/kg; allopurinol (ALLO), 100 mg/kg iv.; ciprofloxacin (CF), 20 mg/kg ip. given one hour before occlusion, 60 mg/kg at the end of the occlusion period; deferoxamine (DFO), 30 mg/kg im. one hour before the occlusion and 30 mg/kg iv. at the end of reperfusion. Vitamin E (VIT E), 250 mg/kg im. 24h before operation. Sulfasalazine (SUL), 1 g/kg and Mesalazine (MES), 1 g/kg, intraluminally 2 h before the occlusion each.

RESULTS

Figure 1 shows the increase of lipid peroxides in tissue contingent on durance of reperfusion with a more than 20 fold increase after 240 minutes of reperfusion. Like demonstrated in previous experiments and even in other ischemia models (Augustin et al., 1990), there is a remarkable (more than tenfold) increase of lipid peroxides during the ischemic period, indicating an oxidative tissue damage already during ischemia. In comparison with these results the myeloperoxidase activity (Fig.1) increased during ischemia 10 fold, after 150 min reperfusion 14 fold (p<0.01) and after 240 min 17 fold (p<0.05 compared to 0 min and 150 min each). The increase of myeloperoxidase shows an immigration of neutrophils. The difference in pattern of increase, compared to lipid peroxides, leads to scavenging mechanisms which are overcharged with oxidative metabolites after a certain reperfusion time.

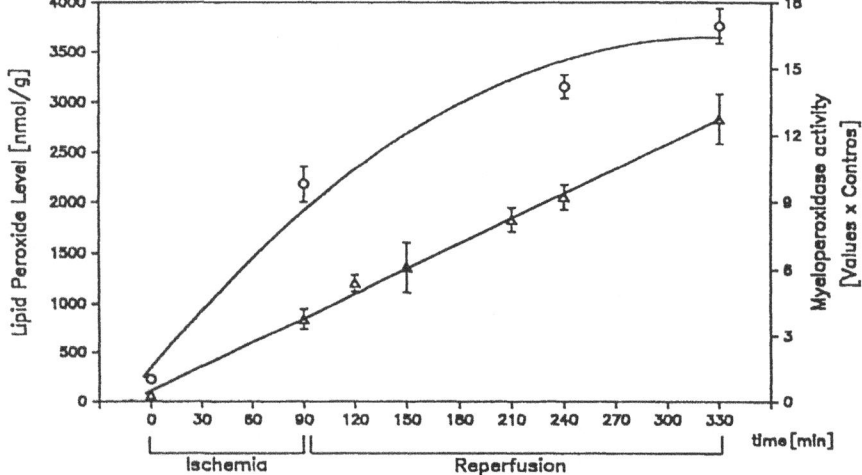

Figure 1. Increase of myeloperoxidase activity (circles) and lipid peroxides (triangles) during ischemia and reperfusion. Regression order: MPO: $y = a_0 + a_1x + a_2x^2$ ($a_0 = 1.54$; $a_1 = 0.091$; $a_2 = -0.00014$); LPO: $y = a_0 + a_1 x$ ($a_0 = 108.62$; $a_1 = 8.25$).

The above mentioned increase of lipid peroxides in intestinal tissue of animals without therapy after 150 min reperfusion was comparable to that shown in previous experiments (Fig.2). VIT E reduced the lipid peroxide level to the range of initial values (p<0.01). With the SOD/CAT-treatment a remarkable reduction of LPO was reached (p<0.01) which may be due to the higher concentration of the drugs used in these experiments (in comparison to previous results). The ALLO-therapy also showed beneficial effects with reduction of LPO to a quarter of values without therapy (p<0.01). DFO-treatment only achieved a reduction which was not significantly different from ischemic values (see Fig.1), but from reperfusion values (p<0.05). These data fit to the myeloperoxidase values (Fig.3) with VIT E showing the largest effect on MPO activity (p<0.01), followed by SOD/CAT (p<0.01), allopurinol (p<0.01) and DFO (p<0.05). The results indicate that acting directly to tissue damaging oxidative metabolites seems to be a better approach than the inhibition of single pathways, which produce free radicals, like the Haber Weiss or Fenton reaction. This does not seem to be valid for a scavenging of harmless radicals (superoxide anion or hydrogen peroxide).

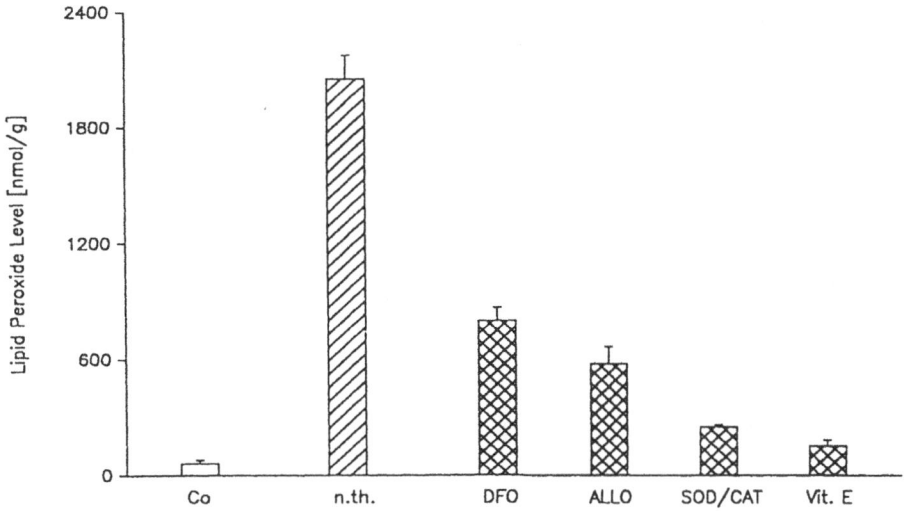

Figure 2. Lipid peroxide levels in intestinal tissue. All values showed a significant difference (p<0.05 for DFO; p<0.01 for the other therapy groups) as compared to the n.th. group. Abbreviations and doses see methods.

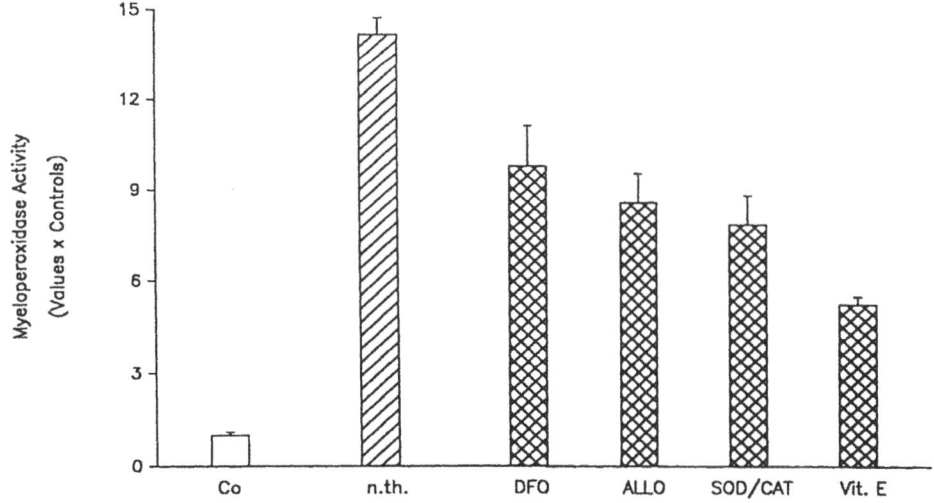

Figure 3. Myeloperoxidase activity expressed as values x controls. All values were significant as compared to n.th. (p<0.05 for DFO; p<0.01 for others).

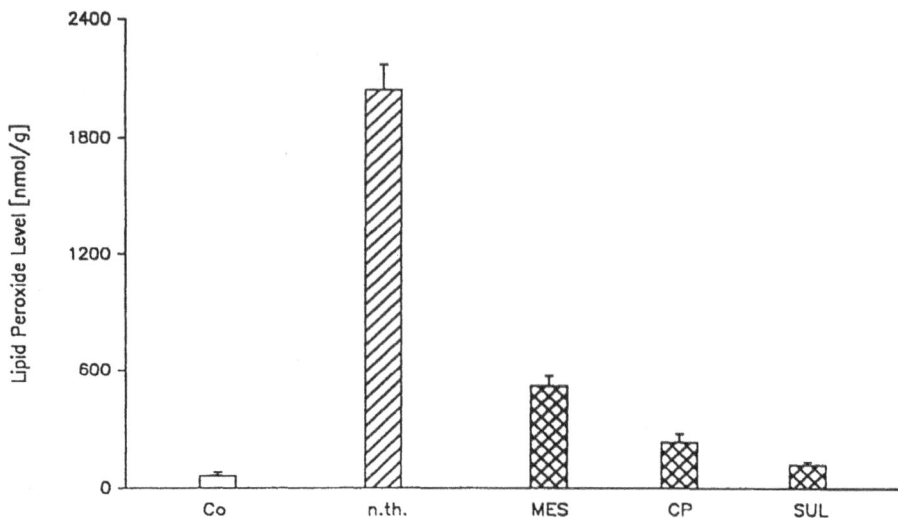

Figure 4. Lipid peroxide level in intestinal tissue under the influence of antiinflammatory drugs. All values are significant (p<0.01) as compared to the n.th. group. Abbreviations and doses see methods.

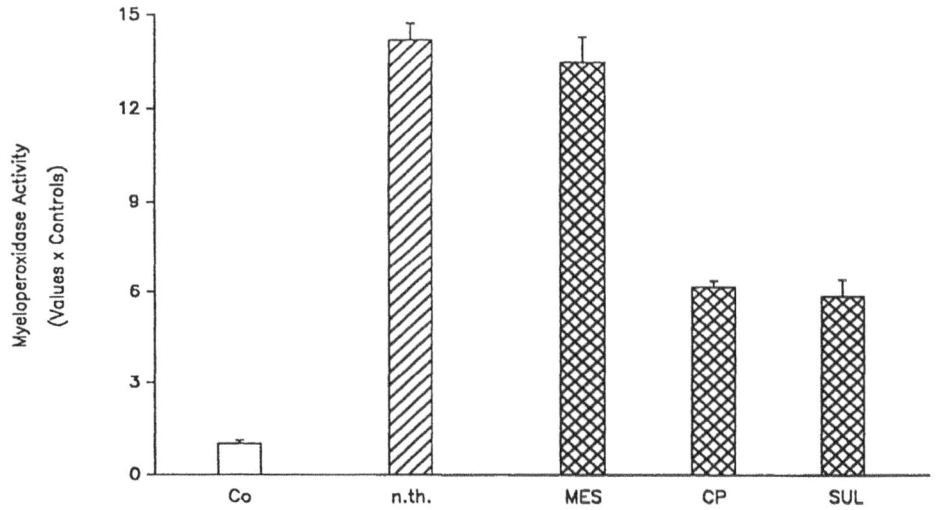

Figure 5. Myeloperoxidase activity in intestinal tissue under the influence of antiinflammatory drugs. CP and SUL are significant (p<0.01), whereas MES is not significant as compared to the n.th. group.

These harmless oxidative metabolites are not only precursors of tissue damaging radicals, like singlet oxygen or hydroxyl radical, but they are also able to form chemoattractants for neutrophils. This is corroborated by the considerable reduction of MPO by SOD/CAT.

CP, a fluorinated quinolone showed an effect on lipid peroxide level and myeloperoxidase activity (Fig.4 and 5, p<0.01 each) which is comparable to that observed under VIT E therapy. The forecoming discussion about a contribution of inflammation in ischemic intestinal diseases seems to be confirmed by the fact of tissue protection by an antibiotic. Sulfasalazine (a combination of an antibiotic and 5-ASA with the active metabolite being still in discussion) also showed a beneficial effect on both determined parameters (p<0.01 each, Fig.4 and 5). These results are comparable to those obtained with CP. With Mesalazine these values were not reached. The lipid peroxides were different from n.th. values (p<0.01), whereas the MPO activity was not significantly reduced. Thus, recent findings which propagate a characteristic action of SUL, former believed as a carrier for 5-ASA, seem to be confirmed.

An assessment of survival in two groups with beneficial effects on lipid peroxides showed a significant improvement in the two experimental groups (Table 1, Fig.6). CP and an antibiotic pretreatment was already presented by us concerning the effect on lipid peroxidation (Lutz and Augustin,

Table 1. Mean survival time and survival rate of two therapy groups and controls. Both were significantly different from controls (n.th.). (pr.tr.=antibiotic pretreatment, p<0.05); CP=ciprofloxacin, p<0.05).

Group	n	survival time (hours)	survival rate (%)			
			12h	48h	120h	168h
n.th.	10	6.54 ± 0.80	0	0	0	0
pr.tr.	7	63.67 ± 22.43	41.7	33.3	33.3	33.3
CP	7	69.15 ± 32.29	41.7	41.7	33.3	33.3

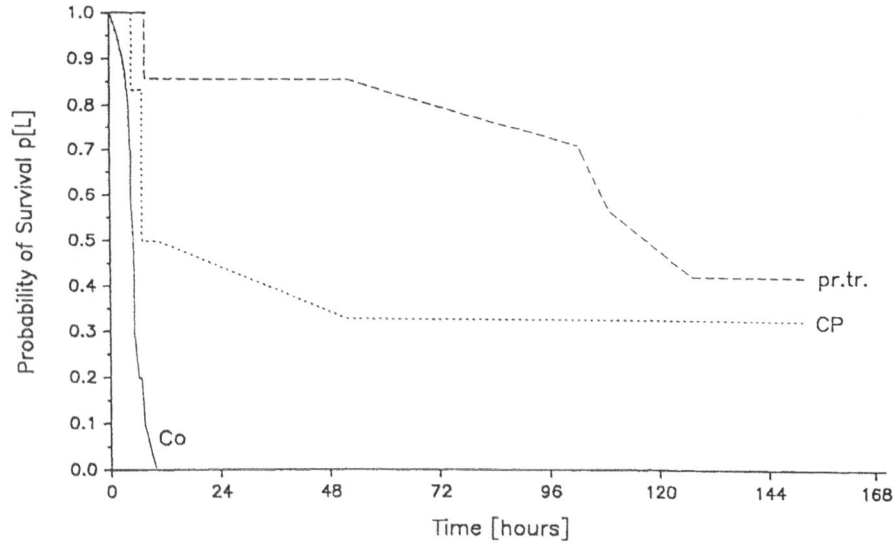

Figure 6. Estimation of survival of animals after 90 minutes of mesenteric artery occlusion (Kaplan and Meier, 1958). Both therapy groups are significant (p<0.01 for pr.tr.; p<0.05 for CP) as compared to the n.th. group.

1989; Lutz et al., 1990). This confirms the theory of oxygen free radical production being one of the major pathophysiological events in ischemic intestinal diseases.

DISCUSSION AND CONCLUSION

Our results in scavenger experiments present highly dosed vitamin E to be most effective. It is believed to become steric opportune fixed into membrane phospholipids that the position of the hydroxyl dyat of the benzene ring is in optimal position for a reaction with potential tissue damaging oxygen metabolites (Sokol, 1989). Therefore vitamin E is named a chain breaking antioxidant and is believed to be the most important lipid soluble one in vivo. It is still unclear, if vitamin C is a free radical scavenger as well, but there is body of evidence that the primary function of body own vitamin C is to react with tocopherones to regenerate vitamin E for further reactions (Hayden and Paniello, 1987). Since vitamin E reduces MPO as well, we conclude that this drug prevents neutrophil immigration. This can be interpreted by a prevention of the production of chemoattractants for neutrophils by reactive oxygen species like the formation of a chemotactical factor by the reaction of $O_2 \cdot$ and arachidonic acid. These experiments indicate the advantage of free radical scavengers, which act against tissue damaging agents. This is confirmed by results obtained in allopurinol therapy, which is believed to be a hypochlorous acid and hydroxyl radical scavenger as well (references see above). The difference in effectiveness may be due to the different application manner (pretreatment vs. reperfusion therapy). Therefore it is remarkable that allopurinol reduces LPO and MPO below values already reached in ischemia. The results of SOD/CAT show that they act against important precursors of the tissue damaging radicals (see above). A combination of SOD with catalase is to be carried out, because the dismutation of SOD leads to H_2O_2 which is not only a substrate of the Haber Weiss mechanism but also can activate proteolytic enzymes by the release of Ca^{++}-ions.

The great importance of free iron in the ischemic disease is shown by the results of the DFO therapy group. The lipid peroxide level could be reduced to a value obtained immediatly after ischemia, whereas the reduction of myeloperoxidase activity was only significant after extension of the group number to six animals. These data lead to the conclusion, that free iron is contributed to propagation rather than initiation, which was proposed by Aruoma et al. (1989). We also conclude from our data that the iron dependent free radical production is one, but not the major source, of oxygen metabolites in this disease, because of the nearly missing reduction of myeloperoxidase activity which indicates that there was no prevention from neutrophil immigration. It should be taken into account that iron acts in different ways to form oxidative metabolites which means decomposing of already originated lipid peroxides from other reactions into peroxyl and alkoxyl (lipid-O$\cdot$) radicals that are capable of abstraction of hydrogen themselves (Gutteridge, 1988). The possibility of iron liberation from transferrine and lactoferrine in the acid ischemic microenvironment was mentioned above already. As shown in Fig.7, intestinal ischemia goes along with a hemorrhage which may lead to an iron overload and stimulate

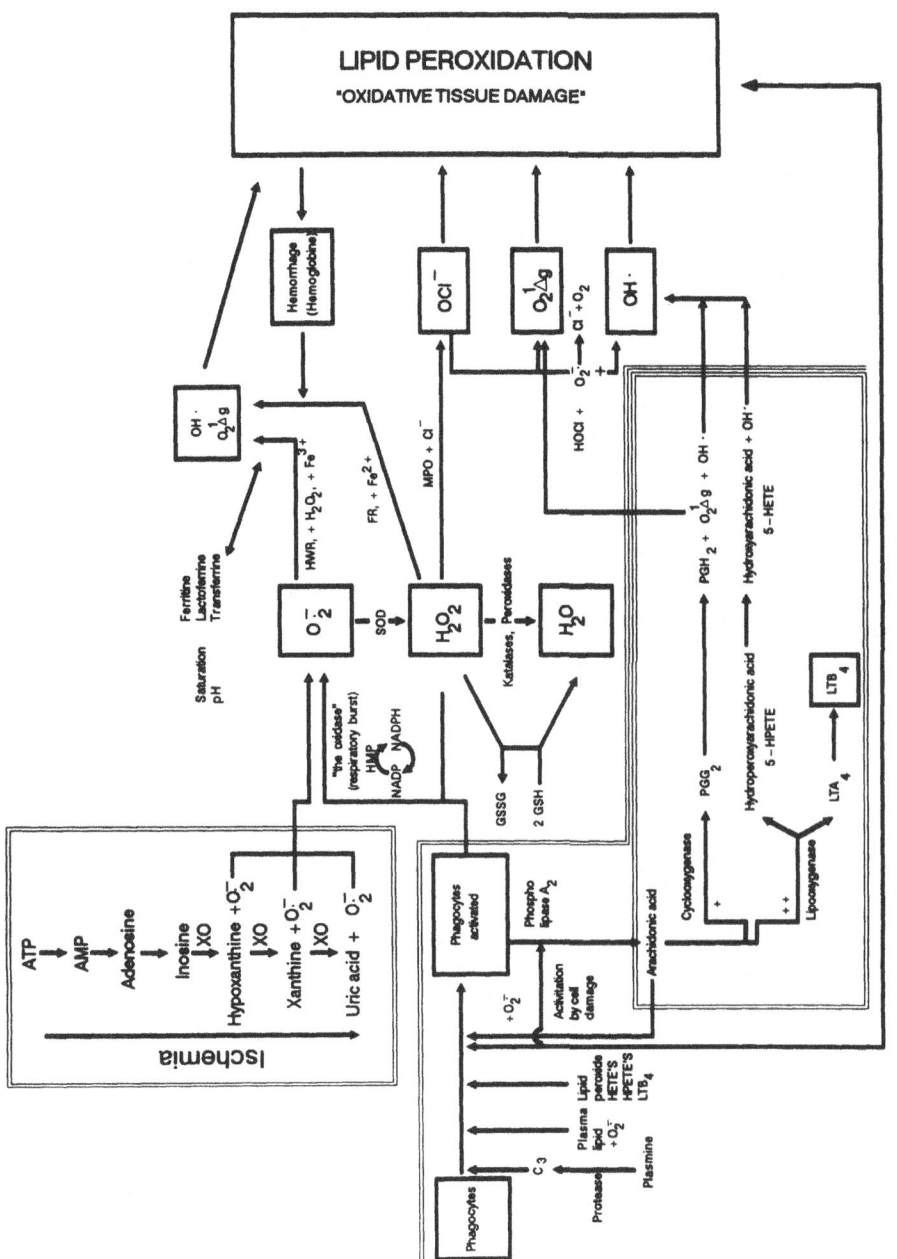

Figure 7. Production of oxidative metabolites by ischemia and inflammation. Upper part: xanthine oxidase and iron dependent reactions. Middle: reduction of oxygen to water (four electron reduction). Left and below: activation of neutrophils; involvement of the arachidonic acid pathway. Abbreviations: FR = Fenton reaction; HWR = Haber Weiss reaction; LT (A,B) = leucotriene (A,B); PG= prostaglandine; C₃ = compl. 3; HMP = hexosemonophospate shunt.

247

iron liberation from these enzymes (Aruoma and Halliwell, 1987). Gutteridge (1986) demonstrated that iron which is released from hemoglobin by hydrogen peroxide or organic hydroperoxides is the generator of hydroxyl radicals in the Fenton reaction.

The effect of the antibiotic CP confirms the fact that ischemic intestinal disease is accompanied with a peritonitis or inflammatory reactions of stimulated neutrophils which can lead to intraabdominal sepsis in the animals. Benefit results of a CP treatment could be obtained from other experimental models as well (Lahnborg and Nord, 1989).

There is still discussion about the active metabolites in the antiinflammatory drugs sulfasalazine and mesalazine (5-ASA). Whereas Ahnfelt-Ronne and Nielsen (1987) showed that 5-ASA is the active metabolite of sulfasalazine, Niel and coworkers (1987) proposed the azo linking group of sulfasalazine as the structural feature of the free radical scavenging activity of this compound. They only could show a slight effect of 5-ASA alone. Recent findings by us confirm these results with no effects of 5-ASA (not presented), given intraluminally, which is to be explained by a rapid resorption and inactivation by acetylation (Ronne and Nielsen, 1987). Therefore 5-ASA should be applicated as a coated compound in order to prevent rapid resorption and to keep the substance at the site of inflammation. The above mentioned results with sulfasalazine as a free radical scavenger itself are not in contradiction with the effects of mesalazine on lipid peroxides. The action of this drug on neutrophils can also be a phospholipase A_2 or lipoxygenase inhibition since these pathways could be shown to be involved in postischemic inflammation (Otamiri and Tagesson, 1989). This fact was also proposed by Aruoma et al. (1987) which could show an hydroxyl radical and OCl^- scavenging activity of sulfasalzine and 5-ASA as well.

SUMMARY

The influence of oxygen free radical scavengers and antiinflammatory drugs on postischemic lipid peroxidation and myeloperoxidase activity was shown. The best results were obtained from vitamin E and the antiinflammatory treatment with CP and SUL, whereas an iron elimination only showed slight effects on myeloperoxidase activity above all. In experiments without therapy a linear increase of lipid peroxides dependent on reperfusion durance was found, whereas myeloperoxidase already showed a remarkable increase during ischemia and early reperfusion. This difference can be interpreted by scavenging mechanisms, which are overcharged after an appointed durance of reperfusion.

KEYWORDS: allopurinol - catalase - ciprofloxacin - deferoxamine - ischemia - lipid peroxides - mesalazine - myeloperoxidase - reperfusion - superoxide dismutase - sulfasalzine -

REFERENCES

Ahnfelt-Ronne I., Nielsen O.H. (1987) The anti-inflammatory

moiety of sulfasalazine, 5-amonosalicylic acid, is a radical
scavenger. Ag. Act. 21, 191-194

Aruoma O.I., Halliwell B. (1987) Superoxide-dependent and as-
corbate-dependent formation of hydroxyl radicals from hydrogen peroxide in the presence of iron. Biochem. J.241, 272-278

Aruoma O.I., Wasil M., Halliwell B., Hoey B.M., Butler J.
(1987) The scavenging of oxidants by sulfasalazine and its
metabolites. Biochem. Pharmac. 36, 3739-3742

Aruoma O.I., Halliwell B., Laughton M.J., Quinlan G.J.,
Gutteridge J.M.C. (1989) The mechanism of lipid peroxidation.
Evidence against a requirement for an iron(II)-iron(III)
complex. Biochem. J. 258, 617-620

Augustin A., Lutz J. (1988). The effect of a temporary occlusion of the superior mesenteric artery on the level of lipid
peroxides in plasma and intestinal tissue. Europ. J. Physiol.
11, R 51

Augustin A., Purucker E., Milz J., Lutz J. (1990) Intestinal
and hepatic lipid peroxidation after aortic occlusion and
reperfusion. Europ. J. Physiol. 415 Suppl.1, R 45

Biemond P., Swaak A.J.G., Beindorf M., Koster J.F. (1986)
Superoxide-dependent and -independent mechanisms of iron
mobilization from ferritin by xanthine oxidase. Biochem. J.
239, 169-173

Bradley P.P., Priebat D.A., Christensen R.D., Rothstein G.
(1982) Mesurement of cutaneous inflammation: Estimation of
neutrophil content with an enzyme marker. J. Invest.
Dermatol. 78, 206-209

Del Maestro R. (1979) The influence of of oxygen derived free
radicals on in vitro and in vivo systems. Acta Univ. Ups.
Uppsala Abstr. 340

Della Corte E., Stirpe F. (1972) The regulation of rat liver
xanthine oxidase. Involvement of thiol groups in the conversion of the enzyme activity from dehydrogenase (type D) into
oxidase (type O) and purification of the enzyme. Biochem. J.
126, 739-745

Granger D.N., Höllwarth M.E., Parks D.A. (1986) Ischemia-
reperfusion injury: role of oxygen derived free radicals.
Acta Physiol. Scand., Suppl. 548, 47-63

Grisham M.B., Hernandez L.A., Granger D.N. (1986) Xanthine
oxidase and neutrophil infiltration in intestinal ischemia.
Am. J. Physiol. 251, G567-G574

Grootveld M., Halliwell B., Moorhouse C.P. (1987). Action of
uric acid, allopurinol and oxypurinol on the myeloperoxidase-derived oxidant hypochlorous acid. Free Rad. Res. Comms.
4, 69-76

Kaplan E.L., Meier P. (1958) Nonparametric estimation from
incomplete observation. J. Am. Stat. Assoc. 53 457-481

Gutteridge J.M.C. (1986) Iron promotors of the fenton reaction and lipid peroxidation can be released from hemoglobin by peroxides. FEBS Lett. 201, 291-295

Gutteridge J.M.C. (1988) Lipid peroxidation: some problems and concepts. In oxygen radicals and tissue injury (ed. B. Halliwell), Kansas; Allen Press. 9-19

Hayden R.E., Paniello R.C. (1987) The effect of glutathione and vitamins A,C, and E on acute skin flap survival. Laryng. 97, 1176-1179

Krawisz JE, Sharon P, Stenson WF (1984) Quantitative assay for acute intestinal inflammation based on myeloperoxidase activity. Gastroenteroloy 87, 1344-1350

Lahnborg G., Nord C.E. (1989) Effect of ciprofloxacin compared to gentamicin in the treatment of experimental intraabdominal infections in rats. Scand. J. Infect. Dis. Suppl. 60, 35-38

Lutz J., Hamar J., Netzer K.O., Stark M. (1985). Survival from mesenteric occlusion shock influenced by different treatment in rats. Int. J. Microcirc. Clin. Exp. 4, 103

Lutz J., Augustin A. (1989). The influence of a temporary cessation and reperfusion of intestinal blood flow on the level of hepatic lipid peroxides. In: Oxygen Transport to Tissue XI (Eds. K. Rakusan et al.) Plenum Pub. N.Y. p. 803-808

Lutz J., Augustin A., Friedrich E. (1990). Severity of oxygen free radical effects after ischemia and reperfusion in intestinal tissue and the influence of different drugs. In: Oxygen Transport to Tissue XII (Eds. J. Piiper et al.) Plenum Pub. N.Y. (in press)

Moorhouse P.C., Grootveld M., Halliwell J.G., Quinlan G., Gutteridge J.M.C. (1987). Allopurinol and oxypurinol are hydroxyl radical scavengers. FEBS Lett. 213, 23-28

Niel T.M., Winterbourn C.C., Vissers M.C.M. (1987) Inhibition of degranulation and superoxide production by sulfasalazine. Biochem. Pharmac. 36, 2765-2768

Ohkawa H., Ohishi N., Yagi K. (1979). Assay for lipid peroxides in animal tissues by thiobarbituric acid reaction. Anal. Biochem. 95, 351-358

Otamiri T., Tagesson C. (1989) Role of Phospholipase A_2 and oxygenated free radicals in mucosal damage after small intestinal ischemia and reperfusion. Am. J. Surg. 157, 562-566

Parks D.A., Williams T.K., Beckman J.S. (1988) Conversion of xanthine dehydrogenase to oxidase in ischemic rat intestine: a reevaluation. Am. J. Physiol. 254, G768-G774

Roy R.S., McCord J.M. (1983) Superoxide and ischemia: conversion of xanthine dehydrogenase to xanthine oxidase. In: Proceedings of the third international conference on superoxide and superoxide dismutase. (Eds. Greenwald and Cohen) New York, Elsevier-North Holland.

Sokol R.J. (1989) Vitamin E and neurologic function in man. Free Rad. Biol. Med. <u>6</u>, 189-207

Weiss S.J. (1989) Tissue destruction by neutrophils. New Engl. J. Med. <u>320</u> 365-376

INFLUENCE OF FREE RADICAL SCAVENGERS ON MYELOPEROXIDASE ACTIVITY AND LIPID PEROXIDATION IN ACUTE SKIN GRAFTS

Ralf K. Goldstein, Albert Augustin*, Johannes
Milz*, Guenther Burg, and Joachim Lutz

Department of Dermatology, *Department of
Physiology, University of Wuerzburg
Roentgenring 9, 8700 Wuerzburg, Germany

Introduction

Oxygen free radicals are known to be partly responsible for tissue necrosis in free skin grafts. This was shown by the positive effect of radical scavenging agents, such as allopurinol (Im et al., 1984) and superoxide dismutase (Manson et al., 1984; Sagi et al., 1986) on the survival rate of island skin flaps. The reported results, however, are difficult to interpret. Anoxia (Tan et al., 1984), inflammatory reactions (Sasaki and Pang, 1981), such as the respiratory burst (Fantone and Ward, 1982), and enzyme leakage from cells (Blake et al., 1987) may contribute to a variable extent to tissue damage after transplantation. Allopurinol may act as a direct superoxide and hydroxyl- scavenger rather than as a xanthin oxidase inhibitor (Moorhouse et al., 1987). Superoxide dismutase reduces the amount of superoxide anions, but even more reactive species, like the hydroxyl radical and hypochlorite acid, can be formed from the resulting hydrogen peroxide. In addition, the commonly used pedicle flap has to be subdivided at least into a lip area and a central zone, which includes the pedicle, due to different oxygen availability. We used a free flap model, which shows no different areas. To rule out anoxia and enzyme leakage as contributing factors, we focused on lipid peroxidation measured by the biochemical method described below rather than by rate of tissue necrosis. In order to estimate the intensity of inflammation in the affected skin and to observe the effect of scavenging agents on the inflammatory process we determined the myeloperoxidase activity in the grafts as a marker of neutrophil content. The influence of the free radical scavengers allopurinol and α -D-tocopherol on myeloperoxidase activity and lipid peroxidation in this model was examined.

Materials and methods

Thirty- nine male wistar rats weighing between 220 and 300 g were used. The animals got a light ether anaesthesia during the operation procedures. A 2.5 to 2 cm abdominal skin flap was dissected, totally lifted and infolded immediately in their donor sites. The grafts were removed and assayed after twelve hours.

Basic skin levels of myeloperoxidase activity and and lipid peroxides were obtained in the first group. The second group of animals received no pretreatment. In the first therapy group one dose of 1000 IU/kg b.wt. of vitamin E (α -D-tocopherol) was applied i.m. 24 hours before the operation. In the second therapy group allopurinol was given for three days (100 mg/kg b.wt. i.v. per day) prior to the operation.

Myeloperoxidase assay was done as described by Bradley (1982). Myeloperoxidase values are expressed as values x basic level. For determination of lipid peroxides a reaction with thiobarbituric acid was used in a modification of the method of Ohkawa et al. (1979) (Lutz et al., 1990).

For statistical evaluation an unpaired students t-test was used. Results are expressed as mean $\pm$ SEM; a p-value of <0.05 was considered to be significant.

Results

After twelve hours, myeloperoxidase activity in skin grafts of animals without pretreatment reached 160 $\pm$ 14 times higher values than in normal rat skin (p<0.001) (Fig. 1), indicating immigration of polymorphonuclear leucocytes (PMNs) of approximately the same extent. In the same time tissue levels of lipid peroxides in this group increased from 103 $\pm$ 15 ($\pm$ SEM) (basic level) to 1020 $\pm$ 131 nmol/g wet weight (p<0.001) (Fig. 1).

After pretreatment with allopurinol myeloperoxidase activity was reduced to 61 $\pm$ 5 times higher values than in controls (p<0.005) (Fig. 2). Vitamin E had a less pronounced effect and myeloperoxidase activity was only reduced to 102 $\pm$

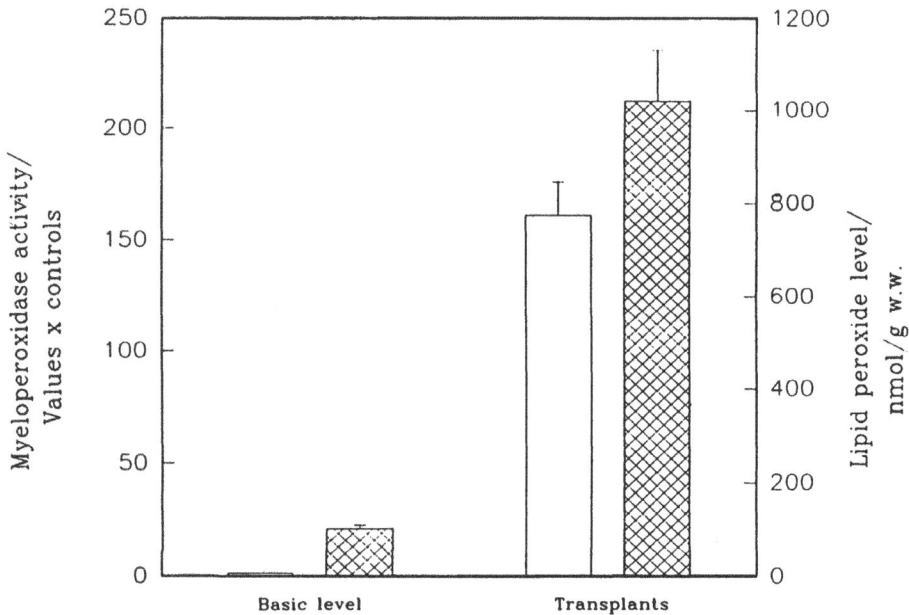

Fig. 1. Basic levels of myeloperoxidase activity and lipid peroxides in rat skin and increased levels in skin grafts.

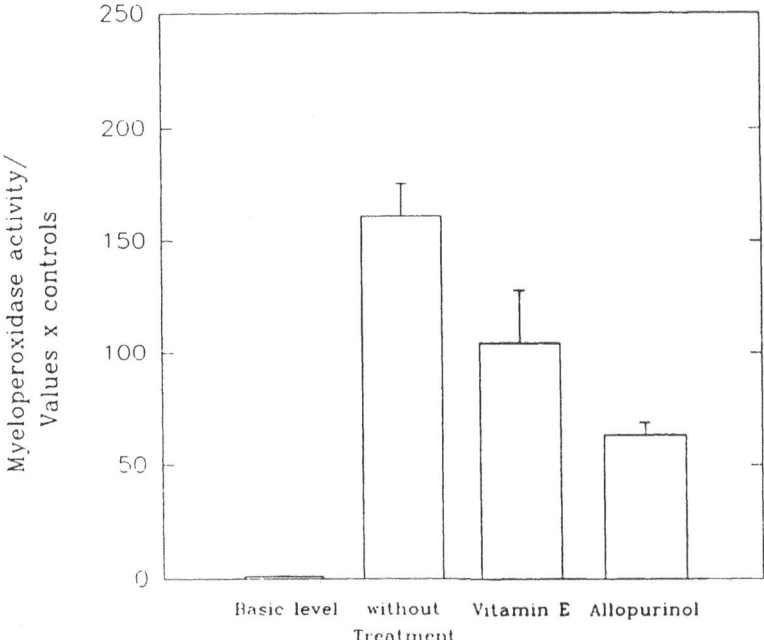

Fig. 2. Myeloperoxidase activity in skin grafts of animals without pretreatment and after treatment with vitamin E or allopurinol . The increase in enzyme activity is a marker for neutrophil content.

22 times basic level (p<0.05) (Fig.2). This is consistent with a significant reduction in neutrophilic infiltration of the transplanted tissue.

Lipid peroxide level was markedly reduced to approximately the same extent in both therapy groups. After pretreatment with allopurinol the level of TBA- reactive substances only amounted to 197 ± 58 nmol/g w.w. (p<0.001) (Fig. 3). Tissue levels of lipid peroxides in flaps of animals that received vitamin E increased to 196 ± 49 nmol/g w.w. (p<0.001) (Fig. 3).

Discussion

Oxygen free radicals are supposed to contribute substantially to necrosis of free skin flaps. Manson et al. (1984) as well as Im et al. (1984) and Sagi et al. (1986) found a beneficial effect of radical scavenging agents, such as allopurinol and superoxide dismutase, on the survival of island skin flaps. However, to which extent tissue necrosis is caused by other damaging mechanisms is unclear. The significant increase in lipid peroxide level after twelve hours indicates a free radical damage, especially of membrane structures, which can not be explained by reperfusion injury, since all vessels were cut during the operation procedure.

The increase of myeloperoxidase activity is interpreted as immigration of polymorphonuclear leukocytes (esp. neutrophils) secondary to the operation trauma, since Bradley et al., (1982) and Lundberg et al., (1983) already showed a direct relationship of myeloperoxidase activity the the neutrophil count in vivo. The extent of 160 time higher values in peroxidase activity is compatible with the results Laiho (1988) obtained in traumatic incision wounds. This accumulation of PMNs due to trauma may be a major cause of free radical damage, because several mechanisms lead to oxygen radical production in activated phagocytes, such as the NADPH- oxidase system (Badwey et al., 1979; Nakamura et al., 1987) and myeloperoxidase (Blake et al., 1987), as well as prostaglandin production (dto.). Oxygen radicals then can themselves serve as chemotactical factors (Oyanagui,

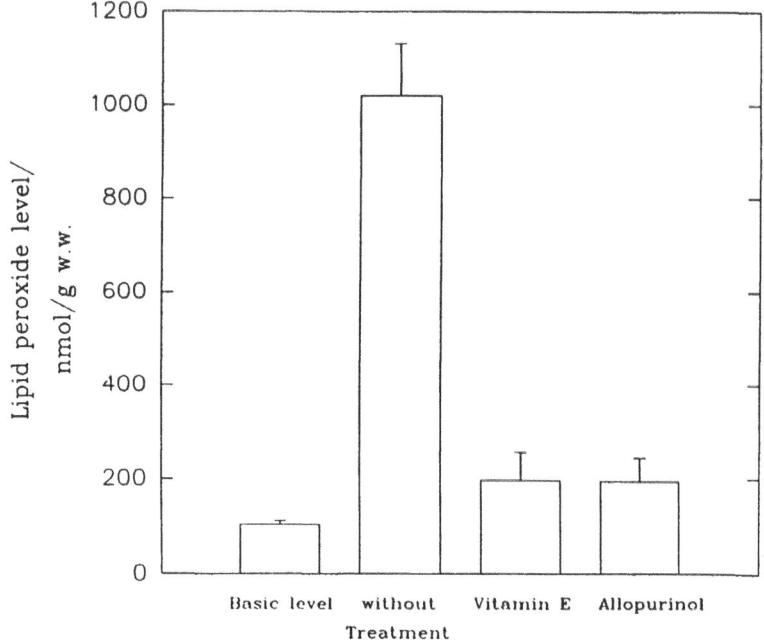

Fig. 3. Level of lipid peroxides in skin grafts of animals without pretreatment and after treatment with vitamin E or allopurinol. Oxygen radical damage, as measured by lipid peroxides, is markedly reduced in therapy groups.

Y., 1987) for further invasion of neutrophil granulocytes (Freeman and Crapo, 1982) and also potentiate prostaglandin production (Blake et al., 1987). In addition the neutrophils are important sources of other soluble mediators (Krawisz, 1984) and also cause the release of chemotactic factors from mast cells (Stendahl et al., 1983). This suggests a self-perpetuating mechanism of inflammation at least for the initial period, until further regulatory factors take action.

Pretreatment with allopurinol and α-D-tocopherol (vitamin E) reduced the amount of lipid peroxides in the skin grafts to almost normal values and PMN immigration was less intense. Allopurinol was thought to block the enzyme xanthine-oxidase, thus preventing the skin graft partly from reperfusion injury and, as a consequence, from necrosis by oxygen radicals (Im et al., 1984 and 1989). Manson et al., (1984) already assumed a contribution of ischema to tissue necrosis, but also reached the conclusion that only the reperfusion injury can be prevented by scavengers of free radicals. We could show however a formation of oxygen radicals in the skin grafts without a period of reperfusion and prove the positive effect of scavengers in this model. As described by Moorehouse et al., (1987), allopurinol and his metabolite oxypurinol could have served as hypochlorite and hydroxyl-radical scavenger, independent of the xanthin-oxidase system.

Lipid peroxidation could also be prevented by a pretreatment with vitamin E, which blocks the formation of peroxy radicals from polyunsaturated fatty acid constituents in cellular membranes (Schulman et al., 1980). It terminates free radical reactions by competing for peroxy radicals, which lead to the formation of tocopherol dimers (Fantone and Ward, 1982). Since vitamin E is fixed in lipid membranes, its effect on lipid peroxidation in our experiments is equal to that of allopurinol. Reactions of radical compounds with other cell components, like proteins and DNA, and extracellular effects may be prevented more effectively by allopurinol. This is a possible explanation for the marked reduction of neutrophil immigration to about 30% after pretreatment with allopurinol, whereas in α-D-tocopherol- group the level was only reduced to about 60%.

This leads to the conclusion that activated neutrophils are a major source of oxygen free radicals in free skin grafts. Since various inflammatory mediators play a role as chemotactical factors after initial trauma, neutrophil immigration can not be totally prevented by radical scavengers, despite effective prevention from lipid peroxidation. But due to the decrease in lipid peroxides, PMN infiltration is less prominent, which, vice versa, also reduces the formation of oxygen free radicals. It is therefore assumed that the level of lipid peroxides and neutrophilic accumulation, and thus the inflammatory response, depends on the effectiveness of the scavenger to interfere with this mechanism. Allopurinol seems to be more effective in this model than vitamin E.

Summary

Activated neutrophil granulocytes are an important source of oxygen free radicals in acute skin grafts. Lipid peroxidation and immigration of PMNs, indicating inflammatory mechanisms, affect each other to a variable extent. The effect of the scavengers allopurinol and α-D-tocopherol on both lipid peroxidation and neutrophilic infiltration was investigated. A possible mechanism for the superior effect of allopurinol compared to vitamin E was discussed.

References

Badwey, J. A., Curnutte, J. T., and Karnovsky, M. L., 1979, The enzyme of granulocytes that produces superoxide and peroxide, N Engl J Med, 300:1157.

Blake, D. R., Allen, R. E., and Lunec J., 1987, Free radicals in biological systems - a review orientated to inflammatory processes, Brit Med Bull, 43:371.

Bradley P. P., Priebat, D. A., Christensen, R. D., and Rothstein, G., 1982, Measurement of cutaneous inflammation: estimation of neutrophil content with an enzyme marker, J Invest Dermatol, 78:206.

Fantone, J. C., and Ward, P. A., 1982, Role of oxygen-derived free radicals and metabolites in leukocyte-dependent inflammatory reactions, A J P, 107:397.

Freeman, B. A., and Crapo, J. D., 1982, Biology of Disease. Free radicals and tissue injury, Lab Invest, 47:412.

Im, M. J., Hoopes, J. E., Yoshimura, Y., Manson, P. N., and Bulkley, G. B., 1989, Xanthine:Acceptor oxidoreductase activities in ischemic rat skin flaps, J Surg Res, 46:230.

Im, M. J., Shen W.-H., Pak, C. J., Manson, P. N., Bulkley, G. B., and Hoopes, J. E., 1984, Effect of allopurinol on the survival of hyperemic island skin flaps, Plast Reconstr Surg, 73:276.

Krawisz, J. E., Sharon, P., and Stenson, W.F., 1984, Quantitative assay for acute intestinal inflammation based on myeloperoxidase activity, Gastroenterology, 87:1344.

Laiho K., 1988, Peroxidase activity in traumatic skin lesions, Z Rechtsmed, 100:65.

Lundberg, C., and Arfors, K. E., 1983, Polymorphonuclear leucocyte accumulation in inflammatory dermal sites as measured by ^{51}Cr-labeled cells and myeloperoxidase, Inflammation, 7:247.

Lundberg, C., Lebel, L., and Gerdin B., 1984, Inflammatory reaction in an experimental model of open wounds in the rat, Lab Invest, 50:726.

Lutz, J., Augustin, A., Friedrich, E., 1990, Severity of oxygen free radical effects after ischemia and reperfusion injury in intestinal tissue and the influence of different drugs, in: Oxygen transport to tissue XII, Piiper J. et al. ed., Plenum Pub., New York.

Manson, P. N., Anthenelli, R. M., Im, M. J., Bulkley, G. B., Hoopes, J. E., 1984, The role of oxygen-free radicals in ischemic tissue injury in island skin flaps, Ann Surg, 198:87.

Moorhouse, P. C., Grootveld, M., Halliwell, B., Quinlan, J. G., and Gutteridge, J. M., C., 1987, Allopurinol and oxypurinol are hydroxyl radical scavengers, FEBS lett, 213:23.

Nakamura, M., Murakami, M., Koga, T., and Minakami, S., 1987, Oxygen activation by phagocyte-specific NADPH oxidase, in: The biological role of reactive oxygen species in skin, Hayaishi, O., Imamura, S., Miyachi, Y., ed., Elsevier, New York, Amsterdam, London.

Oyanagui, Y., 1987, Active oxygens and glucocorticoids in animal inflammation models, in: The biological role of reactive oxygen species in skin, Hayaishi, O., Imamura, S., Miyachi, Y., ed., Elsevier, New York, Amsterdam, London.

Sagi A., Ferder, M., Levens, D., and Strauch, B., 1986, Improved survival of island skin flaps after prolonged ischemia by perfusion with superoxide dismutase, Plast Reconstr Surg, 77:639.

Sasaki, G. H., and Pang, C. Y., 1981, Experimental evidence for involvement of prostaglandins in viability of acute skin flaps: effects on viability and mode of action, Plast Reconstr Surg, 67:335.

Schulman, J. D., Mudd, S. H., Schneider, J. A ., Spielberg, S. P., Boxer, L., and Oliver, J., 1980, Genetic disorders of glutathione and sulfur amino-acid metabolism, Ann Int Med, 93:330.

Stendahl, O., Molin, L., and Lindroth, M., 1983, Granulocyte-mediated release of histamine from mast cells, Int Archs Allergy appl Immun, 70:277.

Tan, C. M., Im, M. J., Meyers, R. A. M., and Hoopes, J. E., 1984, Effects of hyperbaric oxygen and hyperbaric air on the survival of island skin flaps, Plast Reconstr Surg, 70:27.

CARDIOVASCULAR AND RESPIRATORY SYSTEMS

MORPHOMETRIC METHODS FOR THE EVALUATION OF

CAPILLARY GROUPING PATTERNS IN RAT HEART

S. Batra and K. Rakusan

Department of Physiology
Faculty of Medicine, University of Ottawa
Ottawa, Ontario, Canada K1H 8M5

INTRODUCTION

Many methods for assessing tissue capillarity from sections taken perpendicular to the axis of anisotropy have been advanced in recent years. The central theme to these methods has generally involved the measurement of capillary density, i.e. the number of capillaries per unit of cross sectional area. This measurement is the primary descriptor of the relative contributions of vascular and tissue components in the plane. Notwithstanding this average measurement, tissue supply of oxygen has been shown to be compromised as the heterogeneity of capillary spacing increases (Turek and Rakusan, 1981); and hence modern approaches have taken into account to the spatial relationships amongst neighbouring capillaries (Kayar et al., 1982; Hoofd et al., 1985). As the local capillary environment, more than the average tissue value for capillarity, describes better the range of geometrical conditions for oxygen transport, the morphometric methods described in this study take into account the spatial relationships amongst neighboring capillaries.

The capillary material studied has been further classified into two groups: arteriolar (AC) and venular (VC) capillary portions, as made distinct with an enzymatic staining technique that distinguished these two regions (Batra et al., 1989). All capillary profiles were digitized and reduced to a bivariate point pattern. Subsequent analyses then took into account the spatial relationship amongst capillaries as a function staining response.

Capillary grouping patterns were assessed by two comparable methods. The first method involved the one dimensional measure of inter-capillary distance between two neighbouring capillaries. The second method took into account all neighbouring capillaries in the measurement of the theoretical tissue supply area or domain. We have previously shown that these two methods may demonstrate different results in the hypertrophied rat heart (Batra and Rakusan, in press). To better define the parameters underlying this discrepancy, simulations of capillary fields with various degrees of grouping were generated, and analyzed by both methods to compare their sensitivity to grouping phenomenon.

METHODS

Twenty male Sprague-Dawley rats were used in this study (BW = 256 ± 5 g; mean ± SE). Littermates were randomly divided into two groups: sham operated controls (CON; n=10) and pressure overload hypertrophy (HYP; n=10). Hypertrophy was induced by aortic constriction in 5 day old rats. A subdiaphragmatic, suprarenal ligature was tied around the abdominal aorta, with a 30G needle serving as the template for the constriction. In sham operated controls, the abdominal aorta was exposed, but the ligature was not tied. At 6 weeks postoperative, animals were anaesthetized with pentobarbital, and the heart was quickly excised, weighed, dissected and frozen in liquid nitrogen. Tissue cross sections (16 μm) of the LV mid-wall were prepared by cutting parallel sections to the principle axis (base to apex) of the heart.

The staining protocol has been described elsewhere (Batra et al., 1989). Briefly, the technique involved two histochemical methods in implemented series. First, tissue sections were exposed to an incubating medium sensitive to Dipeptidyl Peptidase IV in the capillary endothelium. This treatment stained the venular portion of capillaries red. Next, sections were transferred to a solution sensitive to Alkaline Phosphatase. This treatment stained the arterial portion of capillaries blue.

Capillary grouping patterns for both the control and hypertrophy group were assessed by two morphometric methods. The first method, the Delaunay triangulation (Rogers, 1964), consisted of the delineation of straight line segments between neighboring capillaries (Fig. 1). To ensure that the process was unequivocal, we used the convention that no two line segments could cross. When more than one possibility existed, the shorter segment was taken. From digitized reconstructions of these fields, intercapillary distance (ICD) was taken to be the length of these segments. These data were grouped into three categories: AC-AC, for line segments connecting to arteriolar capillaries; VC-VC, for line segments connecting to venular capillaries; and AC-VC for line segments connecting one arteriolar and one venular capillary. From the ICD, the area of the Krogh cylinder could be determined as a circle with radius equal to ½ICD. The second method, capillary domains (Fig. 1), involved the delineation of a polygon around each capillary profile such that the area within any given polygon was closer to that capillary than any other. The capillary domain, previously referred to as a Dirichlet tessellation (Rogers, 1964) partitions the tissue plane into an irregular lattice pattern. The area of individual polygonal tissue regions surrounding capillary profiles is taken as the respective tissue supply region.

Coincident to these morphometric approaches, all fields were tested for randomness, i.e., the null hypothesis of an underlying Poisson process. To determine whether the distribution of red and blue capillary profiles was random, a statistical test was employed (Venema, 1988). The test statistic was based upon the number of unlike (red-blue) neighboring pairs in the plane. When capillary type grouping existed, the number of neighbors of dissimilar histochemical type was less than what would be expected from a randomly distributed pattern. As the number of unlike neighboring pairs follows a Gaussian distribution (Cliff and Ord, 1973), a test of significance (Z statistic), comparing the actual and expected number of unlike neighboring pairs in the field was used. For a two-tailed test at $\alpha = 0.01$, the critical Z value was ± 2.56. Values < -2.56 were indicative of a non-random grouping by histochemical type. Values > +2.56 indicate a repulsion of capillaries of the same histochemical type.

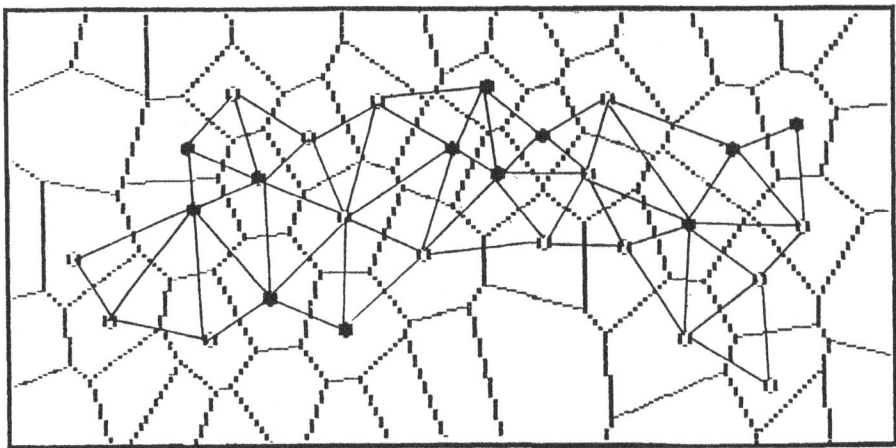

Fig. 1 Delineations of Capillary Domains (thick lines) and Triangulations (thin lines) in the tissue plane for 28 capillary profiles. Domains that are not fully enclosed within the field are excluded. Open circles = arteriolar capillaries; filled circles = venular capillaries.

We have reduced the problem to model form, by generating 5 simulations of capillary distribution patterns with pre-defined levels of grouping by histochemical type (Figures 3a-3e). From these constructs, both the triangulation and domain method were applied. The differences in these morphometric methods in terms of tissue supply regions for arteriolar and venular capillaries are illustrated in figures 4 and 5.

RESULTS

The present model of aortic constriction allowed for a gradual onset pressure overload, which resulted in a substantial increase in cardiac mass. Left ventricular weight increased by 92.5% (CON = 602 ± 9 mg; HYP = 1157 ± 65 mg, P<p.01). The relative index, left ventricular weight to body weight ratio, increased by 109% (CON = 2.3 ± 0.1 mg/g; HYP = 4.6 ± 0.3 mg/g, p<0.01).

From the triangulation method, in control hearts, intercapillary distance was shown to be shorter on the venular side of neighboring capillaries (AC = 25.3 ± 0.3 μm; VC = 22.9 ± 0.2 μm, p<0.01). When expressed as an average area of the Krogh cylinder corresponding values for AC and VC regions were 503 ± 4 μm and 412 ± 3 μm, respectively. This propensity towards smaller tissue supply areas on the venular side of neighboring capillaries was also shown by the capillary domain method (AC = 499 ± 3 μm^2; VC = 456 ± 5 μm^2, p<0.01). Both morphometric approaches gave rise to similar findings in control hearts: greater capillary proliferation on the venular side of neighbouring capillaries, resulting in smaller capillary tissue supply regions (Figure 2; left).

In hypertrophied hearts, the distinction between arteriolar and venular intercapillary distance was not preserved (AC = 27.2 ± 0.4 μm; VC = 27.0 ± 0.2 μm, p=0.56). When expressed as an average area of the Krogh cylinder corresponding values for AC and VC were 581 ± 6 μm and 573 ± 4 μm, respectively. The capillary domain method, however, demonstrated that the AC

Table 1. Results of a test for randomness for control hearts, hypertrophied hearts, and simulations A-E.

Cond	N	Nr	Nb	Nt	N-br	EN-br	SD EN-br	Z
CON	183±7	69±5	213±4	799±22	191±8	296±18	10.4±0.5	-9.8±0.6
HYP	192±7	53±3	139±7	543±20	177±9	216±9	9.3±0.3	-4.1±0.3
SIM-A	212	62	150	577	58	294	10.0	-18.2
SIM-B	212	62	150	577	58	294	10.0	-12.6
SIM-C	212	62	150	577	58	294	10.0	- 5.7
SIM-D	212	62	150	577	58	294	10.0	- 1.9
SIM-E	212	62	150	577	58	294	10.0	+ 5.2

N=number of capillaries per field; r=red; b=blue; Nt=total number of neighbors; N-br=number of mixed neighbors; EN-br=expected number mixed neighbors; SD= standard deviation; Z=test statistic.

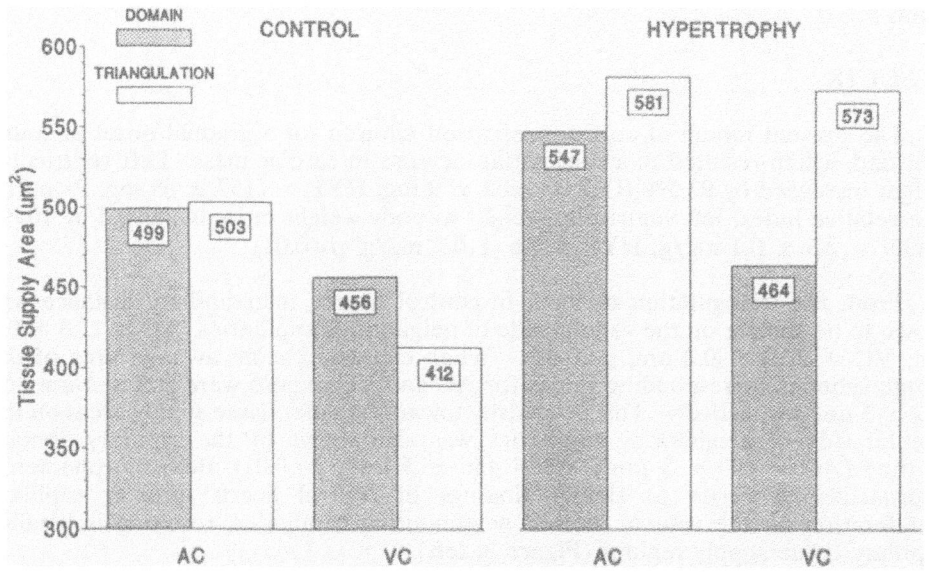

Fig. 2 Comparison of tissue supply regions (μm^2) as determined by the domain and triangulation methods in control and hypertrophied heart. Results from the triangulation method (ICD) are represented as area of the Krogh cylinder (πr^2)

domain area was larger than the VC domain area (AC = 547 ± 6 μm^2; VC = 464 ± 5 μm^2, p<0.01). Thus, in hypertrophied hearts, these methods gave rise to disparate results: the triangulation method did not detect differences between arteriolar and venular capillary regions, whereas the capillary domain method demonstrated significantly smaller tissue supply areas on the venular side of neighboring capillaries (Figure 2; right).

The test for spatial randomness of red and blue capillary profiles indicated significant type grouping in all control hearts. The mean Z score was -9.81 ± 0.61 (range -6.41 to -12.47; n=10). In hypertrophied hearts, capillary grouping by histochemical type was also prevalent in all hearts. The mean Z score was -4.13 ± 0.27 (range -2.72 to -5.58; n=10). Although capillary type grouping was exhibited in both control and hypertrophied hearts (Table 1), the magnitude of grouping was significantly less in hypertrophied hearts (p<0.01; Student's t-test).

The simulations of capillary grouping patterns (Figs. 3A-E) indicated that the correlation between both morphometric approaches: the triangulation method and the capillary domain method, was contingent upon the degree of capillary grouping. In figures 3A, 3B, and 3C, where the distribution of AC and VC profiles was highly non-random (Z = -18.12, -12.56 and -5.65, respectively), both methods displayed similar results: reduced capillary supply regions for venular capillaries (AC > VC). In figure 3D, where the distribution of capillary profiles was random (Z = -1.9) the first discrepancy in the two morphometric methods was observed: the domain method detected differences in tissue supply region (AC > VC) while the triangulation method did not show differences in tissue supply region (AC = VC). In figure 3E, where the pattern was for a repulsion of capillary profiles of the same histochemical type (Z = +5.6), the domain method still displayed the same trend (AC > VC) while the triangulation method showed a reversal (VC > AC). Figure 4 illustrates the tissue supply regions generated by both morphometric methods for all simulations. The triangulation method expressed as percentage of the domain method is shown in figure 5.

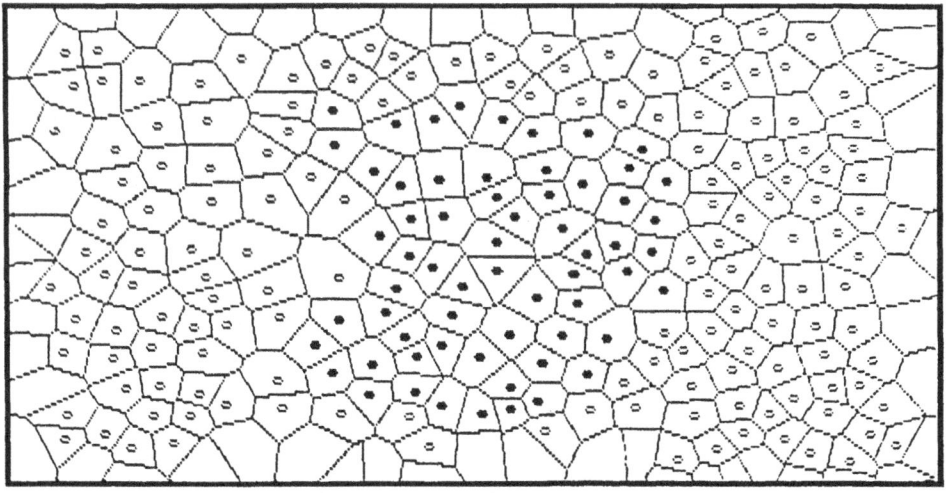

Fig. 3a

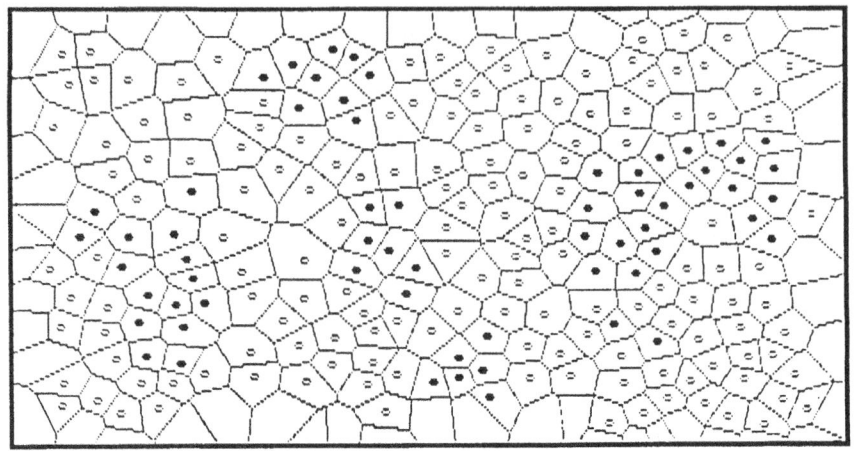

Fig. 3b

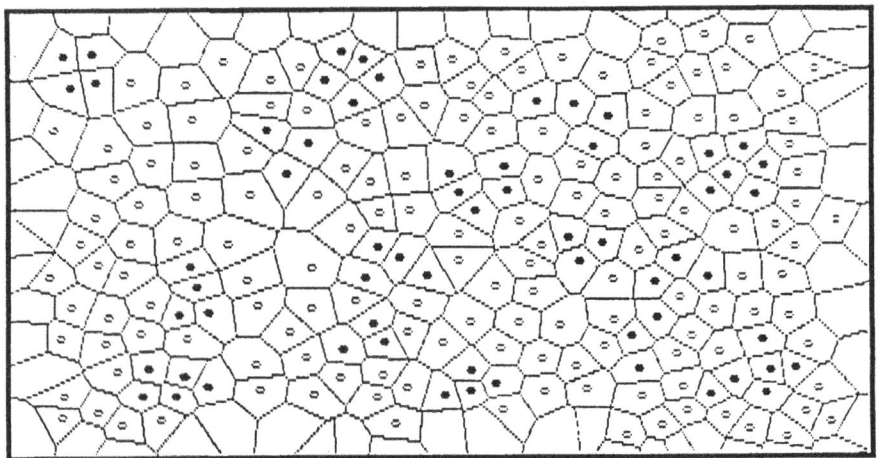

Fig. 3c

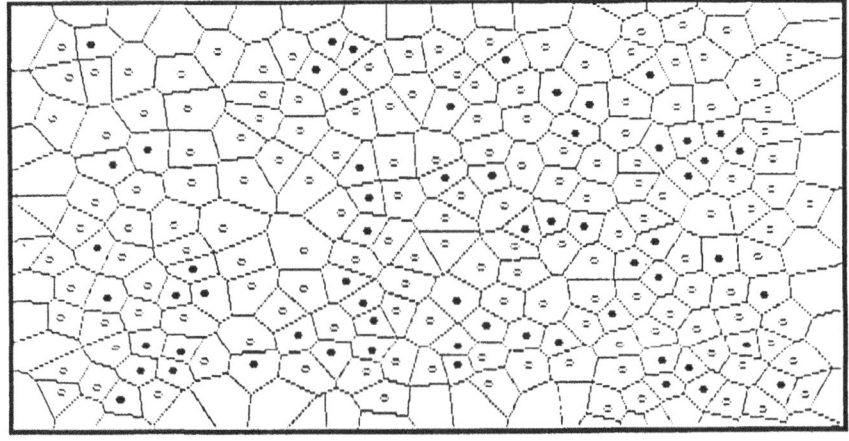

Fig. 3d

266

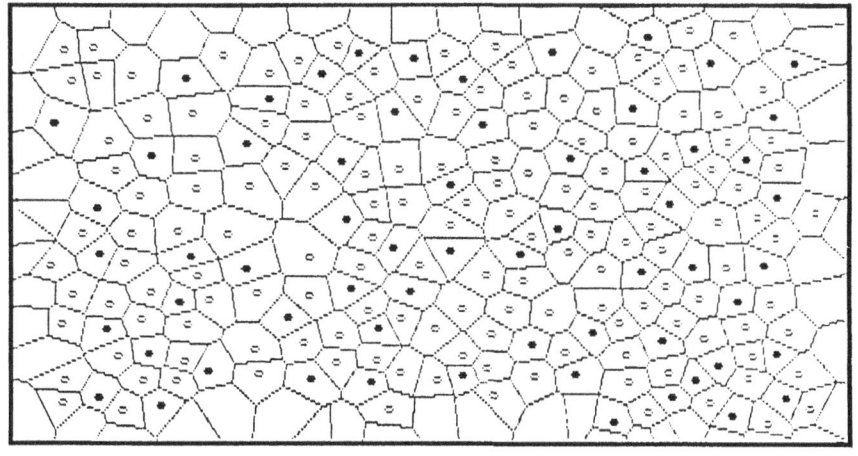

Fig. 3e

Figs. 3a-3e Simulations of capillary profiles in the tissue plane with various degrees of grouping by histochemical type.

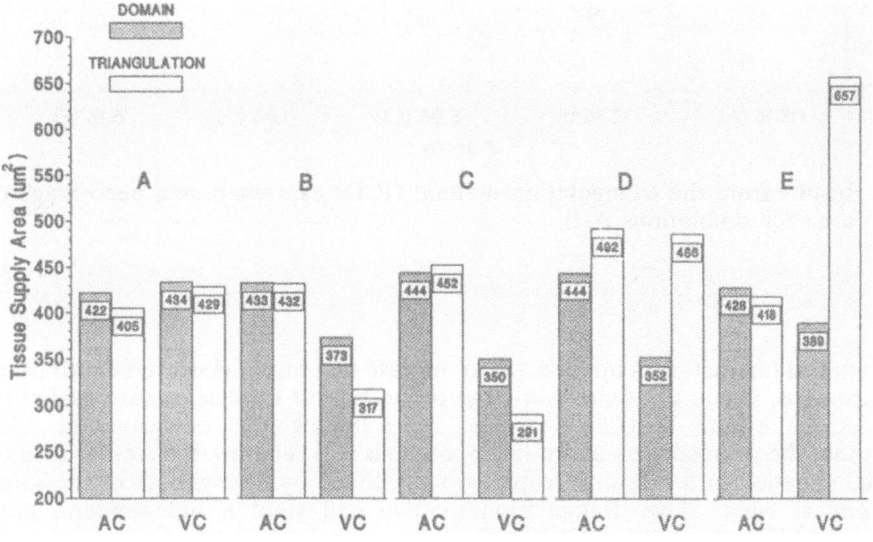

Fig. 4   Comparison of Tissue Supply Regions (μm^2) as determined by the triangulation and domain methods for simulations A-E.

DISCUSSION

In control hearts, the finding of smaller intercapillary distances and domain areas on the venular side of neighboring capillaries was consistent. However, in hypertrophied hearts, intercapillary distances obtained by the triangulation method were not different between the arteriolar and venular side, whereas domain areas exhibited significantly smaller values for venular capillaries. This discrepancy led us to further consideration of the attributes of these two morphometric methods. Both methods are similar in that their calculation of tissue supply regions (Krogh cylinder area) takes into account the local capillary environment, i.e. the distance between neighboring capillaries. In a field consisting of entirely one capillary type, both methods will yield similar results. However, in the case of a bivariate point pattern, the results are influenced by the degree of association between capillaries of differing histochemical type.

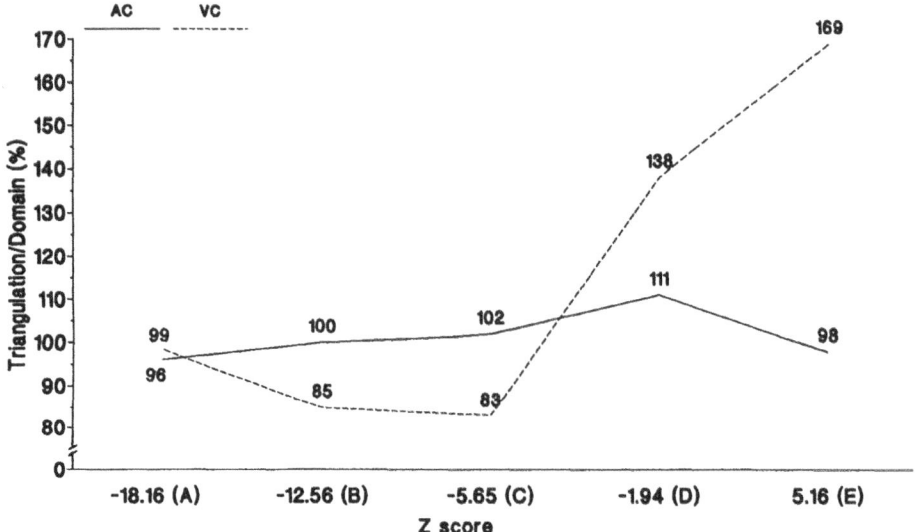

Fig. 5 Results from the triangulation method (ICD) expressed as a percentage of domain area for simulations A-E.

When this interaction is minimal, as in the case of a highly clustered distribution (Simulations A, B and C), there is a large proportion of capillaries surrounded by neighbors of the same histochemical type, i.e., a homogenous environment. This ensures that the triangulation and domain methods will generate the similar results. Take the situation of a reference capillary surrounded by 6 neighbors of the same histochemical type. The triangulation method will yield 6 independent, one-dimensional ICD values (AC-AC). The domain method will yield an averaged, two-dimensional value for domain area from the same 6 neighbors (AC). which will be an integration of the 6 individual results of the triangulation method. When the interaction between capillaries of differing histochemical type increases (simulations D and E), the attributes of these two methods begin to digress. The case of a reference arteriolar capillary surrounded by 6 neighbors, 4 AC and 2 VC, will render 4 typical AC-AC values. But the domain method will be influenced by the penetration of the 2 VC profiles resulting in a modified domain area, the product of a heterogeneous local capillary environment.

In control hearts, a high degree of capillary grouping by histochemical type allowed for a predominantly homogenous population of similar neighbors surrounding a given capillary. As a result, both methods yielded similar results, which agreed with the results of simulations B and C. These data would support a model where transverse arterioles run at an oblique angle to the axis of anisotropy. The arteriole furnishes capillaries that are staggered at various levels in the heart, as a function of the angle of the transverse arteriole and the position that the capillary emanates. Capillaries leaving at any one level would be in register in terms of their arteriolar and venular regions. From tissue cross sections, this arrangement would be manifested by grouping of capillaries of the same histochemical type.

In hypertrophied hearts, where the degree of capillary type grouping was significantly reduced, the triangulation method did not detect differences between the arteriolar and venular tissue supply regions, while domain areas exhibited significantly smaller values for venular capillaries. These results were similar to those generated in simulation D. In hypertrophied hearts, the observation of less grouping by histochemical type may indicate a sharper angle of the transverse arteriole, which then furnishes capillaries at more dispersed positions in the tissue. Alternatively, this may represent a reduction in arteriolar unit size, i.e. the number of capillaries furnished by a given arteriole in hypertrophied heart. This trend towards less capillary grouping by histochemical type is likely an attempt of the microcirculation to ensure homogeneous distribution of oxygen to tissue. From cross sections, large zones of venular capillaries, with presumably lower PO_2 values, would be encroached by arteriolar capillaries with higher PO_2 values to ensure favorable conditions for the diffusion of oxygen in the face of significant cardiac hypertrophy.

To consider which method may provide a more realistic picture of the capillary geometry from arteriole to venule, we have measured the distances between capillaries from longitudinal sections (Batra and Rakusan, in press). The results from these measurements, taken along transect lines at 20% intervals from terminal arteriole to collecting venule, have shown that the tissue supply region decreases in a stepwise manner along the A-V pathlength in both normal and hypertrophied hearts. Accordingly, the method of capillary domains which has demonstrated a similar trend, appears to be preferable to the triangulation method. The method of capillary domains is more sensitive to the local capillary environment and is better able to extract critical information in the face of changes in capillary grouping phenomenon.

We conclude that these two morphometric methods do not furnish uniform results over a diverse range of capillary grouping patterns when capillary profiles are reduced to a bivariate point configuration. The underlying variable is the degree of capillary grouping by histochemical type. Additional factors still to be tested include the proportion of capillaries of either histochemical type; the number of neighbors surrounding each capillary; and systematic differences in capillary supply area.

ACKNOWLEDGEMENTS

Supported by the Medical Research Council of Canada. The authors wish to thank Jimmy Gao and Ching Kuo for their expert technical assistance.

REFERENCES

Batra, S., and Rakusan, K. (in press). Capillarization of the hypertrophic heart: discrepancy of the results obtained by the triangulation and domain methods. *J. Cardiovasc. Pharmacol.*

Batra, S., Rakusan, K., and Kuo, C. (1989). Spatial distribution of coronary capillaries: A-V segment staggering. *In*: "Oxygen Transport to Tissue-XI" (K. Rakusan, G.P. Biro, T.K. Goldstick, Z. Turek, Eds.), pp. 241-247. Plenum Press, New York and London.

Cliff, A.d., and Ord, J.K. (1973). "Spatial Autocorrelation." London, Pion.

Green, P.J. and Sibson, R. (1978). Computing Dirichlet tessellations in the plane. *Computer Journal*, **21**, 168-173.

Hoofd, L., Turek, Z., Kubat, K., Ringnalda, B.E.M., Kazda, S. (1985). Variability of intercapillary distance estimated on histological sections of rat heart. *In*: "Oxygen Transport to Tissue VII" (F. Kreuzer, S.M. Cain, Z. Turek, T.K. Goldstick, Eds.) pp. 239-247, Plenum Press, New York and London.

Kayar, S.R., Archer, P.G., Lechner, A.J. and Banchero, N. (1982). The closest-individual method in the analysis of the distribution of capillaries. *Microvasc. Res.*, **24**, 326-341.

Rogers, C.A., (1964). Packing and Covering, Cambridge Mathematical Tracts #54, Cambridge University Press.

Turek, Z., and Rakusan, K. (1981). Lognormal distribution of intercapillary distance in normal and hypertrophic rat heart as estimated by the method of concentric circles: its effect on tissue oxygenation. *Pflügers Arch.* **391**, 17-21.

Venema, H.W. (1988). Spatial distribution of fibre types in skeletal muscle: test for a random distribution. *Muscle & Nerve.* **II**:301-311.

ESTIMATION OF THE OXYGEN GRADIENT ACROSS PHOSPHOLIPID BILAYERS OF MITOCHONDRIA FROM REPERFUSED RABBIT HEARTS AFTER ISCHEMIA

T. Koyama and M.-Y. Zhu

Research Institute of Applied Electricity, Hokkaido University, 060 Sapporo, Japan

INTRODUCTION

Oxygen supply to mitochondria has been studied theoretically by Jones and Kennedy (1986) and Clark et al. (1985, 1987). However, the effects of the phospholipid bilayers of mitochondria on oxygen transport remains to be further studied. Oxygen transport through phospholipid bilayers of myocardial mitochondria after ischemia-reperfusion is particularly important, because serious dysfunctions of hearts often occur when hearts are reperfused after a transient ischemia. There seems to be no published study available to this problem. We have tried to approximate the oxygen gradient across the membrane lipid layers of mitochondria isolated from hearts exposed to reperfusion after ischemia in anesthetized open chest rabbits.

In a preceding study (Koyama & Araiso 1989) the membrane viscosity of mitochondria from normally perfused rat heart was measured by time-resolved fluorometry. The diffusion coefficient of oxygen (DO_2) in the membrane was estimated with the drastic assumption that the ratio of the membrane viscosity against water corresponded to the reciprocal ratio of DO_2 value in the membrane against water. Using the obtained DO_2 value and assuming further a cylindrical diffusion, the possible maximal oxygen gradient was estimated to be only 3.9 nM across two phospholipid bilayers of mitochondrial membranes. This small gradient, however, is an overestimation, since DO_2 is underestimated because of the above drastic assumption. Oxygen molecules which are smaller than phospholipid molecules diffuse through the ordered structures of phospholipid layers of membranes. The actual DO_2 is larger than the one estimated in the preceding study.

Fischkoff and Vanderkooi obtained large DO_2 values in liposomes and erythrocyte membranes by measuring the pyrene fluorescence which was quenched by diffusing oxygen. In the present study DO_2 values in lipid bilayer of cardiac mitochondria from the control area and the ischemic-reperfused area of myocardium were estimated by comparing the membrane viscosity with the DO_2 values given by Fischkoff and Vanderkooi. Using approximated values for DO_2 the oxygen gradients were calculated across phospholipid bilayers of myocardial mitochondria.

METHODS

Rabbits were anesthetized with i.v. Nembutal (25mg/kg), maintained with i.p. Urethan (1g/kg) and ventilated via thoracheal canula with a Harvard ventilator. The aortic pressure was monitored with an electronic manometer through a catheter placed in the left common carotid artery. Physiological saline solution was continuously infused through a catheter placed in the right jugular vein, to minimize the decrease in aortic pressure caused by the thoracotomy and dehydration.

The left thorax was opened and the heart was exposed. The left anterior descending artery of the coronary artery was occluded for 30 minutes and then the occlusion was released to allow reperfusion of the ischemia-exposed myocardium. After 30 minute reperfusion the left ventricular myocardium which was perfused by the anterior descending artery (ischemic-reperfused area) and the myocardium which was perfused by the left circumferencial artery (control area) were separately sampled. Both muscle samples were separately cut into small pieces with scissors and homogenized with a Polytron (Kinematica, Switzerland) at 0 °C. Mitochondria were obtained by repeated centrifugation and resuspension of the mitochondrial pellets. The amount of peroxidized phospholipid was measured by the thiobarbiturate reaction and expressed as the produced amount of malondialdehyde (MDA).

The dynamic microstructure of mitochondrial membranes was studied with a home-made nanosecond time-resolved fluorometer using the fluorescence of the hydrophobic fluorophore, diphenyl hexatriene (DPH). Isolated myocardial mitochondria in which the membrane lipid bilayers were made to contain DPH (Koyama et al. 1990a) were suspended in KCl-Tris buffer solution and illuminated with polarized pulsed light. The fluorescence decay curves of two directions, parallel and perpendicular to the polarization of the pulsed light for excitation, were measured with a single photon counting circuit controlled with a personal computer. The fluorescence anisotropy decay curve was constructed from the fluorescence decay curves. The fluorescence anisotropy decay curves deviated from monoexponential ones probably because of the inhomogeneity of mitochondrial membranes. We assumed that they represent average values for fluorescence and anisotropy parameters. Mean values for the membrane viscosity (η) and the wobbling angle of phospholipids (θ_w) were calculated from the obtained fluorescence life time (τ), steady state anisotropy (r_s), anisotropy at infinite time (r_∞) and anisotropy decay time (i.e. rotational correlation time, ϕ).

CALCULATION

Approximation of DO_2

Fischkoff and Vanderkooi (1975) using the pyrene fluorescence quenching method measured DO_2 in liposomes of dimyristoyl phosphatidylcholine (DMPC) and dipalmitoyl phosphatidylcholine (DPPC) which are main constituents of phosphatidylcholine in biomembranes. The DO_2 values were shown to be 6.7 and 4.1 x $10^{-5} cm^2$/sec, respectively, at 37 °C, including the high absorption coefficient of oxygen to lipids. Using the time-resolved fluorometry, the viscosities of liposomes of DMPC and DPPC were measured to be 0.35 and 2.7 poise, respectively, using DPH fluorescence in our study (Araiso et al. 1990). The above DO_2 values are plotted against membrane viscosity in Fig. 1. DO_2 seems to be related to the membrane viscosity of the phospholipid bilayer. In human erythrocyte membrane ghosts (RBC) whose viscosity was found to be 1.0 poise in our previous study (Koyama & Araiso 1988), DO_2 was given to be 3.2 x $10^{-5} cm^2$/sec (Fischkoff & Vanderkooi, 1975). Its plotting in Fig. 1 is deviated from the

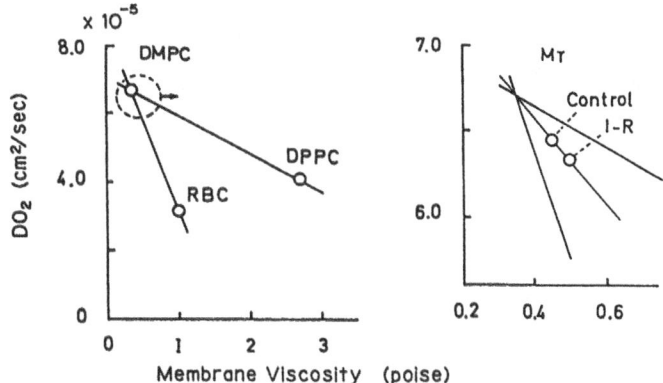

Fig. 1. A diagram showing the relation between reported DO_2 values and membrae viscosity measured with a time-resolved fluorometer in DMPC, DPPC and erythrocyte membranes. The encircled portion was expanded on the right side diagram. Membrane viscosities obtained in the mitochondria (M_T) from control and ischemic-reperused (I-R) areas are plotted, and the points on the bisector are shown for the estimation of DO_2.

line connecting DMPC and DPPC liposomes. This deviation may be caused by the presence of membrane proteins and the coexistence of various phospholipids in membrane ghosts. There is an ambiguity but it was assumed in the present study that the relation was distributed in the area between the two straight lines in Fig. 1. The viscosity obtained in mitochondria from the control and the ischemic-reperfused myocardium were plotted and the midpoints for two lines were used to approximate DO_2 values.

Estimation of Oxygen Gradient

Estimation of oxygen gradient was made by the model of cylindrical diffusion of the preceding study. Briefly a long cylinder, of which diameter was 2 μm, was assumed as a model of the mitochondrion. All the mitochondria are arranged end to end for a long cylinder for the ease of calculation. The total length of the cylinder of all mitochondria contained in 1 ml tissue, l, is given by; l = total inner surface of mitochondria including cristae (11×10^4 cm^2/cm^3 in rats, Page & MacCaster 1973)/$2\pi r$ = 1.75×10^8 cm. The flux of oxygen, Q, is estimated by dividing the oxygen consumption rate of the hearts (10 ml/100 g/ min = 9.35 ml/100 ml/min = 1.79×10^{-3} ml/ml/sec (converted by the assumption that the density of myocardial tissue is 1.07) in cats (Millican 1939) with the total length of mitochondrial cylinder = 1.02×10^{-11} ml/ml/sec. The membrane protein occupies 80 w% of the mitochondrial inner membrane. Oxygen diffusion is slower through the protein because of its tight structure than through lipid bilayers, reducing the surface area effective for oxygen diffusion. We assume that 80 % of the total surface area of mitochondrion is occupied with protein and that the length of the mitochondrial cylinder should be reduced to 20 % of the length mentioned above. The oxygen flux and the oxygen gradient, therefore, must be increased five times of the mentioned values.

The obtained DO_2 values were put into Eq. (1) of the preceding study (Koyama & Araiso, 1989) with the multiplication factor of 5 because of the above consideration :

$$C_2 - C_1 = 5 \times 1000 \times ln[b/a] \times Q / 2\pi DO_2,$$

where a, b and ln represent the inner, outer radius of the mitochondrial cylinder (1000 and 1010nm, respectively) and natural logarithm, respectively. The difference of 10 nm between the radius a and b represents the total thickness of the mitochondrial outer and inner lipid bilayers of the membranes.

RESULTS

Examples of recorded fluorescence and anisotropy decay curves are shown in Fig. 2. The anisotropy decay curve, r(t), in the ischemic-reperfused area reached a nearly constant value higher than in the control area at infinite time (r_∞). The steady state anisotropy (r_s) was also found higher in the ischemic-reperfused area than in the control area. Calcula-

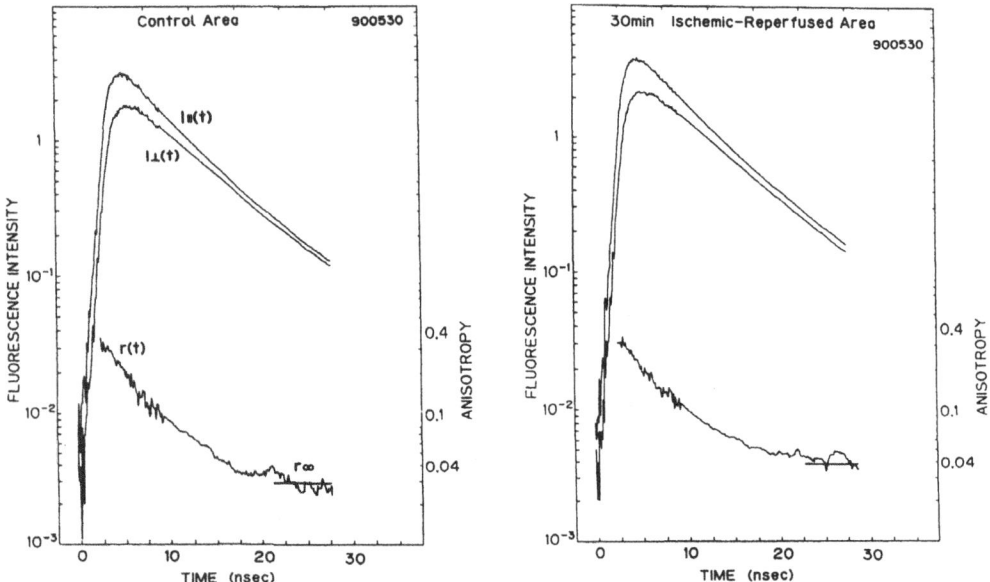

Fig. 2. Recordings of DPH fluorescence and anisotropy decay curves in myocardial mitochondria isolated from control and ischemic-reperfused area.

tions using these values showed that the wobbling angle (θ_w) was reduced and the viscosity (η) increased by the 30 minute ischemia-reperfusion. Obtained parameters are listed in Table 1. MDA was increased from 0.53 to 4.30 nmol/mg-protein by the ischemia-reperfusion.

The viscosity values, 0.45 and 0.50 poise for the control and ischemic-reperfused area, respectively, were plotted in Fig. 1 to obtain DO_2 values. DO_2 values were approximated to be 6.5 and 6.3 x 10^{-5}cm^2/sec, respectively.

Putting these DO_2 values into the equation above, oxygen gradients across the mitochondrial lipid layers were calculated to be 0.055 and 0.057 nM for the mitochondria from the control area and the ischemic-reperfused area, respectively.

DISCUSSION

As surgical manipulation of hearts in open chest rabbits often caused decreases in arterial blood pressure, only fifteen minute ischemia and reperfusion of the anterior descending artery was made in a preceding study (Koyama et al. 1991). Significant differences in the membrane viscosity and wobbling angle of phospholipids were observed in mitochondria obtained from the control area and from the ischemic reperfused area. In the present study, the period of the ischemia and reperfusion was made longer by a continuous infusion of saline with an intention to induce stronger changes in the dynamic microstructure of mitochondrial membranes. However, only a similar increase in membrane viscosity and a similar decrease in wobbling angle of phospholipids were observed.

The present estimation showed much smaller oxygen gradients across the phospholipid bilayers than in the last estimation (Koyama & Araiso 1989) where the bulk viscosity and DO_2 of water were used as the standard for the approximation of DO_2 values. This difference suggests the importance of the ordered microenvironment for oxygen diffusion.

Table 1.   Parameters for DPH fluorescence, anisotropy decay and dynamic microstructure in rabbit myocardial mitochondria. I-R represents mitochondria from ischemic-reperfused area.

	r_s	r_∞	τ (ns)	ϕ (ns)	θ_w (°)	η (poise)
Control	0.121	0.03	5.80	2.41	65	0.45
I-R	0.130	0.04	5.88	2.50	62	0.50

The membrane viscosity is increased by the presence of proteins (Koyama et al. 1990b). The relation between DO_2 and membrane viscosity may deviate from the lines shown in Fig. 1. The membrane viscosity in the inner lipid bilayer was found higher than that in the outer one (Koyama et al. 1990b). These effects were ignored for the ease of the calculation in the present study. Therefore, the present result must be considered with reservation, but while the estimated oxygen gradient is a first order approximation, the result that the oxygen gradient increases somewhat in the 30 minute ischemia-reperfusion test may not be subjected to substantial alterations. The phospholipid bilayers of mitochondrial membranes provide no diffusion barrier to oxygen even after the ischemia-reperfusion. This is probably due to the large total surface area of mitochondria and the high absorption coefficient of lipid layers to oxygen.

A high solubility of oxygen in erythrocyte membranes has been studied in detail (McCabe 1986). MDA value, the index of the lipid peroxidation, however, increased in the present study. Lipid peroxidation produces a hydrophilic -OOH group on the acyl chains of phospholipids. If mitochondrial membranes are hydrated, the absorption coefficient of mitochondrial membranes is reduced and its effects on the oxygen gradient must be taken into consideration. The present measurements provided no clear information on the question whether membranes are hydrated and their absorption coefficient to oxygen is reduced after the 30 minute ischemia-reperfusion. The life time of DPH fluorescence, τ, increased also in the present experiment (see Table 1) as in the case of 15 minute ischemia-reperfusion (Koyama et al. 1991). The fluorescence of the hydrophobic fluorophore is readily quenched by dipole of the microenvironment (Wratten et al. 1989). The value of τ was found significantly reduced by strong peroxidation of myocardial mitochondria incubated with oxidative agents (Koyama et al. 1990c). Thus, it may be said that the present result shows no indication of the possibility of membrane hydration and reduction in the absorption coefficient to oxygen.

SUMMARY

Mitochondria isolated from myocardium exposed to 30 minute ischemia followed by 30 minute reperfusion showed an increase in membrane viscosity and a decrease in wobbling angle of phospholipids, compared with those from the normally perfused myocardium in anesthetized open-chest rabbits. The values for the membrane viscosity were used to estimate the oxygen gradient across the lipid bilayers of mitochondrial membranes with a model of cylindrical diffusion. The effective diffusion coefficient for oxygen, DO_2, was approximated to be 6.5 and 6.3×10^{-5} cm^2/sec in the control and ischmic-reperfused area, respectively, by comparing reported DO_2 values with values for membrane viscosity. For the surface area of the inner mitochondrial membrane including cristae and for the oxygen consumption rate of the myocardium, reported values for rats and cats, respectively, were employed. Using these values, oxygen gradients across the lipid bilayers of mitochondrial lipid membranes were estimated to be only 0.055 and 0.057 nM in the control and 30 minute ischemic-reperfused myocardium, respectively. If the mitochondrial membranes are hydrated because of the ischemia-reperfusion, the absorption coefficient of the membrane to oxygen will decrease and the oxygen gradient will be increased. In the present study, however, the fluorescence life time of DPH, the hydrophobic fluorophore, showed no shortening despite the ischemia-reperfusion. Hence, no indication of membrane hydration was obtained.

REFERENCES

Araiso T., Saitoh H., Shirahama H. & Koyama T. 1990, Analysis for the molecular motion of phospholipid bilayer with picosecond fluorometry. Biorheology 27, 375-387.

Clark A. Jr. & Clark P. A. A. 1985, Local oxygen gradients near isolated mitochondria. Biophys. J. 48, 931-938.

Clark A. Jr., Clark P. A. A., Connett, R. J., Gayeski T. E. J. & Honig C. R. 1987, How large is the drop in PO2 between cytosol and mitochondrion ? Am. J. Physiol. 252, C583-587.

Fischkoff S. & Vanderkooi J. M. 1975, Oxygen diffusion in biological and artificial membranes determined by the fluorochrome pyrene. J. Gen. Physiol. 65, 663-676.

Jones D. P. & Kennedy F. G. 1986, Analysis of intracellular oxygenation of isolated adult cardiac myocytes. Amer. J. Physiol. 250 (Cell Physiol. 19): C384-390.

Koyama T. & Araiso T. 1989, Oxygen diffusion through mitochondrial membranes. Oxygen Transport to Tissue XI. Ed. Rakusan K., Biro G. P.,

Goldstick T. K. & Turek Z., Plenum Press, New York and London, pp. 763 -767.

Koyama T. & Araiso T. 1988, Effects of α-tocopherol-nicotinate administration on the microdynamics of phospholipids of erythrocyte membranes in human subjects. J. Nutr. Sci. Vitaminol. 34, 449-457.

Koyama T., Zhu M.-Y., Kinjo M. & Araiso T. 1990a, Dynamic microstructure of plasma and mitochondrial membranes from bullfrog myocardium. Jap. J. Physiol. 40, 65-78.

Koyama T., Zhu M.-Y., Kinjo M. & Araiso T. 1990b, Microdynamics of outer and inner membranes of mitochondria from bullfrog myocardium. Jap. J. Physiol. 40, 551-560.

Koyama T., Zhu M.-Y., Kinjo M. & Araiso T. 1990c, Dynamic microstructure and hydration of peroxidized membrane of rat cardiac mitochondria and effects of adriamycin. Jap. J. Physiol. 40, 635-649.

Koyama T., Zhu M.-Y., Kinjo M. & Araiso T. 1991, Dynamic microstructure of mitochondrial membranes from rabbit heart subjected to reperfusion after ischemia. Jap. H. J. 32, 247-253.

McCabe M. 1986, The solubility of oxygen in erythrocyte ghosts and the flux of oxygen across the red cell membrane. Oxygen Transport to Tissue vol. VIII. Ed. Longmuir I. S., Plenum Press, New York and London, pp. 13-20.

Millican G. A. 1939, Muscle hemoglobin. Physiol. Rev. 19, 505-523.

Page E. & McCallister L. P. 1973, Quantitative electron microscopic description of heart muscle cells. Amer. J. Cardiol. 31, 172-181.

Wratten M. L., Gratton E., Ven M. v. d. and Sevanian, A. 1989, DPH lifetime distributions in vesicles containing phospholipid hydroperoxides. Biochem. Biophys. Res. Comm., 164, 169-175.

THE OXYGEN DEPENDENCE OF MITOCHONDRIAL OXIDATIVE PHOSPHORYLATION

AND ITS ROLE IN REGULATION OF CORONARY BLOOD FLOW

William L. Rumsey, Cindy Schlosser, E. Matti Nuutinen[*], Michael Robiolio, and
David F. Wilson

Department of Biochemistry and Biophysics
Medical School
University of Pennsylvania
Philadelphia, PA 19104

[*]Department of Pediatrics
University of Oulu
Oulu, Finland

INTRODUCTION

It is well known that coronary flow is proportional to myocardial oxygen consumption, resulting in a nearly constant arterial-venous oxygen concentration difference over a wide range of cardiac work rates. To assure adequate oxygen delivery to the cells, the control of coronary blood flow occurs by an, as yet, undefined mechanism that senses local oxygen concentration, a phenomenon first described in 1925 by Hilton and Eichholtz. We have proposed previously that mitochondrial oxidative phosphorylation is an oxygen sensor linking myocardial oxygen consumption to coronary flow (Nuutinen et al, 1982; Nuutinen et al, 1983; Starnes et al, 1985). It was argued that the vasodilatory responses to metabolic inhibitors (amytal, dinitrophenol) or to alterations of substrate utilization were a result of metabolic changes occurring in the cardiac myocytes from which a message was somehow sent to the coronary vasculature. It could not be ruled out, however, that similar metabolic alterations transpired within the cells of the vasculature. In the latter case, vasoactive mediators are now known to be released from endothelial cells and these effectors may be important in the regulation of blood flow. On the other hand, the oxygen concentration will likely be greater within the cells of the vasculature than in the cardiac myoctes of the intact heart. Since it is now widely believed that the mechanism responsible for sensing local oxygen concentration in the heart and adjusting coronary vascular tone involves mitochondrial oxidative phosphorylation (Olson & Bunger, 1987; Sparks & Bardenheuer, 1986), we have measured the oxygen dependence of respiration in isolated cardiac myocytes and compared these measurements to those made with cell types not of cardiac origin.

The oxygen dependence of respiration has been measured in isolated cardiac myocytes by indirect methods previously (Kennedy and Jones, 1986, Wittenberg and Wittenberg, 1985) with conflicting results. In coupled myocytes, half-maximal oxidation of cytochrome a_3 in the study of Kennedy and Jones was reported to be 8.0 uM oxygen whereas the latter authors observed that half-maximal oxidation of cytochrome oxidase was about 0.2 uM oxygen. The discrepancy in the values is quite large and has been attributed to differences in methodology which resulted in large gradients of oxygen between the bulk medium in which the cells were suspended and the plasmalemmal membrane (Wittenberg and Wittenberg, 1985). It is necessary therefore to have a method available for measuring oxygen which avoids the generation of these oxygen concentration differences. We have developed an optical method for measuring oxygen which has both high sensitivity (10^{-3} M to 10^{-8} M) and rapid response (< 1 msec) (Vanderkooi and Wilson, 1986; Vanderkooi et al, 1987; Wilson et al, 1987; 1988). Because oxygen is not continuously added to the medium suspending the cells, this method also avoids large pericellular gradients of oxygen.

METHODS

The in vitro measurement of oxygen concentration using the quenching of phosphorescence has been described previously (Robiolio et al, 1989, Wilson et al, 1988). In brief, cells were suspended in oxygenated buffer and added to a sample chamber (1 cm square, glass cuvette mounted on a magnetic stirrer) containing similar buffer, the oxygen probe, Pd-coproporphyrin (2 uM) and catalase. Addition of a small aliquot of H_2O_2 provided a means for reintroducing oxygen into the assay system. A xenon flash lamp was used to excite the lumiphore with light of wavelengths between 380 and 440 nm and the phosphorescence lifetime measured by the decay of light emission following the flash (for more detail, see Vanderkooi et al, 1987). A photomultiplier tube detected the light emitted (> 665 nm) from the sample chamber. The signal was digitized, phosphorescence lifetime calculated and the data stored with the aid of a microcomputer and a software program written in the C language. Further data analysis was performed using Asystant by Macmillan Software Co.

Calcium tolerant ventricular myocytes were prepared from hearts of adult rats according to the procedure of Wittenberg and Robinson (1981). In brief, hearts from Sprague Dawley rats (300 g) were perfused retrogradely for 4 minutes with a low calcium Hepes-Ringer buffer equilibrated with 100% oxygen and then perfused via recirculation with the same medium plus 0.1% collagenase. The ventricles were minced and incubated in similar medium (collagenase = 0.15%, $CaCl_2$ = 300 uM, 0.7% BSA) for 10 min at 37 C. Three to four additional incubations were usually necessary to harvest most of the cells. The cells were washed, separated by density gradient centrifugation in isotonic Percoll and assessed for viability by morphometric analysis. The cell suspension contained 85-90% rod-shaped, quiescent cells. Metabolite determinations were made as described previously (Rumsey et al, 1988). Mitochondria were isolated from rat heart according to the method of Fuller et al (1985).

RESULTS

Oxygen dependence of the respiration of cardiac myocytes

Measurements of the rate of oxygen consumption by myocytes isolated from rat heart showed that above about 20 uM, oxygen was without significant effect on cellular respiration (Rumsey et al, 1990). The oxygen concentration required for the half-maximal rate of respiration (P_{50}) was 2.23 $\pm$ 0.13 uM when the maximal velocity (V_{max}) was 0.28 $\pm$ 0.02 uM/sec. The latter value is a measure of the concentration of respiratory proteins within the cuvette and the metabolic state of the cells. Although the addition of more cells to the cuvette raised the rate of respiration, as much as threefold, the value of P_{50} remained unaffected. The quiescent nature of these cells resulted in a high energy state, i.e., [ATP]/[ADP] = 10.1 $\pm$ 1.1, Creatine Phosphate/Creatine = 8.2 $\pm$ 1.7.

The metabolic state of the cardiac myocytes was altered in two ways; by raising the concentration of calcium in the suspending medium from 0.3 to 1.3 mM, and by addition of the uncoupler of oxidative phosphorylation, FCCP, to the assay cuvette. These additions were made after reoxygenating the samples with H_2O_2. Calcium is a well known effector of the activity of key dehydrogenases of the Krebs cycle which are largely responsible for establishing the redox state of the intramitochondrial pyridine nucleotides, i.e., the intramitochondrial [NAD^+]/[NADH] ratio. By increasing the calcium concentration to 1.3 mM, the value of P_{50} was increased by about 48% with only a marginal rise in the V_{max}, about 10%. In separate experiments the [ATP]/[ADP] ratio was not changed by calcium addition but a 10% decrease was found in the intramitochondrial [NAD^+]/[NADH] ratio. The uncoupler, FCCP, is known to decrease the cellular energy state and, in response, to increase the respiratory rate. By addition of FCCP which maximally stimulated cellular respiration, the [ATP]/[ADP] ratio decreased to less than 1 and respiration rose by over ninefold, from 13.2 ± 1.4 to 127 ± 9.6 nmol O_2/min/g dry wt. The latter change resulted in nearly a fourfold increase in the value of P_{50}. If the rate of respiration by the uncoupled cells was inhibited by addition of amytal, a respiratory chain inhibitor, to a level equivalent to that observed in the coupled state, the P_{50} declined to 1.37 ± 0.05 uM.

The P_{50} for respiration of mitochondria isolated from rat heart was 0.6 uM with a V_{max} of 0.2 uM/sec when oxidizing glutamate/succinate/malate in the presence of 0.8 mM ATP. The P_{50} was decreased to less than 0.05 uM for uncoupled mitochondria. Thus, for normal cells in the presence of calcium, the mean oxygen concentration difference between the plasma membrane and the mitochondrial cytochrome oxidase is approximately 1.3 uM.

<u>Oxygen dependence of the respiration of extracardiac cells</u>

Previous work with suspensions of human neuroblastoma cells (Robiolio et al, 1989) showed that the rate of oxygen utilization was constant down to about 13 uM. As the oxygen concentration declined below this value, the respiratory rate decreased progressively with decreasing oxygen concentration. In two different cell lines (CHP-404 and IMR-5 cells), the P_{50} values were 1.3 uM and 1.2 uM, with corresponding Vmax values of 4.8 ± 0.6 nmol O_2/min/mg dry wt and 4.1 ± 0.3 nmol O_2/min/mg dry wt, respectively. It is important to note that the Vmax values expressed as uM/sec, increased proportionally to the concentration of cells in the measurement system, however, the P_{50} values were independent of the cell concentration.

The energy level of the neuroblastoma cells was high when measured under conditions similar to those of the heart cells. The measured [ATP]/[ADP] and [Creatine Phosphate]/[Creatine] ratios were 6.4 ± 0.4 and 0.77 ± 0.05 (n = 5) for CHP-404 cells, and 8 ± 1 and 0.54 ± 0.05 (n = 3) for IMR-5 cells, respectively.

Addition of FCCP resulted in nearly a fivefold increase in V_{max} and the P_{50} decreased from 1.3 to 0.1.1 uM. Increasing the concentration of uncoupled cells in the assay cuvette up to a factor of four produced a proportional elevation in V_{max} without a change in the P_{50}. Inhibiting the rate of respiration of the uncoupled cells with amytal decreased the V_{max} and the P_{50}. Under the latter conditions, the P_{50} approached that of uncoupled mitochondria (< 0.05 uM, Wilson et al, 1988) as the respiratory rate became quite low with increasing additions of amytal. This behavior indicates that for uncoupled cells, in contrast to normal cells, the P_{50} is due to a diffusion induced oxygen pressure difference between the extracellular medium and the mitochondria.

Lastly, the oxygen dependence of respiration was also measured using suspensions of cultured smooth muscle cells from pulmonary artery. The P_{50} of these smooth muscle cells was 0.70 uM ± 0.02 (n = 2) with a V_{max} of 0.2 uM/sec. The oxygen dependence of respiration in these cells was also a function of the metabolic state, i.e., the P_{50} increases to 1.2 ± 0.09 uM but decreased to 0.16 uM when the cells were uncoupled with FCCP and inhibited with amytal, $V_{max} = 0.14$ uM/sec.

Table 1. Comparison of P_{50} values (uM) from different cell types.

	Cardiocyte (6)	CHP-404 (6)	IMR-5 (3)	Smooth Muscle (2)
Coupled (low Ca^{++})	2.3 ± 0.1	ND	ND	ND
plus Ca^{++} (1.3 mM)	3.47 ± 0.2	1.3 ± 0.05	1.2 ± 0.06	0.7 ± 0.02
Uncoupled	9.53 ± 0.7	1.1 ± 0.04	ND	1.2 ± 0.09

The values represent means $\pm$ S.E. for the number of experiments in ( ). ND = not determined. The cardiocytes were freshly isolated from adult rat heart, whereas the CHP-404 and IMR-5 were cultured lines of neuroblastoma cells provided by the Cell Center, University of Pennsylvania, and the smooth muscles cells were cultured from freshly isolated cells from fetal bovine pulmonary artery provided by Dr. Macarak of the Connective Tissue Research Institute, University of Pennsylvania.

DISCUSSION

It may be important that the P_{50} was markedly different between the cardiac myocytes and the extracardiac cells (Table 1). The differences are largely due to the intrinsic metabolic characteristics of each cell type. This is quite clear in the experiment in which calcium was added to the suspension of cardiac myocytes resulting in a substantial increase of P_{50} with essentially no change in V_{max}. The results also show that the P_{50} for oxygen is near the levels of oxygen reported for the cytosol of the cardiac myocytes, in situ, during normoxic conditions (Coburn et al, 1973). Thus even modest changes in oxygen delivery to the cells will have a significant impact on the ability of the mitochondria to synthesize ATP. This may, in turn, provide a possible metabolic signal for alteration of vessel tone. By contrast, the P_{50} of the smooth muscle cells was fivefold less than the cardiocytes. If coronary smooth muscle or endothelial cells show a comparable oxygen dependence as those from pulmonary artery, then this dependence may be too insensitive to changes in the intravascular oxygen concentration to serve as an effective oxygen sensor. Recent work by Mertens and co-workers (1990) shows that the P_{50} of endothelial cells isolated from rat coronary vessels was about 1.3 uM. In the presence of glucose, the levels of ATP and ADP remained constant when measured between 0.16 and 160 uM oxygen, suggesting that the energetic requirements of these cells are low and that these cells are resistant to hypoxia. The oxygen dependence of respiration in the latter cells appears to be similar to the extracardiac cells discussed above. Since the concentration of oxygen in the vessels of the intact heart will likely be greater than that affecting the rate of mitochondrial oxidative phosphorylation, even in severe ischemia, it is unlikely that the oxygen sensor responsible for adjusting coronary vessel tone is located in cells other than the cardiac myocytes.

The diffusion induced oxygen concentration difference between plasma membrane and mitochondria may contribute to the measured P_{50}, since the latter is based on measurements of the extracellular oxygen concentration. This contribution to the measured P_{50} has been determined by adding uncoupler to the cells. When mitochondria are uncoupled, the respiratory rate is independent of oxygen concentration to less than 0.1 uM. Thus the measured P_{50} for uncoupler treated cells is determined only by the oxygen pressure difference induced by oxygen diffusion from the extracellular medium to the mitochondria. When the respiratory rate of the uncoupled cells is inhibited until it is the same as that for the coupled state, the P_{50} is a direct measure of the oxygen diffusion gradient in the coupled cells. In normal cells (coupled), the measured P_{50} is substantially larger than that for uncoupled cells respiring at the same rate, indicating that diffusion is not a major factor in determining the P_{50} for oxygen in normal cells.

In Summary

The oxygen dependence of mitochondrial oxidative phosphorylation measured in isolated cells of cardiac and non-cardiac origin are affected by the metabolic state of the cells. The contribution of oxygen diffusion to the measured P_{50} value in resting cells is small. In cardiac myocytes, and possibly in the other cells as well, this contribution may become significant near maximal levels of respiration. The influence of cellular energy metabolism on the oxygen dependence of respiration in cardiac myocytes suggests strongly that mitochondrial oxidative phosphorylation in these cells is an oxygen sensor for adjusting coronary vascular tone during normal cardiac function.

Acknowledgements

This research was supported by a grant, GM-21525, from the National Institutes of Health.

REFERENCES

Coburn, R.F., Ploegmakers, F., Gondrie, P., and Abboud, R., 1973, Myocardial myoglobin oxygen tension. Am. J. Physiol. 224: 870-876.

Fuller, E.O., Goldberg, D.I., Starnes, J.W., Sacks, M. Delavoria-Papadopoulos, M., Mitochondrial respiration following acute hypoxia in the perfused rat heart. 1985, J. Mol. Cell. Cardiol. 17: 71-81.

Hilton,R. and Eichholtz, F., 1925, The influence of chemical factors on the coronary circulation. J.Physiol. 59:413-425.

Kennedy, F,G, and Jones, D.P., 1986, Oxygen dependence of mitochondrial function in isolated rat cardiac myocytes. Am. J. Physiol. 250: C374-C383.

Mertens, S., Noll, T., Spahr, R., Krutzfeldt, A., and Piper, H.M., 1990, Energetic response of coronary endothelial cells to hypoxia. Am. J. Physiol. 258: H689-H694.

Nuutinen, E.M., Nelson, D., Wilson, D.F. and Erecinska, M., 1983, Regulation of coronary blood flow: effects of 2,4-dinitrophenol and theophylline. Am. J. Physiol. 244: H396-H405.

Nuutinen, E.M., Nishiki, K., Erecinska, M., and Wilson, D.F., 1982, Role of mitochondrial oxidative phosphorylation in regulation of coronary blood flow. Am. J. Physiol. 243: H259-H169.

Olson, R.A., and Bunger, R., 1987, Metabolic control of coronary blood flow. Prog. Cardiovasc. Dis. 24: 369-387.

Robiolio, M., Rumsey, W.L., and Wilson, D.F., 1989, Oxygen diffusion and mitochondrial respiration in neuroblastoma cells. Am. J. Physiol. 256: C1207-C1213.

Rumsey, W.L., Kilpatrick, L., Wilson, D.F., and Erecinska, M. 1988, Myocardial metabolism and coronary flow: effects of endotoxemia. Am. J. Physiol. 255: H1295-H1304.

Rumsey, W.L., Schlosser, C., Nuutinen, E.M., Robiolio, M., and Wilson, D.F., 1990, Cellular energetics and the oxygen dependence of respiration in cardiac myocytes isolated from adult rat. J. Biol. Chem. In Press.

Sparks, H.V., and Bardenheuer, H., 1986, Regulation of Adenosine formation by the heart. Circ. Res. 58: 193-201.

Starnes, J. W., Wilson, D.F., and Erencinska, M., 1985, Substrate dependence of metabolic state and coronary flow in perfused rat heart. Am. J. Physiol. 249: H799-H806.

Vanderkooi, J.M., Maniara, G., Green, T.J., and Wilson, D.F., 1987, An optical method for measurement of dioxygen based on quenching of phosphorescence. J.Biol.Chem. 262: 5476-5482.

Vanderkooi, J.M., and Wilson, D.F., 1986, A new method for measuring oxygen in biological systems. In:"Oxygen transport to tissue VIII, Longmuir, I.A. ed., Plenum Press, New York, p. 189-193.

Wilson, D.F., Rumsey, W.L., Green, T.J., and Vanderkooi, J.M., 1988, The oxygen dependence of mitochondrial oxidative phosphorylation measured by a new optical method for measuring oxygen concentration. J.Biol.Chem. 263: 2712-2718.

Wilson, D.F., Vanderkooi, J.M., Green, T.J., Maniara, G., DeFeo, S.F. and Bloomgarden, D.C., 1987, A versatile and sensitive method for measuring oxygen. In:"Oxygen transport to tissue IX". Silver, I.A. and Silver, A., ed., Plenum Press, New York and London, (<u>Adv. Exp. Med. Biol.</u> 215, 71-77).

Wittenberg, B.A., and Robinson, T.F. 1981, Oxygen requirement, morphology, cell coat and membrane permeability of calcium-tolerant myocytes from hearts of adult rats. <u>Cell Tiss. Res.</u> 216: 231-251.

Wittenberg, B.A., and Wittenberg, J.B., 1985, Oxygen pressure gradients in isolated cardiac myocytes. <u>J. Biol. Chem.</u> 260: 6548-6554.

OXYGEN PENETRATION IN AND RELEASE FROM LUNG SURFACTANT

Erna Ladanyi and Karlheinz Stalder

Department of Occupational Health
University of Göttingen
Federal Republic of Germany
D-3400 Göttingen, Windausweg 2

INTRODUCTION

The transport of inhaled oxygen to the lung tissue is a rather complicated process. The first steps of this process include the penetration of the oxygen through the so-called lung surfactant surface layer LSSL lining the alveoles at the air / water interface, its diffusion through the underlying aqueous hypophase containing the lipid and protein precursors of the surfactant, and finally, its release from this subphase towards the next cell wall. Additionally, a kind of interaction of oxygen with one or more of the hypophase components can not be excluded.

We already described in detail in the pages of this journal the structure and role of the lung surfactant system in general and that of the LSSL in particular (Ladanyi, 1988). We also reported that different properties of the LSSL *in vitro* could be studied by using electrochemical methods such as d.c. and a.c. polarography. The results concerning surface activity (Ladanyi, Zugravu and Tomoaia, 1974; Ladanyi and Stalder, 1979), composition (Ladanyi, 1980), double-layer capacitance (Ladanyi et al., 1988), the influence of environmental and occupational noxious agents on different LSSL-properties (Ladanyi et al., 1974; Ladanyi, 1980; Stalder and Ladanyi, 1980; Ladanyi and Stalder, 1983; Ladanyi, 1986; Ladanyi, 1987; Ladanyi and Stalder, 1987) illustrate the usefulness of these techniques.

We started to learn more about the behaviour of oxygen with the lung surfactant. It was reported that LSSL represents a certain barrier for the crossing oxygen. The phospholipid and protein components of LSSL seemed to play an antagonistic role concerning the height of this barrier: the phospholipids hindered oxygen penetration while the hydrophilic Sp-A diminished the energetic barrier the crossing oxygen had to overcome (Ladanyi, 1988). On the other hand the hypophase seemed to be a place of storage for the oxygen since bronchoalveolar lavages BAL accumulated more oxygen than saline did (Ladanyi, 1988), ande the BAL of a healthy human accumulated more than patients claiming some pulmonary disorders (Ladanyi and Stalder, 1989). A screening of the hypophase components for enhanced oxygen accumulation showed some affinity of oxygen to Sp-A, which allowed to speculate that Sp-A may be involved in the oxygen storage in BAL (Ladanyi and Stalder, 1989).

We report now a study of the kinetics of oxygen penetration and release in lung surfactant *in vitro*. A possible difference between them would give an additional explanation for oxygen accumulation in BAL.

The data we present below show that there is indeed a difference between the kinetics of oxygen penetration from the air into an oxygen free BAL and of the kinetics of its release from the oxygenated BAL into an oxygen free environment.

EXPERIMENTAL

Oxygen uptake and release kinetics was measured by using the Direct Current Tast Polarographic DCTP technique for evaluating the time-dependent changes in the height of the second oxygen reduction wave. The polarograph was a Polarecord E 506 (Metrohm, Herisau, Switzerland) equipped with an x-y recorder. A dropping mercury electrode DME was used as a working electrode (dropping time mechanically adjusted to 1 sec), a KCl saturated Ag/AgCl microelectrode as reference and a Pt tip as an auxilliary electrode, respectively. The measurements were carried out at 25°C. The polarographic vessels could be fixed alternatively under one lid equipped with the electrodes and the de-aerating tube and under another hermetically closing lid. The contact surface between the gaseous and aqueous phases was equal to 4 sqcm.

Oxygen uptake and its release were studied consecutively on the same sample. The uptake measurement consisted basically in de-aerating the sample with nitrogen, in exposing the sample to the air for free uptake, in reconnecting the vessel with the lid holding th electrodes and in recording the values of the stepwise re-oxygenation. Oxygen release measurement consisted in pipetting the oxygen saturated sample in a nitrogen filled vessel having a light nitrogen stream and in recording the values of the stepwise diminishing remaining-oxygen in the sample. Oxygen release values were calculated as the difference between the value of the oxygen uptake at its maximum and the remaining-oxygen values.

The position of the HMDE was kept constant, penetrating and releasing oxygen had to make the same distance from the interface to the electrode in the bulk and *vice versa*.

The biological material consisted of human natural BAL obtained by rinsing with 100 ml saline the lung (one lobe) of patients with indication for bronchoscopy. (Gift of Prof.Hüttemann, Lenglern, FRG). The BAL was free of cells and cell debries.

Saline was prepared from NaCl p.a (Merck, Germany) in double distilled de-ionized water.

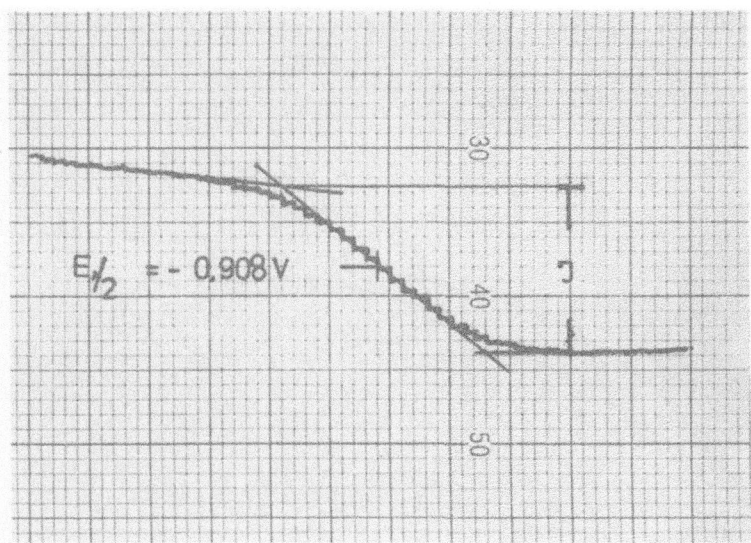

Fig.1.   The   second   reduction   wave   of   the   oxygen   at   a
dropping mercury electrode.   Supporting   electrolyte:
saline

RESULTS AND DISCUSSION

    The   second DCTP wave of oxygen reduction (Fig.   1) is   an
excellent indicator for the presence and behaviour of oxygen in
a liquid sample. As known,   its height is proportional to   the
concentration   of   the   dissolved oxygen   (and   therefore   the
values recorded in current units can be any time converted into
concentration units), while the half-wave potential $E_{1/2}$ of the
reduction   wave   has   a   characteristic   value   in   the   given
supporting electrolyte.

    $E_{1/2}$ of the second wave of oxygen reduction in saline   is
equal to - 0.908 V. A shift of this value towards more negative
values   would   indicate   adsorption   of   some   species   at   the
electrode/electrolyte interface hindering the transport of   the
oxygen to the electrode.

    As   expected,   in   the   BAL samples   the $E_{1/2}$ of   oxygen
reduction   was   shifted   towards   more   negative   values   thus,
surfactant   components adsorbed at the electrode.   Under   these
conditions   the $E_{1/2}$ was around - 1.2 V.   The BAL samples   also
stored more oxygen than the saline did. While the oxygen uptake
limit   (before de-aeration) in saline corresponded to 0.90   µA,
BAL   showed   a   value   of   1,22   µA,   which   represents   a   30%
augmentation compared to the saline.

    Oxygen   uptake   curves   (Fig.   2) of BAL   and   saline   had
different shapes. Although oxygen penetration into BAL was much
faster than into saline, the uptake in BAL continued long after
saline   achieved   the   equilibrium value of   0.9   µA.   In   other
words,   oxygen uptake is doublefold favored in BAL compared   to
saline: it is faster and reaches a higher absolute value.

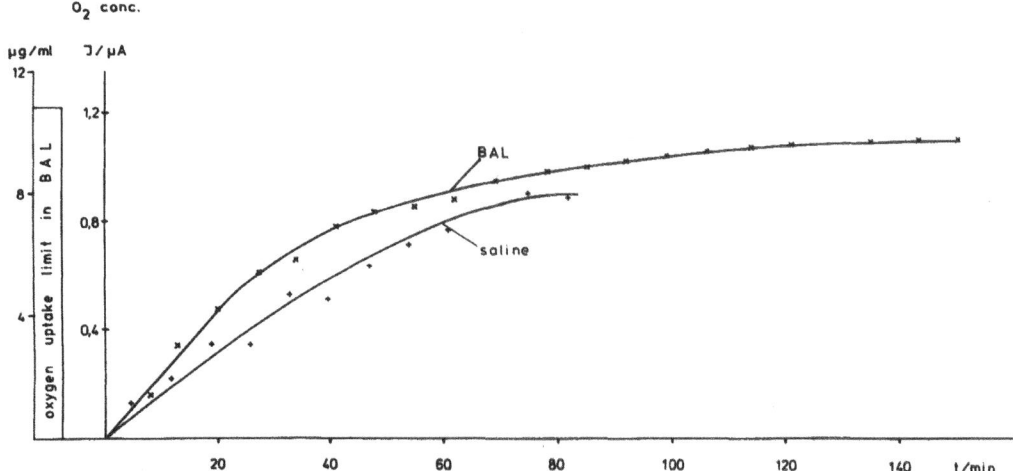

Fig. 2   Kinetics of oxygen uptake from the air into de-
aerated BAL and saline measured as reduction current

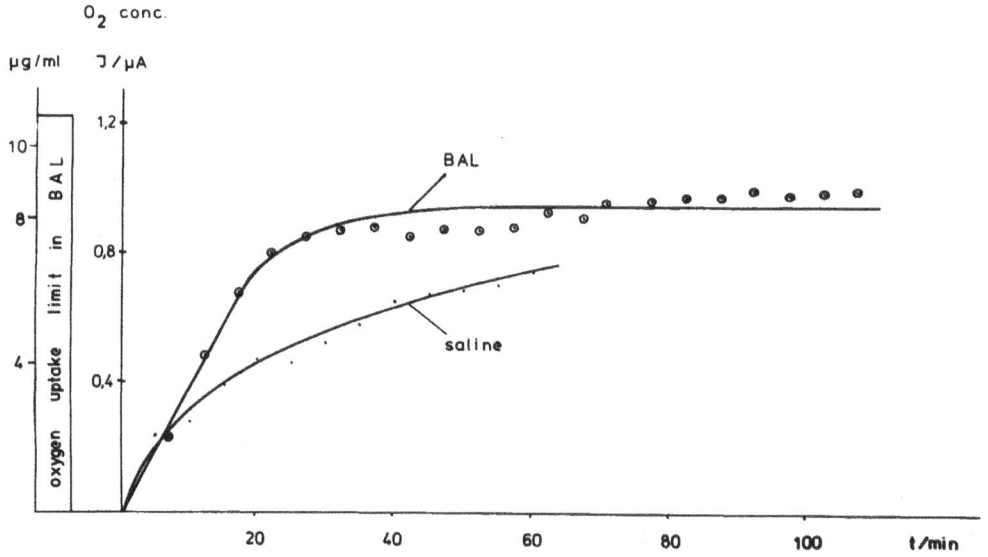

Fig. 3   Kinetics of oxygen release from oxygen-saturated BAL
and saline into a nitrogen environment. Polarographic
reduction current data

The release curves of oxygen from BAL and saline were also different (Fig. 3). Again the release was faster in BAL than in saline but while in BAL the release achieved after 30 minutes a constant value, in saline it continued to rise. It seemed interesting that this release equilibrium value in BAL was equal to the maximum of oxygen uptake in saline. It should be mentioned in this place that oxygen uptake measurements could be performed with a good reproducibility both for BAL and saline while the release measurements in saline (probably because some instability of the unbuffered system) showed a pronounced scattering of the values and a rather poor reproducibility.

Comparing now uptake and release curves of BAL (Fig. 4) it is evident that uptake and release kinetics of oxygen in BAL were different. While the uptake was a relatively slow, basically non-linear process, the release had a fast linear part. After this linear part the release achieved a plateau. The level of this plateau was equal to the uptake maximum in saline that is, it was about 20% under the level of maximal oxygen uptake and 30% under the maximal uptake limit. Thus, even prolonged releasing conditions do not lead to complete release of the oxygen from the system.

We suggest that the differences in oxygen uptake and release kinetics might contribute to explain the accumulation of oxygen in BAL. The positive difference between the release and uptake kinetics at the beginning of the curve is consistent with a steady oxygen penetration into the BAL. The difference in the maximal oxygen uptake level and the maximal release level indicates a partial retention of the oxygen in the system.

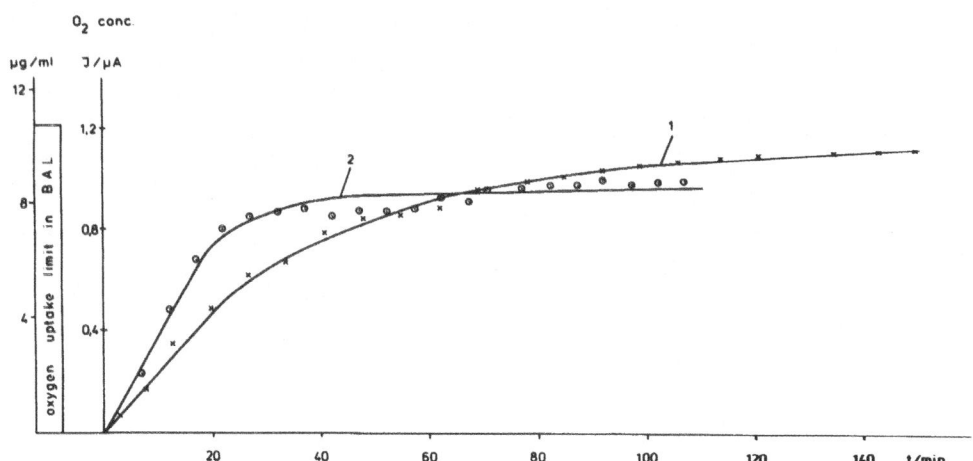

Fig.4   Oxygen uptake (curve 1) and release (curve 2) kinetics in BAL measured as polarographic reduction current

SUMMARY

The kinetics of oxygen penetration from the air into an oxygen free human BAL/saline and of its release from the oxygenated BAL/saline into a nitrogen environment was studied.

Time-dependent oxygen concentration at constant temperature was monitored by recording the direct polarographic current of the second reduction wave of oxygen at a dropping mercury electrode.

The obtained kinetic curves showed that not only uptake and release were quicker in BAL than in saline but also the corresponding equilibrium values were higher. Release and uptake curves in BAL were markedly different. The release was faster than the uptake but its maximal value was about 30% under the maximal uptake level.

We suggest that the differences in oxygen uptake and release kinetics might contribute to explain the previously found accumulation of oxygen in BAL. The positive difference between the release and uptake kinetics at the beginning of the curves is consistent with a steady oxygen penetration through the BAL. The difference in the maximal oxygen uptake level and the maximal release level indicates a partial retention of the oxygen in the system.

REFERENCES

Ladanyi, E., Zugravu, E., Tomoaia,M.,1974, Electrochemical Methods in Surface-Activity Studies of Lung Surfactant.I. Polarographic Maximum Suppressing Ability of Lung Surfactant, Int. Arch. Arbeitsmed., 33:245

Ladanyi, E., Stalder,K., 1979, Alternating current-tast-polarographic determination of surface activity of lung surfactant, J. Electroanal.Chem., 99:321

Ladanyi, E., 1980, Polarographische Elektrosorptions-analyse des oberflächenaktiven Systems der lunge (Lung surfactant), Dissertation, Technische Universität Clausthal

Ladanyi, E., Stalder, K., 1980, Changes occuring in the lung surfactant under the action of inhalative occupational substances, Verh. Dtsch. Ges. Arbeitsmed., 20:519

Ladanyi, E., Stalder, K.,1983, Contribution to the medical importance of dusts resulting by the use of agricultural machines, Verh.Dtsch.Ges.Arbeitsmed., 23:523

Ladanyi, E., 1986, Inhalative Noxen und das Surfactant-System der Lunge, Prax.Klin.Pneumol., 40:465

Ladanyi, E., Möbius, D., Stalder, K., von Wichert, P., 1987, Structure of isolated lung surfactant monolayer, Symposium on Membrane Lipids, 20-21 March,Sintra, Portugal, ACTAS do INSTITUTO de BIOQIMICA (in press)

Ladanyi, E., Stalder, K., 1987, Modellversuche zum Einfluß von Formaldehyd auf das Lungensurfactant, Verh.Dtsch.Ges.Arbeitsmed.,27:545

Ladanyi, E., 1987, Present knowledge in the field of lung surfactant electrochemistry, J.Bioel.Bioenerg., (in Press)

Ladanyi, E., Miller, I., Popovitz-Biro, R., Marikovsky, J., von Wichert, P., Müller, B., Stalder, K., 1988, Molecular structure of the extracellular surface-layer of the human lung surfactant, in Progr.Resp.Res. P. v. Wichert, Ed., Karger, Basel

Ladanyi, E., 1989, in "Oxygen Transport to Tissue XI", Rakusan, Ed.,Plenum, New York

Ladanyi, E., K. Stalder 1990, in "Oxygen Transport to Tissue XII", Piiper, Ed., Plenum, New York

Stalder,K., Ladanyi, E., 1980, Changes occurring in the lung surfactant under the action of inhalative occupational substances, Verh. Dtsch. ges. Arbeitsmed., 20:519

IMPROVEMENT OF PULMONARY GAS EXCHANGE AFTER SURFACTANT REPLACEMENT
IN RATS WITH PNEUMOCYSTIS CARINII PNEUMONIA

E.P. Eijking, G.J. van Daal, R. Tenbrinck, J.F. Sluiters*, E. Hannappel[@], W. Erdmann
and B. Lachmann

Depts. of Anesthesiology and *Clinical Microbiology, Erasmus University, Rotterdam,
The Netherlands, [@]Institute for Biochemistry, University of Erlangen, Erlangen, FRG

INTRODUCTION

Infectious diseases are of great concern in intensive care health service in immunocompromized patients; especially infectious diseases of the lung. In these patients pneumocystis carinii is a major cause of pneumonia leading to severe respiratory distress (CDC Update, 1984; Walzer et al, 1974). Infectious diseases of the lung can cause an outpouring of edema fluid into the alveoli leading to disruption of the surfactant system, either by plasma proteins or by specific enzymes. It has been demonstrated in rats that pneumocystis carinii pneumonia (PCP) decreases the total amount of phospholipids in bronchoalveolar lavage (BAL) fluid whereas phospholipase activity in BAL fluid increases (Sheehan et al, 1986; Kernbaum et al, 1983). Furthermore, lung compliance also decreases in PCP (Sheehan et al, 1986). These findings suggest that in the pathophysiology of respiratory failure in PCP the pulmonary surfactant system may be involved.

To test this hypothesis we designed a study to investigate the effect of surfactant replacement therapy on pulmonary function in rats with PCP. If surfactant replacement would increase gas exchange, then one could conclude that a disturbed surfactant system is the main factor in the pathophysiology of respiratory failure in PCP.

MATERIALS AND METHODS

Surfactant

The surfactant used in these experiments is a natural surfactant isolated from bovine lungs in basically the same manner as previously described (Metcalfe et al, 1980).

Animal model

This protocol was approved by the Animal Care and Use Committee of the Erasmus University Rotterdam, The Netherlands.

The studies were performed in 31 male Wistar rats (initial body weight: 110-120 g), placed in two groups. The first group (n=23) was treated with cortisone acetate, based on the model described by Frenkel and Chandler (1966; 1979). These animals received 12.5 mg cortisone acetate s.c. four times weekly over 8-12 weeks and 10 μg/ml doxycycline in drinking water to reduce a possible bacterial superinfection.

The second group (n=8) received no treatment and served as controls. All animals were housed in standard plastic cages and received standard food *ad libitum* and were weighed at weekly intervals.

After 8-12 weeks the cortisone acetate treated animals developed respiratory failure as judged by clinical symptoms (e.g. tachypnoea, cyanosis and fast decrease in body weight). Thirteen animals died before the experiments with surfactant treatment started. Ten rats which suffered clinically from respiratory failure were anesthetized with pentobarbital, tracheotomized and paralyzed with pancuronium. A catheter was inserted into the carotid artery for blood sampling. Animals were ventilated pressure-controlled with a Servo Ventilator 900 C (Siemens-Elema, Solna, Sweden) at the following settings: $FiO_2 = 1.0$, ventilation frequency=30/min, peak airway pressure=22 cm H_2O, PEEP=2 cm H_2O and I:E=1:2.

After a ± 20 min stabilization period, 7 animals received 200 mg/kg bovine surfactant suspended in 0.7 ml of a 0.6% NaCl solution intratracheally, whereas 3 received the same amount of 0.6% NaCl-solution. Blood samples were taken from the carotid artery immediately before surfactant instillation and at 30 and 60 min post treatment. Blood gases were measured with the ABL 330 Acid-Base Laboratory (Radiometer, Copenhagen, Denmark).

Histological examination

At the end of the experiments the animals were sacrificed with an overdose of intra-arterially administered pentobarbital. The lungs were removed and the left lobe was minced. From this minced lobe a cytospin preparation was made. This preparation was stained with a modified Gomori stain (Musto et al, 1982), which allows visualization of the cysts (not the intra-cystic structures) and Giemsa stain, and examined to confirm presence of PCP. The right lobes were fixed in 10% formalin for light microscopy examination, dehydrated, embedded in paraffin and 6 μm sections of the right middle lobe were stained with hematoxylin and eosin, Grocott-Gomori methenamine silver nitrate and periodic acid-Schiff stain.

Statistical analysis

Statistical evaluation of data was performed using the Wilcoxon signed rank test for paired observations or the Mann-Whitney-U test (two-tailed). All data are expressed as mean ± SEM.

RESULTS

Animal study

During an 8-12 week period the animals developed respiratory insufficiency, which could be reversed by surfactant instillation, as demonstrated by a significant improvement of PaO_2 to near normal values (Table).

Figure 1 presents an example of one rat with PCP with the lowest PaO_2 values in the group before surfactant instillation. After treatment with surfactant the pulmonary function increased dramatically.

Histological examination

Figure 2 shows the lungs of a rat with PCP which received no exogenous surfactant; the alveoli are filled with a characteristic foamy edema. In Figure 3 the alveoli are stabilized and fully aerated after surfactant treatment, but a more or less emphysematous structure is obvious.

DISCUSSION

The results demonstrate that surfactant instillation in rats with respiratory failure due to pneumoncystis carinii infection leads to restoration of pulmonary gas exchange by allowing resorption of

Table. 1. PaO_2 values before and after surfactant treatment in PCP rats

	n =	PaO_2 (mmHg)
PCP before surfactant/NaCl	10	164.7 ± 49.3
PCP 30' after NaCl-solution	3	67.2 ± 17.4
PCP 60' after NaCl-solution	3	59.5 ± 15.6
PCP 30' after surfactant	7	482.9 ± 33.2*
PCP 60' after surfactant	7	464.2 ± 20.8*
Control (healthy animals)	8	513.8 ± 10.7*

Data given as mean ± SEM; *significantly different from PCP before surfactant (p<0.01).

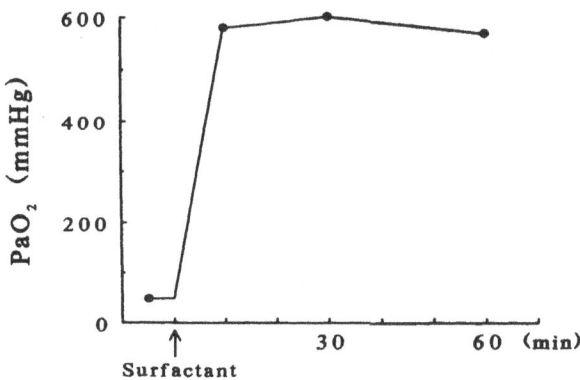

Figure 1. PaO_2 over time in a rat with PCP having the lowest PaO_2 of the whole group before surfactant treatment; In this animal PaO_2 was also measured 10 min after surfactant instillation. The arrow indicates the time of surfactant instillation.

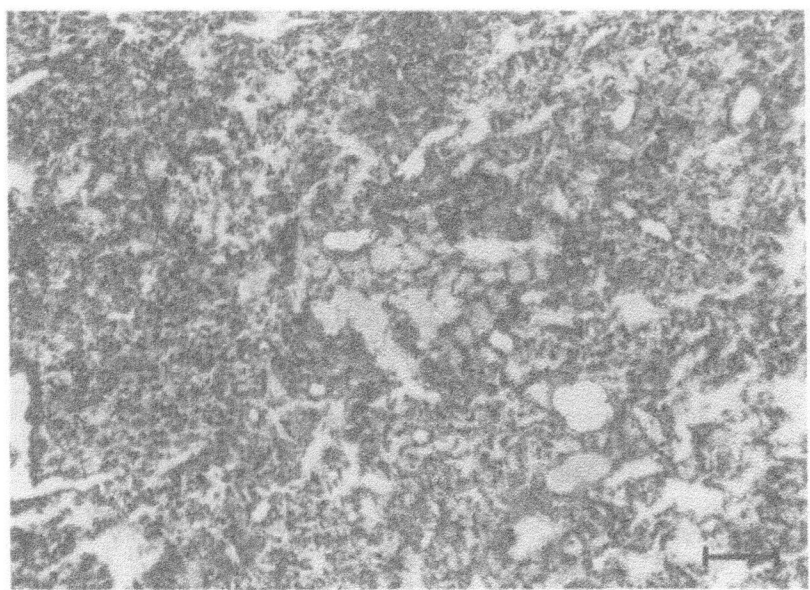

Figure 2. Typical histological findings in a rat with PCP which did not receive exogenous surfactant. Hematoxylin and eosin stain; the scale-line equals 100 μm.

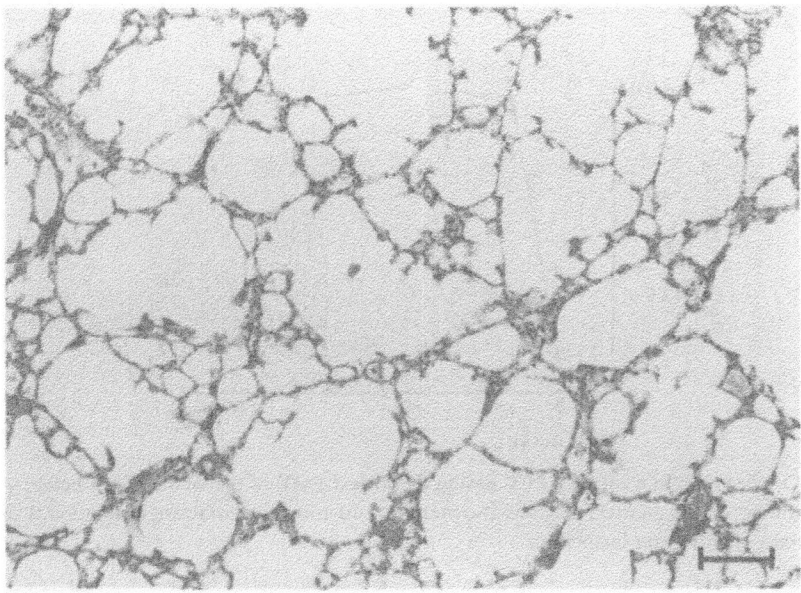

Figure 3. Typical histological findings in rats with PCP after treatment with surfactant; note the emphysematous structure. Hematoxylin and eosin stain; the scale-line equals 100 μm.

pulmonary edema (Guyton et al, 1984), which could be demonstrated by histological examination (Fig 3). However, in animals treated with exogenous surfactant it was also found that many alveolar septa were damaged, leading to an emphysematous lung structure, which suggests that in these lungs proteolytic enzymes may be over-represented. These results underline our hypothesis that the pulmonary surfactant system is mainly responsible for the respiratory failure caused by PCP.

It is established that immunosuppressed rats, as is the case in humans, can develop pneumocystis carinii infection of the lungs, leading to respiratory distress. The current treatment of choice is parenteral administration of trimethoprim-sulfamethoxazole or pentamidine-isothionate. Both drugs have about the same rate of success of around 75% (percentages vary depending on the study) but have high incidences of systemic adverse reactions (for review see: Levine and White, 1988). To prevent these adverse reactions it is necessary to lower the systemic concentration, which led to studies on aerosolized pentamidine. Results from these aerosol studies have shown that it is possible to get high concentrations in the lungs with low systemic concentrations (Montgomery et al, 1987).

For the future one could use surfactant also as a carrier-substance for antimicrobial drugs, thereby serving two treatment goals. First to overcome the respiratory failure and second to get a high intra-alveolar drug concentration and a low systemic concentration. The advantages would include not only quick restoration of pulmonary function, but also reduced drug-induced systemic adverse reactions as a result of lower systemic antimicrobial drug concentrations. We have recently demonstrated that pentamidine-isothionate and trimethoprim-sulfamethoxazole, antimicrobial drugs used for treatment of PCP, do not inhibit the exogenous surfactant used in these experiments (unpublished observations).

From our histological observations one may also speculate that proteinase inhibitors, given to the patient before manifestation of pneumocystis carinii infection, could prevent morphological damage to the lungs.

To conclude: in rats with severe respiratory distress due to PCP surfactant replacement fully restored pulmonary function. Surfactant therapy could be of great advantage in future treatment of PCP, both by direct restoration of pulmonary function and as a carrier for antimicrobial drugs.

SUMMARY

The effect of intratracheal surfactant instillation on pulmonary function in rats with pneumocystis carinii pneumonia (PCP) was investigated. In these animals which developed PCP with severe respiratory failure after s.c. administration of cortisone acetate over 8-12 weeks, pulmonary function could be improved by surfactant instillation, as measured by an increase in PaO_2. Histological examination showed that alveoli of rats with PCP which received no surfactant treatment are filled with foamy edema, whereas after surfactant treatment alveoli are stabilized and well-aerated. These results indicate that surfactant therapy could be used in patients with severe PCP to overcome an acute stage of respiratory distress while at the same time surfactant could serve as a carrier substance for antimicrobial drugs to attain high intra-alveolar and low systemic antimicrobial drug concentrations.

ACKNOWLEDGEMENT

This work was in part supported by The Dutch Foundation for Medical Research (SFMO).

REFERENCES

CDC Update: acquired immuno-deficiency syndrome (AIDS)-United States, MMWR 1984, 32: 688.

Chandler, F.W., Frenkel, J.K. and Campbell, W.G., 1979, Animal model of human disease: pneumocystis pneumonia, Am. J. Path., 95: 571.

Frenkel, J.K., Good, J.T. and Schultz, J.A., 1966, Latent pneumocystis infection of rats, relapse, and chemotherapy, Lab. Invest., 15: 1559.

Guyton, A.C., Moffatt, D.S. and Adair, T.H., 1984, Role of alveolar surface tension in transepithelial movement of fluid, In: "Pulmonary surfactant," B. Robertson, L.M.G. van Golde and J.J. Batenburg, eds., Elsevier Science Publishers, Amsterdam, 171.

Kernbaum, S., Masliah, J., Alcindor, L.G., Bouton, C. and Christol, D., 1983, Phospholipase activities of bronchoalveolar lavage fluid in rat pneumocystis carinii pneumonia, Br. J. Exp. Path., 64: 75.

Levine, S.J. and White, D.A., 1988, Pneumocystis carinii, Clin. Chest. Med., 9 (3): 395.

Metcalfe, I.L., Enhorning, G. and Possmayer, F., 1980, Pulmonary surfactant-associated proteins: their role in the expression of surface activity, J. Appl. Physiol., 49: 34.

Montgomery, A.B., Debs, R.J., Luce, J.M., Corkery, K.J., Turner, J., Brunette, E.N., Lin, E.T. and Hopewell, P.C., 1987, Aerosolized pentamidine as sole therapy for pneumocystis carinii pneumonia in patients with acquired immunodeficiency syndrome, Lancet, 2: 480.

Musto, L., Flanigan, M. and Elbadawi, A., 1982, Ten-minute silver stain for pneumocystis carinii and fungi in tissue sections, Arch. Pathol. Lab. Med., 106: 292.

Sheehan, P.M., Stokes, D.C., Yeh, Y. and Hughes, W.T., 1986, Surfactant phospholipids and lavage phospholipase A$_2$ in experimental pneumocystis carinii pneumonia, Am. Rev. Respir. Dis., 134: 526.

Walzer, P.D., Perl, D.P., Krogstad, D.J., Rawson, P.G. and Schultz, M.G., 1974, Pneumocystis carinii pneumonia in the United States: epidemiologic, diagnostic, and clinical features, Ann. Intern. Med., 80: 83.

ATTENUATION OF HYPOXIC PULMONARY VASOCONSTRICTION IN ACUTE OLEIC ACID LUNG

INJURY - SIGNIFICANCE OF VASODILATOR PROSTANOIDS

Kazuhiro Yamaguchi, Masaaki Mori, Akira Kawai, Koichiro
Asano, Tomoaki Takasugi, Akira Umeda and Tetsuro Yokoyama

Department of Medicine, School of Medicine, Keio University
Tokyo 160, Japan

INTRODUCTION

In order to assess the physiological abnormalities and the pathogenesis of adult respiratory distress syndrome (ARDS), especially that
associated with pulmonary fat embolism, acute lung injury caused by mono-
unsaturated nonsterified fat oleic acid (cis-9-octadecenoic acid) has been
widely used in animal experiments. This lung injury results in an extensive, multifocal, and heterogeneously distributed lung damage with alveolar
flooding, interstitial edema and microatelectasis. These morphological
changes have generally been considered to develop significant shunt flow
causing severe hypoxemia and loss of lung compliance. Based on the microscopic examinations, however, Schoene et al. (1984) have failed to demonstrate a quantitative relationship between the morphological alterations
and the amount of shunt in dogs with lung damage induced by oleic acid
administration. Further, Brigham et al. (1983) have shown that gas exchange abnormality in patients with ARDS does not correlate with extent of
pulmonary edema estimated from extravascular lung water but significantly
with vascular permeability-surface area. They have inferred from their
experimental findings that the ability of the lung circulation to reduce
perfusion of damaged and edematous areas is essentially important in preserving blood oxygenation in ARDS. Such studies may indicate that local
mechanism of regulating the distribution of blood flow to diseased areas is
considerably attenuated in acute lung injury and this disturbance may
partly be responsible for its impediment in gas exchange represented by an
augmented shunt flow.

In normal lungs, hypoxic pulmonary vasoconstriction (HPV) is commonly
accepted as the main factor to modulate the distribution of pulmonary perfusion and to serve for maintaining gas exchange, as pulmonary blood flow
from hypoxic region is directed to better oxygenated areas of the lung
(Euler and Liljestrand, 1946; Fishman, 1976). If HPV is deteriorated at
hypoxic regions, gas exchange efficiency would be lowered because of a
relative increase in blood flow to worse oxygenated areas.

Eicosanoids have been implicated as mediators of a variety of afflic-
tions to the lungs. Cyclooxygenase and lipoxygenase products of arachidonic acid appear in high concentrations in lung lymph and blood after
infusion of either oleic acid or endotoxin (Olanoff et al., 1984; Ball et
al., 1989; Chang et al., 1989). Among them, locally elaborated vasodilator
prostanoid such as prostacyclin (PGI_2) may oppose vasoconstriction, resulting in inhibiting the vigor of hypoxic pressure response in injured

areas. It is, therefore, possible that local accumulation of vasodilator prostanoids yields the attenuation of HPV predominantly in diseased areas of the lung and evokes the impaired gas exchange in acute lung injury. The present study was undertaken to test the hypothesis that HPV would be significantly diminished especially in diseased areas of acute lung injury induced by administrating oleic acid. Additionally, a possible role of locally accumulating vasodilator prostanoids in modifying the vascular reactivity responding to hypoxic spell under these circumstances was examined by inhibiting prostaglandin biosynthesis with indomethacin, an inhibitor of cyclooxygenase activity.

METHODS

Preparation and instrumentation

Twenty four mongrel dogs of either sex weighing 12.2±3.0 kg were anesthetized with a slow injection of pentobarbital sodium (25 mg/kg), placed in a supine position and had their trachea intubated. Ventilation was maintained with a tidal volume of 10-15 ml/kg without positive end-expiratory pressure (Model 613, Harvard Apparatus, Southnatick, MA). Respiratory rate was adjusted to maintain arterial PCO_2 ($PaCO_2$) between 30 and 40 Torr, and diaphragmatic paralysis was secured with pancuronium bromide (4 mg iv, repeated as necessary). The temperature of the animal at the aortic arch was kept at 37±1°C with a thermostatically regulated heating pad. The right femoral artery was cannulated for sampling blood for gas analysis as well as recording systemic arterial pressure. A 5-Fr catheter with thermistor and conductivity electrodes in close proximity (HE-2900, Electro-Catheter Corp., Rahway, NJ) was passed through the left femoral artery to the arch of the aorta, and was used to quantitate the extravascular lung water in terms of a standard double-indicator-dilution technique (Nobel and Severinghaus, 1972). A 7-Fr double-lumen Swan-Ganz catheter was placed just inside the main pulmonary artery via the right femoral vein to measure cardiac output ($\dot{Q}_T$) and pulmonary arterial pressures as well as to take mixed venous blood, while a venous catheter was inserted into the left femoral vein for injection and infusion of various indicators.

Vascular pressures were measured relative to the level of the mid-right atrium. $\dot{Q}_T$ was obtained by the method of thermodilution injecting 5 ml of 3% saline kept at zero degree into the proximal lumen of Swan-Ganz catheter located in the right atrium. This injectate simultaneously allowed one to estimate the amount of extravascular lung water using heat as a diffusible indicator and sodium ion as a nondiffusible indicator, respectively (Nobel and Severinghaus, 1972). Pulmonary vascular resistance (PVR) in the unit of dyne·sec/cm^5 was calculated from the formula, 80·(P$\overline{pa}$ — P$\overline{pw}$)/$\dot{Q}_T$, where P$\overline{pa}$ and P$\overline{pw}$ were mean pulmonary artery pressure and pulmonary arterial wedge pressure, respectively. PVR was used as an index of the flow-resistive properties of the pulmonary circulation. Minute ventilation ($\dot{V}_E$) and respiratory frequency were monitored by means of a pneumotachograph of Fleisch (LTV-132T, Nihon Kohden Corp., Tokyo) with a differential pressure transducer (TP-602T, Nihon Kohden Corp., Tokyo). PO_2, PCO_2 and pH in blood samples were analyzed with the method of electrodes (Micro 13, Instrumentation Laboratory, Lexington, MA).

After base-line measurements of hemodynamic and gas-exchange parameters, the animal was given an injection of oleic acid (Wako Pure Chemical Industries Ltd., Tokyo) at a dose of 0.07 ml/kg into the femoral vein. A decline in arterial PO_2 (PaO_2) as well as changes in $\dot{Q}_T$, P$\overline{pa}$ and extravascular lung water were subsequently observed every 30 min. A quasi steady state concerning hemodynamics and arterial blood gases were established nearly 90 min later so that necessary measurements were commonly performed between 90 and 180 min after the administration of oleic acid.

Before the measurements, the lungs were hyperinflated with a volume of 30 ml/kg to minimize the formation of atelectasis.

Inert gas measurement

To evaluate the gas exchange efficiency in an canine model with or without lung damage, sulfur hexafluoride (SF_6), ethane, cyclopropane, halothane and diethyl ether dissolved in normal saline was infused intravenously at a constant rate of 2 ml/min for at least 30 min. After a steady state was attained, the samples of both arterial and mixed venous blood were simultaneously taken and the expired gas, which was maintained over 40°C before analysis to avoid condensation and loss of highly soluble gases, was collected. The concentrations of inert gases in the samples were measured on a gas chromatograph (model 163, Hitachi Ltd., Tokyo) equipped with an electron capture detector for SF_6 and a flame ionization detector for remaining indicator gases. Blood-gas partition coefficients for the inert gases were determined for each animal by applying the extraction method of Wagner et al. (1974).

After correcting for experimental error, retention (R) of the inert gas, defined as a quotient between the partial pressure of the indicator gas in arterial blood and that in mixed venous blood (Wagner et al., 1974), was utilized to appreciate the impaired gas exchange in the lungs. In theory, increase in R value is indicative of deterioration in gas exchange efficiency regardless of cause (Hlastala and Robertson, 1978).

Experimental protocol

Before administration of oleic acid, hemodynamic parameters including extravascular lung water, blood gases and R values for five indicator gases were examined at an inspired O_2 concentration (FIO_2) of 0.15, 0.21 and 0.60. After the stable lung injury was achieved, the same measurements were made at varied FIO_2 as described above. Indomethacin (5 mg/kg) were subsequently infused through the peripheral venous catheter over a 10-min period (Sprague et al., 1986). 30 min was allowed after giving the cyclooxygenase inhibitor for re-establishment of stability (Sprague et al., 1986). The entire sequence of measurements and sample collections were then repeated at FIO_2 of 0.60.

Statistical analysis

Significance of differences between the two conditions was assessed either by a t-test for paired observations or by a Wilcoxon signed-ranks test. Data are presented as means $\pm$ SD, and p value less than 0.05 was deemed significant.

RESULTS

Extravascular lung water determined from the method of double indicator dilution increased by 125% after infusion of oleic acid. Although a significant correlation between the amount of lung edema and the severity of hypoxemia was found for all cases including normal and injured dogs, neither was there a mutual relation between these two variables as far as the data obtained from the animals with lung damage were solely considered (Figure 1).

Oleic acid caused a increase in pulmonary vascular resistance, PVR at an average of 103% of the value before its administration under a condition of FIO_2 0.21. PVR in injured lungs decreased as FIO_2 was higher, the trend being similar to that observed in normal lungs before injection of oleic acid (Figure 2). Alterations in FIO_2 did not produce any difference either in $\dot{Q}_T$ or in extravascular lung water but did in mixed vanous PO_2 ($P\bar{V}O_2$),

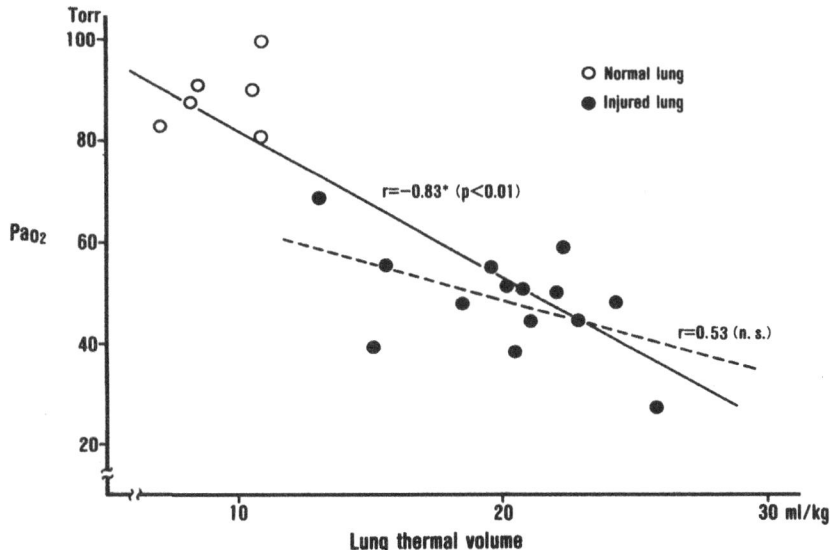

Fig. 1. Relationship between PaO₂ and extravascular lung
water. LTV: extravascular lung water. No correlation
can be found as far as the data obtained from the
animals with lung damage are solely considered
(dotted line).

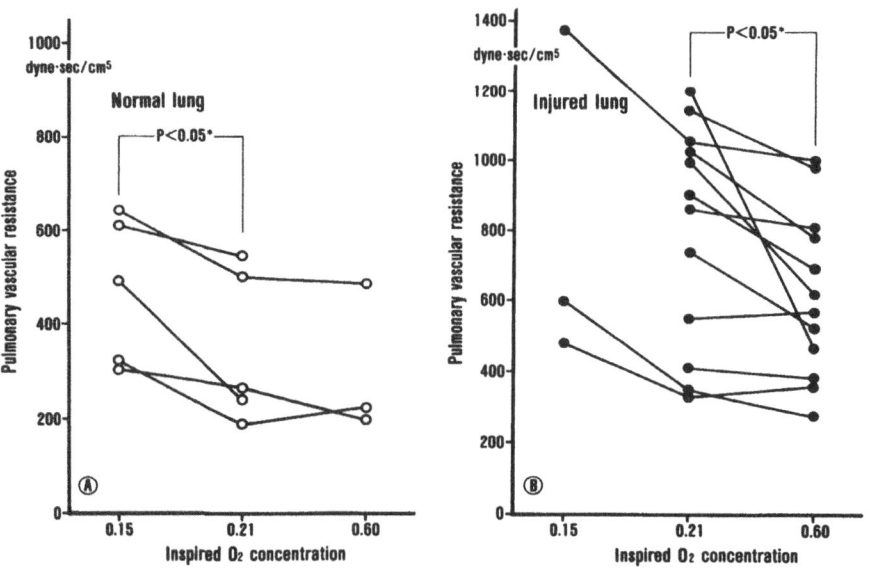

Fig. 2. Effect of inspired O₂ concentration on PVR. (A):
normal dogs. (B): dogs with lung injury.

which was 33 Torr at FIO_2 of 0.21 but 45 Torr at FIO_2 of 0.60 in the animals with lung damage.

R estimate for SF_6 (RSF_6), an indicator of the intrapulmonary shunt and the areas with very low ventilation-perfusion ratios ($\dot{V}_A/\dot{Q}$), did considerably decline in injured lungs during hyperoxic gas breathing, whereas RSF_6 in normal lungs was insignificantly changed comparing the results with hypoxic gas breathing to those investigated during exposure to normoxic or hyperoxic gas mixture. Qualitatively similar findings were obtained for other four inert gases with varied solubility (Figure 3).

The pressor response in pulmonary circulation after indomethacin infusion was markedly augmented compared with the preinfusion response. i.e. PVR increased by 50% from 712 dyne·sec/cm^5 to 1071 dyne·sec/cm^5 (Figure 4). Although neither $\dot{Q}_T$ nor extent of lung edema was different before and after administrating indomethacin, $P\bar{V}O_2$ was a little lower in injured animals with cyclooxygenase inhibited.

Indomethacin produced a significant reduction in R estimate of each indicator gas in the animals with injured lung (Figure 5). Mean RSF_6 declined by 21% from 0.35 to 0.27 accompanied by a considerable rise in PaO_2 from 84 Torr up to 99 Torr under a condition of FIO_2 0.60 (Figure 6).

DISCUSSION

Pulmonary fat embolism is a frequent clinical complication following major trauma, particularly that involving long-bone fracture (Sherr et al., 1974). This often causes ARDS within 24-72 hours after traumatic event (Sherr et al., 1974). Fat emboli appear to consist largely of triglycerates, which is hydrolyzed to fatty acids by pulmonary lipases (Sherr et al., 1974; Shier and Wilson, 1980; Jones et al., 1982). Experimental study (Sherr et al., 1974) showed that oleic acid comprised about 50% of the total fatty acids present in pulmonary emboli subjected to long-bone trauma, indicating that oleic acid might be one of the major substances to induce ARDS associated with pulmonary fat embolism. Thus, the model of lung injury produced by injection of oleic acid may be useful to study the pathophysiological abnormalities accompanied with ARDS induced by fat emboli.

Intravenous administration of oleic acid causes a leakage of protein-rich fluid into the interstitium and alveoli. The injury is multifocal and tends to favor peripheral sites while sparing normal parenchyma (Schoene et al., 1984). Pulmonary edema may be the result of oleic-acid-induced free radicals assaulting microvascular endothelium, because the increased permeability can be attenuated by antioxidant pretreatment (Townsley et al., 1985). The most prominent pathophysiological disturbance after oleic acid injection is severe hypoxemia which has generally been attributed to a significant amount of shunt flow, $\dot{Q}_S/\dot{Q}_T$ caused by the lung damage including alveolar flooding and interstitial edema (Dueck et al., 1977; Dantzker et al., 1979; Montaner et al., 1986). Interestingly however, Schoene et al. (1984) could not demonstrate the distinct association between $\dot{Q}_S/\dot{Q}_T$ and extent of lung injury developed by oleic acid. Furthermore, Brigham et al. (1983) failed to show the correlation between the severity of hypoxemia and the amount of pulmonary edema in patients with ARDS of varied etiologies. Experimental findings given in Figure 1 are qualitatively similar to those reported by Brigham et al.(1983), i.e. the extent of pulmonary edema estimated from extravascular lung water did not correlate with arterial hypoxemia in oleic-acid lung injury as well. These results suggest that lung edema per se following parenchyma damage is a necessary but not sufficient condition to elucidate severe hypoxemia mainly caused by an augmented $\dot{Q}_S/\dot{Q}_T$ in acute lung injury involving that by oleic acid.

The distribution of pulmonary blood flow is predominantly determined by the resistances of small arteries, which are regulated by the local vascular tones sensitive to lung-tissue hypoxia (Fishman 1976). In normal

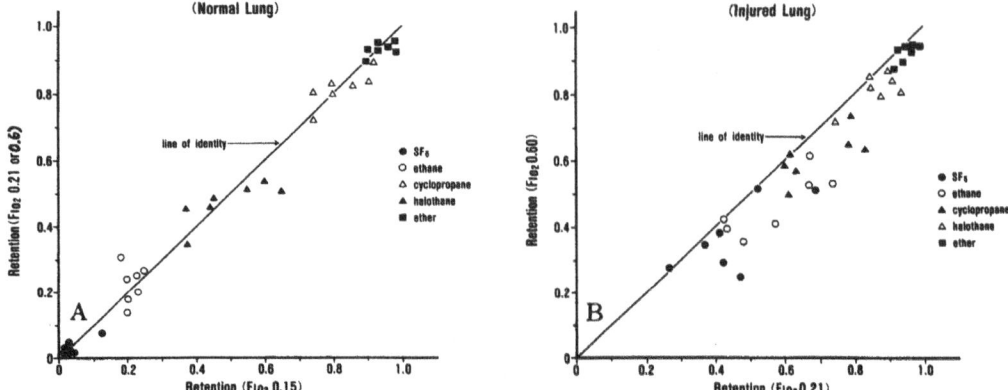

Fig. 3. Effect of inspired O_2 concentration on gas exchange efficiency. (A): normal dogs. (B): dogs with lung injury. See text for further details.

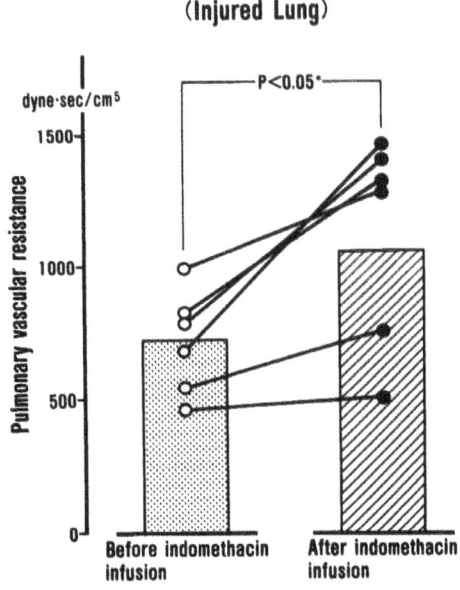

Fig. 4. Effect of inhibiting cyclooxygenase activity on PVR in injured lungs.

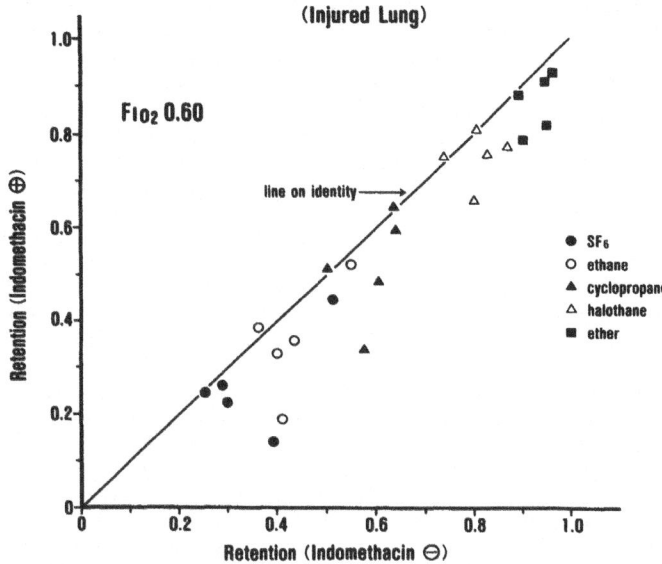

Fig. 5. Effect of inhibiting cyclooxygenase activity on gas
exchange efficiency in the animals with lung damage.

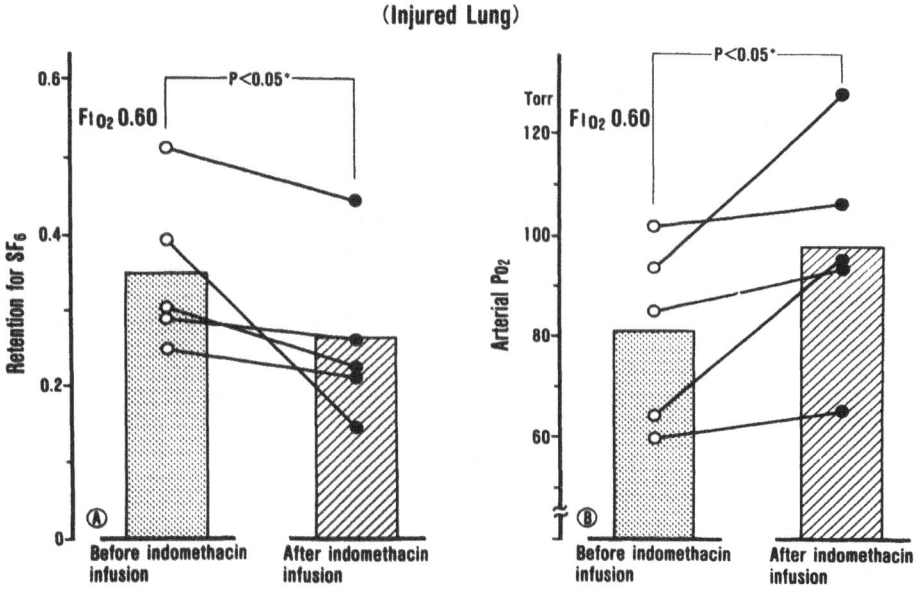

Fig. 6. Reduction in the shunt-like effect after cyclooxygenase
inhibition in injured lungs. (A): retention for SF_6,
a rationale index for shunt flow. (B): arterial PO_2
before and after indomethacin administration.

lungs, hypoxic regions with low $\dot{V}_A/\dot{Q}$ ratios elicit vasoconstriction, i.e. HPV shifting the blood flow there to better oxygenated areas, thus maintaining gas exchange efficiency as a whole (Euler and Liljestrand, 1946). If HPV is deteriorated in injured lungs, vascular beds in hypoxic areas fail to be contracted, leading to a relative increase of perfusion in injured and edematous areas of the lung. This causes the augmentation of $\dot{Q}_S/\dot{Q}_T$ and worsens the gas exchange efficiency. Diseased lungs damaged by oleic acid yielded a decrease in PVR without an appreciable change in either $\dot{Q}_T$ or extravascular lung water as an increase in FIO_2 from 0.21 to 0.60 (Figure 2). Such overall hemodynamic behavior is similar to that observed in normal lungs (Figure 2). However the change from normoxic gas breathing to hyperoxic gas breathing resulted in a significant reduction of the R estimate of each indicator gas in injured animals but did not in the animals without administrating oleic acid (Figure 3). The findings on the untreated animals suggested that, when exposing to hyperoxic gas, vascular tone within the pulmonary microcirculation diminishes nearly in an uniform manner in normal areas, preventing a considerable redistribution of pulmonary blood flow and thus maintaining the gas exchange efficiency in the lung. Different results obtained from injured animals indicate that, at a higher FIO_2, there appears to be a nonuniform reduction in the vascular tone of pulmonary microvessels in the lung and a sizable shift in perfusion from diseased areas to those preserved from tissue damage, leading to a considerable improvement in $\dot{Q}_S/\dot{Q}_T$ as well as gas exchange efficiency. Combining the experimental findings observed for both normal and injured animals (Figures 2 and 3), the vascular responsiveness to O_2 tension in diseased areas of oleic-acid lung injury may appreciably be attenuated. This prevents the vasoconstriction in hypoxic regions evoked by tissue damage and results in the augmented blood flow in the areas with low $\dot{V}_A/\dot{Q}$ ratios including intapulmonary shunt.

Considerable increase in concentrations of thromboxane, prostaglandins as well as leukotrienes was seen both in bronchoalveolar lavage fluid and in plasma after oleic acid injection (Olanoff et al., 1984; Ball et al., 1989). These products derived from arachidonic acid are potent mediators either constricting or dilating smooth muscles of airway tracts and of vascular trees (Snapper et al., 1983; Ahmed et al., 1986; Coggeshall et al., 1988). Among them, prostacyclin (PGI_2), one of the cyclooxygenase products of arachidonic acid, is a noble vasodilator and its local accumulation may cope with the vasoconstriction initiated by alveolar hypoxia in injured areas of the lung treated with oleic acid (Schulman et al., 1988), thus yielding a paralysis of vascular response to O_2. Abolition of PGI_2 biosynthesis by the administration of indomethacin caused a significant increase of PVR with concomitant change neither in $\dot{Q}_T$ nor in extravascular lung water (Figure 4). This was accompanied by a distinct improvement in gas exchange efficiency estimated from R values of five inert gases and arterial PO_2 (Figures 5 and 6). The findings may support the idea that reduction of vasodilator prostaglandins such as PGI_2 restores local hypoxic pressor response in injured areas and diminishes blood flow to the areas having venous-admixture effect with a resultant increase in perfusion to the portions relatively preserving the normal structure. Inhibition of cyclooxygenase activity not only reduces the production of PGI_2 but also diverts arachidonic acid into the lipoxygenase pathway, resulting in increased generation of leukotrienes (Morris et al., 1980). In addition to the decrease in PGI_2, the enhanced production of leukotrienes may, in part, contribute to vasoconstriction after indomethacin infusion. This possibility, however, requires a further investigation because of the uncertainty concerning how much the concentration of leukotrienes increases locally by inhibiting cyclooxygenase in acute lung injury particularly induced by oleic acid.

Restraint on cyclooxygenase pathway also attenuates the genesis of thromboxane, which is accepted as one of the important mediators for alterations in lung mechanics such as resistance to airflow in injured

lungs (Snapper et al., 1983; Ahmed et al., 1986; Coggeshall et al., 1988).
Previous study showed that indomethacin could completely block the
endotoxin-induced changes in lung mechanics (Ahmed et al., 1986). Although
the parameters available for estimating lung mechanics were not examined in
the present study, qualitatively the same phenomena as described above
would be expected to occur also in oleic-acid-induced lung injury. Namely,
severe bronchoconstriction caused by thromboxane in diseased areas may be
effectively prevented by the administration of indomethacin, resulting in
restoring the ventilation to diseased areas. Cooperating with a decrease
in perfusion to injured areas given by the inhibition of vasodilator
prostaglandins, an increase in ventilation due to blocking the genesis of
thromboxane further improves $\dot{V}_A/\dot{Q}$ inhomogeneity in injured lungs after
inhibiting cyclooxygenase activity.

SUMMARY

    To assess a significant role of hypoxic pulmonary vasoconstriction,
HPV, on maintaining the gas exchange efficiency in acute lung injury, 24
mongrel dogs were treated with intravenously injecting 0.07 ml/kg of oleic
acid. Hemodynamic and gas-exchange parameters were investigated at varied
inspired O_2 concentration, FIO_2. To know a possible contribution of
vasoactive prostanoids in regulating vascular reactivity under these
circumstances, observations were repeated after infusion of indomethacin.
The impairment of gas exchange in injured lungs was examined by measuring
the fractional retention, R, of the gas in arterial blood. For this
evaluation, a normal saline containing five foreign inert gases such as
sulfur hexafluoride, SF_6, ethane, cyclopropane, halothane and diethyl ether
was infused at a constant rate through a peripheral vein. After a steady
state was established, the expired gas was collected and the samples of
both arterial and mixed venous blood were simultaneously taken for the
inert-gas analysis. The concentrations of the indicator gases in the
samples were measured in terms of a gas chromatograph equipped with an
electron capture detector for SF_6 and a flame ionization detector for the
other four gases. Although pulmonary vascular resistance, PVR, after
injecting oleic acid at FIO_2 0.60 was significantly smaller than that
obtained at FIO_2 0.21, cardiac output, $\dot{Q}_T$ as well as extravascular lung
water were not different between the two conditions. R value for the
indicator gas was consistently lower at FIO_2 0.60 irrespective of the gas
species. As increasing FIO_2, R estimate concerning SF_6, RSF_6, rational
index of the fractional blood flow perfusing shunt area, decreased
significantly. Administration of indomethacin caused the rise in PVR
without an apperciable change in either $\dot{Q}_T$ or extravascular lung water but
a considerable diminution in R value for the inert gas. RSF_6 after
infusion of indomethacin decreased from 0.35 to 0.27, accompanied by a
significant rise in arterial PO_2 from 84 to 99 Torr. The findings are
highly compatible with the idea that HPV is distinctly attenuated in
diseased areas induced by oleic acid probably due to a local accumulation
of vasodilator prostanoids. Inhibiting prostanoid biosynthesis may
selectively enhance the vascular reactivity to O_2 in shunt vessels and may
redistribute the perfusion from shunt to relatively normal areas, thereby
improving gas exchange at alveolar region without altering the total amount
of extravascular lung water.

REFERENCES

Ahemed, T., Wassermann, M.A., Muccitelli, R., Tucker, S., Gazeroglu, H.,
    and Marchette, B., 1986, Endotoxin-induced changes in pulmonary
    hemodynamics and respiratory mechanics, <u>Am. Rev. Respir. Dis.</u>,
    134:1149.

Ball, H.A., Cook, J.A., Spicer, K.M., Wise, W.C., and Halushka, P.V., 1989, Essential fatty acid-deficient rats are resistant to oleic acid-induced pulmonary injury, J. Appl. Physiol., 67:811.

Brigham, K.L., Kariman, K., Harris, T.R., Snapper, J.R., Bernard, G.R., and Young, S.L., 1983, Correlation of oxygenation with vascular permeability-surface area but not with lung water in human with acute respiratory failure and pulmonary edema, J. Clin. Invest., 72:339.

Chang, S., Westcott, J.Y., Pickett, W.C., Murphy, R.C., and Volkel, N.F., 1989, Endotoxin-induced lung injury in rats: role of eicosanoids, J. Appl. Physiol., 66:2407.

Coggeshall, J.W., Christman, B.W., Lefferts, P.L., Serafin, W.E., Blair, I.A., Butterfield, M.J., and Snapper, J.R., 1988, Effects of inhibition of 5-lipoxygenase metabolism of arachidonic acid on response to endotoxemia in sheep, J. Appl. Physiol., 65:1351.

Dantzker, D.R., Brook, C.J., Dehart, P., Lynch, J.P., and Weg, J.G., 1979, Ventilation-perfusion distributions in adult respiratory distress syndrome, Am. Rev. Respir. Dis., 120:1039.

Duek, R., Wagner, P.D., and West, J.B., 1977, Effects of positive end-expiratory pressure on gas exchange in dogs with normal and edematous lungs, Anesthesiol., 47:359.

Euler, U.S., and Liljestrand, G., 1946, Observations on the pulmonary arterial blood pressure in the cat, Acta Physiol. Scand., 12:301.

Fishman, A.P., 1976, Hypoxia and pulmonary circulation, Cir. Res., 38:221.

Hlastala, M.P., and Robertson, H.T., 1978, Inert gas elimination characteristics of the normal and abnormal lungs, J. Appl. Physiol., 44:258.

Jones, J.G., Mintz, B.D., Reeley, J.M., Royston, R.D., Crow, J., and Grossmann, R.F., 1982, Pulmonary epithelial permeability is immediately increased after embolization with oleic acid but not neutral fat, Thorax, 37:169.

Montaner, J.S.G., Tsang, J., Evans, K.G., Mullen, J.B.M., Burns, A.R., Walker, D.C., Wiggs, B., and Hogg, J.C., 1986, Alveolar epithelial damage. A critical difference between high pressure and oleic acid-induced low pressure pulmonary edema, J. Clin. Invest., 77:1786.

Morris, H.R., Piper, P.J., Taylor, G.W., and Tippins, J.R., 1980, The role of arachidonate lipoxygenase in the release of SRS-A from guinea pig chopped lung, Prostaglandins 19:371.

Nobel, W.H., and Severinghaus, J.W., 1972, Thermal and conductivity dilution curves for rapid quantitation of pulmonary edema, J. Appl. Physiol., 32:770.

Olanoff, L.S., Reines, H.D., Spicer, K.M., and Halushka, P.V., 1984, Effects of oleic acid on pulmonary capillary leak and thromboxanes, J.Surg. Res., 36:597.

Sherr, S., Montemurno, R., and Raffer, P., 1974, Lipids of recovered pulmonary fat emboli following trauma, J. Trauma, 14:242.

Shier, M.R., and Wilson, R.F., 1980, Fat embolism syndrome: traumatic coagulopathy with respiratory distress, Surg. Annu., 12:139.

Schoene, R.B., Robertson, H.T., Thorning, D.R., Springmeyer, S.C., Hlastala, M.P., and Cheney, F.W., 1984, Pathophysiological patterns of resolution from acute oleic acid lung injury in the dog, J. Appl. Physiol., 56:472.

Schuman, L.L., Lennon, P.F., Ratner, S.J., and Enson, Y., 1988, Meclofenamate enhances blood oxygenation in acute oleic acid lung injury, J. Appl. Physiol., 64:710.

Snapper, J.R., Hutchison, A.A., Ogletree, M.L., and Brigham, K.L., 1983, Effects of cyclooxygenase inhibitors on the alterations in lung mechanics caused by endotoxemia in the unanesthetized sheep, J. Clin. Invest., 72:63.

Sprague, R.S., Stephenson, A.H., Dahms, T.E., and Lonigro, A.J., 1986, Effects of cyclooxygenase inhibition on ethchlorvynol-induced acute lung injury in dogs, J. Appl. Physiol., 61:1058.

Townsley, M.I., Taylor, G.E., Korthuis, R.J., and Taylor, A.E., 1985, Promethazine or DPPD pretreatment attenuates oleic acid-induced injury in isolated canine lungs, <u>J. Appl. Physiol</u>., 59:39.

Wagner, P.D., Saltzmann, H.A., and West, J.B., 1974, Measurement of continuous distributions of ventilation-perfusion ratios: theory, <u>J. Appl. Physiol</u>., 36:588.

^{99m}Tc-DTPA CLEARANCE: A SENSITIVE METHOD FOR EARLY DETECTION OF AN

IMPENDING DISTURBANCE IN GAS EXCHANGE

J.A.H.Bos, P.Wollmer[*], W.Erdmann and B.Lachmann

Dept. of Anaesthesiology, Erasmus University, Rotterdam, The Netherlands, [*]Dept. of Clinical Physiology, University of Lund, Sweden

INTRODUCTION

The barrier between the alveolar spaces and pulmonary capillary vasculature allows rapid exchange of oxygen and carbon dioxide between the blood and the alveolar air, but it severely restricts the diffusion of large solutes, such as albumin and other blood proteins, from the lung capillaries into the alveoli. The integrity of this barrier is altered in a wide variety of acute and chronic lung diseases as, for instance, the adult respiratory distress syndrome (ARDS) [Lachmann et al., 1982]. These pathological states of the lung are always accompanied by a severe impairment of gas exchange [Lachmann et al., 1982].

An indication of the state of the permeability of the alveolo-capillary barrier can be of major diagnostic and, possibly, therapeutic importance and is often measured by means of the clearance rate of hydrophylic solutes. A much used solute in this respect is ^{99m}Technetium-labeled-diethylenetriaminepentaacetic acid (^{99m}Tc-DTPA) [O'Brodovich and Coates, 1987; Barrowcliffe and Jones, 1987]. The pulmonary clearance of ^{99m}Tc-DTPA allows a noninvasive assessment of the integrity of the alveolo-capillary barrier. A part of this barrier is the surfactant lining layer, which plays a major role in maintaining adequate gas exchange by preventing alveolar collapse at end-expiration and also preventing alveolar flooding [Lachmann, 1987].

The purpose of this study was to investigate whether pulmonary clearance of ^{99m}Tc-DTPA is a useful and sensitive parameter for early detection of a disturbed surfactant system which, in turn, can lead to impairment of pulmonary gas exchange.

MATERIALS AND METHODS

Fourteen New Zealand adult rabbits (CPB, Zeist, The Netherlands) (2.5 ± 0.5 kg body weight) were used. The animals were anaesthetized with pentobarbital (50-60 mg kg^{-1} i.v.) with additional doses, as required, to maintain anaesthesia. Following tracheotomy a carotid artery was cannulated for arterial blood pressure monitoring and blood sampling. Neuromuscular block was induced with pancuronium (0.3-0.4 mg kg^{-1} i.m.). All animals were ventilated using a Servo Ventilator 900 C (Siemens Elema AB, Solna, Sweden) with, initially, the following ventilator settings: frequency (f) was 30 b.p.m, the inspiratory:expiratory (I/E) ratio was maintained at 1:2, FiO$_2$ was 1.0 and minute ventilation was set to maintain PaCO$_2$ at 30-40 mm Hg. Animals were divided into two groups (7 per group). In group 1, a very mild respiratory failure was induced by lung lavage according to Lachmann [Lachmann et al., 1980]. In brief, lavage was performed with 30 ml kg^{-1} of isotonic saline at 37°C equal to the gas volume needed to inflate the healthy lungs to about 40 cm H$_2$O. Each volume of saline was administered through the tube at a pressure not exceeding 40 cm H$_2$O. Two such lavages were performed. After induction of respiratory failure, group 1 was volume controlled ventilated with the following ventilator settings: f = 30 min^{-1}, tidal volume (V$_T$): 10-15 ml kg^{-1} and positive end-expiratory pressure of 3 cm H$_2$O. Group 2 (n = 7) had no lavage and served as healthy controls. These animals were also ventilated according to the mentioned settings. The animals were ventilated for 120 min during which blood gases were measured (15, 60 and 140 min) (ABL-2, Radiometer, Denmark). A solution of ^{99m}Tc-DTPA, which was prepared from a commercial kit (Technescan DTPA, Mallincrodt Diagnostica, Petten, The Netherlands), was nebulized into the inspiratory line of the ventilator using an air jet nebulizer (Ultravent, Mallinkrodt Diagnostica, The Netherlands). This type of nebulizer produces an aerosol with fine particles (mean size 1.7 μm, as measured by laser light scattering), favouring alveolar deposition [Brain and Valberg, 1979]. The supply of pressurized air to the nebulizer was controlled by a pneumatic valve connected to the ventilator via an electronic circuit (for set-up see Figs. 1 and 2). The nebulizer operated only during expiration, filling tube I with aerosol particles. One side of tube 1 was connected to a one-way valve (A) through which the aerosol was injected into the inspiratory line of the ventilation circuit. This in order to administer the particles with the ensuing insufflation. The other side of tube 1 was connected to a bacterial filter (B) and a pressure valve (C) set at an opening pressure of 25 cm H$_2$O. Whenever the pressure in tube I rose above 25 cm H$_2$O, the valve opened allowing air to escape, this keeping the pressure in the whole system (tube 1, ventilation circuit and animal lungs) during nebulizing below the threshold pressure and thus avoiding pneumothorax. The bacterial filter (B) was placed in front of the pressure valve to collect the ^{99m}Tc-DTPA particles in the aerosol, preventing contamination of the laboratory air.

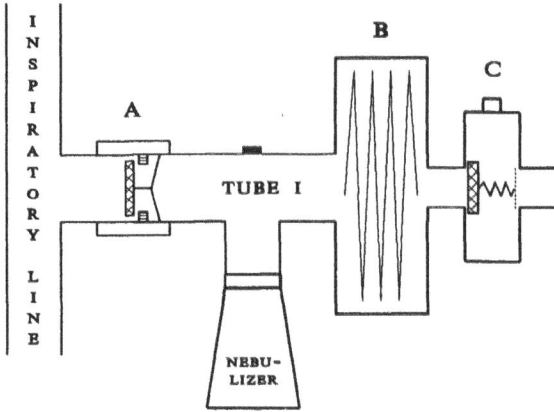

Fig 1. Set-up of nebulizer
A = one-way valve; B = bacterial filter; C = pressure valve

This whole system was placed in a plexiglas box to contain any other possible leakages. The box was protected with heavy lead shielding. Any particles not retained in the lungs of the animal were trapped in a lead shielded filter placed in the expiratory line of the ventilation line.
Before each run of the nebulizer, the ventilatory system was checked carefully for air leaks. During nebulizing, in both groups pressure controlled ventilation was used with an insufflation pressure of 2 kPa, an I/E ratio of 1:1 and a frequency of 30 b.p.m in order to reach a high activity over the lungs in a short

time. When a count of approximately 200-300 counts per second over the lungs was reached (after 2-3 min), aerosol administration was stopped, normal ventilation was resumed and clearance measurement started. Gamma camera images were obtained in successive 1-min frames for 20 min and stored in a 64x64 images matrix in a computer (Digital RT 11, Maynard, USA). Data from the clearance measurements were analysed by selecting a region of interest, namely both lungs of all animals, and generating a time-activity curve. A mono-exponential function was fitted to the experimental data and a half-life time (T½) of the tracer in the lungs calculated. When the clearance measurement was completed, animals were sacrificed by an i.v. overdose of pentobarbital.

Statistical analysis

Values are given as mean ± SD, unless stated otherwise. Differences between the means were tested by the Mann-Whitney-Wilcoxon test for unpaired samples.

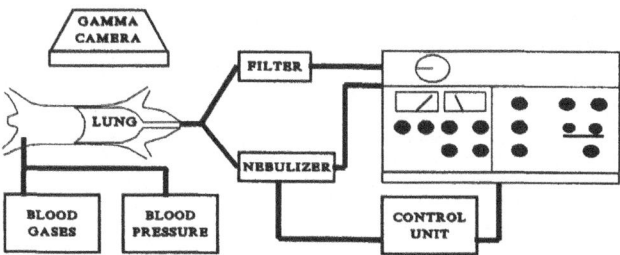

Fig. 2. Schema of experimental set-up.

RESULTS

Distribution of the inhaled ^{99m}Tc-DTPA aerosol was uniform in the lungs of all animals. The calculated T½ of ^{99m}Tc-DTPA in group 1 was 19 ± 4.7 min and in group 2, 93 ± 13 min. The relative count rates of the two groups are shown in Figure 3. The blood gases values during the study period are presented in Table 1. In group 1, all blood gases returned to normal values after 140 min indicating that the induced surfactant deficiency was mild (the difference between blood gas values of groups 1 and 2 was not statistically significant after 140 min; it was, however, after inducing respiratory failure). The corresponding permeability of the alveolo-capillary barrier, as measured by the clearance rates of group 1, however, did not recover to baseline values. Animals with normal blood gas values after 140 min had maximum T½ of the tracer in the lungs of approximately 30 min. Arterial blood pressure remained stable in all animals of both groups during the entire study.

DISCUSSION

The measurement technique for pulmonary clearance of ^{99m}Tc-DTPA (molecular weight 492 daltons, molecular radius ± 0.57 nm) is a relatively new, easy to perform, noninvasive method for assessment of the permeability of the alveolo-capillary barrier. This barrier, which consists of several structural layers, plays a major role in gas exchange in that it enables optimal conditions for the process of oxygen and carbon dioxide transport.

Table 1.   PaO$_2$ and PaCO$_2$ (in mm Hg). Significant differences (P<0.05, Mann-Whitney-Wilcoxon test for unpaired samples): # group 1 versus group 2. Values are mean ± SD.

Group	1			2	
Time (min)	PaO$_2$	PaCO$_2$ lavage		PaO$_2$	PaCO$_2$ no lavage
control	537	31		549	32.2
	±18	±4.9		±13	±6.3
after lavage					
15'	379#	33.2		557	33.4
	±148	±6.5		±20	±5.2
60'	435#	34.1		576	38.4
	±180	±5.7		±34	±3.4
140'	579	31.3		587	36.7
	±34	±4.0		±19	±4.5

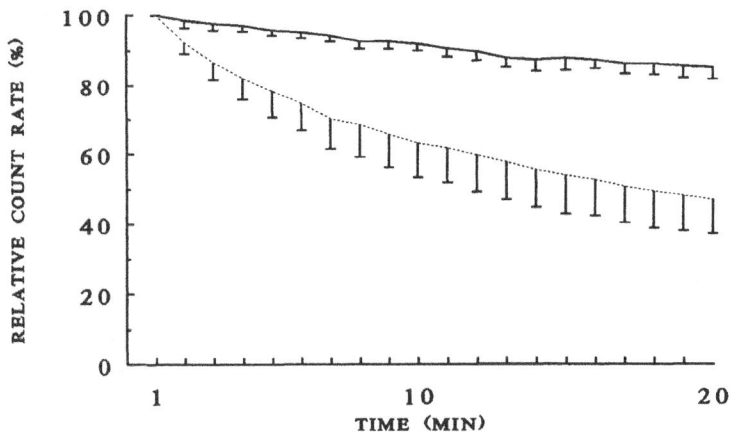

Fig 3.   Mean time-activity curves of ^{99m}Tc-DTPA in animal lungs.
(dashed line = group 1; solid line = group 2). Values are mean ± SD.

Although the technique has been well investigated over the last decade, the mechanisms which influence or dictate the clearance rate of ^{99m}Tc-DTPA are not yet fully understood. Most theories focus only on the integrity of the alveolar epithelium as the main factor of influence for the diffusion rate across the alveolo-capillary barrier. The mechanism behind this theory is that the hydrophylic solute molecule must pass through the aqueous pores between the alveolar epithelial cells, the so-called intercellular junctions. These pores have a median diameter of 0.6-1.0 nm. When compared to the median diameter of a ^{99m}Tc-DTPA molecule, which is $\pm$ 0.57 nm, it is highly probable that the main route of diffusion of the tracer across the alveolo-capillary layer is in fact through these pores. Studies, however, by Evander and Wollmer showed that it is not only the alveolar epithelium that dictates the diffusion rate but also the pulmonary surfactant system [Evander et al., 1987]. In a rabbit model they demonstrated that when the surfactant layer in the alveoli is removed by lung lavage [Lachmann et al., 1980], the clearance rate of ^{99m}Tc-DTPA across the alveolo-capillary membrane greatly increases. That this increase in clearance rate could not be attributed to damage of the alveolar epithelium by lavage was demonstrated by the fact that surfactant replacement after lung lavage restored the pulmonary clearance rate of ^{99m}Tc-DTPA to almost normal values. Further evidence for the effect that the surfactant system has on the diffusion rate of ^{99m}Tc-DTPA is the greatly increased pulmonary clearance rate in patients with ARDS and neonates with hyaline membrane disease [Jefferies et al., 1985; Jones et al., 1983]. The established pathophysiological factor in both these syndromes is a significant surfactant deficiency with its concomitant results. In experimental ARDS models, e.g., by infusion of intravenous oleic acid or by prolonged ventilation with 100% oxygen resulting in oxygen toxicity, results also showed an increased diffusion rate [Huchon et al., 1981; Matalon and Egan, 1981]. The increased permeability of the alveolo-capillary barrier, as measured with ^{99m}Tc-DTPA clearance technique, is also seen in case of pneumocystis carinii pneumonia, pulmonary sarcoidosis and fibrosing alveolitis, etc [Mason et al., 1985; Jones and Higenbottam, 1985; Jacobs et al., 1985; Rinderknecht et al., 1980].

Another factor important to the outcome of the pulmonary clearance of ^{99m}Tc-DTPA is the pattern of mechanical ventilation that is applied before, during and after nebulizing of a ^{99m}Tc-DTPA containing solution with the subsequent deposition of the tracer in the alveoli. Positive pressure ventilation increases the pulmonary clearance in neonatal lambs when compared to unventilated controls [Ramanathan et al., 1990]. The degree to which clearance increases is directly related to the duration of ventilation and to the height of positive pressure used during ventilation. Another feature of mechanical ventilation is the fact that the frequency used does not influence the clearance rate, but the use of a positive end-expiratory pressure (PEEP) does. Whenever a PEEP above 10 cm H_2O was used, a marked acceleration of clearance of ^{99m}Tc-DTPA was found [O'Brodovich et al., 1986]. This effect of PEEP seems to be largely mediated through an effect of PEEP on total lung volume. Especially when lung volume is elevated above the functional residual volume, clearance of ^{99m}Tc-DTPA is increased [Marks et al., 1985].

The mechanisms involved in the lung volume-induced increased clearance of the tracer are not yet fully understood. An increase in lung volume could result in stretching of the interepithelial junctions. The diameter of these pores should thereby be increased and, subsequently, the diffusion rate of ^{99m}Tc-DTPA through these pores. Another theory focuses on the possible effects of mechanical ventilation on the surfactant system and its function. An increase of lung volume could adversely affect the surfactant layer and its integrity, thus influencing the permeability of the alveolo-capillary barrier. Studies done by Evander and Wollmer with the ^{99m}Tc-DTPA clearance technique also indicate that mechanical ventilation can be deleterious to the surfactant system. Slow ventilation with large tidal volumes produce an increase in ^{99m}Tc-DTPA clearance compared to shallow ventilation with small tidal volumes [Evander et al., 1990].
Faridy [Faridy et al., 1966] and McClenahan [McClenahan and Urtnowski, 1967] showed that ventilation decreases lung compliance of ventilated lungs and that this effect is directly related to the tidal volume used. These findings and those of many others confirm that mechanical ventilation can adversely affect the pulmonary surfactant system and that possible alterations in permeability of the alveolo-capillary barrier by artificial ventilation, as measured the ^{99m}Tc-DTPA technique, can indicate changes in the integrity of the surfactant system.

The purpose of this study was evaluate to the usefulness of the ^{99m}Tc-DTPA clearance technique for very early detection of a disturbance in gas exchange and compare the results of this technique with blood gas values. The results show that the pulmonary clearance rate of ^{99m}Tc-DTPA in group 1 is abnormal compared to the untreated controls (group 2); this despite the fact that after 120 and 140 min, blood gases in group 1 had returned to normal. This observation that there is no good correlation between blood gases and clearance rate of ^{99m}Tc-DTPA for detection of a disturbed pulmonary surfactant system has already been reported by Evander [Evander et al., 1988] who administered the detergent dioctyl sodium sulfosuccinate in aerosol to rabbits. This detergent is known to cause no structural damage to the alveolar cells, nor does it increase the pulmonary microvascular membrane permeability to macromolecules. Their results showed no changes in gas exchange and lung mechanics 30 min after detergent administration, but the clearance rate of ^{99m}Tc-DTPA was immediately abnormal showing a highly increased permeability to the tracer molecule. The authors opined that the detergent acted on the alveolar surfactant layer and disrupted its normal function.

This study, and that of Evander [Evander et al., 1988], shows that blood gases and lung mechanics are not sensitive enough to detect an early disturbance of the pulmonary surfactant system which, in turn, could lead to severe impairment of oxygen transport across the alveolo-capillary barrier. It is suggested that pulmonary clearance of ^{99m}Tc-DTPA may be of great value in the diagnosis and monitoring of the response to treatment in, for example, pneumocystis carinii pneumonia.

ACKNOWLEDGEMENT

This study was in part financially supported by The Dutch Foundation for Medical Research (SFMO)

REFERENCES

Barrowcliffe, M.P. and Jones. J.G., 1987, Solute permeability of the alveolar capillary barrier, Thorax 42:1.

Brain, J.D., and Valberg, P.A., 1979, Deposition of aerosol in the respiratory tract. Am Rev Respir Dis., 120:1325.

Evander, E., Wollmer, P., Jonson, B., and Lachmann, B., 1987, Pulmonary clearance of inhaled ^{99m}Tc-DTPA: effects of surfactant depletion by lung lavag, J Appl Physiol, 62:1611.

Evander, E., Wollmer, P., and Jonson, B., 1990, Pulmonary clearance of inhaled ^{99m}Tc-DTPA: effects of ventilation pattern, Clin Physiol., 10:199.

Evander, E., Wollmer, P., and Jonson, B., 1988, Pulmonary clearance of inhaled ^{99m}Tc-DTPA: effect of the detergent dioctyl sodium sulfosuccinate in aerosol, Clin Physiol., 8:105.

Faridy, E.E., Permutt, S., and Riley, R.L., 1966, Effect of ventilation on surfactant forces in excised dogs' lungs. J Appl Physiol., 21:1453.

Huchon, G.J., Little, J.W., and Murray, J.F., 1981, Assessment of alveolar-capillary membrane permeability of dogs by aerosolization, J Appl Physiol., 51:955.

Jacobs, M.P., Baughman, R.P., Hughes, J., and Fernandez-Ulloa, M., 1985, Radioaerosol lung clearance in patients with active pulmonary sarcoidosis, Am Rev Respir Dis., 131:687.

Jefferies, A.L., Coates, G., and O'Brodovich, H., 1985, Pulmonary epithelial permeability in hyaline membrane disease, N Engl J Med., 311:1075.

Jones, J.G., Royston, D., and Minty, B.D., 1983, Changes in alveolar-capillary barrier function in animals and humans, Am Rev Respir Dis., 127:S51.

Jones, D.K., and Higenbottam, T.W., 1985, Pneumocystis pneumonia increases the clearance rate of inhaled ^{99m}Tc-DTPA from lung to blood, Chest., 88:631.

Lachmann, B., Danzmann, E., Haendly, B., and Jonson, B., 1982, Ventilator settings and gas exchange in respiratory distress syndrome. In: Prakash, O., ed., Applied physiology in clinical respiratory care. Martinus Nijhoff, The Hague/Boston/London, 141.

Lachmann, B., 1987, The role of pulmonary surfactant in the pathogenesis and therapy of ARDS, In: Vincent,J.L., ed., Update in Intensive Care and Emergency Medicine, Berlin, Springer-Verlag, 123.

Lachmann, B., Robertson, B., Vogel, J., 1980, In-vivo lung lavage as an experimental model of the respiratory distress syndrome, Acta Anaesthesiologica Scandinavica, 24:231.

Marks, J.D., Luce, J.M., Lazar, N.M., Ngao-Sun Wu, J., Lipavsky, A.J.A. and Murray, J.F., 1985, Effect of increases in lung volume on clearance of aerosolized solute from human lungs, J Appl Physiol 59:1242.

Mason, G.R., Effros, R.M., Mena, I. and Duane, G., 1985, Pulmonary clearance of a small solute is increased by pneumocystis pneumonia in patients with acquired immune deficiency syndrome [Abstract], Am Rev Respir Dis, 131:A116.

Matalon, S., and Egan, E.A., 1981, Effects of 100% O_2 breathing on permeability of alveolar epithelium to solute, J Appl Physiol: Respirat. Environ. and Exercise Physiol, 50:859.

McClenahan, J.B., and Urtnowski, A., 1967, Effect of ventilation on surfactant and its turnover rate, J Appl Physiol, 23:215.

O'Brodovich, H. and Coates, G., 1987, Pulmonary clearance of ^{99m}Tc-DTPA: a noninvasive assessment of epithelial integrity, Lung, 165:1.

O'Brodovich, H., Coates, G. and Marrin, M., 1986, Effect of inspiratory resistance and PEEP on ^{99m}Tc-DTPA clearance, J Appl Physiol, 60:1461.

Ramanathan, R., Gregory, R.M. and Usha Raj, J., 1990, Effect of mechanical ventilation and barotrauma on pulmonary clearance of ^{99m}Tc-DTPA in lambs, Pediatric Research, 47:70.

Rinderknecht, J., Shapiro, L., Krauthammer, M., Taplin, G., Wasserman, K., Uszler, J.M. and Effros, R.M., 1980, Accelerated clearance of small solutes from the lungs in interstitial lung disease, Am Rev Respir Dis, 121:105.

ACUTE RESPIRATORY FAILURE DURING PNEUMONIA INDUCED BY

SENDAI VIRUS

GJ van Daal, EP Eijking, KL So, RBAM Fievez,
MJW Sprenger*, DW van Dam, W Erdmann, and B Lachmann

Depts of Anesthesiology and *Virology and WHO influenza center
Erasmus University Rotterdam, The Netherlands

INTRODUCTION

Although there is no universal agreement on the criteria for the diagnosis of the adult respiratory distress syndrome (ARDS), the syndrome is characterized by distinct clinical and pathophysiological features (Ashbaugh et al., 1967; Petty and Ashbaugh 1971). ARDS can be the outcome of a number of disorders including shock of any etiology, infectious causes, trauma, drug overdose and many others (Hopewell and Murray, 1977). During the past 20 years little has changed in the possibilities for the treatment of ARDS and the mortality rate is still extremely high (50 - 70%) (Shale, 1987).

In this study a model of acute respiratory failure due to viral pneumonia in rats, closely resembling ARDS, is presented. This model permits evaluation of different therapeutical approaches for improving gas exchange during ARDS. Furthermore, preliminary results of surfactant substitution therapy in this model are presented.

MATERIALS AND METHODS

Animals

Male Sprague-Dawley rats (SPF, 170-200 g, 6 - 8 weeks old) were used. Rats were housed under filter bonnets and autoclaved food and water were available ad libitum. Infected animals were removed and kept separately in another facility.

Virus

Sendai virus (Myxovirus parainfluenza type 1) is a single-stranded RNA virus (150-200 nm diameter) belonging to the family of the paramyxoviruses, to which several important human viruses also belong (e.g. mumps, measles and respiratory syncytial virus). Sendai virus generally replicates in respiratory epithelium (Ito et al., 1982; Ito et al., 1983; Tyrrell and Coid, 1970), primarily in bronchial and bronchiolar epithelium (Blandford and Heath, 1972). Sendai virus was propagated in 11-day embryonated chicken eggs. The hemagglutination (HA) titre of the stock solution was determined to be approximately 1:3000.

Animal inoculation

Twenty-four non-anesthetized animals were randomly assigned to four groups and were exposed for 90 min in an 11 liter aerosol chamber, through which an aerosol (1 : 2 dilution in PBS of stock solution) flow of 5 l/min was led. Aerosol was produced with an air jet nebulizer (Ultravent, Malinckrodt Diagnostica, The Netherlands). This device produces small particles, size 0.6 - 2 μm (Dahlbäck et al., 1986), which allows for alveolar deposition of the aerosol (Brain and Valberg 1979). Two additional animals were infected in the same way for histological examination of the lungs.

Evaluation of arterial blood gases and surfactant substitution

Each day 6 animals were anesthetized (pentobarbital sodium 60 mg/kg i.p.), tracheotomized and a metal cannula was inserted into the trachea. A catheter (0.8 mm outer diameter) was inserted into the right carotid artery for drawing arterial blood samples. Animals were paralyzed (pancuronium bromide 0.1 mg/kg i.m.) and mechanically ventilated with a Siemens 900c ventilator, in a pressure controlled mode, at a rate of 35/min, P_{peak} = 15 cm H_2O (15/0), I/E ratio 1:2 and FiO_2 = 1. Then peak-airway pressure was increased to 20 cm H_2O (20/0). Next, peak-airway pressure was increased to 25 cm H_2O and positive-end-expiratory-pressure (PEEP) of 4 cm H_2O (25/4) was introduced. At each different ventilator setting 15 min was allowed for stabilization before 0.3 ml blood was collected and arterial blood gases were determined (ABL 330, Radiometer, Copenhagen, Denmark). On the second day after infection, two animals received 1.5 ml exogenous natural, bovine surfactant (phospolipids 50 mg/ml) intratracheally and were monitored for two hours after surfactant substitution; airway pressures 25/4, rate 35/min, I/E ratio 1:2 and FiO_2 = 1.

Thorax-lung compliance registration

After evaluation of the influence of different ventilator settings on blood gases, animals were transferred to a body-plethysmograph (for details see Lachmann et al., 1980) and thorax-lung compliance was registered at a rate of 6/min, P_{peak} = 15 cm H_2O, I/E ratio 1:2 and FiO_2 = 1.

Statistical evaluation of data

Data are presented as mean ± standard deviation (SD). Statistical analysis of data was done with the Wilcoxon test for within-group comparison and with the Mann-Whitney U test for between-group comparison. Statistical significance was accepted at $p < 0.05$ (two-tailed).

RESULTS

Bodyweight of infected animals (182.8 ± 35.3 g, day 3 and day 4 group, n = 12) increased during the first day after infection (202.1 ± 52.3 g); animals suffered significant weightloss during the second (197.3 ± 51.3 g) and the third day (173.0 ± 44.8 g) after infection. Furthermore, animals showed clinical signs of illness like ruffled fur, tachypnea, rhinitis and reduced motility from the first day after infection.
Figure 1 shows mean arterial oxygen (P_aO_2) and carbon dioxide (P_aCO_2)-tension of control animals and three groups of infected animals, on three successive days after infection, during artificial ventilation (P_{peak} = 15 cm H_2O, I/E ratio 1:2 and FiO_2 = 1). With these ventilator settings P_aO_2 dramatically decreased from the first day after infection whereas P_aCO_2 significantly increased. Arterial pH (table 1) was significantly decreased in all groups of infected animals whereas arterial bicarbonate concentration (HCO_3^-) was not significantly different compared to controls. The remaining six animals (day 4 group) died spontaneously between the third and the fourth day after infection and blood gases and thorax-lung compliance could, therefore, not be evaluated.

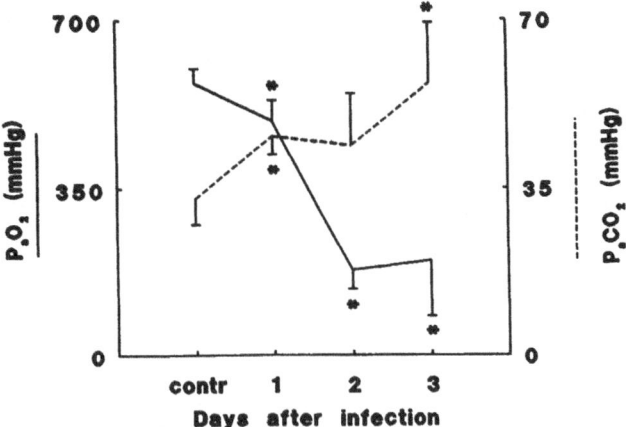

Figure 1. Mean arterial oxygen (P_aO_2) and carbon dioxide (P_aCO_2) tension of control animals and three groups of rats, infected with Sendai virus, on three successive days after infection, during artificial ventilation (P_{peak} = 15 cm H_2O, I/E ratio 1:2 and FiO_2 = 1).
* $p < 0.05$, vs controls, n=6/group.

Table 1. Influence of different ventilator settings on arterial PCO_2, pH and HCO_3^-, during the course of infection in rats infected with Sendai virus. Animals were mechanically ventilated with a Siemens 900c ventilator, pressure controlled, at different pressures at a rate of 35/min, I/E ratio 1:2 and FiO_2=1. Data are presented as mean (SD), n=6 per group. # $p < 0.05$, MW U-test, infected animals vs controls. * $p < 0.05$, Wilcoxon test, vs controls.

	P_aCO_2 (mmHg)			pH			HCO_3^- (mmol/l)		
	15/0	20/0	25/4	15/0	20/0	25/4	15/0	20/0	25/4
contr.	32.7 (5.6)			7.49 (0.03)			24.4 (2.6)		
day 1	45.7# (3.9)	36.1* (3.7)	38.6* (3.0)	7.31# (0.02)	7.40* (0.03)	7.36* (0.03)	22.6 (1.3)	21.6 (1.1)	21.3 (1.2)
day 2	44.3# (10.8)	43.9 (10.4)	38.3 (8.1)	7.32# (0.1)	7.32 (0.09)	7.35 (0.07)	21.5 (1.0)	21.5 (1.5)	19.9 (1.5)
day 3	56.8# (12.3)	54.9 (12.6)	57.6 (14.0)	7.29# (0.07)	7.30 (0.08)	7.29 (0.08)	26.3 (2.0)	25.8 (1.9)	26.7 (1.7)

Thorax-lung compliance was significantly decreased on the second and the third day after infection (Fig. 2) whereas protein concentrations of bronchoalveolar lavage (BAL) fluids were significantly increased on all three days after infection (Fig. 2).

The effect of increased peak airway pressures on P_aO_2 is shown in figure 3. Only on the second day after infection P_{peak} of 25 cm H_2O with PEEP of 4 cm H_2O significantly increased P_aO_2. Increased airway pressures did not result in significantly increased P_aO_2 on days one and three after infection. However, increased airway pressures significantly reduced P_aCO_2 and increased pH on the first day after infection whereas levels of significance were not reached on days two and three (table 1). Increased airway pressures did not significantly influence HCO_3^-(table 1).

The increase of airway pressures from 15/0 to 25/4 only slightly increased P_aO_2, whereas instillation of surfactant restored P_aO_2 almost to normal (Fig. 4).

Histological examination of lungs of two infected animals two days after infection showed alveolar flooding with protein rich fluid and atelectatic areas.

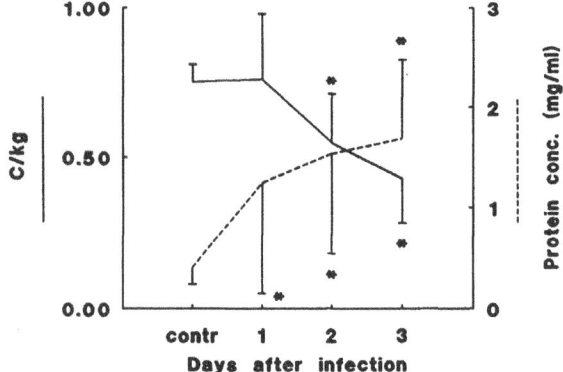

Figure 2. Thorax-lung compliance (C/kg) and protein content (mg/ml) of BAL fluid of rats infected with Sendai virus. * p < 0.05, vs controls, n=6/group.

DISCUSSION

We wanted to investigate whether we could induce severe respiratory failure in rats by infection with Sendai virus. Studies of intranasal inoculation of rats with high dose Sendai virus, as described by Giddens et al. (Giddens et al., 1987), performed earlier in our laboratories, did not result in hypoxemia or acidosis. This in spite of the fact that animals showed clinical signs of illness like ruffled fur, rhinitis, reduced motility and loss of bodyweight. However, animals spontaneously recovered within one week after infection. Therefore, we decided to employ concentrated live Sendai virus aerosol during an extended period of time. In this way we were able to induce severe respiratory failure with lethal outcome in four days. During the progression of the infection animals suffered significant loss of bodyweight and showed severe signs of illness.

At ventilator settings which resulted in hyperventilation and respiratory alkalosis in control animals, infected animals showed hypoventilation and respiratory acidosis as a result of mismatching between ventilation and perfusion of the lungs; P_aCO_2 of infected animals increased to 135% (day 2) and 173% (day

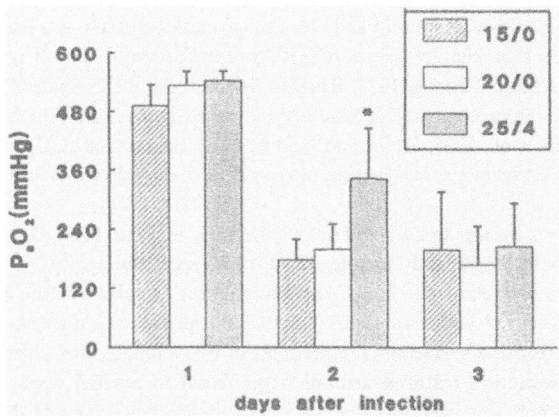

Figure 3. Effect of increased airway pressures on arterial oxygen content (P_aO_2) during artificial ventilation of rats infected with Sendai virus. * $p < 0.05$, vs 15/0, n=6/group.

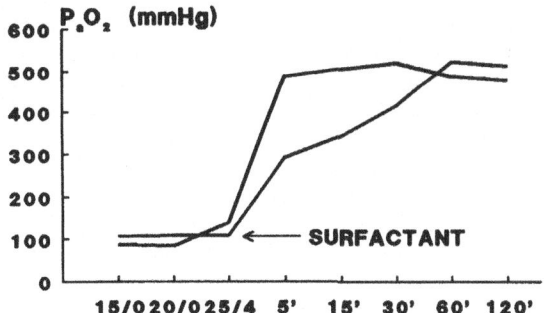

Figure 4. Effect of increased airway pressures and surfactant instillation on arterial oxygenation during artificial ventilation of 2 rats infected with Sendai virus.

3) of P_aCO_2 of control animals. Furthermore, infected animals developed severe hypoxemia as a result of atelectasis, flooded alveoli and, as mentioned above, mismatching between ventilation and perfusion; P_aO_2 of infected animals decreased to 31% (day 2) and 35% (day 3) of P_aO_2 of control animals in three days.

An intact surfactant system is indispensable for the maintenance of proper lung function and therefore any type of surfactant deficiency, whether primary or secondary, will contribute to the development of severe pulmonary pathology. It appears that loss of surfactant function is not the primary pathogenic factor for the occurrence of the severe respiratory failure in this study, but it may contribute significantly. Infection with Sendai virus may lead to destruction of type I cells and loss of integrity of the alveolar-capillary membrane, allowing albumin and other macromolecules to enter the alveoli, as observed in this study. One of the common characteristics of ARDS is the accumulation of protein-rich edema in the alveolar space (Hopewell and Murray, 1977; Rinaldo and Rogers, 1982; Shale, 1987; Bradley, 1987). Edema components like plasma proteins and cell membrane constituents inhibit the biophysical activity of the surfactant system (Holm et al., 1985; Holm and Notter, 1987; Ikegami et al., 1984; Fuchimukai et al., 1987; Seeger et al., 1985). Furthermore, destruction of type II cells may decrease surfactant synthesis and secretion.

Increasing ventilatory pressures to 20/0 did not significantly increase P_aO_2. Only $P_{peak} = 25$ cm H_2O with PEEP (4 cm H_2O) could significantly increase P_aO_2 of infected animals (92% increase) on day 2, which is direct evidence for a surfactant deficiency. Any tendency for alveolar collapse during endexpiration - which can be compensated only by counterpressure e.g. PEEP - speaks for a surfactant deficiency in these areas. To prove this, we performed surfactant replacement in two animals with severely reduced arterial oxygenation. Surfactant substitution restored arterial oxygenation to normal whereas increased airway pressures (25/4) alone did not; this indicates that surfactant deficiency is one of the major pathogenic causes of respiratory failure in this model of ARDS. However, since the applied ventilator settings were not very effective in restoring gas exchange, higher pressures or different modes of ventilation might have improved arterial oxygenation and reduced acidosis in infected animals.

Although there is no overall consensus on the criteria for the diagnosis of ARDS, the essential features of ARDS are met in this animal model (Pepe et al., 1982; Fowler et al., 1983; Stevens and Raffin, 1984). Infection of rats with Sendai virus led to acute respiratory failure with severe hypoxemia, increased intrapulmonary shunting, stiff lungs and pulmonary edema.

Surfactant replacement therapy seems to be a promising approach for the treatment of ARDS-like syndromes caused by viral pneumonia, as can be concluded from the preliminary results following surfactant replacement in this study. Several animal models of ARDS have demonstrated some form of surfactant deficiency or dysfunction and in seven of these models the beneficial effects of exogenous surfactant has been demonstrated: pulmonary oxygen toxicity (Holm et al., 1985), pulmonary damage induced by xanthine oxidase instillation (Saugstadt et al., 1984), in vivo lung lavage (Lachmann et al., 1980), bilateral cervical vagotomy (Berry et al., 1986), anti-lung serum infusion (Lachmann et al., 1987), induced damage to bronchial surfactant by artificial ventilation combined with saline instillation (Lachmann, 1985), and finally influenza A virus pneumonia in mice (Lachmann and Danzmann, 1984).

Although most of these models show characteristic features of ARDS, inevitable variations exist concerning the technique employed for the induction of pulmonary damage.

Therefore, we opine that this virus model may help clarify the underlying principles of ARDS and from the various therapeutic approaches available elucidate which approach is most appropriate.

SUMMARY

In this study a model of acute respiratory failure due to viral pneumonia in rats, closely resembling ARDS, is presented. Severe respiratory failure with lethal outcome in four days was induced by infection concentrated Sendai virus aerosol. This model permits evaluation of different therapeutical approaches for improving gas exchange during ARDS. Furthermore, preliminary results of surfactant substitution therapy in this model are presented.

ACKNOWLEDGEMENT

This work was supported by the Dutch Foundation for Medical Research (SFMO).

REFERENCES

Ashbaugh, D.G., Bigelow, D.B., Petty, T.L., et al., 1967, Acute respiratory distress in adults, Lancet, 2: 319.

Berry, D., Ikegami, M., Jobe, A., 1986, Respiratory distress and surfactant inhibition following vagotomy in rabbits, J Appl Physiol, 61: 1741.

Blandford, G., Heath, R.B., 1972, Studies on the immune reponse and pathogenesis of Sendai virus infection of mice. I. The fate of viral antigens, Immunology, 22: 637.

Bradley, R.B., 1987, Adult respiratory distress syndrome, Focus Crit Care, 14: 48.

Brain, J.D., Valberg, P.A., 1979, Deposition of aerosol in the respiratory tract, Am Rev Resp Dis, 120: 1325.

Dahlbäck, M., Nerbrinck, O., Arborelius, M., et al., 1983, Output characteristics from three medical nebulizers, J Aerosol Sci, 17: 563.

Fowler, A.A., Hamman, R.F., Good, J.T., et al., 1983, Adult respiratory distress syndrome: risk with common predispositions, Ann Intern Med, 98: 593.

Fuchimukai, T., Fujiwara, T., Takahashi, A., et al., 1987, Artifical pulmonary surfactant inhibited by proteins, J Appl Physiol, 58: 326.

Giddens, W.E., Van Hoosier, G.L., Garlinghouse Jr, L.E., 1987, Experimental Sendai virus infection in laboratory rats II. Pathology and immunohistochemistry, Lab Anim Sci, 37: 442.

Holm, B.A., Notter, R.H., Siegle, J., 1985, Pulmonary physiological and surfactant changes during injury and recovery from hyperoxia, J Appl Physiol, 59: 1402.

Holm, B.A., Notter, R.H., 1987, Effects of hemoglobin and cell membrane lipids on pulmonary surfactant activity, J Appl Physiol, 63: 1434.

Holm, B.A., Notter, R.H., Finkelstein, J.N., 1985, Surface property changes from interactions of albumin with natural lung surfactant and extracted lung lipids, Chem Phys Lipid, 38: 287.

Hopewell, P.C., Murray, J., 1977, The adult respiratory distress syndrome, in: "Respiratory emergencies", E.M. Shibel, K.M. Moser, eds., C.V. Mosby, St. Louis.

Ikegami, M., Jobe, A., Jacobs, H., et al., 1984, A protein from airways of premature lambs that inhibits surfactant function, J Appl Physiol, 57: 1134.

Ito, Y., Yamamoto, F., Takano, M., et al., 1982, Comparative studies on the distribution mode of orthomyxovirus and para-myxo-virus receptor possessing cells in mice and birds, Birds Med Microbiol Immunol, 171: 59.

Ito, Y., Yamamoto, F., Takano, M., et al., 1983, Detection of cellular receptors for Sendai virus in mouse tissue sections, Arch Virol, 75: 103.

Lachmann, B., Robertson, B., Vogel, J., 1980, In vivo lung lavage as an experimental model of the respiratory distress syndrome, Acta Anaesth Scand, 24: 231.

Lachmann, B., 1985, Possible role of bronchial surfactant, Eur J Respir Dis, 67: 49.

Lachmann, B., Hallman, M., Bergmann, K.C., 1987, Respiratory failure following anti-lung serum: study on mechanisms associated with surfactant system damage, Exp Lung Res, 12: 163.

Lachmann, B., Danzmann, E., 1984, Acute respiratory distress syndrome. in: "Pulmonary surfactant," B. Robertson, L.M.G. van Golde, J.J. Batenburg (eds), Elsevier, Amsterdam.

Pepe, P.E., Potkin, R.T., Reus, D.H., et al., 1982, Clinical predictors of the adult respiratory distress syndrome, Am J Surg, 144: 124.

Petty, T.L., Ashbaugh, D.G., 1971, The adult respiratory distress syndrome, Chest, 60: 233.

Rinaldo, J.E., Rogers, R.M., 1982, Adult respiratory distress syndrome: Changing concepts of lung injury, N Engl J Med, 15: 900.

Saugstadt, O.D., Hallman, M., Becher, G., et al., 1984, Protective effect of superoxide dismutase on severe lung damage caused by xanthine oxidase, <u>Pediatr Res</u>, 18: 802.

Seeger, W., Stohr, G., Wolf, H.R.D., et al., 1985, Alteration of surfactant function due to protein leakage: Special interaction with fibrin monomer, <u>J Appl Physiol</u>, 58: 326.

Shale, D.J., 1987, The adult respiratory distress syndrome - 20 years on, <u>Thorax</u>, 42: 642.

Stevens, J.H., Raffin, T.A., 1984, Adult respiratory distress syndrome - 1. Aetiology and mechanisms, <u>Postgrad Med</u>, 60: 505.

Tyrrell, D., Coid, C., 1970, Sendai virus infection of rats as a convenient model of acute respiratory infection, <u>Vet Rec</u>, 86: 164.

ESTIMATION OF RESPIRATORY MECHANICS IN DOGS WITH ACUTE LUNG INJURY

Zhang Li-fan, Han Li-ping, Wu Xing-yu,
Zhang Rong, Sun Xi-qing and Li Xiang-yu

Laboratory of Applied and Systems Physiology,
Department of Aerospace Physiology, The Fourth
Military Medical University, Xi'an, 710032, China

INTRODUCTION

In the pathogenesis of adult respiratory distress syndrome (ARDS), changes of pulmonary mechanics could lead to ventilation-perfusion ratio mismatch and refractory hypoxemia (Snapper, 1985). Real-time monitoring of pulmonary mechanics of patients with ARDS would be helpful for early detection and diagnosis, evaluation of the severity of the disease, assessment of the efficacy of each therapy developed, and in the future, the realization of the control and the optimization of ventilatory treatment (Suter, 1985; Koller et al., 1985).

However, the traditional methods used at present to study and monitor the pulmonary mechanics of patients with ARDS in Intensive Care Unit (ICU) have their limitations (Suter, 1985; Zhang and Zhang, 1990), especially being not suitable for tracking the slowly varying parameter changes. Thus it is highly desirable to assess the efficacy of the recently reported identification methods (Avanzolini and Barbini, 1984) on dogs with acute lung injury. In view of the value of measuring the pressure-volume (PV) characteristics in monitoring the progression of ARDS (Matamis et al., 1984), it is also attempted to develop a practical method for monitoring changes of PV characteristics without any interruption of the mechanical ventilation (Koller et al., 1985), and to evaluate it on animal models. Accordingly, the aim of the present work was to compare in canine oleic acid-induced acute lung injury model, the dynamic changes of the pulmonary mechanics measured by using the interrupter technique (Gottfried et al., 1985), least-squares (LS) estimation, identification technique, and tracing the PV curve, with a view to evaluating the efficacy and practicality of each method to be applied in monitoring the pulmonary mechanics of ARDS patients.

MATERIALS AND METHODS

Animal Preparation

Twelve mongrel dogs of either sex were equally divided into two groups, weighing 12.3 ± 1.9 and 10.7 ± 2.9 (SD) kg, respectively. The first group was assigned for assessing different parameter estimation methods, and the second for evaluating the method of obtaining dynamic PV curve and observing

the dynamics of pulmonary mechanics. The experimental animals were anesthetized by intravenous administration of pentobarbital sodium with an initial
dose of 30 mg/kg of body weight and subsequent doses of 4-6 mg/kg/h to ensure
adequate anesthesia. The animals were paralyzed with a dose of 10 mg of α-
tubocurarine chloride, followed by subsequent doses of 10 mg/h. They were
placed in a supine position on a well-fitted animal board made of plastics
with their relaxed forefeet hanging loosely at their sides. A tracheostomy
was performed and a tracheal intubation was made to connect the ventilator
through the resistance tube of a Fleisch pneumotachogragh (MFP-1200, Nihon
Kohden, Japan). The tidal volume of ventilation was regulated to maintain
end-tidal CO_2 ($Fetco_2$) ranging between 4 and 5%. The ventilator frequency
was 15 breaths/min and the inspired oxygen fraction (FIO_2) about 50%. The
esophageal pressure (Pes), which is an approximation to pleural pressure
(Ppl), was recorded by a standard technique, in order to measure transpulmonary pressure (Ptp = Pao-Ppl, Pao-airway pressure). A 7F Swan-Ganz catheter was inserted via the left external jugular vein to obtain the pulmonary
arterial pressure (Ppa)/ pulmonary capillary wedge pressure (Ppw) and to
collect mixed venous blood sample. An arterial catheter was inserted via
the right femoral artery with its tip located at abdominal aorta to measure
arterial pressure (Psa) and to collect arterial blood sample. A venous catheter, inserted via the right femoral vein, was positioned near the inlet
of the right atrium and subsequently used for injections of drugs and infusion of saline (15-20 ml/kg/h).

Experimental Techniques and Data Processing

Electric analog signals of respiratory gas pressures (Pao, Ptp), flow
rate ($\dot{V}$), gas fractions (Fo_2, Fco_2) and circulatory physiological variables
(Psa, Ppa, Ppw, ECG) were all recorded on an 8-channel polygraph (RM-6000,
Nihon Kohden, Japan) and stored by an 8-channel digital tape recorder. Mixed
venous and arterial blood samples were analysed for blood gases on an ABL 3
acid-base analyser (Radiometer, Denmark).

Interrupter technique: After disconnecting the animal from the ventilator, about 350-400 ml of air was inflated into the lungs and the airway
opening was occluded immediately. The airway was then rapidly reopened and
a series of about 7 interruptions were performed manually during the subsequent relaxed expiration. During the period of interrupted expiratory flowing, Pao, $\dot{V}_E$ and V_E were recorded simultaneously (see Fig.2). An average
of 5 interrupted expirations was analysed, to obtain the numerical values
for each set of parameters. The Crs was obtained by linear regression of
Pao and V_E, and Rrs was obtained by curvilinear regression of Pao and $\dot{V}_E$.
The pressure-flow relationship for the endotracheal cannula and its connecting tube was curvilinear, being described by a power function $P = 2.85\ \dot{V}^{1.06}$.

Least-squares (LS) fitting: The analog signals of P and $\dot{V}$ from 3 consecutive respiratory cycles were sampled at a frequency of 50 Hz. Subsequently the sampled signals were filtered by a 5-point moving average filter.
Then the respiratory mechanical parameters were estimated by use of least-
squares fitting. The one-compartment model may be written as a differential
equation

$$P(t) = P_z(t) - P_1(t) = C^{-1}V(t) + R\dot{V}(t) = DV(t) + R\dot{V}(t) \qquad (1)$$

where P(t) is the pressure difference exerted on the system; V(t) is the respiratory volume, $\dot{V}(t)$ is the flow rate; C, D, and R are compliance, elastance,
and resistance parameters, respectively. Depending on sites of pressure
measurement we can obtain the estimates of parameters for the lung and total
respiratory system while using the data of transpulmonary pressure (Ptp) and
tracheal pressure (Pao), respectively. According to the least-squares estimation, we can obtain the estimates, $\hat{D}(\hat{C})$ and $\hat{R}$, by solving the equations

$$\hat{D} \sum_{i=1}^{n} V(i)\,\dot{V}(i) + \hat{R} \sum_{i=1}^{n} \dot{V}(i)^2 = \sum_{i=1}^{n} P(i)\,\dot{V}(i)$$

$$\hat{D} \sum_{i=1}^{n} V(i)^2 + \hat{R} \sum_{i=1}^{n} \dot{V}(i)\,V(i) = \sum_{i=1}^{n} P(i)\,V(i)$$

(2)

Recursive extended least-squares (RELS) algorithm. The pretreatment of P and $\dot{V}$ signals is the same as stated in the previous paragraph. The equation (1) may be written as a difference equation

$$y(k) \equiv P(k) - P(k-1) = b_0 \dot{V}(k) + b_1 \dot{V}(k-1)$$

(3)

where P(k) is the respiratory pressure at time k, $\dot{V}(k)$ is the flow rate at time k, b_0 and b_1 are parameters to be estimated. These parameters are linked to the discrete parameters by simple relationship

$$\begin{cases} R = b_0 \\ C = T/(b_0 + b_1) \end{cases}$$

(4)

where T is the sampling interval. The RELS algorithm calculate the coefficients from these equations at each point in time k by minimizing a cost function

$$J = \sum_{i=0}^{k} \lambda^{k-i} \{ y(i) - y_M(i) \}^2$$

(5)

where $y(i)$ and $y_M(i)$ are the real and the model-predicted value at time i, respectively; and λ is a forgetting factor incorporated so that slowly varying parameters may be tracked. Estimates of the vector of coefficients and its covariance matrix are updated based on their previous values and a vector of new data. Finally, the estimates $\hat{C}$ and $\hat{R}$ of the continuous-time model were obtained through the step invariance transform.

Tracing static PV curve: Before each measurement, the lungs were inflated 2-3 times with air to a value of 2.45-2.94 kPa for Pao, to establish a standard volume history. The lungs were then inflated with air in 200-ml increments until a Pao of approximately 2.94 kPa was reached. Immediately after inflation, the air was removed in a similar series of steps down to FRC. At each step, V, Pao and Ptp were recorded, to obtain data for tracing a PV curve. The PV curves characterizing the elastic property of lungs and the respiratory system were obtained, respectively, from Ptp and Pao data, and the mean PV curve was obtained by interpolation.

Tracing dynamic PV curve: The dynamic PV curve of the total respiratory system was traced by making full use of the PEEP (positive end-expiratory pressure) mode of the ventilator. Before each measurement, the lungs were inflated 2-3 times with air to a Pao value of about 2.94 kPa. Then the lungs were at first inflated in a stepwise manner by increasing the PEEP value breath-by-breath, until a Pao of approximately 2.94 kPa was reached. Immediately after inflation, the lungs were deflated by a stepwise decrease of the PEEP value, until Pao returned to atmospheric pressure. During both inflation and deflation, the pressure and flowrate were simultaneously recorded and stored for off-line data processing to trace a dynamic PV curve.

Measurement of functional residual capacity: After 3 deep inflations to a value of 2.45-2.94 kPa for Pao, the animal was disconnected from the ventilator and a specially-made 320 ml syringe containing a known volume of helium in air, at a concentration of 8-10%, was attached to the airway. The

animal was ventilated from the syringe 10 times. The helium concentration
of the gas in the syringe was analysed using the helium analyser of a computerized spirometer (OSP-Ⅲ, Chest, Japan), and the FRC was calculated by
means of the standard dilution equation.

Experimental Protocol

After the control measurements of physiological variables, PV characteristics, and the control analysis of blood gases, 0.1 ml/kg oleic acid (OA)
was infused slowly (over 2-3 min) via the venous catheter. Sequential physiological measurements and blood gases analyses were then obtained up to
4-5 hours after infusion of OA (as shown in the relevant Figures and Tables
below). After all the measurements had been obtained, the animals were
heparinized and exsanguinated to death. The right and left lungs were
excised and the wet/dry weight ratios of lungs were determined.

Statistical Analysis

Analysis of variance (ANOVA) and Newman-Keuls test were performed to
evaluate changes over time, a P value of 0.05 being regarded as statistically significant.

RESULTS

Blood Gases and Hemodynamics

The data for blood gases are summarized in Table 1. There was a significant decrease (P<0.01, ANOVA) in Pao_2 and $P\bar{v}o_2$, and a significant increase
(P<0.01) in $P(A-a)o_2$ and $\dot{Q}s/\dot{Q}t$ in the 4-hour period after infusion of OA.
There was no significant change in Fro_2, $Petco_2$ and $Paco_2$ (P>0.05) over the
course of the experiment. The hemodynamic data are listed in Table 2. During the 4 hour period after infusion of OA, $\bar{P}pa$ and $\bar{P}sa$ increased and remained elevated at 2- to 4- hour (P<0.01), but $\bar{P}pw$ and HR did not show any
statistically significant change (P>0.05) throughout the experiment.

Table 1.　Changes of Blood Gases after OA Infusion (mean± SE. n=6)

		control	1h	4h	ANOVA
Pao_2	(kPa)	33.52±2.15	16.04±2.40	15.55±2.51	P<0.01
$P\bar{v}o_2$	(kPa)	7.13±0.55	5.35±0.63	5.80±0.51	P<0.01
$Paco_2$	(kPa)	4.99±0.09	5.64±0.32	6.08±0.52	P>0.05
$P(A-a)o_2$	(kPa)	6.31±2.20	22.49±2.40	22.24±2.11	P<0.01
$\dot{Q}s/\dot{Q}t$	(%)	4.90±2.30	10.30±2.10	10.50±2.00	P<0.01

Table 2.　Changes of Hemodynamics after OA Infusion (mean± SE. n=6)

	control	1h	2h	3h	4h	ANOVA
HR (min⁻¹)	166 ±9	153 ±16	155 ±15	165 ±15	166 ±16	P>0.05
$\bar{P}sa$ (kPa)	13.48 ±0.89	14.60 ±1.52	17.20 ±1.63	17.60 ±2.08	17.24 ±1.85	P<0.01
$\bar{P}pa$ (kPa)	1.65 ±0.12	1.81 ±0.31	1.96 ±0.14	2.17 ±0.14	2.43 ±0.17	P<0.01
$\bar{P}pw$ (kPa)	0.61 ±0.09	0.57 ±0.09	0.63 ±0.08	0.63 ±0.06	0.71 ±0.06	P>0.05

Functional Residual Capacity (FRC) and Wet/Dry Weight Ratios of Lungs

Fig.1 shows the change of FRC with time (ANOVA, P<0.01). During the first 2 hours after infusion of OA, FRC promptly decreased from its control value of 702.3 (mean)±101.3 (SE) to 588.5 ±74.6 ml (at 2h), although during the following 2 hours FRC tended to decrease, but the differences among values obtained at 2-, 3-, and 4- hour, as confirmed by the Newman-Keuls test, were not significant. Four hours after infusion of OA, FRC decreased to a value of 538.7±58.0 ml.

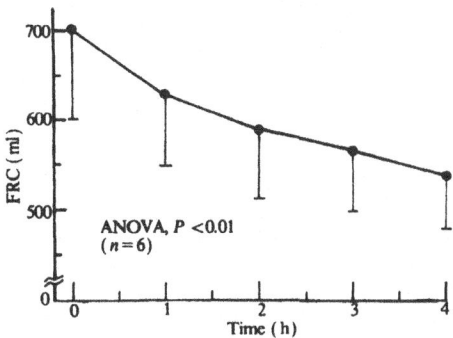

Fig.1. Measurement of FRC over time in 6 dogs given oleic acid
Note: vertical bars, 1 SE.

The wet/dry weight ratios of right, left, and both lungs of the first group 4 hours after infusion of OA were 8.04±0.19, 8.36±0.29, and 8.20±0.24, respectively.

Pulmonary Mechanical Parameters Measured with Interrupter Technique

A piece of the original tracings of Pao, V̇E and VE recorded during a series of interruptions throughout a passive expiration is given in Fig. 2. Fig.3, depicting the results obtained with the interrupter technique, shows the changes in the pulmonary mechanical parameters over time. The control

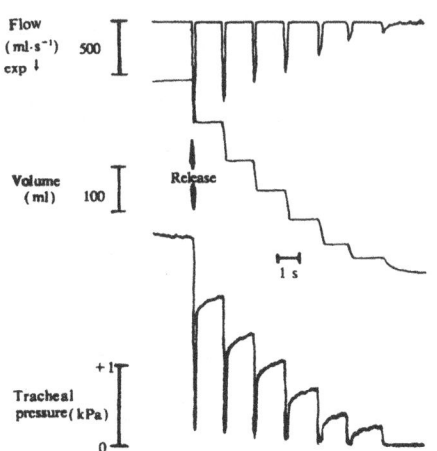

Fig.2. Tracings of flow, volume and airway pressure throughout a series of expiratory interruptions in a dog (2#)

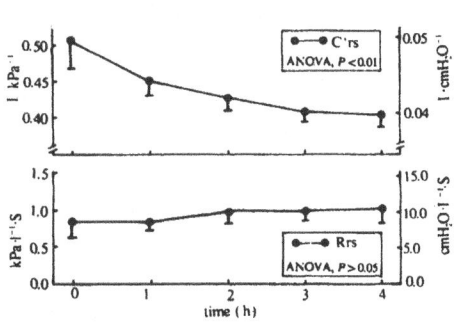

Fig.3. Effect of oleic acid on pulmonary mechanical parameters in 6 ventilated dogs —interrupter technique
Note: vertical bars, 1 SE.

value of $\hat{R}rs$ was 0.86 ± 0.21 kPa$\cdot$l$^{-1}\cdot$s and after OA infusion $\hat{R}rs$ tended to increase, but the changes were not significant ($P>0.05$). The control value of $\hat{C}rs$ was 0.51 ± 0.04 l$\cdot$kPa^{-1}. Immediately after OA infusion, $\hat{C}rs$ decreased rapidly to 0.45 ± 0.02 l$\cdot$kPa^{-1} at 1 hour ($P<0.01$); during the following 3 hours, $\hat{C}rs$ showed a gradual tendency to decrease, but the changes were not significant ($P>0.05$). The value of $\hat{C}rs$ obtained at 4 hours after OA infusion was 0.41 ± 0.02 l$\cdot$kPa^{-1}.

Pulmonary Mechanical Parameters Estimated with LS Fitting

Pieces of the original tracings of respiratory flow rate ($\dot{V}$) and pressure (P) under mechanical ventilation taken before and at different time intervals after OA infusion are given in Fig 4. Note that the amplitudes of both Ptp and Pao increased progressively after OA infusion. The estimated results by LS fitting for both $\hat{C}$ and $\hat{R}$ at various time intervals are summarized in Fig 5.

The trends of changes in compliance for both inspiratory and expiratory phases were very similar (see the upper two panels in Fig.5). For the inspiratory phase, 4 hours after OA infusion, the estimated lung compliance ($\hat{C}_{L,I}$) and total compliance ($\hat{C}rs,_I$) decreased from their control values of 0.99 ± 0.38 and 0.46 ± 0.05 to 0.41 ± 0.07 and 0.28 ± 0.02 l$\cdot$kPa^{-1}, respectively (ANOVA, $P<0.01$). During the first hour after OA infusion, both $\hat{C}_{L,I}$ and $\hat{C}rs,_I$ decreased sharply and exponentially. During the following 1- to 4- hours period, however, they tended to decrease slowly, but the changes were statistically nonsignificant. For the expiratory phase, the trends of changes in compliance over time were similar to those for the inspiratory phase. Four hours after OA infusion, both $\hat{C}_{L,E}$ and $\hat{C}rs,_E$ decreased from their control values of 1.17 ± 0.45 and 0.49 ± 0.05 to 0.34 ± 0.04 and 0.27 ± 0.02 l$\cdot$kPa^{-1}, respectively (ANOVA, $P<0.01$).

With regard to the changes in airflow resistance over time, trends for the two respiratory phases were quite different (see the lower two panels in Fig.5). For the inspiratory phase, the control value of $\hat{R}_{L,I}$ was higher than that of $\hat{R}rs,_I$, being 1.40 ± 0.15 and 0.73 ± 0.22 kPa$\cdot$l$^{-1}\cdot$s, respectively, and neither $\hat{R}_{L,I}$ nor $\hat{R}rs,_I$ showed any significant change during the 4-hour period after OA infusion (ANOVA, $P>0.05$). For the expiratory phase, the control value of $\hat{R}_{L,E}$ was lower than that of $Rrs,_E$, being -0.06 ± 0.13 and 0.32 ± 0.05 kPa$\cdot$l$^{-1}\cdot$s, respectively. Both $\hat{R}_{L,E}$ and $\hat{R}rs,_E$ increased and at 4-hour after OA infusion, reached values of 0.42 ± 0.05 and 0.67 ± 0.08 kPa$\cdot$l$^{-1}\cdot$s, respectively (ANOVA, $P<0.01$).

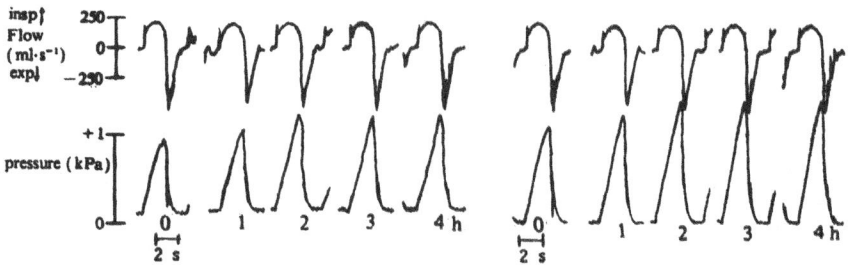

Fig.4. Tracings of flow and pressure (left panel—transpulmonary P., right panel—airway P.) of a ventilated dog (8#), showing the effect of oleic acid

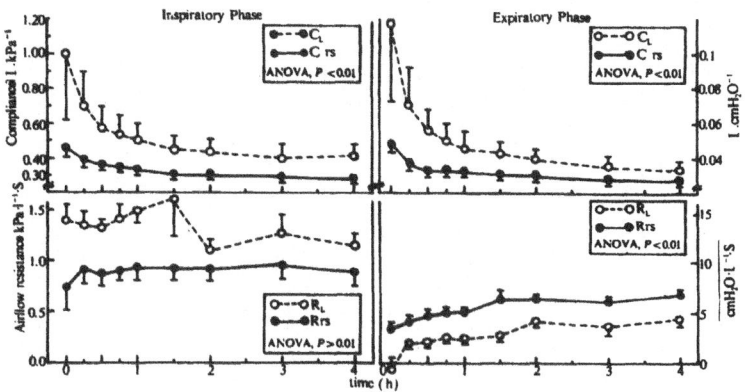

Fig.5. Effect of oleic acid on pulmonary mechanical parameters
in 6 ventilated dogs—LS fitting
Note: vertical bars, 1 SE.

Pulmonary Mechanical Parameters Estimated with RELS Algorithm

Fig.6 presents the actual transpulmonary waveform of five consecutive respiratory cycles of an animal together with the estimates for both the resistance ($\hat{R}$) and compliance ($\hat{C}$) obtained by the RELS algorithm. Note that after about 50 steps of computation, the estimates converged at a stabilized level with the estimated resistance for the expiratory phase being consistently larger than that for the inspiratory phase, and the estimated compliances of the two phases being nearly the same. The general time courses of the averaged estimates for the six dogs were similar and are summarized in Fig.7.

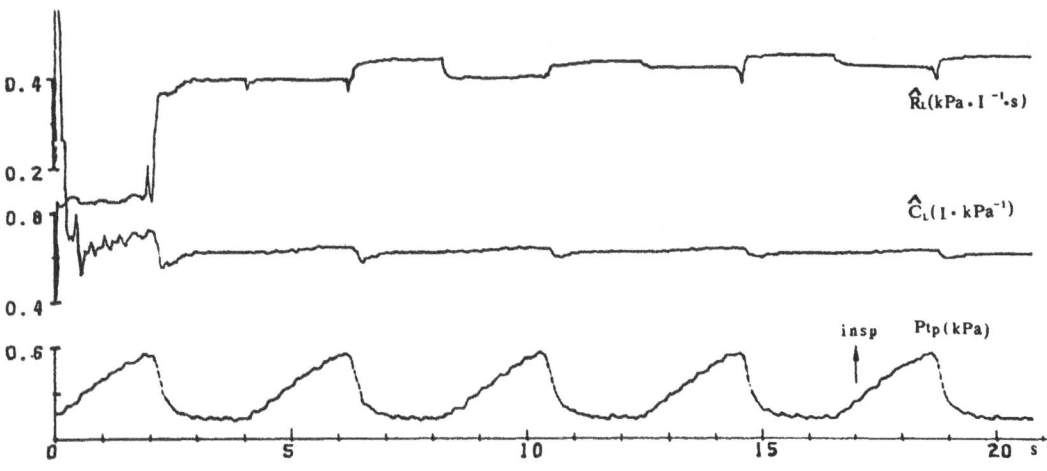

Fig.6. Tracings of actual transpulmonary pressure waveform (Ptp)
and estimates of resistance ($\hat{R}_L$) and of compliance ($\hat{C}_L$)
for five respiratory cycles of a ventilated dog (9#)

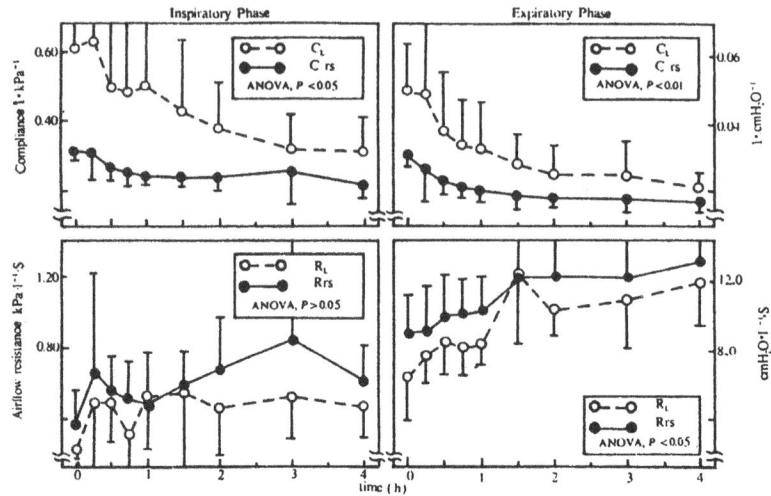

Fig.7. Effect of oleic acid on pulmonary mechanical parameters
in 6 ventilated dogs—RELS algorithm
Note: vertical bars, 1 SE.

The trends of changes in compliance for both the inspiratory and expiratory phases were very similar (see the upper two panels in Fig.7). For the
inspiratory phase, 4 hours after OA infusion, the mean estimated lung compliance ($\hat{C}_{L,I}$) and total compliance ($\hat{C}_{rs,I}$) decreased from their control values of 0.61 ± 0.19 and 0.31 ± 0.02 to 0.31 ± 0.09 (ANOVA, $P<0.05$) and $0.22\pm
0.04$ ($P<0.01$) $l\cdot kPa^{-1}$, respectively. As regards the expiratory phase, 4
hours after OA, $\hat{C}_{L,E}$ and $\hat{C}_{rs,E}$ decreased from their control values of $0.49
\pm0.13$ and 0.30 ± 0.03 to 0.21 ± 0.04 and 0.17 ± 0.02 $l\cdot kPa^{-1}$, respectively
(ANOVA, $P<0.01$).

The time courses of group mean of estimated resistance and the relation-
ship in magnitude between $\hat{R}_L$ and $\hat{R}_{rs}$ for both the two respiratory phases were
similar except that the values for the expiratory phase were consistently
higher than those for the inspiratory phase (see the lower two panels in
Fig.7). For the inspiratory phase, the control values of group mean esti-
mates for $\hat{R}_{L,I}$ and $\hat{R}_{rs,I}$ were 0.22 ± 0.06 and 0.37 ± 0.18, and tended to increase in a biphasic pattern to values of 0.46 ± 0.17 and 0.60 ± 0.21 $kPa\cdot l^{-1}\cdot s$
at 4-hour after OA, respectively, but both the changes were statistically
nonsignificant (ANOVA, $P>0.05$). As regards the expiratory resistance, the
mean estimates, $\hat{R}_{L,E}$ and $\hat{R}_{rs,E}$, increased steadily from their control values
of 0.63 ± 0.25 and 0.81 ± 0.28 to 1.16 ± 0.23 (ANOVA, $P<0.05$) and 1.29 ± 0.21
($P<0.01$) $kPa\cdot l^{-1}\cdot s$ at 4-hour after OA infusion, respectively.

Static and Dynamic PV curves

Fig.8 presents mean static PV curves for both the total respiratory
system (Pao-V curve) and the lungs (Ptp-V curve) obtained before and 4 hours
after OA infusion, demonstrating changes in PV characteristics of the chest-
lung system and lungs, respectively. It shows that 4 hours after OA infusion, there appeared an inflection on the inflation limb of the PV curve,
corresponding to a mean pressure value of 0.98 kPa for both Pao-V and Ptp-V
curves. Further, the mean half-volume width of Pao-V and Ptp-V curves increased from its control values of 0.69 and 0.36 to 1.05 and 0.81 kPa 4 hours
after OA infusion, respectively, indicating a larger hysteresis. Together
with the changes in FRC, these changes resulted in a downward and rightward
shift of the PV curves.

334

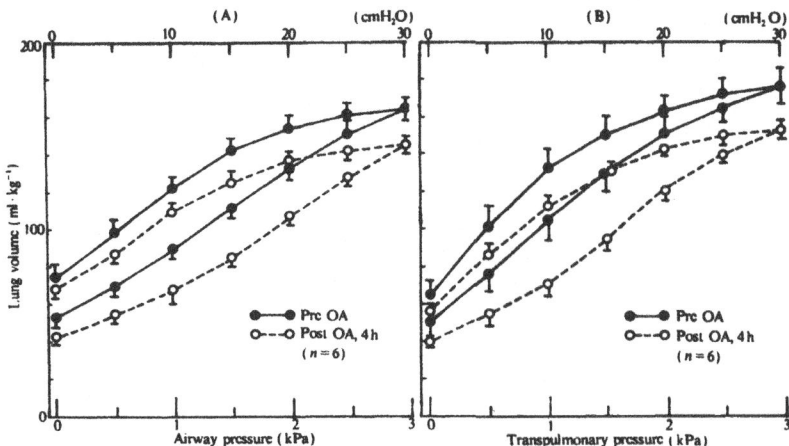

Fig.8. Mean pressure-volume curves pre- and post (4 h)- oleic
acid in 6 dogs. (A) Pao-V curves (chest-lung system),
(B) Ptp-V curves (lungs); vertical bars, 1 SE.

A set of compound dynamic PV curves of the total respiratory system of
a dog obtained before and at 1-, 3- and 5-hour after OA is given in Fig.9,
showing serial changes in PV characteristics of the chest-lung system after
OA infusion. The averaged data of the parameters characterizing these changes for the six dogs are listed in Table 3. Analysis of variance (ANOVA)
confirmed the following main changes caused by OA infusion, such as an increase in the airway pressure corresponding to the point of inflection, a
larger increase in both the ratio and degree of hysteresis, an increase in
the area enclosed by the outer envelope of the compound PV loop, and a decrease in the slope of the loop.

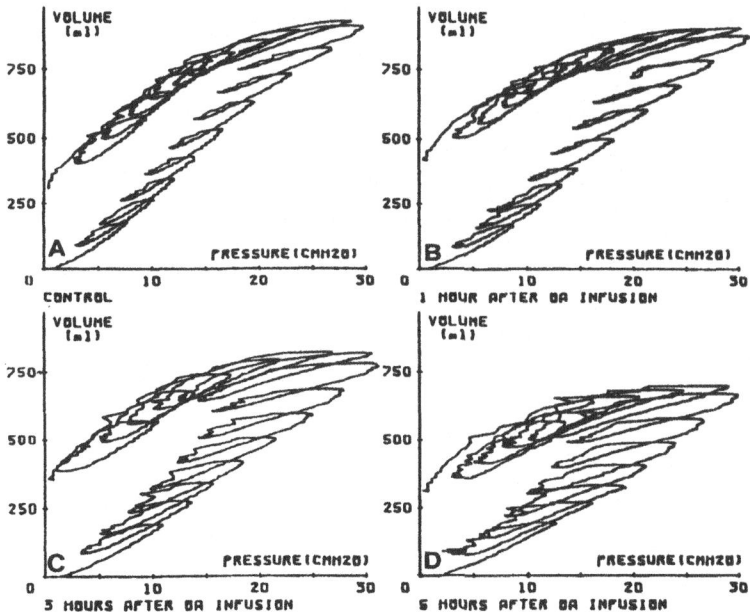

Fig.9. Compound dynamic PV curves obtained from a ventilated dog (7#)
before(A) and after oleic acid at 1-(B), 3-(C), and 5-hour(D)

Table 3. Changes in Parameters of Compound Dynamic PV Curve after OA
          Infusion (mean ± SE, n=6)

	Control	1h	3h	5h	ANOVA
Area enclosed by inner envelope (l · kPa)	0.496 ±0.020	0.465 ±0.016	0.381 ±0.025	0.356 ±0.029	P<0.01
Area enclosed by outer envelope (l · kPa)	1.090 ±0.069	1.281 ±0.125	1.356 ±0.173	1.451 ±0.185	P<0.05
Point of inflection (kPa)	0.602 ±0.042	0.755 ±0.059	1.037 ±0.151	1.311 ±0.191	P<0.01
Half-volume width (kPa)	1.359 ±0.057	1.614 ±0.088	1.725 ±0.062	1.928 ±0.067	P<0.01
Ratio of hysteresis	0.35 ±0.01	0.41 ±0.01	0.44 ±0.01	0.45 ±0.03	P<0.01
Degree of hysteresis (kPa)	1.052 ±0.035	1.258 ±0.051	1.381 ±0.046	1.445 ±0.066	P<0.01
Slope of loop (ml/kPa)	360.51 ±20.61	318.16 ±21.02	308.78 ±26.63	309.39 ±33.27	P<0.01

DISCUSSION

Viewing the data summarized in Fig.1 and Table 1, and the data obtained for the wet/dry weight ratios of the lungs, we consider that in our canine oleic acid model some typical pathophysiological changes of acute lung injury actually tookplace. The fact that no pulmonary hypertensive response occurred in the anesthetized dogs during the early phase after OA infusion might be explained by the adequate oxygen delivery to them by artificial ventilation. with Pao$_2$ maintained at a level above 13.3 kPa throughout the experiment.

The interrupter technique has been introduced by Gottfried et al. (1985) as an rapid quantitative approach to examine the passive elastic and flow-resistive properties of the total respiratory system in mechanically venti-lated patients using a simple, noninvasive equipment. We therefore believe it should remain a direct standard technique against which parameter esti-mation techniques should be compared. For all the three methods tested, it was found that the trends of changes in pulmonary mechanics after OA assessed by using each of them were basically similar. Let us discuss this in further detail.

With regard to changes in pulmonary elasticity over time, we have shown by the use of each of the three techniques that immediately after OA the compliance of lungs as well as chest-lung system decreased markedly and approached a very low level within two hours, followed by a very gradual decline, which was not statistically significant in most cases, until the end of the experiment (see Fig. 3, 5 and 7). The above results are in close agreement with those obtained for the FRC and static PV curve in this experi-ment (see Fig. 1 and 8) and the results previously reported (Snapper, 1985; Suter, 1985). However, the results in respect of pulmonary resistance as obtained by each of the three techniques tested were inconsistent with each other. When using the interrupter technique the results obtained for $\hat{R}rs,E$ after OA infusion were statistically not significant. Although LS fitting yielded more efficient estimates and satisfactory results for $\hat{C}_I$, $\hat{C}_E$ and $\hat{R}_E$, yet the results obtained for $\hat{R}_I$ ($\hat{R}_{L,I}$ and $\hat{R}rs,I$) were statistically nonsig-nificant and physiologically unreasonable. Only using the identification technique with RELS algorithm, we got more satisfactory results demonstrat-ing the changes in both the compliance and resistance parameter for the ex-piratory as well as the inspiratory phases.

Therefore, we put forward our argument for using the identification technique to monitor the lung mechanics of ventilated patients with ARDS. The reasons are as follows: (1) the inherent noise and statistical incorrect-ness of the isolated points method are overcome; (2) the apparatus used for interrupting airflow could be omitted; (3) there is no interference with me-chanical ventilation; (4) this method could be used on-line and to track the slow changes; and (5) it is a necessary step for the realization of the con-trol and optimization of the mechanical ventilation.

However, the following considerations are of crucial importance when adopting the identification technique for monitoring of lung mechanics.

(1) Model. According to the results of the present work and that re-ported by Chapman and Newell (1989), the first-order one-compartment model would be more appropriate for clinical use, yielding unequivocal results re-flecting the significant changes during the progression of acute lung injury. Although the second-order model has been claimed as a rather satisfactory re-presentation of pulmonary mechanics by some authors (Brusasco et al., 1980; Avanzolini and Barbini, 1984), yet its clinical usefulness is limited because of the difficulties in obtaining satisfactory results. Chapman and Newell (1989) reported that in tests on dogs developing OA-induced unilateral hemor-

337

rhagic pulmonary edema, the two-compartment estimates were seldom physiologically plausible.

(2) Algorithm. Besides algorithms based on LS, the maximum likelihood
method (Brusasco et al., 1980) and an adaptive tracking algorithm based on
steepest-descent (Linkens, 1985) also have been applied in this area. Among
them the recursive algorithms introduced by Avanzolini and Barbini (1985) are
of potential value to solve the problem. Our results indicate that after
about 50 steps of computation by use of the RELS algorithm, we could get a
stabilized estimate of the vector of unknown parameters, with expiratory resistance being larger than inspiratory resistance for each respiratory cycle
(see Fig. 6). The results shown in Fig. 3, 5 and 7 clearly indicate that the
RELS algorithm has yielded more satisfactory estimates reflecting the changes
of respiratory mechanics after OA. Therefore, it is attractive for clinical
monitoring of ventilated patients with ARDS.

(3) Separate estimation of inspiratory and expiratory mechanical parameters. Our finding shown in Fig. 5, 6 and 7 that the estimates obtained for
different respiratory phases were different, agrees with the results reported
by Chapman and Newell and most of the other authors (see Lorino, H., et al.,
1982; Zhang and Zhang, 1990). There is no satisfactory explanation yet.
Considering that the inspiratory lung mechanics may be more complicated under
mechanical ventilation, we support the suggestion made by Chapman and Newell
that lung mechanics can be monitored by tracking expiratory mechanics only.
It is acceptable and clinically worthwhile at least for monitoring ventilated
patients with ARDS.

(4) Assessment of the lung mechanics by the estimates of the total respiratory system. Transpulmonary pressure is needed to compute pulmonary mechanics. The measurement of pleural pressure is usually approximated by
the measurement of esophageal pressure. Nevertheless, this technique is not
suitable for clinical monitoring of severe patients with ARDS. Moreover, the
value of esophageal pressure is debatable in supine patients. To obviate
this drawback of clinical systems for monitoring lung mechanics, the tendency is to assess the mechanics of the total respiratory system by the data
of pressure and flow measurements at airway opening (Lorino, A.M., et al.,
1988). This attempt is supported by our finding shown in Fig.5 and 7, that
the trends of changes in estimates after OA infusion for either the lungs
or the total respiratory system are closely in agreement. However, it must
be emphasized that this assessment should be limited to paralysed ventilated
patients.

Tracing the PV curve could describe more comprehensively the changes
of pulmonary mechanics during the progression of ARDS. It is generally accepted that in the early stage of ARDS alveolar edema is associated with an
increased hysteresis ratio due to rightward shift of the inflation limb, and
the presence of an inflection on the inflation limb of the PV curve. The
inflection is explained by the inflational reopening of the airspaces closed
during the period of deflation (Slutsky et al., 1980). The inflection pressure is presently used as an indicator for adjusting the level of PEEP in
patients with the ARDS, since it has been demonstrated that when PEEP is
raised higher than inflection pressure the efficiency of oxygen transport at
lungs is markedly improved (Matamis et al., 1984). In addition, Matamis et
al, comparing total respiratory PV curve to roentgenographic changes obtained
in 19 patients, showed that the patterns of PV curves were correlated with
the stages of ARDS and to the pattern of the chest x-ray film. Thus tracing
the PV curve would be of potential value in monitoring the progression of
ARDS in ventilated patients. In this respect, the most important result of
this study is that the trends of changes after OA for both the pulmonary and
total respiratory PV curve were quite similar (see Fig.8). Therefore, it is
suggested that in clinical monitoring the characteristics of lung mechanics

could be assessed by tracing the PV curve of the total respiratory system, so as to obviate difficulties encountered in the measurement of esophageal pressure of ventilated patients with ARDS. For the sake of keeping the mechanical ventilation uninterrupted, based on the concept of "pneumoloop" raised by Koller et al (1985), we have developed our method for obtaining dynamic PV curve of the total respiratory system and defined some parameters to characterize the pattern of the PV curve. The results obtained (see Fig. 9 and Table 3) show that by tracing the dynamic compound PV curve and measuring its parameters, we can follow the sequential changes of the degree of hysteresis and inflection pressure, with the ventilation uninterrupted.

SUMMARY

In mechanically ventilated anesthetized dogs with acute lung injury induced by intravenous infusion of oleic acid (OA), changes in airflow, lung volume, and tracheal (Pao), transpulmonary (Ptp) pressures were measured. Changes in lung mechanics were studied before and after OA infusion at intervals, during an observation period lasting 4-5 hours, by using the interrupter technique and identification technique, and by measuring the pressure-volume (PV) characteristics.

The main results are listed as follows: (1) compliance and FRC showed a marked decrease, Pao and Ptp showed a marked increase within 2 hours after OA. (2) airflow resistance estimated by the identification technique showed a marked increase after OA. (3) static PV characteristics of both the lungs and total respiratory system showed similar changes 4 hours after OA, characterized by the presence of an inflexion on the inflation limb, increased hysteresis and a rightward and downward shift of the loop. (4) the trend of changes of dynamic PV characteristics obtained by changing the PEEP value in a stepwise manner, with the ventilation uninterrupted, was similar to that of the static PV curve.

It is suggested that an on-line identification technique with one-compartment model to track the slow changes in combination with serial measurements of the dynamic PV characteristics of the total respiratory system would be more appropriate for monitoring lung mechanics of ventilated patients with ARDS.

REFERENCES

Avanzolini, G., and Barbini, P., 1984, A versatile identification method applied to analysis of respiratory mechanics, IEEE Trans. Biomed. Eng., 31: 520.

Avanzolini, G., and Barbini, P., 1985, A comparative evaluation of three on-line identification methods for a respiratory mechanical model, IEEE Trans. Biomed. Eng., 32: 957.

Brusasco, V., Tiano, A., Ramoino, R., and Lamedica, G., 1980, Identification and parameter estimation of the mechanical ventilatory system, Respiration, 39: 75.

Chapman, F.W., and Newell, J.C., 1989, Estimating lung mechanics of dogs with unilateral lung injury, IEEE Trans. Biomed. Eng., 36: 405.

Gottfried, S.B., Rossi, A., Higgs, B.D., Calverley, P.M., Zocchi, L., Bozic, C., and Milic-Emili, J., 1985, Noninvasive determination of respiratory system mechanics during mechanical ventilation for acute respiratory failure, Am. Rev. Respir. Dis., 131: 414.

Koller, W., Aloy, A., Mutz, N., and Pauser, G., 1985, Computer systems in artificial ventilations (new techniques), in: Computers in Critical Care and Pulmonary Medicine, P.M. Osswald, ed., 212-220, Springer-Verlag, Berlin.

Linkens, D.A., 1985, Identification of respiratory and cardiovascular
    systems, IFAC Symp. Ident. & Syst. Par. Est., York, UK, 55-66.
Lorino, A.M., Benichou, M., Macquin-Mavier, I., Lorino, H., and Harf, A.,
    1988, Respiratory mechanics for assessment of histamine bronchopulmo-
    nary reactivity in guinea pigs, Respiration Physiology, 73: 155.
Lorino, H., Lorino, A.M., Harf, A., Atlan, G., and Laurent, D., 1982, Linear
    modeling of ventilatory mechanics during spontaneous breathing, Comp.
    Biomed. Res., 15: 129.
Matamis, D., Lemaire, F., Harf, A., Brun-Buisson, C., Ansquer, J.C., and
    Atlan, G., 1984, Total respiratory pressure-volume curves in the adult
    respiratory distress syndrome, Chest, 86: 58.
Slutsky, A.S., Scharf, S.M., Brown, R., and Ingram, Jr., R.H., 1980, The
    effect of oleic acid-induced pulmonary edema on pulmonary and chest
    wall mechanics in dogs, Am. Rev. Respir. Dis., 121: 91.
Snapper, J.R., 1985, Lung mechanics in pulmonary edema, Clinics in Chest
    Medicine, 6: 393.
Suter, P.M., 1985, Assessment of respiratory mechanics in ARDS, in: Lung
    Biology in Health and Disease, vol 24, W.M. Zapol and K. Falke, eds.,
    507-519, Dekker, N.Y..
Zhang, L.F., and Zhang, R., 1990, Model and identification of respiratory
    mechanical system, Space Med. Med. Eng., 3: 58 (in Chinese).

EFFECT OF HYPERVENTILATION ON OXYGENATION OF THE BRAIN CORTEX OF NEONATES

David F. Wilson, Anna Pastuszko, Roy Schneiderman, Jane E. DiGiacomo
Marek Pawlowski and Maria Delivoria-Papadopoulos

Departments of Biochemistry and Biophysics, Physiology and of Pediatrics
Medical School, University of Pennsylvania, Philadelphia, PA 19104

INTRODUCTION

It is well known that even brief periods of hypoxia lead to serious disruption of neural function. Neuronal properties are modified within seconds of the onset of anoxia and within 2.5-3 minutes anoxic depolarization begins (see for example Silver, 1977a,b), characterized by massive loss of K^+ and diffusion of Cl^- into the cells. If reoxygenation occurs within a few minutes the cellular events and integrated neural function recovers, but as the period of oxygen deprivation is extended there is progressive irreversible loss of function.

In an attempt to evaluate the effects of hypoxia on the newborn brain, a number of indicators, including cerebral blood flow (Lou et al, 1979a,b) and Lactic dehydrogenase in the cerebral spinal fluid (Hall et al, 1980) have been studied. However, these parameters have not provided an accurate measure of the time course or degree of cerebral injury. Hyperventilation is often used clinically to increase pulmonary blood flow in infants with pulmonary hypertension or to decrease cerebral edema following perinatal asphyxia. Because the decrease in $PaCO_2$ might be expected to decrease cerebral blood flow, hyperventilation may result in brain tissue hypoxia. Therefore, in order to determine the physiologic changes in cortical oxygenation induced by this therapy, we have examined the effects of hyperventilation on cerebral oxygen pressure as well as the $PaCO_2$ and pH of the systemic blood in newborn piglets. The oxygen pressure in the blood of the cerebral cortex was measured using a new optical method based on the oxygen dependent quenching of phosphorescence (see Vanderkooi et al, 1987; Wilson et al, 1988; Rumsey et al, 1988; Wilson and Pawlowski, this volume).

METHOD OF APPROACH

Newborn piglets were anesthetized and ventilated with a small animal respirator. The blood pressure was monitored continuously throughout the experiments and periodically blood was removed from the femoral artery and analyzed for $PaCO_2$, pH, and PaO_2. The head placed in a stereotaxic unit and the scalp removed from the left parietal region to expose the skull. A hole approximately 2 cm in diameter was made in the skull over the parietal hemisphere and the dura carefully reflected. A cranial window was placed in the hole, sealed with bone wax and cemented into place with dental acrylic. The surface of the brain was superfused with artificial CSF and observations were made using a Wild Macrozoom microscope with epifluorescence attachment. The images were taken using a Xybion intensified CCD camera and recorded on an All-Tronics Medical Systems Super VHS video recorder. The illuminating light for the epifluorescence attachment was a 45 watt xenon flashlamp mounted in a Leitz lamp housing. The flash lamp was controlled by a 80386 microcomputer which determined the timing of the flashes and the gating of the video camera intensifier. Excitation was through an interference filter with 540 nm center and a bandwidth at half height of 30 nm; phosphorescence was observed through a long pass cutoff filter with 50% transmission at 630 nm. A typical protocol for measuring phosphorescence lifetimes was as follows: Number of flashes averaged for each delay time, 16; delay times after the flash, 10 usec, 40 usec, 75 usec, 125 usec, 225 usec 350 usec, 450 usec and 550 usec; gate width in all cases, 2,500 usec. The video images were collected and processed using the Universal Imaging Corp. Image 1/AT image processing system.

The phosphorescent oxygen probe used in these studies was the Pd complex of tris (p-carboxyphenyl) porphyrin. The probe was infused (20-30 mg/kg) through the femoral artery as a 1:1 molecular complex with bovine serum albumin dissolved in physiological saline at pH 7.4. The ventilation protocols used in the experiments are given in the appropriate Figure Legends.

RESULTS

Images of the phosphorescence of the oxygen probes in the vascular system of the cortex

Images of the surface of the cortex show a distribution of phosphorescence with the brightest emission associated with the veins. This is shown in Figure 1 as a three dimensional graph with phosphorescence intensity as the vertical axis. There is also substantial phosphorescence from the capillary beds while the arterioles are generally not seen or appear as darker vessels as they cross the veins. The phosphorescence distribution is very sensitive to vascular integrity and any extravasation of albumin from the vessels appears

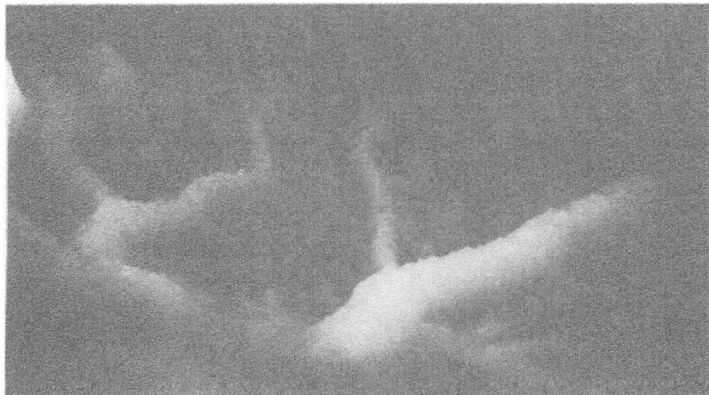

Figure 1. An image of the phosphorescence of oxygen probe in the blood of the cortex of newborn piglets. The piglet was anesthetized and a cranial window prepared. The surface of the brain was superfused with artificial CSF and observations were made using a Wild Macrozoom microscope with an epifluorescence attachment. The image was taken using a Xybion intensified CCD camera and recorded on a Super VHS video recorder. The Pd complex of tris (p-carboxyphenyl) porphyrin was infused through the femoral artery as a 1:1 complex with bovine serum albumin in physiological saline at pH 7.4. When the probe was injected, the arteriolar system brightened first followed by the capillary bed and then the venous system, with the transit of the initial dye front requiring only a few seconds. The phosphorescence is presented as a three dimensional graph with the intensity on the ordinate.

as a region of greatly intensified phosphorescence. This occurs because of the Pd-porphyrin accompanies the albumin into the interstitial space where it is in an environment with an oxygen pressure lower than that of the blood. The presence of probe in the interstitial space and the low oxygen pressure combine to cause a marked increase in phosphorescence in the region of extravasation.

Changes in oxygen pressure in the cortex of a piglet induced by hyperventilation

Hyperventilation increased the intensity of phosphorescence from all areas of the brain surface. The VCR recording is a sequence of 8 images of decreasing intensity each recorded for approximately 1.0 seconds. The images were analyzed by identifying regions of interest and generating a digital file of the average intensity of each region as a function of time using the image processor. The values of the average intensity in each region were then plotted as the logarithm of intensity against the delay time for which the image was collected. In each case there was a good fit to a straight line, indicating the decay of most of the phosphorescence approximates a single exponential. The measured decay constants (T) were substituted into the Stern-Volmer equation (equation 1):

$$T^0/T \ = \ 1 \ + \ k_Q * T^0 * PO_2 \qquad\qquad (1)$$

where T^0 and T are the lifetimes at zero oxygen and at the oxygen pressure PO_2 and k_Q is the quenching constant for the probe determined at 38^0 C. The oxygen pressure in each region of interest was calculated from the measured value of T.

The effect of hyperventilation on the oxygen pressure in the vessels of the cortex is shown in Figures 2a,b. In the experiment, the gated sets of video images were recorded just prior to beginning hyperventilation and at the indicated later times during periods of hyperventilation and recovery. Figure 2a indicates the observed changes in phosphorescence lifetime for two regions (approximately 100 um square), one in the capillary bed and the other in a vein. The changes were followed during a 25 minute period in which the ventilation rate was increased to two different levels for 5 minutes each, separated by a recovery period. The lifetimes were about 75 usec during normal ventilation and increased with hyperventilation. With doubling of the ventilator rate the phosphorescence lifetimes increased to approx. 125 usec with incomplete recovery during the following the six minute period of normal ventilation. Increasing the ventilation rate to three times normal further increased the phosphorescence lifetime to almost 200 usec. Measurements of PaO_2 showed the arterial oxygen pressure was constant at approximately 100 torr during both normal and hyperventilation while $PaCO_2$ and pH had the values given at the top of the figure. As expected, hyperventilation decreased $PaCO_2$ and increased pH of the arterial blood.

The cortical oxygen pressures were calculated with equation 1 using the quenching constant (k_Q) of 324 $Torr^{-1}sec^{-1}$ and lifetime at zero oxygen (T^0) of 595 usec at 38^0 C (see Pawlowski and Wilson, this book). The normal cortical oxygen pressure was approximately 25-35 Torr.

Time dependence of the effect of hyperventilation on the oxygen pressure in the cortex

The cortical oxygen pressure observed in the experiment shown in Figure 2 was still changing slightly at the end of the 5 minutes of hyperventilation. In the experiment shown in Figure 3a,b, the degree of hyperventilation was increased (the respiratory rate was increased to 3.5 times normal and the tidal volume was increased two fold over normal) and extended to a full hour. The phosphorescence lifetimes were measured in four different regions of the brain in order to allow comparisons of the behavior patterns. As may be seen[1] in Figure 3a, the animal had a low normal CO_2 at zero time ($PaCO_2$ of 32) and as a consequence the phosphorescence lifetimes were longer than for zero time in Figure 2. The increase in phosphorescence lifetimes was more extensive (to values of > 350 usec) and the values varied widely from one region of the brain to another. The variations were directly attributable to regional variations in oxygen pressure.

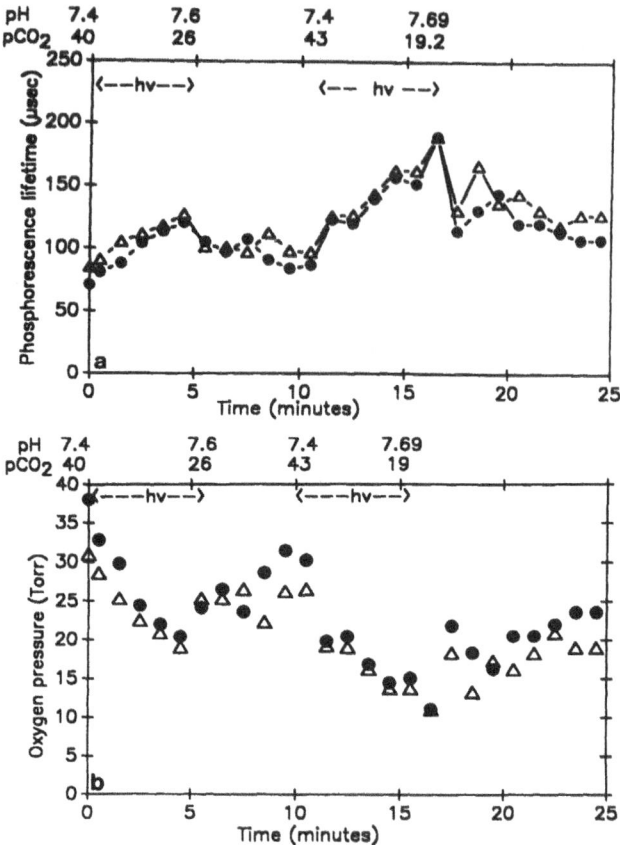

Figure 2a,b. The effect of hyperventilation on the oxygen pressure in blood of the cortex. The experimental conditions were described in the legend of Figure 1. The phosphorescence images were obtained at the indicated times. Samples of arterial blood were taken at the ends of the hyperventilation and recovery periods and the $PaCO_2$, PO_2 and pH measured. The PO_2 remained constant at 100 Torr and the values of the other parameters are given along the top of the figure. a. Two regions of interest approximately 100 um square were selected, one on a vein and the other in the capillary bed, and the digitized data analysed to give the phosphorescence lifetimes for each point in time (ordinate). b. The oxygen pressure was calculated for each point and presented on the ordinate.

Hyperventilation caused a rapid decline in the oxygen pressures in each region of the cortex, from an initial value of about 20 Torr to 5 Torr in less than 6 minutes. From 6 minutes to 60 minutes the average oxygen pressure remained approximately 5 Torr but fell as low as 2.5 Torr at one time point. The oxygen pressure in each individual region fluctuated widely and independently of the other regions, ranging from essentially zero to about 10 Torr. When the ventilation rate was returned to normal the oxygen pressure began recovering immediately, reaching values greater than 30 Torr at 15 minutes recovery and then declining again to normal.

DISCUSSION

Acute hypocapnia decreases cerebral blood flow (Kety and Schmidt, 1946; Reivich, 1964) and causes tissue hypoxia as determined by polarographic and mass spectrophotometric methods (Clark et al, 1958; Kennealy et al, 1980). Our data confirm those observations; more importantly, we are able to quantitate the effects of graded hypocapnia on the oxygenation of the cortex. The data presented in this paper indicate the oxygen pressure in the cortex falls nearly linearly with decreasing $PaCO_2$. Among the observations relevant to the cortical oxygen pressure during periods of hyperventilation are:

1. The oxygen pressure is not uniform across the viewing field, but increases or decreases in each region independently of the other regions. Thus, at any point in time the oxygen pressure in a region may be substantially above or below the average value.

2. The average oxygen pressure decreases strikingly as the $PaCO_2$ falls and blood pH increases in spite of the fact that the systemic arterial oxygen pressure remained high (100 torr). The decrease in oxygen pressure in the cortex is approximately proportional to the decrease in $PaCO_2$.

3. The decrease in local regions is not a smooth transition but may show large random fluctuations relative to an averaged curve.

4. Depending on the experimental conditions, return of the ventilation rate to normal may initiate an overshoot in the vascular oxygen pressure which then returns toward normal.

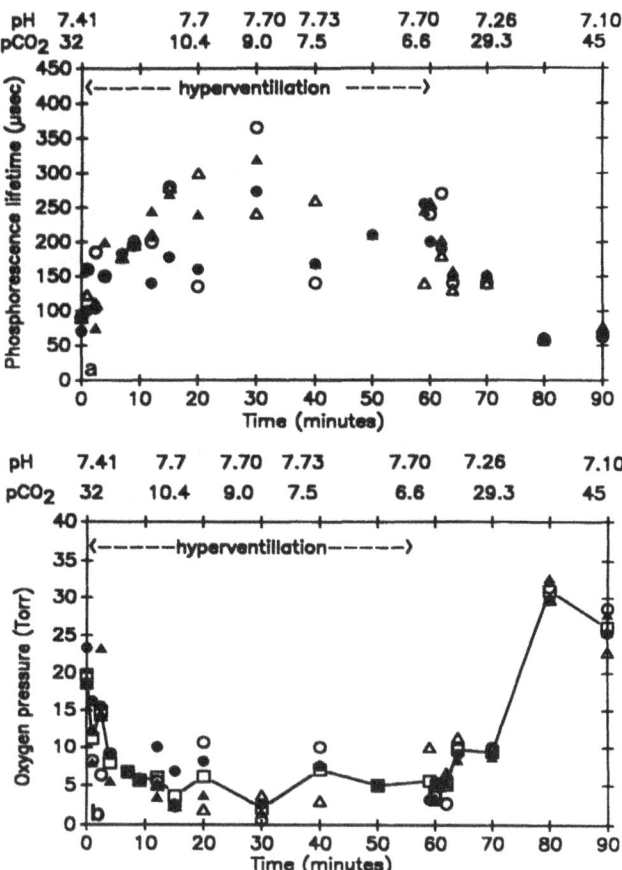

Figure 3a,b. The time dependence of the hyperventilation induced changes in oxygen pressure in the cortex. The experimental conditions were as described in the legend of Figure 1. Hyperventilation (increase in rate by 3.5 fold and in volume of two fold) was begun at time zero and continued for 60 minutes. a. The phosphorescence lifetimes were calculated for four regions of interest (two on veins and two on capillary beds) and the values are presented on the ordinate. b. The oxygen pressure was calculated for each region of interest and the values are presented on the ordinate. The average value for the four regions was calculated and these are graphed with the points connected with a line.

SUMMARY

A new phosphorescence imaging method (Rumsey et al, Science (1988) 1649) has been used to continuously monitor the oxygen pressure in the blood of the cerebral cortex of newborn pigs. The animals' blood pressure was continuously measured and $PaCO_2$, PaO_2 and arterial blood pH were measured periodically. The oxygen pressure in the blood was quantitatively determined for regions of about 100 um square within the image (from a total field of about 3 mm diameter). It was observed that during hyperventilation, which lowered $PaCO_2$ and increased pH of the blood, oxygen pressure decreased in proportion to the decrease in $PaCO_2$. For example, hyperventilation which decreased $PaCO_2$ from its normal value of 40 Torr to 10 Torr caused a rapid (within 5 minutes) decrease in oxygen pressure in the blood of capillaries and veins to approximately 1/4 of normal.

REFERENCES

Clark, L.C., Misrahy, G., and Fox, R.P. (1958) Chronically implanted polarographic electrodes, J. Appl. Physiol. 13: 85-91.

Hall, R.T., Kulkarni, P.B., Sheehan, M.B., and Rhodes, P.G. (1980) Cerebrospinal fluid lactate dehydrogenase in infants with perinatal asphixia, Dev. Med. Child Neurol. 22: 300-307.

Kennealy, J.A., McLennan, J.E., Loudon, R.G., and McLaurin, R.L (1980) Hyperventilation-induced cerebral hypoxia, Amer. Rev. Res. Dis. 122: 407-412.

Kety, S.S. and Schmidt, C.F. (1946) The effects of active and passive hyperventilation on cerebral blood flow, cerebral oxygen consumption, cardiac output, and blood pressure of normal young men, J. Clin. Invest. 25: 107-119.

Reivich, M. (1964) Arterial PCO_2 and cerebral hemodynamics, Am. J. Physiol. 206: 25-35.

Lou, H.C., Lassen, N.A., and Friis-Hansen, B. (1979a) Management of interstitial emphysema by high-frequency low positive pressure hand ventilation in the neonate, J. Pediatr. 94: 118-121.

Lou, H.C., Skov, H., Pedersen, H. (1979b) Low cerebral blood flow: a risk factor in the neonate, J. Pediatr. 95: 606-608.

Rumsey, W.L., Vanderkooi, J.M., and Wilson, D.F. (1989) Imaging of phosphorescence: a novel method for measuring oxygen distribution in perfused tissue, Science 241: 1649-1651.

Silver, I.A. (1977a) Changes in PO_2 and ion fluxes in cerebral hypoxia-ischemia, Adv. Exptl. Med. Biol. 77: 325-334.

Silver, I.A. (1977b) Ion fluxes in hypoxic tissues, Microvascular Res. 13: 409-420.

Vanderkooi, J.M., Maniara, G., Green, T.J. and Wilson, D.F. (1987) J. Biol. Chem. 262: 5476-5482.

Wilson, D.F., Rumsey, W.L., Green, T.J., and Vanderkooi, J.M. (1988) The oxygen dependence of mitochondrial oxidative phosphorylation measured by a new optical method for measuring oxygen concentration, J. Biol. Chem. 263: 2712-2718.

HYPOXIA AND THE "REACTION THEORY" OF CENTRAL RESPIRATORY CHEMOSENSITIVITY

H. Kiwull-Schöne and P. Kiwull

Department of Physiology
Ruhr-University
D-4630 Bochum, FRG

INTRODUCTION

As early as 1911, Winterstein postulated his famous "reaction theory" reducing the hypoxic and hypercapnic ventilatory drives to one and the same mechanism at chemoreceptors, namely protons dissociated either from carbonic acid or from fixed acids generated during oxygen deficiency. However, after the discovery of the O_2-sensitive arterial chemoreceptors, this fascinating idea had to be restricted to the central chemosensitivity, e.g. to the alternative whether H^+ or molecular CO_2 is the adequate stimulus.

As has been reviewed by Loeschcke, 1982, there is a considerable amount of experimental evidence that the extracellular fluid (ECF) pH in the brainstem is the central chemical signal determining pulmonary ventilation. On the other hand, controversial papers appeared being unable to prove ventilation as a unique function of ventral medullary ECF-pH either changed by CO_2-inhalation or by infusion of strong mineral acids (Eldridge et al., 1985; Shams, 1985; Teppema et al., 1983).

In order to achieve a more physiological kind of metabolic acidosis, in the present study intermittent periods of hypoxia were applied to peripherally chemodenervated and partly vagotomized animals. The respiratory effects of the hypoxia induced endogeneous lactacidosis were compared to those following lactic acid infusions at comparable concentrations in the arterial blood. All respiratory changes were related to the brainstem extracellular fluid pH. The ECF pH-sensitivity of the respiratory system during endogenously and exogenously induced (metabolic) lactacidosis was then tested for coincidence with CO_2-induced (respiratory) acidosis. In order to avoid sources of misinterpretation, special attention was paid to possible side-effects of transient CO_2-release following acid infusion and to shifts in the CO_2-threshold, which could falsify the CO_2-sensitivity.

METHODS

The experiments were performed in 12 cats (2.7 ±0.1 kg) and 15 rabbits (2.9 ±0.2 kg). The cats were anaesthetized by 15 mg/kg Ketanest® i.m. and by 17 mg/kg pentobarbital sodium i.v. initially, followed by continuous infusion of 1.7 mg/kg/h. The pentobarbital loading dose in

Oxygen Transport to Tissue XIII, Edited by T.K. Goldstick *et al*.
Plenum Press, New York, 1992

rabbits was 43 mg/kg and the maintenance dose 6.3 mg/kg/h. All animals were peripherally chemodenervated and partly vagotomized. They were either spontaneously breathing or paralysed (Alloferin®) and artificially ventilated at constant volume within the normal range of tidal volume.

In the spontaneously breathing animals, tidal volume (V_T) and inspiratory/expiratory durations (T_I, T_E) were continuously measured by pneumotachography and in the artificially ventilated animals, the integrated compound potential (IPNA) of one desheathed phrenic nerve was taken as parameter of central nervous respiratory output. Additionally, the end-tidal PCO_2 ($P_{ET}CO_2$) was continuously recorded by infrared absorption and the ECF-pH at the exposed ventral surface of the medulla oblongata by a balanced pH-electrode (Kiwull-Schöne and Kiwull, 1983). From blood samples, the oxygen partial pressure (P_aO_2) and O_2-Hb saturation were measured by electrode and Lex O_2 Con apparatus, respectively, as well as the CO_2 partial pressure, CO_2-buffer capacity, actual and standard bicarbonate concentration by the indirect equilibration method (Astrup). For correction of the Haldane effect under hypoxic conditions, see Kiwull-Schöne et al., this volume.

In the series with intermittent periods of hypoxia, the animals initially inhaled O_2-enriched air ($PaO_2 > 20$ kPa) until all measured variables had reached a steady state. After that, CO_2 was added to the inspired gas-mixture (10-15 min) so that the $PaCO_2$ reached a new steady state of about 1.3 kPa above control, then the CO_2-admixture was removed again. Subsequently, the inspired gas was switched to a moderate hypoxic mixture ($PaO_2 = 9.6 \pm 0.37$ kPa, about 72 mmHg) for another 10 min during which the transient changes in ventilation ($\dot{V}$), $P_{ET}CO_2$ and ECF-pH were followed up. Then the inspiratory CO_2 was elevated and removed in much the same way as described before. Upon returning to hyperoxia, again the transient changes of respiratory variables were observed for ten minutes before applying the next CO_2-step. In 11 spontaneously breathing cats, each of these hyperoxic and hypoxic runs lasting for 30-45 min was carried out twice and averaged first for each animal before group mean computation. The same experimental procedure was applied to one artificially ventilated cat, to get besides CO_2-sensitivity also information about the CO_2-threshold under hyperoxic and hypoxic conditions. The thresholds were estimated in terms of those values of $PaCO_2$ or ECF pH, at which the phasic phrenic nerve activity disappeared.

In the series with intravenous lactic acid infusion, the animals started also with inhaling O_2-enriched air ($PaO_2 > 20$ kPa). Again the CO_2-sensitivity was tested by step-wise increases and decreases of $PaCO_2$. After that, 1M lactic acid was infused intravenously for 10 min at a rate of 1 ml/min. The transient changes of ventilatory variables, blood gases and acid-base conditions in blood and brainstem ECF were followed up at minimum intervals of one min, during and up to 60 min after infusion. At the end of observation time, the CO_2-sensitivity was tested once more and compared to that under control conditions before lactic acid application. This kind of experiment was carried out in spontaneously breathing and artificially ventilated rabbits, but, since they exclude feedback effects of pulmonary ventilation on blood gases, only experiments with artificial ventilation will be presented here.

RESULTS

1. Hypoxia induced brainstem acidosis

Base-line ventilation and CO_2-responses of spontaneously breathing cats are shown by Fig. 1 (left diagram) as functions of brainstem ECF-pH.

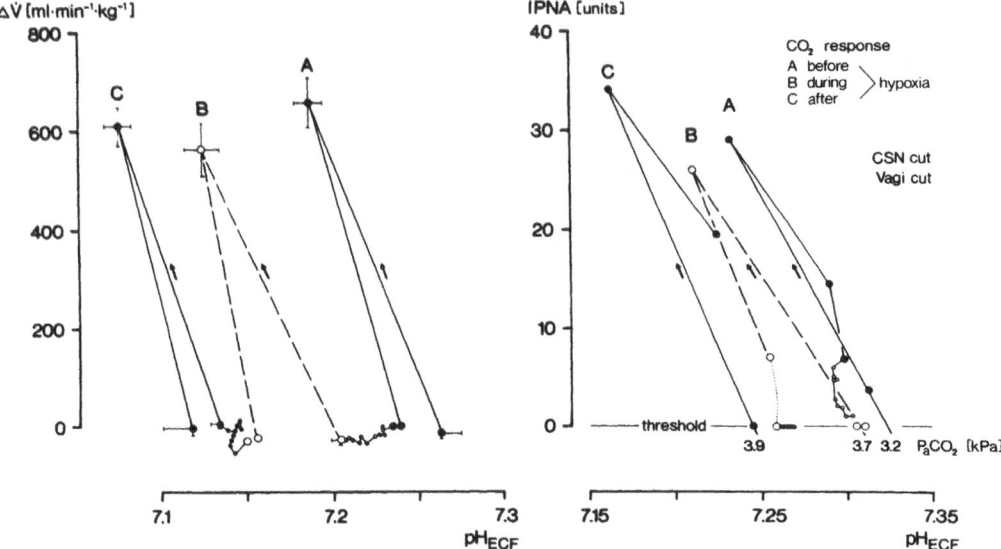

Fig. 1. Effect of hypoxia induced metabolic brainstem acidosis on central
         respiratory chemosensitivity.
         Steady state and transient (smaller symbols) respiratory responses
         as functions of the extracellular fluid (ECF) pH at the ventral
         medullary surface in carotid chemodernervated cats during hyper-
         oxia (•——•) and hypoxia (o---o), before and after step rises in
         PaCO₂ of about 1.3 kPa (A, B, C).
         Left diagram: Changes of pulmonary ventilation ($\Delta\dot{V}$) in spontane-
         ously breathing cats (Means $\pm$SEM, N = 11).
         Right diagram: Integrated phrenic nerve activity (IPNA) in one
         artificially ventilated cat.

The initial minute ventilation under hyperoxic conditions (PaO₂ >20 kPa)
before hypoxia and before CO₂-inhalation was 304 $\pm$15 ml/min/kg and ECF-pH
7.263 $\pm$0.026. During 10 min of moderate hypoxia, ventilation was observed
in one-min intervals as indicated by the smaller symbols in Fig.1. The
hypoxic ventilatory depression did not exceed 10%, and was not prevented
by the accompanying fall in ECF-pH of −0.030 $\pm$0.005 units. Upon return to
hyperoxia, ventilation widely recovered, but in spite of an unchanged
arterial PCO₂, ECF-pH further decreased, at least reaching levels by 0.195
units below control. In contrast to this lack of ventilatory reaction to
metabolic brainstem acidosis, the reaction to CO₂-inhalation still reached
100 $\pm$10.9 ml/min/kg compared to 110 $\pm$14.1 ml/min/kg per 0.01 units
decrease in ECF-pH under control conditions.

     Under comparable conditions, the apneic thresholds in the artificially
ventilated cat (Fig.1, right diagram) could be discerned as values of
P$_a$CO₂ and ECF-pH, at which the phasic phrenic nerve discharge disappeared.
During hypoxia, the threshold values for arterial PCO₂ increased by 0.5
kPa (3.8 mmHg) and those for ECF-pH decreased by 0.017 units. During the
ensuing hyperoxic period, the CO₂-threshold remained elevated by 0.7 kPa
(5.4 mmHg) and the ECF-pH threshold remained in the acid range as well.
However, this shift of the CO₂-threshold towards brainstem ECF acidity was
not accompanied by a reduction in CO₂-sensitivity.

2. Brainstem acidosis induced by intravenous lactic acid infusion

   The experiments with intravenous lactic acid infusion were carried
out to reveal possible differential effects of exogenous and endogenous
lactacidosis on the respiratory system. Carotid chemodenervated rabbits
were used because this species is able to accumulate rather high lactic
acid blood levels under hypoxic conditions, which persist also during the
ensuing hyperoxic recovery period (Table 1). Approximately the same steady
state lactic acid concentrations could be reached 10 to 60 min after
lactic acid infusion. In correspondence with the cats, no ventilatory
drive could be discerned in response to metabolic acidosis, as pulmonary
ventilation in the spontaneously breathing rabbits was 360 $\pm$ 25 and 371
$\pm$12 ml/min/kg before and after hyoxia, respectively.

   For the study with acid infusion artificially ventilated animals were
taken to eliminate feedback effects of ventilation on blood gases due to
the expected CO_2 release from carbonic acid. During the lactic acid infusion, arterial pH was drastically reduced by 0.446 $\pm$0.044 units, but
only about 25% of this acidotic change were reflected in the brainstem
ECF-pH. About 15% of the arterial pH-change could be attributed to the
concomitant rise in PCO_2, reaching a maximum of 0.78 $\pm$0.08 kPa (about 6
mmHg) two min after the start of infusion. When the arterial PCO_2 was
raised by decreasing the pumping rate of the respirator, ECF pH fell on an
average by 0.045 units per kPa, which means that during the first two min
of infusion at least 89% of the ECF pH deflection are caused by the
release of CO_2 and not by lactic acid.

   As can be seen from Fig. 2, the early response of the central inspiratory drive (IPNA/T_I) to acid infusion by up to 40.9 $\pm$12.7% therefore

Table 1.   Arterial acid-base conditions in carotid chemodenervated vagoto-
           mized rabbits before and after induction of a metabolic acidosis.
           Mean values $\pm$SEM of arterial PCO_2 and pH, as well as of standard
           bicarbonate and lactate$^-$ concentration in 9 spontaneously
           breathing rabbits before and after a 10 min period of severe
           hypoxia (P_aO_2 = 3.22 $\pm$0.11 kPa) and in 6 artificially venti-
           lated rabbits before and after a 10 min period of 1M lactic acid
           infusion (1 ml/min).
           * Significant difference against control before either treat-
             ment, P_D< 0.05

	ENDOGENOUS LACTACIDOSIS		EXOGENOUS LACTACIDOSIS	
	before	10 min after	before	10 min after
	h y p o x i a		i n f u s i o n	
P_aCO_2 (kPa)	4.60 $\pm$0.15	4.47 $\pm$0.12	3.12 $\pm$0.15	3.50 $\pm$0.36
pH_a	7.392 $\pm$0.021	7.236* $\pm$0.029	7.442 $\pm$0.032	7.289* $\pm$0.024
$HCO_3^-{}_{st}$ (mmol$\cdot$l^{-1})	21.5 $\pm$1.1	15.1* $\pm$1.0	19.2 $\pm$1.2	14.8* $\pm$0.8
Lactate$^-$ (mmol$\cdot$l^{-1})	6.2 $\pm$1.2	10.9* $\pm$1.0	4.6 $\pm$1.0	11.8* $\pm$1.1

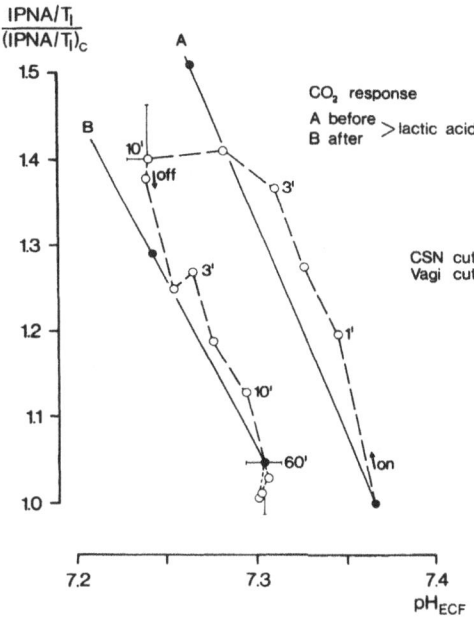

Fig. 2. Effect of metabolic brainstem acidosis
induced by lactic acid infusion on central
respiratory chemosensitivity.
Relative change of central inspiratory
drive (IPNA/T_I) as a function of the
extracellular fluid (ECF) pH at the
ventral medullary surface in artificially
ventilated carotid chemodenervated and
vagotomized rabbits (Means $\pm$SEM, N=6).
Transient changes in response to lactic
acid infusion (o---o) and steady state
changes following rises in P_aCO_2 by the
respirator ($\bullet$—$\bullet$).

closely coincides with that to the rise in P_aCO_2 induced by the respirator,
being on an average 43.3 $\pm$12.2% per 0.1 units decrease in ECF pH.
10 min after stop of infusion, neither P_aCO_2 (Table 1) nor IPNA/T_I (12.7 $\pm$
7.0%) were any longer elevated significantly above control. 60 min after
stop of infusion, the acid-base balance in the arterial blood was largely
restored by at least 75% of the maximum deviation, whereas a delayed
lactacidosis (by up to 5.9 $\pm$1.0 mM) further developed in the brainstem
ECF. As for the hypoxia induced lactacidosis, there were also no ventilatory reactions to exogenously induced lactacidosis under steady state
conditions, although both arterial and ECF-pH were still significantly
reduced by 0.114 and 0.061 units, respectively. Thereby, the base-line
value for P_aCO_2 60 min after lactic acid infusion (3.24 $\pm$0.47 kPa) was not
significantly different from that before infusion (Table). However, the
endtidal PCO_2 was significantly lower than control by as much as 0.30 $\pm$
0.13 kPa (about 2.5 mmHg), indicating an increased arterio-alveolar CO_2-
difference. Nevertheless, the sensitivity of the central inspiratory drive
response to CO_2 was statistically not distinguishable from control.

DISCUSSION

The main finding presented here was that the CO_2-responses of pulmonary ventilation, or of its central nervous respiratory drive analogue, were shifted towards the acid range of brainstem ECF-pH, disregarding whether a metabolic acidosis was either developing endogenously during hypoxia or whether it was induced exogenously by lactic acid infusion. Thereby, the sensitivities to changes in arterial PCO_2 were largely persisting or only slightly reduced, indicating intact chemosensitivity against respiratory acidosis. In case of hypoxia, it could be shown that the apneic thresholds of the phasic phrenic nerve activity were shifted towards lower ECF-pH and remained lower than control also during the ensuing recovery period. Concomitantly, the arterial threshold PCO_2 was elevated, which means that higher CO_2 levels were required to maintain central respiratory rhythmic activity. From this behaviour can be judged that ventilation (or its central nervous drive analogue in case of constant artificial ventilation) is no unique function of the brainstem extracellular fluid pH, as was predicted from the "reaction theory", but strikingly depends on whether ECF pH-changes are either induced by CO_2 or by fixed acids.

After having proposed the "reaction theory", Winterstein, 1915, performed an extended study in rabbits, in which he compared the ventilatory responses to changes in arterial pH induced either by infusion of NaCl solutions saturated with 100% CO_2 or by infusion of various anorganic and organic acids, in order to substantiate arterial H^+ as being the adequate stimulus. A quantitative re-evaluation of these data clearly shows, however, that the rise in ventilation related to the same fall in pH_a is smaller in case of the weaker acids (like acetic or lactic acid) than in case of the strong mineral acids. This indicates that the nature of the infused acid is of importance for the magnitude of effects and could be an explanation for the discrepancy between our findings with endogenous lactacidosis, leading to no respiratory drive at all, and those with strong acid infusion, reported by others (Eldridge et al., 1985; Shams, 1985; Teppema et al., 1983).

On the other hand, pulmonary ventilation had never could be related uniquely to arterial pH, since there were generally observed dramatic effects of CO_2-inhalation compared to rather small effects with acid infusion. This led to the idea that the central chemosensor might be less accessible to the blood than to the cerebrospinal fluid (CSF) compartment. Loeschcke et al., 1958, were able to drive ventilation by perfusing the third brain ventricle with mock CSF at low pH due to lowered HCO_3^- but unchanged PCO_2. Thereby, the fluid passed to the ventral surface of the medulla oblongata. However, a decrease of CSF-pH by as much as 0.73 units did lead to but 21% rise in ventilation, corresponding to less than a tenth of the CO_2-sensitivity. Thus, it was argued again that neither the blood nor the CSF would represent the acid-base environment of the central chemosensor. Several models concerning HCO_3 gradients along a pathway between blood and CSF were established to estimate the pH close to the sensing element in a certain depth below the ventral medullary surface (Loeschcke, 1982).

Meanwhile, it became possible to measure the pH in the brainstem ECF directly by non-invasive balanced surface pH-electrodes, by which could be shown that the dynamic responses of ventilation and ECF-pH to step changes in inhaled CO_2 indeed coincided very closely (Ahmad and Loeschcke, 1982). However, several investigators (Eldridge et al., 1985; Shams, 1985; Teppema et al., 1983) using this method to study responses to either acid infusion or CO_2 inhalation, as well as our group comparing hypercapnia and hypoxia induced lactacidosis (Kiwull-Schöne and Kiwull, 1983), came to the same general conclusion, namely that ventilation was no unique function of ECF-pH under the conditions of respiratory and metabolic acidosis. There are, however, discrepancies in so far as we do not find any ventilatory reaction to endogenous metabolic brainstem acidosis at all (Fig. 1), whereas during intravenous acid infusion, a respiratory sensitivity against ECF-pH during metabolic acidosis seems to exist in principle, being only quantitatively smaller than that during respiratory CO_2-induced acidosis. As common feature, in these studies strong mineral acids were used for experimental induction of metabolic acidosis, such as hydrochloric acid (Eldridge et al., 1985, Teppema et al., 1983) and sulfuric acid (Shams, 1985).

The strength of H^+-ion dissociation should determine the amount of CO_2 to be released from carbonic acid. This offers an explanation for the higher ventilatory responses to strong acids than to weak acids, as can be re-evaluated from the data of Winterstein, 1915 (see above). During spontaneous breathing, this effect in response to sulfuric acid was discussed to be only transient, since the end-tidal PCO_2 very soon was lowered by hyperventilation (Shams, 1985). Furthermore, in this study the vagus nerves were left intact so that influences from aortic chemoreceptors and/or pulmonary vagal afferents could not be ruled out. In a more recent paper, the author addressed the essential side effects of strong acid infusion on lung perfusion and lung mechanics mediated by the release of eicosanoids (Shams and Scheid, 1990), which in turn may elicit pulmonary vagal reflex effects on ventilation. Consequently, three main requirements should be fulfilled to elucidate thouroughly the role of fixed acids in the control of breathing. First, the animals should be peripherally chemo-denervated and vagotomized, second, they should be artificially ventilated to avoid feedback effects of respiratory control, and third, isocapnia should be maintained. Additionally, to exclude the possible influence of the chemical nature of the acid involved, whether strong or weak, lactic acid was taken in the present study for direct comparison with the endogenously developing lactacidosis during hypoxia.

In the spontaneously breathing cats undergoing alternate periods of hypoxia and hyperoxia, we observed a less pronounced hypoxic ventilatory depression than described by Javaheri and Teppema, 1987, as well as no significant increase of $\dot{V}$ above control during hyperoxic recovery, in spite of a progressive acidification of brainstem ECF. The smaller hypoxic depression may be due to the less severe hypoxia which we applied. Interestingly, certain influences by vagal pulmonary reflexes do play a role, since before vagotomy a less severe hypoxic depression and a more effective hyperoxic recovery could be discerned. A more quantitative study under these conditions, dealing with differential vagal cooling, revealed a significant role of myelinated vagal afferents from the lungs (Kalhoff et al., 1990). Besides vagally mediated effects on ventilation, which could be misinterpreted as being due to the concomitant brainstem acidosis, the Haldane effect should be considered as well (Eldridge and Kiley, 1987). Indeed, transient decreases and increases of end-tidal and arterial PCO_2 could be observed on going from hyperoxia to hypoxia and vice versa, provided pulmonary ventilation was kept constant by the respirator and lung perfusion did not change. However, the hypoxic depression and hyperoxic recovery of the phrenic nerve signal did not follow the CO_2-characteristic accordingly expected (Fig. 1, right diagram). Moreover, hypoxia led to a shift of the apneic threshold towards brainstem acidity being restored to some degree during the ensuing period of hyperoxia. This is not compatible with the idea of a respiratory drive by acidosis itself, but quite on the contrary, the basic central respiratory output is much closer to the apneic threshold in the acid than in the alkaline range. The same phenomenon has been observed by Eldridge et al., 1985, during hydrochloric acid infusion, but has been ignored for a final conclusion.

These authors, as well as Teppema et al., 1983, focussed on isocapnic conditions by servo-controlling end-tidal PCO_2. In artificially ventilated, peripherally chemodenervated cats they found that phrenic nerve activity responded at a seemingly lower sensitivity to acidification of the medullary surface pH when caused by infusion of hydrochloric acid rather than if caused by increased CO_2. In our infusion experiments we observed, however, that the arterial PCO_2 returned to control from the 5th min after infusion stop, whereas the end-tidal PCO_2 significantly decreased below, indicating an increased arterio-alveolar CO_2-difference, possibly due to a reduced lung perfusion. A servo-controlled $P_{ET}CO_2$ would thus have led to an unintentional CO_2-load, being interpreted as "isocapnic" respiratory response to brainstem ECF pH-changes by acid infusion. Thus, there are no conclusive data supporting a ventilatory drive by acute non-volatile acid infusion under true "isocapnic" conditions in the arterial blood.

Moreover, it is surprising that severe acid-base disturbances in the arterial blood are rather effectively restored without the assistance of respiratory control during constant artificial ventilation. As far as the brain ECF is concerned, we don't agree that acute experimental metabolic acid-base disturbances in the blood are rapidly reflected beyond the blood-brain-barrier, as being concluded from ventilated endtidal "isocapnic" dogs and rabbits (Javaheri et al., 1981; Davies and Nolan, 1982). We followed up the early transient changes in arterial PCO_2 and were able to conclude that the initial acid deflection in brainstem ECF-pH following intravenous acid infusion was nearly entirely due to a release of CO_2. Only in the later phase of arterial acid-base recovery there was a slow progressive acidification of the brainstem ECF by accumulation of lactic acid, then being rather uneffective on pulmonary ventilation.

The lack of ventilatory reaction to ventral medullary ECF acidosis, if induced by non-volatile acids, is speaking in favour of a specific role for the CO_2/carbonic acid/HCO_3^--system. The basic mechanism by which hypercapnia stimulates ventilation is unknown. To get further insight into the system, we started experiments with intravenous infusion of acetazolamide sufficient to block the enzyme carbonic anhydrase throughout all sites of occurrence. Interestingly, the phrenic nerve activity responded about equally sensitive to either rise in P_aCO_2, disregarding whether caused by acetazolamide or by inhalation of CO_2 (Fig. 3). The concomitant fall in brainstem ECF-pH may also be attributed to the distinct rise in arterial PCO_2 (and probably brain tissue PCO_2, cf. Bickler et al., 1988). On the contrary, Teppema et al., 1990, did not find significant changes of PCO_2, neither in the blood nor in the brainstem ECF, upon treatment with 50 mg/kg acetazolamide i.v. in spontaneously breathing cats, but a distinct and persistent hyperventilation, which was not correlated with the progressive fall in medullary surface pH. In constantly ventilated rabbits, however, 25 mg/kg of the drug led to a rise in PCO_2 by about 10 mmHg in both the arterial blood and the brain cortex (Bickler et al., 1988). By using NMR-techniques, neither in artificially ventilated rabbits (Bickler et al., 1988) nor in spontaneously breathing humans (Vorstrup et al., 1989) considerable changes in brain intracellular pH could be detected as a consequence of acetazolamide administration. On the other hand, acetazolamide caused a distinct acidification of the brain extracellular fluid, which could not even be prevented by adjusting the brain CO_2 tension to the control level (Bickler et al., 1988). Considering that lactic acid concentrations remained unchanged under these conditions, the latter authors proposed acetazolamide to be responsible for increases in brain ECF H^+ by a combination of elevated PCO_2 and possibly delayed dehydration of metabolically produced carbonic acid. This again is pointing to a dominating role of CO_2 over fixed acids in the central chemical control of pulmonary ventilation.

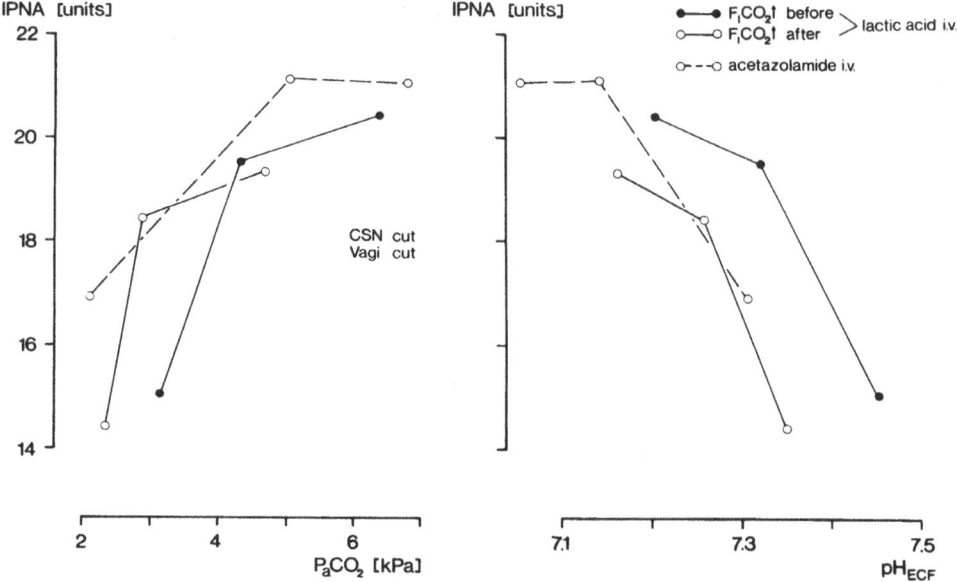

Fig. 3. Effect of brainstem acidosis induced by acetazolamide infusion on
central respiratory chemosensitivity.
Left diagram: Integrated phrenic nerve activity (IPNA) as a
function of arterial PCO_2 in an artificially ventilated, carotid
chemodenervated and vagotomized rabbit. Under control conditions
(●———●) and 60 min after lactic acid infusion (o———o), the P_aCO_2
was varied by CO_2-inhalation. After i.v. infusion of 30 mg/kg
acetazolamide, the P_aCO_2 rose spontaneously (o---o).
Right diagram: IPNA as a function of the extracellular (ECF)
fluid pH at the ventral surface of the medulla oblongata under
the same conditions.

A similar conclusion may be drawn from our collected results, since
neither during endogenously induced lactacidosis nor in response to infusions of lactic acid, the brainstem extracellular fluid pH could be
discerned as the essential determinant for pulmonary ventilation.
Disregarding the metabolic, non-volatile acid-base condition, the central
chemosensitivity remains remarkably stable against changes in PCO_2, speaking in favour of a predominant role for CO_2 or at least carbonic acid.

SUMMARY

In peripherally chemodenervated and vagotomized cats and rabbits,
either spontaneously breathing or artificially ventilated, we studied the
reaction of the respiratory control system to changes in the extracellular
fluid (ECF) pH at the ventral surface of the medulla oblongata. The brainstem ECF-pH was varied either by alternating periods of hypoxia and hyperoxia or by intravenous infusion of lactic acid to achieve endogenous or
exogenous lactacidosis, respectively. Additionally, the arterial PCO_2 was
changed by varying the inspiratory CO_2-fraction or the respirator's
pumping rate. When pulmonary ventilation or central respiratory drive (in
terms of phrenic nerve activity) was related to brainstem ECF-pH, no
unique function resulted for respiratory (CO_2-induced) and metabolic
(lactic acid induced) acid-base changes, thus contradicting the "reaction

theory" for central respiratory chemosensitivity. Under steady state conditions, there was no ventilatory reaction to endogenous or exogenous metabolic brainstem acidosis at all. However, the apneic threshold was shifted towards the acid range, although the sensitivity of the respiratory system to CO_2 remained nearly unchanged, no matter whether CO_2 was inhaled or increased by acetazolamide. This points to a dominating role of CO_2 or at least carbonic acid over fixed acids for the central chemosensitive control of pulmonary ventilation.

ACKNOWLEDGEMENT

We would like to thank Ms F. Werkmeister for her expert technical assistance and Ms S. Adler for her excellent help with the drawings.

REFERENCES

Ahmad, H.R., and Loeschcke, H.H., 1982, Transient and steady state response of pulmonary ventilation to the medullary extracellular pH after approximately rectangular changes in alveolar PCO_2, Pflügers Arch., 395:285.

Bickler, P.E., Litt, L., Banville, D.L., and Severinghaus, J.W., 1988, Effects of acetazolamide on cerebral acid-base balance, J. Appl. Physiol., 65:422.

Davies, D.G., and Nolan, W.F., 1982, Cerebral interstitial fluid acid-base status follows arterial acid-base pertubations, J. Appl. Physiol., 53:1551.

Eldridge, F.L., Kiley, J.P., and Millhorn, D.E., 1985, Respiratory responses to medullary hydrogen ion changes in cats: Different effects of respiratory and metabolic acidoses, J. Physiol., 358:285.

Eldridge, F.L., and Kiley, J.P., 1987, Effects of hyperoxia on medullary ECF pH and respiration in chemodenervated cats, Respir. Physiol., 70:37.

Javaheri, S., and Teppema, L.J., 1987, Ventral medullary extracellular fluid pH and PCO_2 during hypoxemia, J. Appl. Physiol., 63:1567.

Javaheri, S., Clendening, A., Papadakis, N. and Brody, J.S., 1981, Changes in brain surface pH during acute isocapnic metabolic acidosis and alkalosis, J. Appl. Physiol., 51:276.

Kalhoff, H., Kiwull-Schöne, H., and Kiwull, P., 1990, Pulmonary vagal afferents involved in the hypoxic breathing without arterial chemoreflexes, in: "Arterial chemoreception", Eyzaguirre, C., Fidone, S.J., Fitzgerald, R.S., Lahiri, S. and McDonald, D., eds., Springer, Berlin-Heidelberg-New York.

Kiwull-Schöne, H. and Kiwull, P, 1983, Hypoxic modulation of central chemosensitivity, in: "Central neurone environment and the control systems of breathing and circulation", Schläfke, M.E., Koepchen, H.P., and See, W.R., eds., Springer, Berlin-Heidelberg-New York.

Loeschcke, H.H., Koepchen, H.P., and Gertz, K.H., 1958, Über den Einfluß von Wasserstoffionenkonzentration und CO_2-Druck im Liquor cerebrospinalis auf die Atmung, Pflügers Arch., 266:569.

Loeschcke, H.H., 1982, Central chemosensitivity and the reaction theory, J. Physiol., 332:1.

Shams, H., 1985, Differential effects of CO_2 and H^+ as central stimuli of respiration in the cat, J. Appl. Physiol., 58:357.

Shams, H., and Scheid, P., 1990, Effects of thromboxane on respiration and pulmonary circulation in the cat: role of vagus nerve, J. Appl. Physiol., 68:2042.

Teppema, L.J., Barts, P.W.J.A., Folgering, H.Th., and Evers, J.A.M., 1983, Effects of respiratory and (isocapnic) metabolic arterial acid-base disturbances on medullary extracellular fluid pH and ventilation in cats, Respir. Physiol., 53:379

Teppema, L.J., Rochette, F, and Demedts, M., 1990, Effects of acetazolamide on medullary extracellular pH and PCO_2 and on ventilation in peripherally chemodenervated cats, Pflügers Arch., 415:519.

Vorstrup, S., Jensen, K.E., Thomsen, C., Henriksen, O., Lassen, N.A., and Paulson, O.B., 1989, Neuronal pH regulation: Constant normal intracellular pH is maintained in brain during low extracellular pH induced by acetazolamide - ^{31}P NMR Study, J. Cereb. Blood Flow Metab., 9:417.

Winterstein, H., 1911, Die Regulierung der Atmung durch das Blut, Pflügers Arch., 138:167.

Winterstein, H., 1915, Neue Untersuchungen über die physikalisch-chemische Regulierung der Atmung, Biochem. Z., 70:45.

TUMORS

OXYGENATION OF MAMMARY TUMORS: FROM ISOTRANSPLANTED RODENT TUMORS TO PRIMA-

RY MALIGNANCIES IN PATIENTS[*]

P. Vaupel[1], K.-H. Schlenger[1], M. Hoeckel[2], and P. Okunieff[3]

[1]Institute of Physiology and Pathophysiology, [2]Dept. of Gyne-
cology and Obstetrics, University of Mainz, D-6500 Mainz,
Germany
[3]Dept. of Radiation Medicine, Mass. General Hospital Cancer
Center, Harvard Medical School, Boston, MA 02114, USA

INTRODUCTION

The role of oxygen in tumor cell proliferation, radiosensitivity, cytotoxicity of anticancer drugs and hyperthermia treatment has been the subject of a series of investigations (for reviews see Hall, 1988; Teicher et al., 1990, Vaupel et al., 1989a; Vaupel, 1990a,b). Despite the apparent importance of tumor oxygenation, data on pO_2 values in solid tumors are mostly derived from experiments on rodents which might not necessarily reflect the variability of the clinical situation. Due to feasible techniques available now, considerable advances have been made in the past few years in the assessment of tumor hypoxia in patients (for reviews see Vaupel et al., 1989a; Vaupel, 1990a). The latter information may be most beneficial for designing specifically tailored treatment protocols (i.e., it can be used as a prognostic indicator of whether or not an individual patient would be expected to be a good candidate for a certain treatment), for assessing early tumor response to treatment, and/or for examining potentially useful tools for prediction of long-term tumor response (Vaupel et al., 1989a). In addition, the patient data obtained from systematic studies allow for the first time a direct comparison between isotransplanted animal tumors, xenografted human cancers, and primary tumors in patients. In the present study, direct pO_2 readings in isotransplanted mammary tumors in mice, xenografted human breast cancers in immunodeficient rnu/rnu-rats, and in primary malignancies in patients are compared.

MATERIAL AND METHODS

Oxygenation of isotransplanted mammary tumors in mice

Experimental animals were 10 - 12 week old C3H/Sed mice. Early generation isotransplants of a spontaneous mammary adenocarcinoma (MCaIV) were
used. Tumor volumes on the day of study were calculated by an ellipsoid
approximation using the three orthogonal diameters. Tumors of different

---

[*]Dedicated to Professor Dr. Dr. Gerhard Thews on the occasion of his 65th
birthday

sizes and at different implantation sites were investigated (Vaupel et al., 1981, 1989b; Kallinowski et al., 1990). Measurements in the normal subcutis and in resting skeletal muscle were used as controls. Oxygen tension readings in tumors were performed either with gold-in-glass microelectrodes (tip diameter: 1 - 5 μm) or with O_2-sensitive needle electrodes (recessed gold-in-glass electrode, shaft diameter 300 μm, diameter of the cathode 12 μm, and computerized pO_2 histography; Eppendorf, Hamburg, FRG; technical details of the latter system have been described by Vaupel et al., 1989c).

Oxygenation of xenografted human breast cancers

Two different human breast cancer lines obtained from primary tumors were investigated. Tumor tissue was directly transplanted to athymic mice (NMRI-nu/nu). From this "tumor bank", small tissue fragments were implanted s.c. into the flank of male nude rats (WAG/Fra-rnu/rnu). In view of a possible hormone-dependency, the rats were castrated before transplantation of the breast tumors. Oxygen partial pressures were measured with O_2 needle electrodes (computerized pO_2 histography) in "low-flow" and "high-flow" breast cancer xenografts (Vaupel et al., 1987; Kallinowski et al., 1989a,b).

Oxygenation of human breast cancer in situ

Oxygen partial pressures were recorded in pre- and postmenopausal, conscious women (age: 30 - 84 years) with ductal or lobular breast cancers (stages T2 - T4). Control measurements were performed in the contralateral

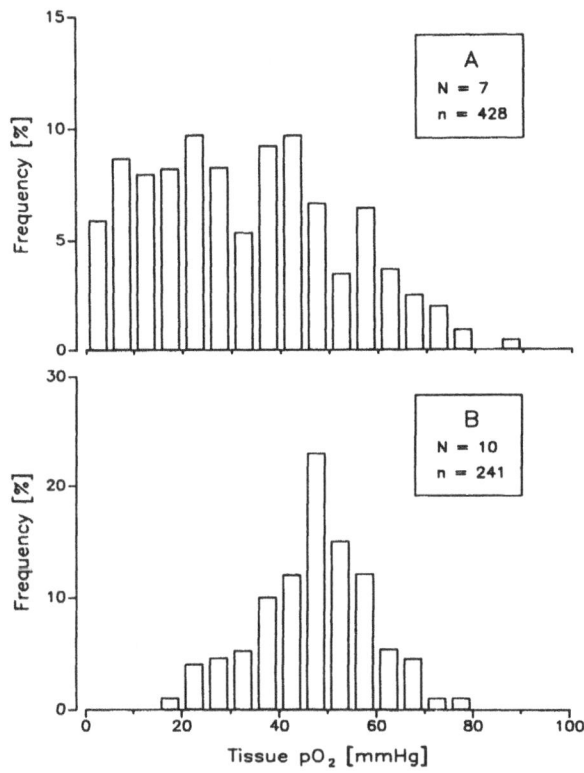

Fig. 1. pO_2 histogram of resting thigh musculature (A) and of the subcutis at the hind foot dorsum (B). N = number of mice, n = number of pO_2 readings.

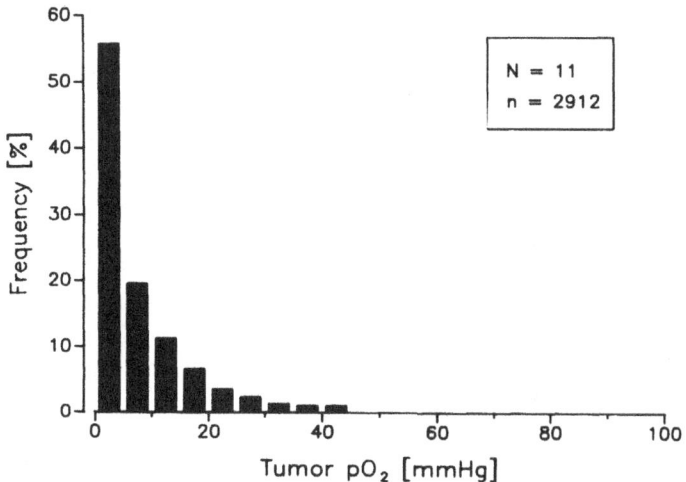

Fig. 2. Oxygen tension distribution in C3H mouse mammary adenocarcinomas
implanted into the subcutis of the hind leg. N = number of tumors,
n = number of pO_2 readings, mean tumor wet weight: 0.5 g, mean pO_2
= 7 mmHg, median pO_2 = 4 mmHg.

normal breast. The pO_2 electrodes were membranized gold-in-glass micro-
cathodes (recessed type, diameter 12 μm) mounted within a 28 gauge needle
(computerized pO_2 histography). All measurements were carried out immediately prior to treatment. A trocar penetrating through the skin was used to
guide the O_2 sensitive electrode to the tumor periphery. In general, 3
defined radial tracks were evaluated in each tumor. Care was taken that the
electrode tracks were either surgically removed together with the tumor
mass or were within the target volume of irradiation. The site and the size
of each breast tumor was localized using either computer tomography, magnetic resonance imaging, sonography or clinical examination. The oxygenation status of the tumors was evaluated in relation to the staging of the
malignancies investigated. Patients had given consent after having been
informed about the experimental nature of the measurements.

RESULTS

pO₂ distribution in isotransplanted mammary tumors in mice

Frequency distributions of measured pO_2 values in resting skeletal
muscle (thigh of mice; upper panel), and in the normal subcutis (hind foot
dorsum; lower panel) are presented in Fig. 1. As expected, there is a
typical scattering of the tissue pO_2 values between 2 and 90 mmHg with a
median pO_2 of 31 mmHg in the high musculature. Values in the mouse subcutis
varied between 16 and 79 mmHg; the median pO_2 value in the subcutis of the
hind foot dorsum was 47 mmHg.
pO_2 histograms measured in MCaIV murine mammary adenocarcinomas are shown
in Fig. 2 (implantation site: subcutis of the hind leg), and in Fig. 3
(implantation site: subcutis of the hind foot dorsum). As a rule, large
areas with radiobiologically hypoxic pO_2 values were found at both implantation sites. Substantial variations of the pO_2 readings were obvious.
Considering individual tumors, between zero and 90% of the pO_2 values were
measured in the class between zero and 2.5 mmHg (this pO_2 class corresponds
to a radiosensitivity which is less than half maximum). A distinct worsening of the tissue oxygenation was observed at larger tumor volumes. Median
oxygen tensions were consistently ⩽ 5 mmHg (see Fig. 3). The volume-de-
pendent changes observed in the tissue oxygenation are the inevitable

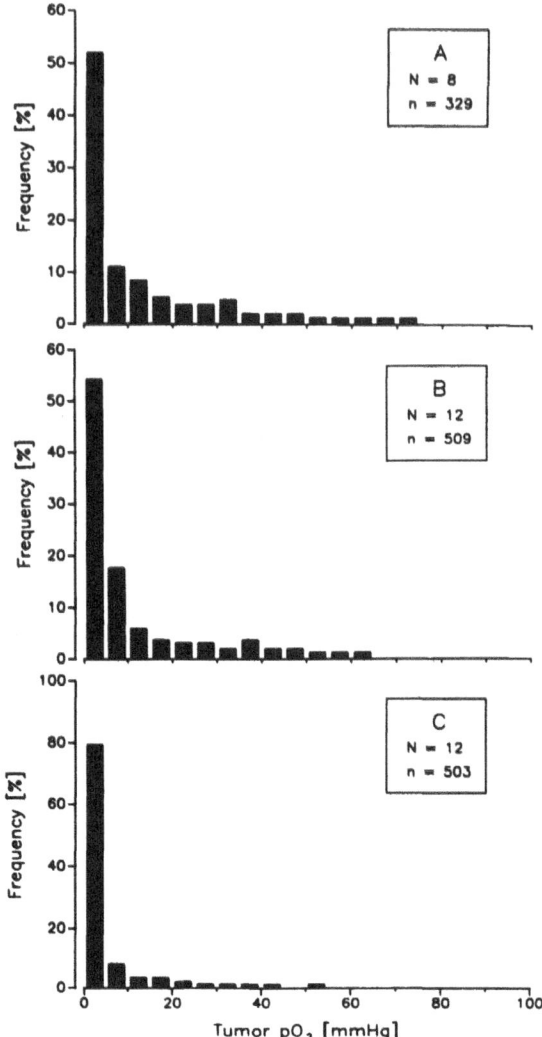

Fig. 3. Oxygen partial pressure distributions in MCaIV mammary adenocarcinomas implanted s.c. into the hind foot dorsum of C3H/Sed mice.
Mean tumor volumes were around 125 mm^3 (A), 175 mm^3 (B), and 350
mm^3 (C). N = number of tumors, n = number of pO$_2$ readings.

consequence of a progressive deterioration and anisotropic distribution of
the tumor perfusion occurring as a result of severe morphological and
functional alterations of the tumor microcirculation (Vaupel et al., 1981).
In MCaIV mammary tumors the mean blood flow decreased exponentially from
0.5 to 0.07 ml/g/min when the tumor volume increased from 200 to 800 mm^3
(Vaupel et al., 1989b). Tumors with volumes of approximately 500 mm^3 and a
mean blood flow rate of 0.2 ml/g/min exhibited median pO$_2$ values of 3 - 5
mmHg.

pO$_2$ distribution in xenografted human breast cancers

Breast cancer xenografts with low perfusion rates (0.1 - 0.2 ml/g/min)
contained large areas of radiobiological hypoxia and anoxia even at small
sizes (Fig. 4). With tumor wet weights of approximately 2.3 g, the median

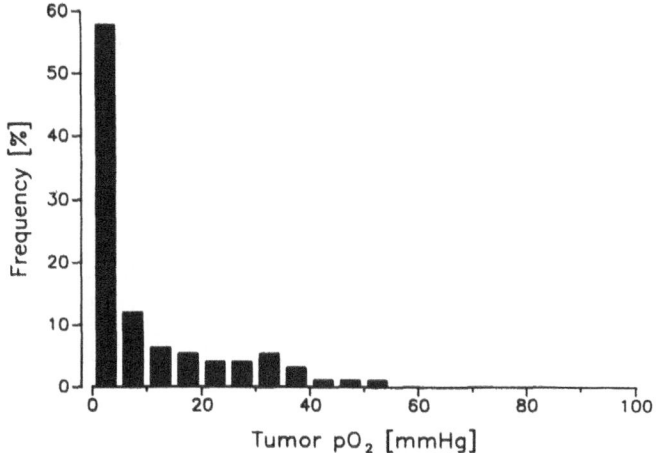

Fig. 4. pO_2 histogram of xenografted "low-flow" medullary breast cancers. Mean tumor volume: 2.3 ml; 1210 pO_2 readings in 10 tumors.

pO_2 values were 2 - 4 mmHg. The tissue oxygenation for breast cancers with high perfusion rates is shown in Fig. 5. This tumor line was well oxygenated, the pO_2 histogram being similar to that of normal liver tissue (Vaupel et al., 1989a). Here, the median pO_2 was 23 mmHg, the mean tumor blood flow being 0.7 ml/g/min. Tissue pO_2 values < 5 mmHg were found in the latter tumor line only at wet weights exceeding 2.5 g.

pO$_2$ distribution in primary breast cancers

Using the computerized pO_2 histography system, pO_2 values were recorded in normal breast and in breast cancers of different clinical stages. As a rule, the mean (and median) pO_2 values were distinctly lower in the malignancies than in the normal tissues. The pO_2 histograms of tumors

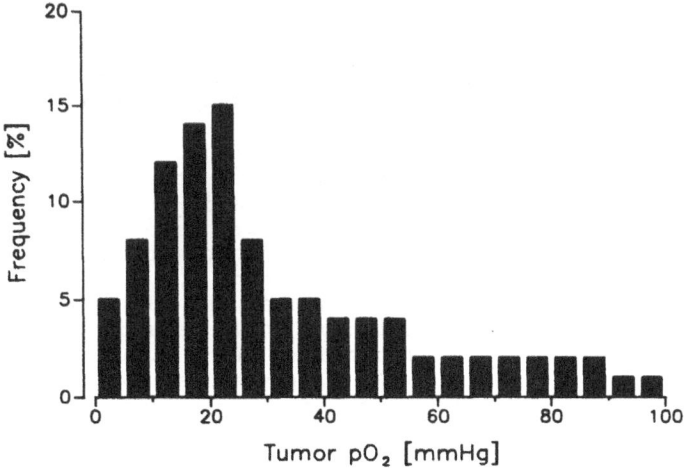

Fig. 5. O_2 partial pressure distribution of s.c. implanted "high-flow" medullary breast cancers. Mean tumor volume: 2.5 ml; 940 pO_2 readings in 9 tumors.

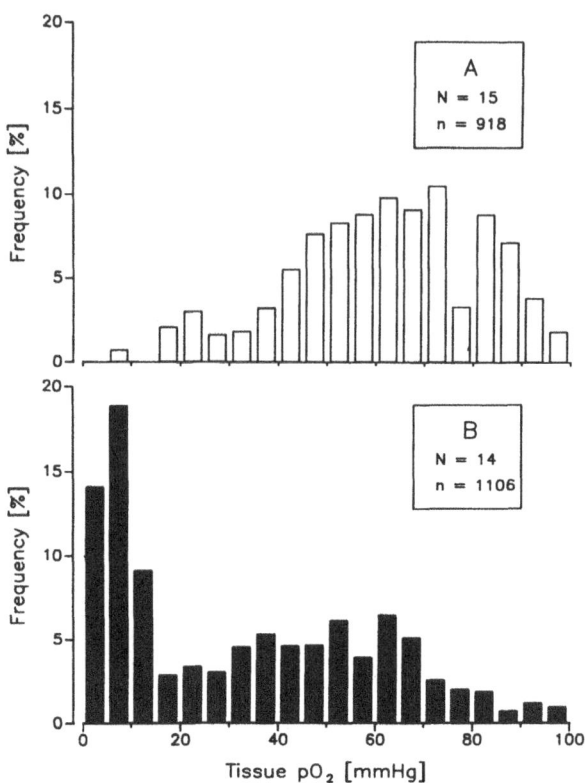

Fig. 6. pO_2 histograms of normal breast tissue (A) and breast cancers (B, pooled data for clinical stages T2 - T4). N = number of patients, n = number of pO_2 readings.

were usually shifted to lower O_2 tensions whereas in the respective normal tissues the pO_2 distribution was more or less Gaussian. O_2 tensions measured in the normal breast of 15 patients (918 pO_2 readings) revealed a median pO_2 value of 64 mmHg (see Fig. 6, top), whereas in 14 cancers of the breast (stages T2 - T4, 1106 pO_2 readings) the median pO_2 was 30 mmHg (Fig. 6, bottom). So far, approximately 1/3 of the cancers investigated exhibited pO_2 values between zero and 2.5 mmHg, i.e., tissue areas with less than half maximum radiosensitivity.

Pooled pO_2 data for all breast cancers of stage T2 are presented in the top panel of Fig. 7, whereas the respective pO_2 histogram for malignancies of stages T3 and T4 are shown in the lower panel. From this compilation there is clear evidence that there are no significant differences between the two groups (median pO_2 in T2 tumors: 27 mmHg, median pO_2 in T3 and T4 tumors: 31 mmHg). This implies that the clinical stages of the breast cancers investigated so far cannot be the paramount factor determining tumor tissue oxygenation. Furthermore, there is experimental evidence that the occurrence of radiobiological hypoxia does not correlate with the stage and/or the grade of the breast tumors investigated. Evidently, both distribution curves are multimodal indicating intra- and inter-tumor heterogeneities.

Besides intra-tumor pO_2 heterogeneities (Fig. 8), there is marked tumor-to-tumor variability, even if tumors of the same clinical stage and grade are compared (Fig. 9). This again emphasizes the urgent need for a

pretreatment characterization of relevant pathophysiological tumor parameters for designing specifically tailored treatment protocols for individual tumors in patients. This necessity for an "individualization" of tumors
is substantiated by the various pO_2 histograms showing considerable tumor-
to-tumor (inter-tumor) heterogeneities.

DISCUSSION

Oxygen partial pressure distributions for <u>isotransplanted mammary tu-
mors</u> have been described in detail (Vaupel et al., 1981; Kallinowski et
al., 1990). In general, as a result of a compromised and anisotropic microcirculation, most of these malignancies reveal hypoxic and anoxic tissue
areas which are heterogeneously distributed within the tumor mass. In
poorly perfused <u>human breast cancer xenografts</u>, hypoxic and anoxic regions
were already present at early growth stages and expanded with tumor growth
(Vaupel et al., 1987; Kallinowski et al., 1989b). In contrast, s.c. breast
cancer xenografts with high perfusion rates have tissue oxygenations comparable to those of most normal organs. This is most probably due to an
adequate vascularization of the latter tumors. Only at larger sizes did
these tumors "outgrow" their vasculature and hypoxic/anoxic tissue areas
develop.

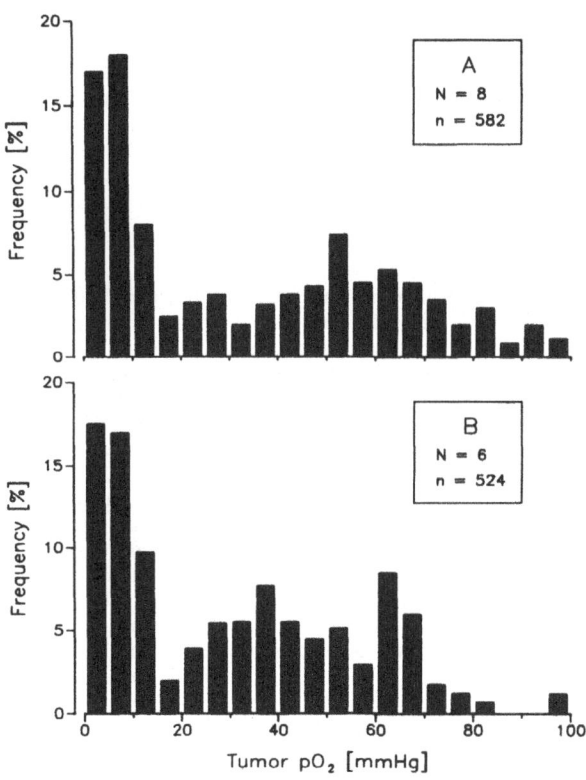

Fig. 7. pO_2 histograms derived from breast cancers of different clinical
stages. Top: T2-tumors, bottom: tumors of stages T3 and T4. N =
number of tumors investigated, n = number of measured pO_2 values.

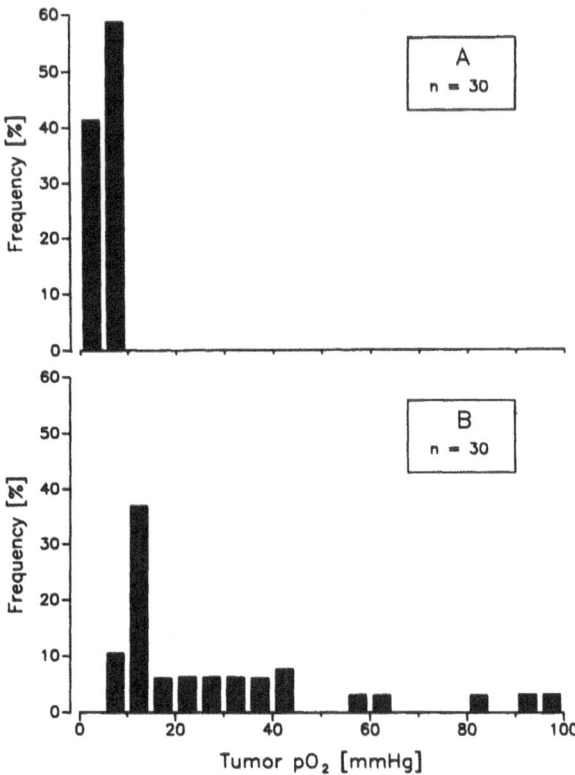

Fig. 8. pO_2 frequency distributions from two different electrode tracks within a breast cancer (stage T2) emphasizing intra-tumor variability in the tissue oxygenation (n = number of pO_2 values recorded).

In the isotransplanted tumors and in the xenografted human breast cancers there is clear experimental evidence that tumor blood flow is the principal modulator of tumor tissue oxygenation (e.g., worsening of the tumor oxygenation with increasing tumor size which parallels a similar drop in the tumor blood flow rate). A close correlation between oxygenation and bioenergetic status suggests that the microcirculation in these tumors yields an O_2-limited energy metabolism (Vaupel et al., 1989c).

In general, the mean (or median) pO_2 values in <u>human breast cancers</u> in situ are distinctly lower than in the normal breast (see Table I). Neither the pO_2 distributions nor the mean and median pO_2 values are characteristic for a certain clinical staging or histological grading. The occurrence of hypoxia and anoxia, thus, does not correlate with the clinical stages of the malignancies investigated. For this reason, the oxygenation status of individual human breast cancers, and the tumor response to irradiation or "O_2-dependent" anticancer drugs cannot be predicted on the basis of the clinical stages or histological grades. Extended and systematic clinical studies also provide evidence for pronounced intra- and inter-tumor heterogeneities in human breast cancers as was also the case with the isotransplanted and the xenografted mammary tumors. The pO_2 distributions found in primary lesions in patients ranged from the pO_2 histograms obtained in "low-flow" isotransplants and xenografts to those in "high-flow" breast cancer xenografts or normal tissues.

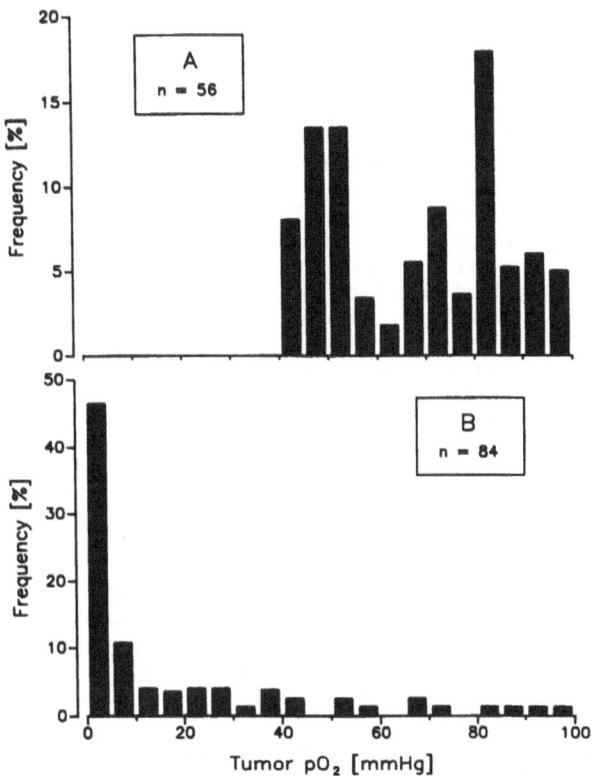

Fig. 9. pO_2 histograms derived from two different breast cancers of the same clinical stage (T2) substantiating marked tumor-to-tumor variability in the oxygenation status (n = number of pO_2 values measured).

Table I. Mean tissue oxygen tension in breast cancers and in the normal breast of patients.

Tumor pO_2 (mmHg)	Normal breast pO_2 (mmHg)	n	References
11	42	3	Cater and Silver (1960), Cater (1964)
32		4	Evans and Naylor (1963)
25*	29*	3	Jamieson and van den Brenk (1965)
21	41	6	Badib and Webster (1969)
33		24	Pappova et al. (1982)
	64	15	this study
27+ (T2)		8	
31+ (T3/T4)		6	

n = number of patients, * estimated mean pO_2 value, + median pO_2 value

SUMMARY

As a result of a compromised and anisotropic microcirculation, iso-transplanted mammary tumors in mice exhibit hypoxic and anoxic tissue areas which are heterogeneously distributed within the tumor mass. Similarly, in poorly perfused human breast cancer xenografts, hypoxia develops at early growth stages and expands with tumor growth. In contrast, breast cancer xenografts with high perfusion rates exhibit an oxygenation status comparable to that of most normal organs. There is clear experimental evidence that the efficiency of tumor blood flow in isotransplanted tumors and in xenografted human breast cancers is the principal modulator of tissue oxygenation. The pO_2 distribution found in primary lesions in patients ranged from the pO_2 histograms obtained in "low-flow" isotransplants and xenografts to those measured in "high-flow" breast cancer xenografts or normal tissues. From our extended and systematic clinical studies there is clear indication that the oxygenation status of human breast cancers in situ does not correlate with the clinical stage and/or histological grade of an individual tumor.

REFERENCES

Badib, A.O., Webster, J.H., 1969, Changes in tumor oxygen tension during radiation therapy, Acta Radiol. Ther. Phys. Biol. 8:247.

Cater, D.B., 1964, Oxygen tension in neoplastic tissues, Tumori, 50:435.

Cater, D.B., Silver, I.A., 1960, Quantitative measurements of oxygen tensions in normal tissues and in the tumours of patients before and after radiotherapy, Acta Radiol., 53:233.

Evans, N.T.S, Naylor, P.F., 1963, The effect of oxygen breathing and radiotherapy upon the tissue oxygen tension of some human tumours, Brit. J. Radiol. 36:418.

Hall, E.J., 1988, Radiobiology for the Radiologist, 3rd edit, Lippincott Co., Philadelphia.

Jamieson, D., van den Brenk, H.A.S., 1965, Oxygen tension in human malignant disease under hyperbaric conditions, Brit. J. Cancer, 19:139.

Kallinowski, F., Schlenger, K.H., Runkel, S., Kloes, M., Stohrer, M., Okunieff, P., Vaupel, P., 1989a, Blood flow, metabolism, cellular microenvironment and growth rate of human tumor xenografts, Cancer Res., 49:3759.

Kallinowski, F., Schlenger, K.H., Kloes, M., Stohrer, M., Vaupel, P., 1989b, Tumor blood flow: The principal modulator of oxidative and glycolytic metabolism, and of the metabolic micromilieu of human tumor xenografts in vivo, Int. J. Cancer, 44:266.

Kallinowski, F., Zander, R., Hoeckel, M., Vaupel, P., 1990, Tumor tissue oxygenation as evaluated by computerized-pO_2-histography, Int. J. Radiat. Oncol. Biol. Phys., in press.

Pappova, N., Siracka, E., Vacek, A., Durkovsky, J., 1982, Oxygen tension and prediction of the radiation response. Polarographic study in human breast cancer, Neoplasma, 29:669.

Teicher, B.A., Holden, S.A., Al-Achi, A., Herman, T.S., 1990, Classification of antineoplastic treatments by their differential toxicity toward putative oxygenated and hypoxic tumor subpopulations in vivo in the FSaIIC murine fibrosarcoma, Cancer Res., 50:3339.

Vaupel, P., 1990a, Oxygenation of human tumors, Strahlenther. Onkol., 166:377.

Vaupel, P., 1990b, Pathophysiological mechanisms of hyperthermia in cancer therapy, in: "Biological basis of oncologic thermotherapy", C. Streffer, P. Vaupel, G. Hahn, eds., Springer, Heidelberg, New York.

Vaupel, P., Frinak, S., Bicher, H.I., 1981, Heterogeneous oxygen partial pressure and pH distribution in C3H mouse mammary adenocarcinoma, Cancer Res., 41:2008.

Vaupel, P., Fortmeyer, H.P., Runkel, S., Kallinowski, F., 1987, Blood flow, oxygen consumption, and tissue oxygenation of human breast cancer xenografts in nude rats, Cancer Res. 47:3496.

Vaupel, P., Kallinowski, F., P. Okunieff, 1989a, Blood flow, oxygen and nutrient supply, and metabolic microenvironment of human tumors: A review, Cancer Res. 49:6449.

Vaupel, P., Okunieff, P., Neuringer, L.J., 1989b, Blood flow, tissue oxygenation, pH distribution, and energy metabolism of murine mammary adenocarcinomas during growth, Adv. Exp. Med. Biol., 248:835.

Vaupel, P., Okunieff, P., Kallinowski, F., Neuringer L.J., 1989c, Correlations between ^{31}P-NMR spectroscopy and tissue O_2 tension measurements in a murine fibrosarcoma, Radiat. Res., 120:477.

MEASUREMENTS OF TUMOR BLOOD FLOW USING INTRAPERITONEAL DEUTERIUM AND

^{2}H-NMR SPECTROSCOPY

Paul Okunieff[*], Junhee Lee[*], Masamitsu Itoh[*],
Peter Vaupel[**]

[*]Dept. of Radiation Medicine
Massachusetts General Hospital Boston, MA 02114
[**]Inst. of Physiology and Pathophysiology Pathophysiology
Division, Univ. of Mainz, D-6500 Mainz, FRG

INTRODUCTION

Tumors usually have a sparse, disorganized, and inefficient
vascular network that leaves a large fraction of the tumor cells in an
oxygen deprived and hostile metabolic microenvironment. Hence tumor
blood flow, or more correctly nutritive perfusion, has important
interactions with treatment efficacy. For example, hypoxic tumor
cells, which occur in tumors with low blood flow, are less susceptible
to radiation and are probably responsible for most radiation treatment
failures (Adams, 1981). Similarly, cytotoxic drug delivery could be
predicted by blood flow measurements, with clear implications
regarding the expected success of chemotherapy. Thus, the ability to
conveniently measure tumor blood flow would have considerable clinical
utility. This report describes a new, simple method of using
deuterium nuclear magnetic resonance spectroscopy (NMR) to determine
tumor blood flow.

MATERIALS AND METHODS

Animals and Tumors

Fibrosarcoma (FSaII) cells in a single-cell suspension were
implanted subcutaneously in the left hind foot dorsum of C3H mice.
Tumor volumes were calculated from the three orthogonal diameters

using an ellipsoid approximation ($V = d_1*d_2*d_3*\pi/6$). Water content of FSaII tumors was determined to be 80.5 ± 0.04 (SE) wt% (n=9) by weighing tumor samples before and after desiccation. Blood flow was measured in tumors using two protocols. First, control tumors were examined before and after delivery of hydralazine (0.01 mg/g i.p.), a vasodilator, to compare the flow changes induced by this drug as measured by deuterium NMR and laser Doppler flowmetry. In the second set of experiments, the inter-tumor heterogeneity in blood flow was examined in control tumors, and in tumors growing in tissue that had received 16 Gy one week prior to tumor implantation. The latter set of tumors simulate the physiology of tumors regrowing after unsuccessful radiation treatment.

Flow Model and Equations

The HOD wash-in technique presented here is analogous to the nitrous oxide method for measuring cerebral blood flow developed by Kety in the 1940's (Kety and Schmidt, 1945 & 1948; Kety, 1949).

The rate at which tracer accumulates in a tumor depends on both the flow rate and the tracer concentration gradient between the tumor and the arterial blood. The tracer concentration in the tumor relates to that in the tumor's arterial blood as follows:

$$d[C_T(t)]/dt = (F/V) [C_A(t)-C_T(t)] \qquad \text{[Equ. 1]}$$

where $C_T(t)$ is the tracer concentration in the tumor as a function of time, $C_A(t)$ is the tracer concentration in arterial blood as a function of time, F is the rate at which blood enters and leaves the tumor (i.e. the flow rate), V is the fluid space of the tumor, and $d[C_T(t)]/dt$ is the derivative of $C_T(t)$ with respect to time (Fig. 1).

In the case of intraperitoneal tracer injection, we found that the arterial tracer concentration $[C_A(t)]$ takes the form of $A_O(1-e^{-t/\tau})$, where A_O and τ are constants. Inserting this expression into the above differential equation and solving for $C_T(t)$ yields the expression:

$$C_T(t) = FA_O\tau/(F\tau-V) [e^{-(F/V)t} - e^{-t/\tau}] - A_Oe^{-(F/V)t} + A_O \qquad \text{[Equ. 2]}$$

which can be used to determine the flow rate F.

D_2O was introduced into the mice by injecting a weight-adjusted volume [0.01 ml/g body weight (bw)] of 10% (or 50%) D_2O doped saline into the peritoneal cavity. The post-injection time course of the arterial deuterium concentration $[C_A(t)]$ was obtained from a cohort of mice by injecting them as above and drawing 0.02 to 0.05 ml blood samples at recorded times thereafter. Blood was drawn from the ocular plexus using heparinized, calibrated capillary tubes. In total, 28 measurements were made. After examining a number of physiologically relevant functions, we chose to fit the data to a saturating exponential with a time constant (τ) of 90 seconds, and a saturation value (A_O) of 0.15% (after delivery of 0.01 ml/g bw of 10% D_2O doped saline). Larger doses of D_2O (e.g., 0.01 ml/g bw of 50% D_2O doped saline) were assumed to scale A_O linearly. The arterial HOD

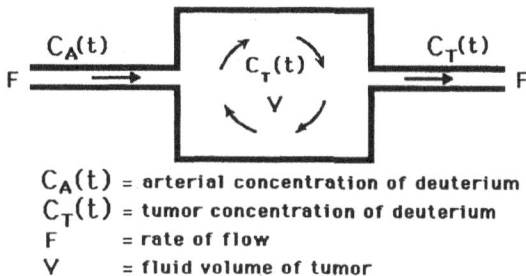

$C_A(t)$ = arterial concentration of deuterium
$C_T(t)$ = tumor concentration of deuterium
F   = rate of flow
V   = fluid volume of tumor

Fig. 1.   Model used for calculation of tumor blood flow.

concentration $[C_A(t)]$ remained nearly constant after saturating due to the long (3 day) half-life of HOD in the mouse (Thomson, 1960; Katz et al., 1962; Peng et al., 1972).

The post-injection time course of the tumor HOD concentration $[C_T(t)]$ was monitored by deuterium NMR. Using the techniques described in the following section, the relative HOD content of the tumor was traced for at least twenty minutes beginning immediately after intraperitoneal injection. Since $C_T(t)$ was traced in relative units, a scaling factor was applied which assumed that the $C_A(t)$ and $C_T(t)$ curves have the same saturation value. Fitting (non-linear least sum of squared errors) of the expression for $C_T(t)$ to the resulting data yielded a value for the flow rate F. As previously noted, the water content of FSaII tumors is 80.5% and F was corrected for this water content.

Deuterium NMR

Deuterium NMR was used to monitor the concentration of deuterated water in the tumors. Measurements were performed in an Oxford 8.5 Tesla magnet at 55.2 MHz. The blood samples for the $C_A(t)$ curve were analyzed in standard NMR test tubes. The HOD concentration of the tumors $[C_T(t)]$ was traced using a custom-built probe that slid into the bore of the magnet. Intraperitoneal delivery of D_2O or hydralazine to the positioned mouse was performed from outside the magnet via two separate catheter tubes (22-gauge Radiopaque Teflon® Catheter, Deseret Medical, Inc., Sandy, UT) fitted with arterial pressure tubing to assure accurate dosaging (High Frequency Pressure Monitoring Lines, NAMIC®, Glens Falls, NY 12801).

Acquisition parameters were as follows: 5 kHz sweep width, 45° excitation pulse (0.01 ms), 512 samples per free induction decay, and 0.1 second recycle delay. Forty-four free induction decays, acquired over 12.5 seconds were summed to produce one spectrum. Relative HOD content was taken as the area under the HOD peak in the resulting spectra.

Laser Doppler Flowmetry

Measurements were made using a Laserflo® Blood Perfusion Monitor 403A (TSI Inc., St. Paul, MN). The mouse was immobilized in a cylindrical tube similar to that used for the NMR analyses, and the fiberoptic probe was positioned centrally over the tumor, no further than 2 mm away, but not in direct contact. A 22-gauge catheter (Radiopaque Teflon® Catheter, Deseret Medical, Inc., Sandy, UT) allowed i.p. delivery of hydralazine to the immobilized mouse. In control animals, saline was injected intraperitoneally.

RESULTS

Our measurements (Fig. 2) indicated an average pre-hydralazine blood flow rate of 42 ± 6.4 (SE) ml/min/100g [15 measurements on 10 tumors with an average volume of 0.11 ± 0.04 (SD) ml]. This rate is comparable to that obtained by Menke and Vaupel (1988) using [85]Kr wash-out and small DS-carcinosarcomas in conscious rats (58 ± 10 ml/min/100g).

Hydralazine was administered at a dose of 0.01 mg/g bw i.p.
Based on the laser Doppler technique, flow drops to a minimum roughly
10 minutes after that dose (Kalmus et al., 1990), and remains at this
level for approximately one hour. Thus, injection of D_2O and
subsequent data acquisition were started 10 minutes after delivery of
hydralazine. This also allowed complete absorption of the injected
drug solution from the abdominal cavity. The slower blood flow rates
induced by hydralazine necessitated monitoring the tumor HOD

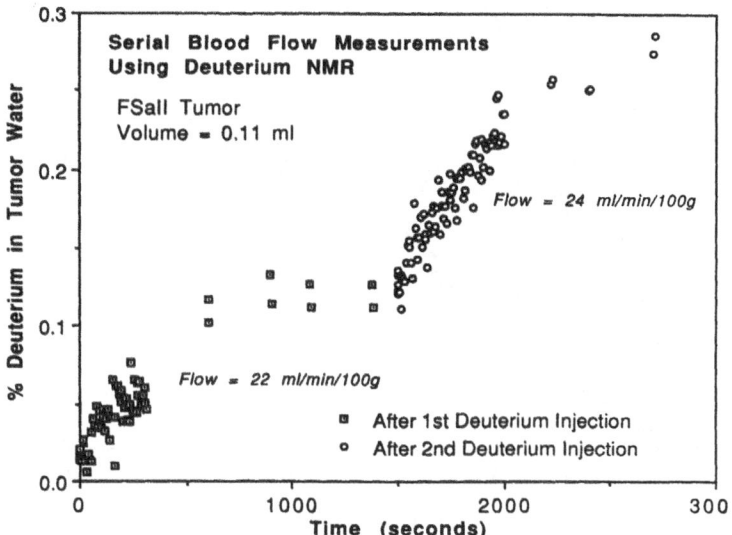

Fig. 2.   Two sequential blood flow measurements made by tumor
deuterium content determinations after i.p. injection of
saline containing 10% D2O. The second D_2O injection was
given at 1500 sec.

concentration for 40 minutes rather than the usual 20 min, since the
slower $C_T(t)$ rise times characteristic of slow flow rates required
longer times to reach the saturation value.

Fig. 3 shows the tumor concentration data acquired pre- and post-
hydralazine in a typical experiment. The tumor volume was 0.073 ml
and the D_2O concentration in the injected saline was 50%. Averaged

over all ten tumors, the post-hydralazine blood flow rate was 4.8 ±
1.5 (SE) ml/min/100g, indicating an 89% decrease from the normal flow
rate (p<0.001). Laser Doppler measurements showed that tumor flow
drops by 93 ± 6.7 (SD) % over the time range of interest (10 to 50
minutes post hydralazine delivery) and were, therefore, in good
agreement with the deuterium NMR measurements.

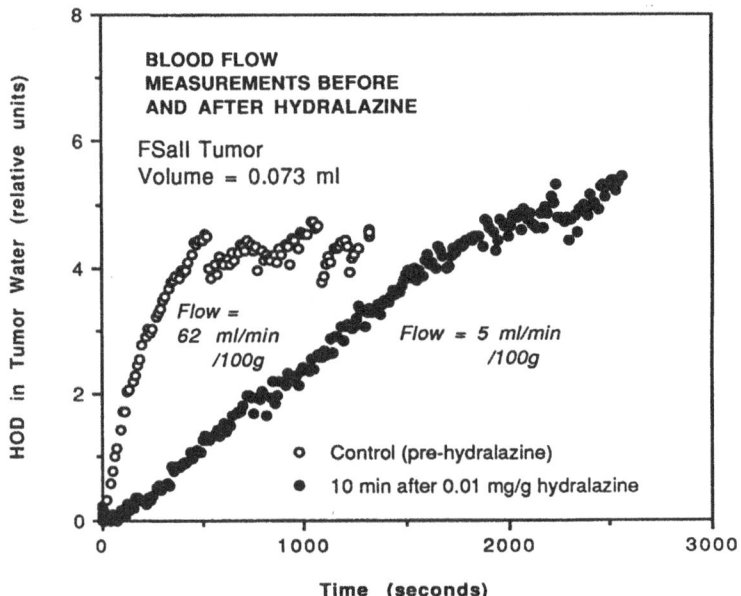

Fig. 3.   Sequential blood flow measurements made before and after
            i.p. delivery of 0.01 mg/g hydralazine. The pre- and post-
            hydralazine tracings are superimposed for clarity.

Fig. 4 shows the tumor blood flow as a function of tumor size for
control tumors and for tumors growing in irradiated tissue. The blood
flow in individual tumors varied greatly as has been previously described for humans (for reviews see Vaupel et al., 1989; Vaupel et
al., 1990) and in animal tumor models (Kim and Ackerman, 1988a; Menke
and Vaupel, 1988; Kallinowski et al., 1989). For control tumors with
volumes of 0.2 ml or less, there was no clear size dependent change in

blood flow, though the flow rate of the largest tumor (0.35 ml) was the lowest value measured (10 ml/min/100g). In contrast, while small tumors (<0.1 ml) growing in irradiated tissue had blood flows similar to control tumors, a differential flow decrease was observed in tumors greater than 0.1 ml (p<0.05).

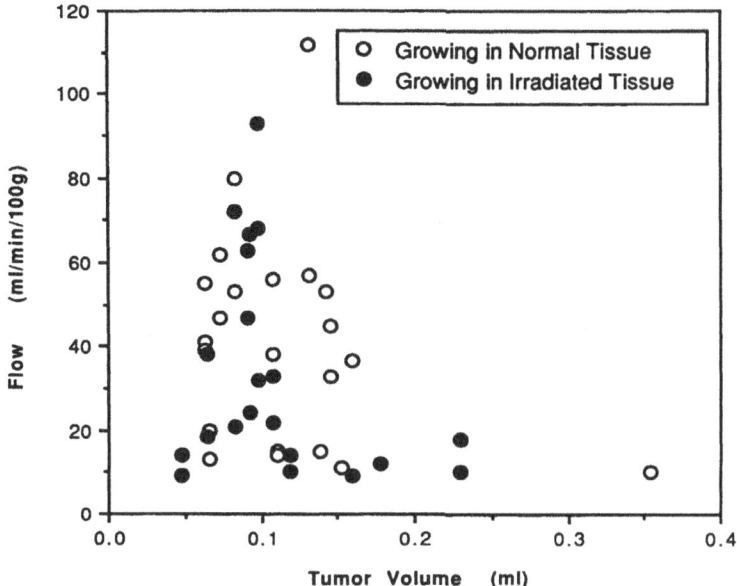

Fig. 4.   Blood flow measurements of tumors growing in normal subcu-
          taneous tissue, or in tissue after pre-treatment with 16 Gy.
          There was substantial inter-tumor variability.   Tumors
          greater than 0.1 ml growing in irradiated tissue had a
          significantly reduced flow rate.

DISCUSSION

     The wash-in method for measuring tumor blood flow presented here has two advantages over the wash-out technique that is more generally used in animal experiments (Ackerman et al., 1987a & 1987b; Kim and Ackerman, 1988a & 1988b; Menke and Vaupel, 1988; Kallinowski et al., 1989; Larcombe McDouall and Evelhoch, 1990).   First, intra-tumor injection of tracer might perturb the flow due to local damage and an increased intratumor pressure (Larcombe McDouall and Evelhoch, 1990). Second, tumors used in this study were approximately a factor of four smaller than was possible to measure using wash-out, since the latter technique requires invasive injection of 0.01 to 0.05 ml of tracer. Finally, our study was performed on fully conscious animals, no sedative was necessary. We believe that these factors illustrate some of the advantages of our wash-in technique over previously reported wash-out techniques.

It should be noted that intraperitoneal delivery of pure saline at a dose of 0.01 ml/g bw has no significant effect on blood flow rate, heart rate, or arterial blood pressure (Okunieff et al., 1988). Furthermore, no significant change in the ^{31}P NMR spectra occurs in C_3H mice after this saline dose. Larger i.p. saline injections (approximately 0.04 ml/g bw) do, however, have mild cardiovascular effects (Okunieff et al., 1988), and thus a minimum of 10 minutes is recommended between successive injections (5 to 10 times the time constant of saline uptake from the peritoneal cavity) to prevent an accumulation of fluid in the peritoneal cavity. The relatively mild cardiovascular effects of intraperitoneal saline administration make it preferable to intravenous bolus administration, which not only has more cardiovascular side effects, but also has a more complex $C_A(t)$ curve.

Extremely high flow rates (> 145 ml/min/100g) are measured less accurately with our technique, and in the rare instances that these may occur, might be better studied using wash-out techniques.

In larger animals, the time course of the arterial HOD concentration could be followed dynamically using arterial catheters, and the accuracy of this technique could be improved since uncertainty in the $C_A(t)$ would be eliminated. This additional data would also prevent miscalculations due to drug-induced modification of the $C_A(t)$ curve.

Another source of error is animal motion. Some minor motion artifact is evident in the pre-hydralazine tumor blood flow measurement presented in Fig. 3. This source of error can be prevented by anesthesia, but anesthesia too can affect tumor blood flow (Menke and Vaupel, 1988) and should be avoided. In three of our measurements, motion was judged to have produced significant errors, which made repeat measurements necessary.

As a final note, the model used to estimate tumor blood flow assumes that the end-capillary concentration of HOD is at equilibrium with the tumor. In a few organs, most notably the brain, water diffusion rates are limited and thus the measurement of blood flow with this technique will only reflect nutritive, and not total, flow. A correction for incomplete H_2O distribution and for the difference in diffusibility of HOD compared to H_2O, therefore, must be made in cases of diffusion impairment.

Ewy et al. (1988) obtained _in vivo_ deuterium images of a cat brain at body concentrations of 4-5%. The images were acquired at a time resolution of roughly 10 seconds and a spatial resolution of 1.7 x 1.7 x 10 mm. The 10 second time resolution is more than adequate for flow calculations. At a 4-5% body concentration, however, serial studies are not possible due to D_2O related toxicity (Hughes and Calvin, 1958; Thomson, 1960; Czajka et al., 1961; Katz et al., 1962; Bachner et al., 1964; Peng et al., 1972). Both the time resolution and spatial resolution might be sacrificed somewhat in order to obtain images at a lower body concentration. If, for example, the time resolution were lowered to 30 seconds (still adequate for flow measurements), and the spatial resolution were lowered to 5.0 x 5.0 x 10.0 mm, the body concentration of deuterium could be decreased to roughly 0.2% per run. This level of imaging resolution is similar to that obtained by positron emission tomography, and the deuterium level required would be safe for extensive serial studies.

SUMMARY

Using i.p. delivered deuterated water, and ^2H NMR spectroscopy, it is possible to non-invasively, serially, and quantitatively measure blood flow through tissue volumes as small as 0.05 ml. Measurements made with this technique show that changes in total tumor blood flow reflect local changes measured superficially by laser Doppler flowmetry, and that tumors of identical histology and size have heterogeneous blood flow rates. Finally, we have shown that tumors growing in an irradiated normal tissue have a lower blood flow rate which is manifest when tumors grow beyond 0.1 ml.

Acknowledgement

The authors would like to thank Leo Neuringer, Ph.D., and Kebede Beshah, Ph.D, for their valuable help. Supported by NIH grants CA48096 and RR00995 as well as the American Cancer Society Career Development Award and Research Grant PDT-313.

REFERENCES

Ackerman, J.J.H., Ewy, C.S., Becker, N.N., Shalwitz, R.A., 1987a, Deuterium nuclear magnetic resonance measurements of blood flow and tissue perfusion employing ^2H_2O as a freely diffusible tracer. Proc. Natl. Acad. Sci. USA 84:4099-4102.

Ackerman, J.J.H., Ewy, C.S., Kim, S.G., Shalwitz, R.A., 1987b, Deuterium magnetic resonance *in vivo*: measurement of blood flow and tissue perfusion. Ann. NY Acad. Sci. 508:89-98.

Adams, G.E., 1981, Development of sensitizers of hypoxic cells. Cancer 48:696-707.

Bachner, P., McKay, D.G., Rittenberg, D., 1964, The pathologic anatomy of deuterium intoxication. Proc. Natl. Acad. Sci. USA 51:464-471.

Czajka, D.M., Finkel, A.J., Fischer, C.S., Katz, J.J., 1961, Physiological effects of deuterium on dogs. Am. J. Physiol. 201(2):357-362.

Ewy, C.S., Ackerman, J.J.H., Balaban, R.S., 1988, Deuterium NMR cerebral imaging *in situ*. Mag. Res. Med. 8:35-44.

Hughes, A.M., Calvin, M., 1958, Production of sterility in mice by deuterium oxide. Science 127: 1445-1446.

Kallinowski, F., Schlenger, K.H., Runkel, S., Kloes, M., Stohrer, M., Okunieff, P., Vaupel, P., 1989, Blood flow, metabolic functions, cellular microenvironment and growth rate of human tumor xenografts. Cancer Res. 49:3759-64.

Kalmus ,J., Okunieff, P., Vaupel, P., 1990, Dose-dependent effects of hydralazine on microcirculatory function and hyperthermic response of murine FSaII tumors. Cancer Res. 50:15-19.

Katz, J.J., Crespi, H.L., Czajka, D.M., Finkel, A.J., 1962, Course of deuteriation and some physiological effects of deuterium in mice. Am. J. Physiol. 203(5):907-913.

Kety, S.S., 1949, Measurement of regional circulation by the local clearance of radioactive sodium. Am. Heart J. 38:321-328.

Kety, S.S., Schmidt, C.F., 1945, The determination of cerebral blood flow in man by the use of nitrous oxide in low concentrations. Am. J. Physiol. 143:53-66.

Kety, S.S., Schmidt, C.F., 1948, The nitrous oxide method for the quantitative determination of cerebral blood flow in man: theory, procedure and normal values. J. Clin. Investigation 27:476-483.

Kim, S.G., Ackerman, J.J.H., 1988a, Multicompartment analysis of blood flow and tissue perfusion employing D_2O as a freely diffusible tracer: a novel deuterium NMR technique demonstrated via application with murine RIF-1 tumors. Mag. Res. Med. 8:410-426.

Kim, S.G., Ackerman, J.J.H., 1988b, Quantitative determination of tumor blood flow and perfusion via deuterium nuclear magnetic resonance spectroscopy in mice. Cancer Res. 48:3449-3453.

Larcombe McDouall, J.B., Evelhoch, J.L., 1990, Deuterium nuclear magnetic resonance imaging of tracer distribution in D_2O clearance measurements of tumor blood flow in mice. Cancer Res. 50:363-369.

Mattiello, J., Larcombe McDouall, J.B., Evelhoch, J.L., 1990, Deuterium NMR imaging of regional tumor blood flow. 38th Ann. Meeting of the Radiation Research Society, Abstract Et-4, New Orleans.

Menke, H., Vaupel, P., 1988, Effect of injectable or inhalational anesthetics and of neuroleptic, neuroleptanalgesic, and sedative agents on tumor blood flow. Radiat. Res. 114:64-76.

Okunieff, P., Kallinowski, F., Vaupel, P., Neuringer, L.J., 1988, Effect of hydralazine-induced vasodilation on the energy metabolism of murine tumors studied by in-vivo ^{31}P-nuclear magnetic resonance spectroscopy. J. Natl. Cancer Inst. 80:745-750.

Peng, S.K., Ho, K.J., Taylor, C.B., 1972, Biologic effects of prolonged exposure to deuterium oxide: a behavioral, metabolic, and morphologic study. Arch. Path. 94:81-89.

Thomson, J.F., 1960, Physiological effects of D_2O in mammals. Ann. NY Acad. Sci. 84:736-744.

Vaupel, P., Kallinowski, F., Okunieff, P., 1989, Blood flow, oxygen and nutrient supply, and metabolic microenvironment of human tumors: A review. Cancer Res. 49:6449-6465.

Vaupel, P., Kallinowski, F., Okunieff, P., 1990, Blood flow, oxygen consumption and tissue oxygenation of human tumors. Adv. Exptl. Med. Biol. (in press).

THE TRUNCATED TCA CYCLE IN HeLa CELL MITOCHONDRIA

Terrence J. Piva* and Edward McEvoy-Bowe

Department of Chemistry & Biochemistry
James Cook University of North Queensland
Townsville, Qld, Australia.

INTRODUCTION

A series of studies were carried out on HeLa cell mitochondria to determine (1) the likely pathways for carbon flux from glutamate, (2) the significance of malic enzyme activity, (3) the significance of pyruvate dehydrogenase activity, and finally (4) the significance of the truncated TCA cycle in HeLa cells (Coleman and Lavietes, 1981).

EXPERIMENTAL

The mitochondria were prepared by a procedure based on that of Moreadith & Fiskum (1984). Throughout the experiments a preparation of bovine heart mitochondria was used as a control. The HeLa cell mitochondria were found to be present in small numbers, and when examined by transmission electron microscopy were shown to have a flattened, round or oval shape similar to the mitochondria isolated from Ehrlich ascites cells (Mitchell, 1967), Morris hepatomas (Petersen et al., 1970), and mouse mammary adenocarcinomas (Sordahl and Schwartz, 1971). The mitochondria lacked the cristal pattern of normal mitochondria.

Oxygen consumption was measured using a Rank Brothers calibrated oxygen electrode in a 1 ml water-jacketed vessel maintained at 30°C. Steady state rates of oxygen consumption were determined as described (Chance and Williams, 1956).

Amino acids were determined by a Waters HPLC using a cation exchange column in association with post-column derivitization with ninhydrin.

The TCA cycle metabolites were determined by standard enzymatic procedures (Bergmeyer, 1985).

RESULTS

Whole cell experiments. In earlier experiments with whole cells it was shown that when the cells were exposed to 3 mM [^{14}] Glutamine for a period of up to 40 min, 90% of the radioactive label in the non-protein fraction of the cells was recovered as glutamine, glutamate and aspartate. Furthermore the specific radioactivity relative to that of the medium glutamine of the three compounds was found to be 1.03, 0.8 and 0.25

---

*Present address is Department of Biochemistry, University of Oxford, South Parks Road, Oxford, U.K.

Table 1. Concentration of free cytoplasmic amino acids in HeLa cells. The results are expressed as the means ±S.E. of 12 separate determinations.

AMINO ACID	[Mm]
Aspartate	8.0 ± 1.4
Glutamate	22.2 ± 2.2
Glutamine	11.3 ± 1.7
Glycine	9.8 ± 1.1
Taurine	2.3 ± 2.5
Alanine	N.D.

respectively, indicating that glutamine synthetase activity was lacking, that the greater part of the glutamate present was derived from glutamine and that the major part of the aspartate carbon appeared to be derived from glucose which was also present in the medium. The cellular concentrations of the major amino acids are shown in Table 1, where it can be seen glutamate was the major amino acid present at a concentration of 20 mM, and for this reason we used a medium concentration 20 mM glutamate in the mitochondrial experiments. Alanine was not detected (< 1 mM).

Respiration studies. Succinate was found to be the only substrate capable of inducing state III respiration, and produced RCR values which ranged from 3.1 to 6.1. Coupled respiration was observed in the HeLa cell mitochondria when malate plus glutamate and malate plus pyruvate were the added respiratory substrate (RCR values being 7.4 and 5.1 respectively). Glutamate derived respiration in the presence of malate was inhibited by aminooxyacetate (Fig. 1).

Mitochondrial Metabolism. Fig. 2 shows that citrate was a major product of pyruvate metabolism, and Fig. 3 shows that both pyruvate and citrate were produced from medium malate. Fig. 4 shows that in the presence of glutamate alone only aspartate and a very small amount of alanine were produced. Fig. 5 shows that the presence of both glutamate and malate in the metabolism resulted in the formation of aspartate, pyruvate, citrate and a small amount of alanine. It is noticeable that the amount of aspartate produced over the first 10 min of incubation time was about double that obtained when glutamate alone was used. The whole cell experiments have shown that the relative specific radioactivity of aspartate is very low compared to glutamine indicating the a major part of the aspartate carbon is glucose derived. The contribution of malate to the rate of formation of aspartate provides at least a partial explanation of how this occurs. On the other hand the rates of pyruvate and citrate formation in the presence of glutamate were about the same as those obtained when malate alone was present. Fig. 6 shows that the presence of aminooxyacetate essentially eliminated aspartate formation when glutamate and malate were present, whereas the rates of formation of pyruvate and citrate remained unchanged.

DISCUSSION

The results from Figs. 1-3 show that exogenous malate gave rise mainly to pyruvate and citrate formation and the most likely pathway for exogenous malate was therefore through the malic enzyme to pyruvate (Moreadith and Lehninger, 1984), and that pyruvate carboxylase and pyruvate dehydrogenase then acted jointly on the pyruvate to form oxaloacetate and acetyl-CoA respectively, which then became available for the synthesis of citrate by citrate synthase. Thus it would appear that citrate derived from exogenous malate was not oxidised (Fig. 1), and that a high proportion of the citrate and pyruvate formed was transported into the medium (Fig. 3). These experiments therefore confirmed the existence of a truncated TCA pathway in HeLa cells with the section between citrate oxidation and α-ketoglutarate formation being deficient, a result which is consistent with the truncated TCA cycle as proposed by Coleman and Lavietes, 1981. Fig. 1 shows that malate alone had little effect on the respiration rate suggesting that it

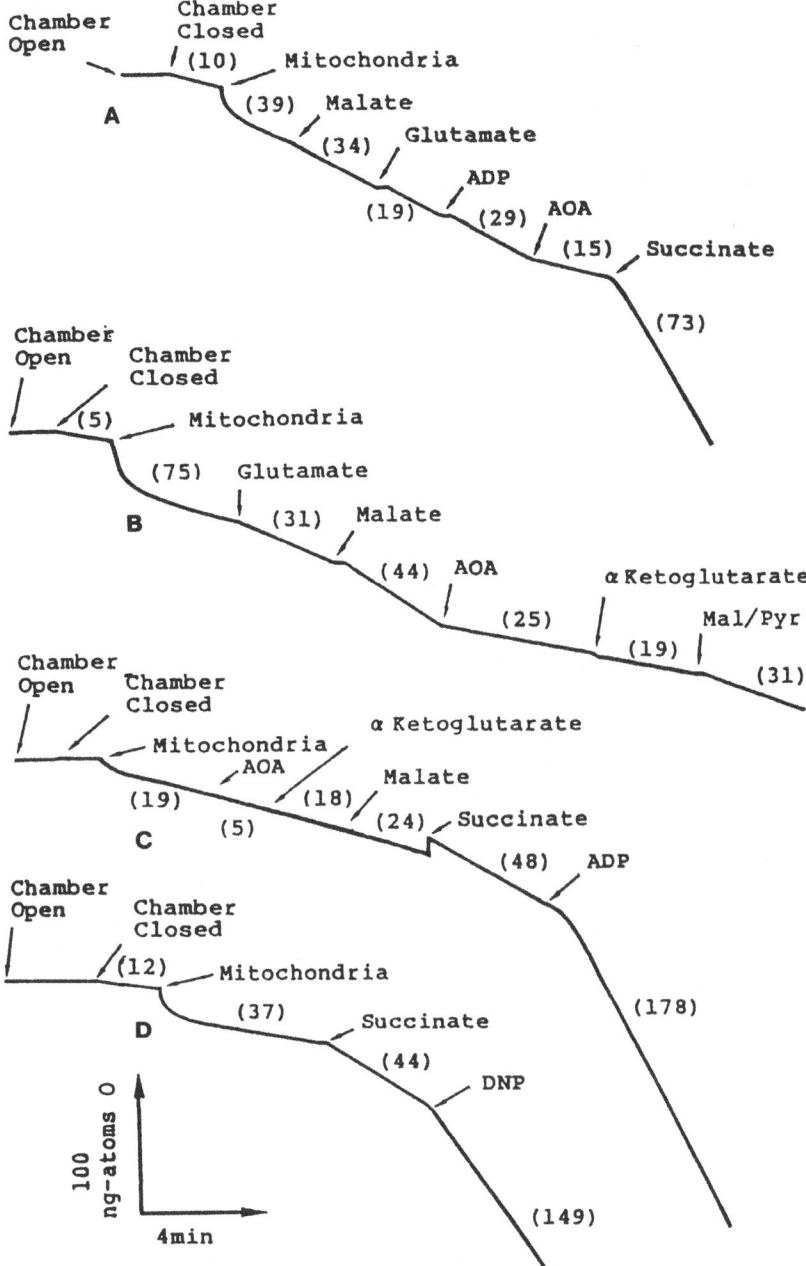

Fig. 1. Effect of various substrates on HeLa cell mitochondria respiration. HeLa cell mitochondria were resuspended at 1mg/ml in a medium containing 130 mM KCl, 10mM Mops, 2 mM $MgCl_2$ and 2 mM KH_2PO_4 at pH 7.2. The substrates were added at 5 mM final concentration except for ADP (0.25 mM), aminooxyacetate (2 mM) and DNP (10 μM). Numbers along the traces represent the amount of oxygen consumed in ng-atoms.

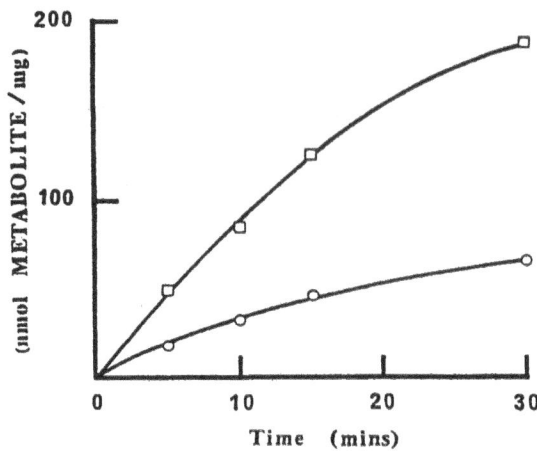

Fig. 2. Metabolic products of pyruvate oxidation in HeLa Cell Mitochondria.

Only those mitochondria possessing an RCR (respirator control ratio) greater than 3 were used. The experimental protocol was the same as that used for the respiration studies except that the concentration of HeLa cell mitochondria in the incubation medium was 1-2 mg protein/ ml.. The Fig. shows citrate efflux. The results are the means of 2 determinations for all the metabolic plots.

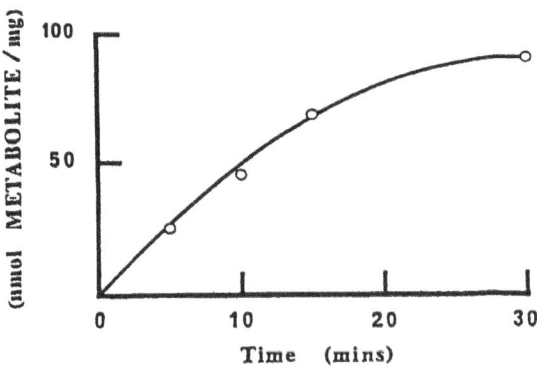

Fig. 3. Metabolic products of malate oxidation in HeLa Cell mitochondria.

The experimental conditions are the same as for Fig. 2 in all the metabolic experiments except that in this experiment malate had a final concentration of 5 mM. (O) citrate efflux, (□) pyruvate efflux.

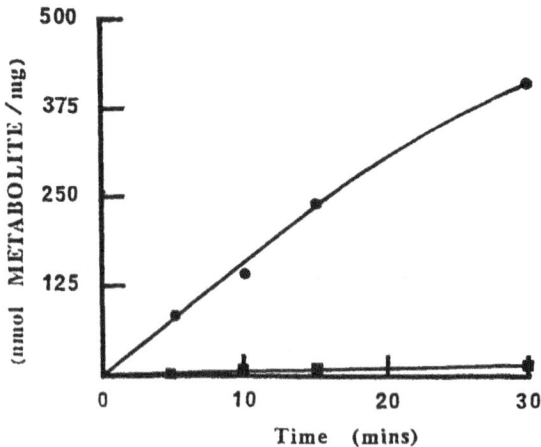

Fig. 4. Metabolic products of glutamate oxidation in HeLa cell Mitochondria.

The concentration for glutamate in this experiment was 20 mM. (●) aspartate, (■) alanine efflux.

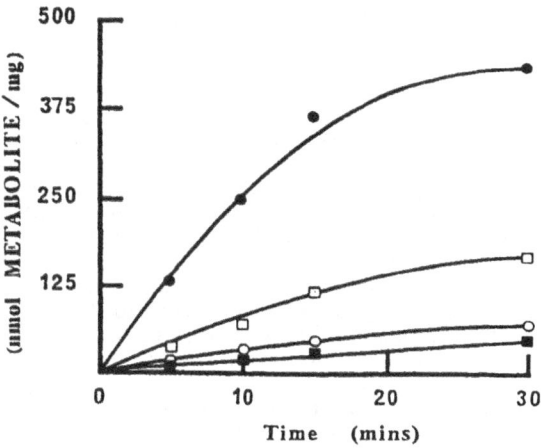

Fig. 5. Metabolic products of glutamate oxidation in HeLa cell mitochondria: effect of malate.

Malate and glutamate were added to the incubation medium to give the concentrations 5 and 20 mM respectively. (●) aspartate efflux, (■) alanine efflux, (○) citrate efflux, (□) pyruvate efflux.

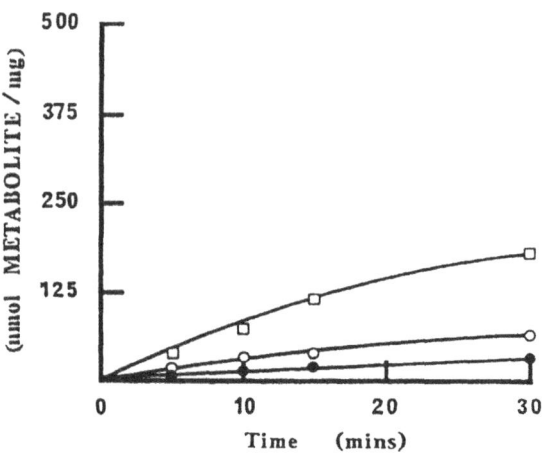

Fig. 6. Effect of aminooxyacetate inhibition on malate and glutamate oxidation in HeLa cell mitochondria.

Experimental conditions were the same as for Fig. 5 except that aminooxyacetate was added to give a concentration of 2 mM. (●) aspartate efflux, (○) citrate efflux, (□) pyruvate efflux.

is the NADP-linked form of the malic enzyme which is involved in the metabolism of medium malate. This enzyme has been previously demonstrated to be characteristic of rapidly growing neoplastic cells (Nagel, Dauchy and Sauer, 1980). The contribution which glutamine makes the to the TCA cycle as a consequence of the high glutaminase activity is clearly related to its ability to partially restore TCA cycle function by acting as an alternative to citrate as the major supplier of α-ketoglutarate. It is not clear why glutamate alone does not promote respiration, whereas malate and glutamate together do. Fig. 5 shows that the presence of malate significantly increased the rate of formation of aspartate, the opposite of that obtained by Moreadith and Lehninger (1984) in their work with Ehrlich tumour mitochondria, whereas the addition of glutamate to malate has little effect on the flow of carbon into citrate, which suggests that the availability of malate derived from α-ketoglutarate has no influence on the rate of formation of pyruvate and citrate derived from exogenous malate. This raises the question as to whether the endogenous malate derived from α-ketoglutarate goes through a separate pathway (malate dehydrogenase) to that of the exogenous malate (malic enzyme). Moreadith and Lehninger have stated that this is what occurred with their Ehrlich tumour mitochondria, though they did not produce any precise evidence for the endogenous route going exclusively through malate dehydrogenase. Thus it appears that the endogenous malate is tightly bound in an enzyme complex whereas the exogenous malate is not or that the oxaloacetate derived from is not. One explanation as to how this occurs may be that the malate dehydrogenase which as a component of the TCA cycle forms part of the proposed TCA cycle metabolon (Srere, 1990) whereas the malic enzyme does not. Figs. 1 and 6 demonstrate that the oxidation of glutamate proceeds through aspartate aminotransferase in that Fig. 1 shows that aminooxyacetate totally inhibited respiration as well as inhibiting the formation of aspartate. Fig. 6 also shows that the elimination of aspartate formation by the aminooxyacetate had no effect on the rate of formation of citrate from exogenous malate, this being further evidence for the hypothesis that the oxidation of endogenous malate goes through malate dehydrogenase .

Thus to return to the four objectives stated at the beginning of this paper, it can be seen that (1) the oxidation of glutamine (glutamate) in HeLa cells results in the formation of aspartate with the glutamate carbon flux being routed successively through α-ketoglutarate, succinate, malate to oxaloacetate, together with some evidence for the possibility that the pathway from malate to oxaloacetate is through malate dehydrogenase rather than malic enzyme followed by pyruvate carboxylase; that (2) the significance of the malic enzyme is is that it acts on exogenous malate to produce pyruvate which then leads to the formation of citrate; that (3) pyruvate dehydrogenase is active but that the pyruvate appears to end up mainly as citrate to be converted subsequently to fatty acids and cholesterol, if adequate amounts of glutamate are present, as aspartate; and that finally evidence has been presented for the existence of a truncated TCA cycle in these cells providing further confirmation that this phenomenon is a characteristic of tumour cells.

ACKNOWLEDGEMENT

We wish to thank Dr. Michael McCabe for his assistance in setting up the oxygen electrode experiments.

REFERENCES

Bergmeyer, H. U. (ed.), 1985, Methods of Enzymatic Analysis, 3rd. edn., Verlag Chemie, Weinheim.
Chance, B. and Williams, G. R., 1956, Adv. Enzymol., 17:65-134.
Coleman, P. S. and Lavietes, B. B., 1981, CRC Crit. Rev. Biochem., 11:341-393.
Mitchell, R. F., 1967, J. Ultrastruct. Res, 18:257-276.
Moreadith, R. R. and Fiskum, G., 1984, Anal. Biochem. 137:360-367.
Moreadith, R. W. and Lehninger, A. L., 1984, J. Biol. Chem., 259:6215-6221.
Nagel, W. O., Duachy, R. T. and Sauer, L. A., 1980, J. Biol. Chem., 255:3849-3854.
Pedersen, P. L., Greenawalt, J. W., Chan T.L. and Morris, H. P., 1970, Cancer Res., 30:2620-2626.
Sordahl, L. A. and Schwartz, A., 1971, Meth. Cancer Res., 6:159-186.

APPARENT HETEROGENEITY BETWEEN LEUKEMIC

LYMPHOCYTE CELL LINES

Kimberley Bondeson and Michael McCabe

Department of Chemistry and Biochemistry
James Cook University of North Queensland
Townsville Queensland 4811 Australia

ABSTRACT

^{31}P-NMR and polarographic techniques were used to investigate glycolytic versus aerobic oxidative activity in normal and leukemic lymphocytes, and to investigate possible heterogeneity in these parameters between two leukemic cell lines. Molt 3 cells showed a 10-fold higher rate of glutamine-dependent respiration than Molt 4 cells, and an increased level of glutamine-uptake.   Molt 3 demonstrated a high intracellular buffering capacity, manifested by constant pHi after addition of glucose, while the same treatment applied to Molt 4 cells induced a change in internal pH of up to 1.23 pH units. This data raises the possibility of heterogeneity of leukemic lymphocytes within the patient from whom the isolation was conducted, or of gross metabolic adaptation by the cell lines in culture.

INTRODUCTION

A general property of malignant tumours appears to be a strong dependence upon exogenous glutamine as an energy source (see, for example, Reitzer et al., 1979). Tumour cells do, however, show a wide range of respiratory activities, dependent upon or reflected by levels of mitochondria and/or structural completeness.   In order to investigate possible metabolic heterogeneity between cell lines representing a single type of cancer, we have determined respiratory parameters for two leukemic lymphocyte lines and their normal counterparts.

MATERIALS AND METHODS

Mixed peripheral blood lymphocytes were isolated using Percoll gradients (McCabe et al., 1984). The leukemic cell lines Molt 3 and Molt 4 were obtained from the ATCC and were grown in RPMI medium containing 2% (v/v) foetal calf serum and antibiotics. Oxygen uptake was monitored using a Rank oxygen electrode. K_m and V_{max} were determined using the Michaelis-Menten relationship in the integrated form of Henri (Longmuir, 1954). Glutamine uptake was measured by a modification of the silicon oil layer technique (Heldt, 1980).

For ^{31}P-measurements, the cells were harvested in mid-exponential growth phase, washed twice in HEPES-saline and resuspended at a density of 1-3 x 10^8 cells/ml. Spectra were obtained on a Bruker Aspect 3000 NMR spectrometer (Spectrospin and Oxford Instrument) operating in the Fourier-transform mode at 121.5 MHz. All experiments were performed at 26°C using 20% D_2O as a lock. Spectra of cultured cells were obtained in 1.5 to 3 minutes, internal pH values were observed within 20 seconds. Typical acquisition parameters were as follows: sweep width 8 kHz, pulse width 12.0 ms (corresponding to a 45° flip angle) data size 4K, repetition rate 0.3735 s. Over 12 hours, cell viability was determined by the trypan blue method and found to be 90%. During NMR experiments, the presence of sugar phosphates in the cell spectra was taken as an indication of cell viability.

To relate the chemical shift of Pi to internal pH, cancer cells were suspended in HEPES saline containing the uncoupler DNA (dinitrophenol) (2 mM) and titrated with acid or base (Ogawa et al., 1981). 85% phosphoric acid was used as an external reference.

The intracellular free Mg^2 concentration was calculated from the chemical shift of the β-ATP peak relative to the α peak of ATP as described by Gupta et al. (1983). The binding constant for magnesium toward ATP is affected by pH. A correction of pH was made by multiplying a factor calculated from: $f = (1 - 10^{6.5-7.4})$ (Vink et al., 1989).

RESULTS

<u>Respiration parameters</u>. Glutamine-dependent respiration rates with Molt 3 cells were approximately an order of magnitude higher than with Molt 4 cells, and were significantly higher than for normal (bovine) lymphocytes (Table 1). In addition, glutamine-uptake rates for Molt 3 were approximately 4-fold higher than for Molt 4, both being higher than for normal (human) lymphocytes. The capacity for high levels of glutamine uptake and oxidation observed with Molt 3 implies that this cell line may predominantly generate ATP by aerobic respiration of glutamine (ie by oxidative phosphorylation), whereas many tumour cell types are thought to be largely dependent upon glycolysis (Reitzer et al., 1979). Though both cell lines displayed Crabtree effects, the effect was much more pronounced with Molt 4 (60%) than Molt 3 (15%), again implying major differences in the relative contributions of glycolysis and oxidative phosphorylation to ATP production between the two lines.

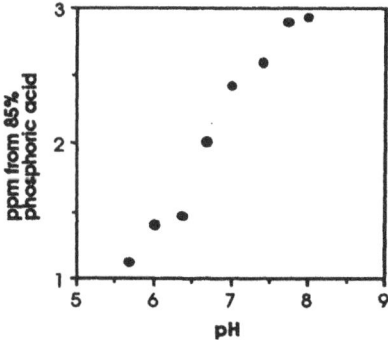

Fig. 1. Dependence of the chemical shift of Pi on pH. The change in chemical shift of Pi in response to acid or base addition was related to spectra accumulation.

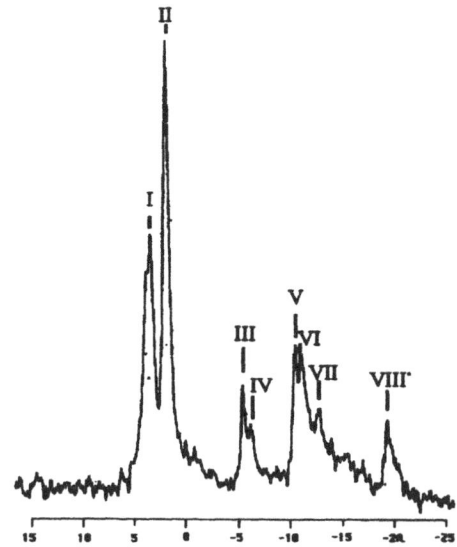

Fig. 2. The 121.5 MHz ^{31}P-NMR spectrum of Molt 3 cancer cells (LB + 20 Hz, 50,900, scans. Peak assignments were as follows: I phosphomonoesters; II Pi; III γATP; IV βADP; V αADP; VI NAD(P); VII Uridine disphophate glucose (UDPG); VIII βATP.

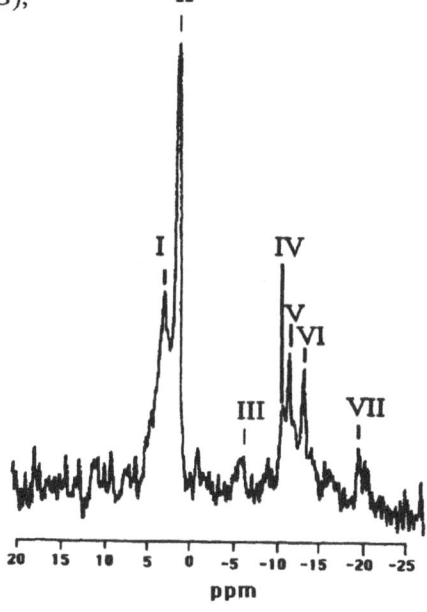

Fig. 3. The 121.5 MHz ^{31}P-NMR spectrum of Molt 4 cancer cells. (LB = 20 Hz, 1,800 scans). Peak assignments were as follows: I phosphomonoesters; II Pi; III γATP; IV αADP and αATP; V Nicotinamide adenine dinulceotide phosphate (NADP); VI Uridine disphosphoglucose (UDPG); VII βATP.

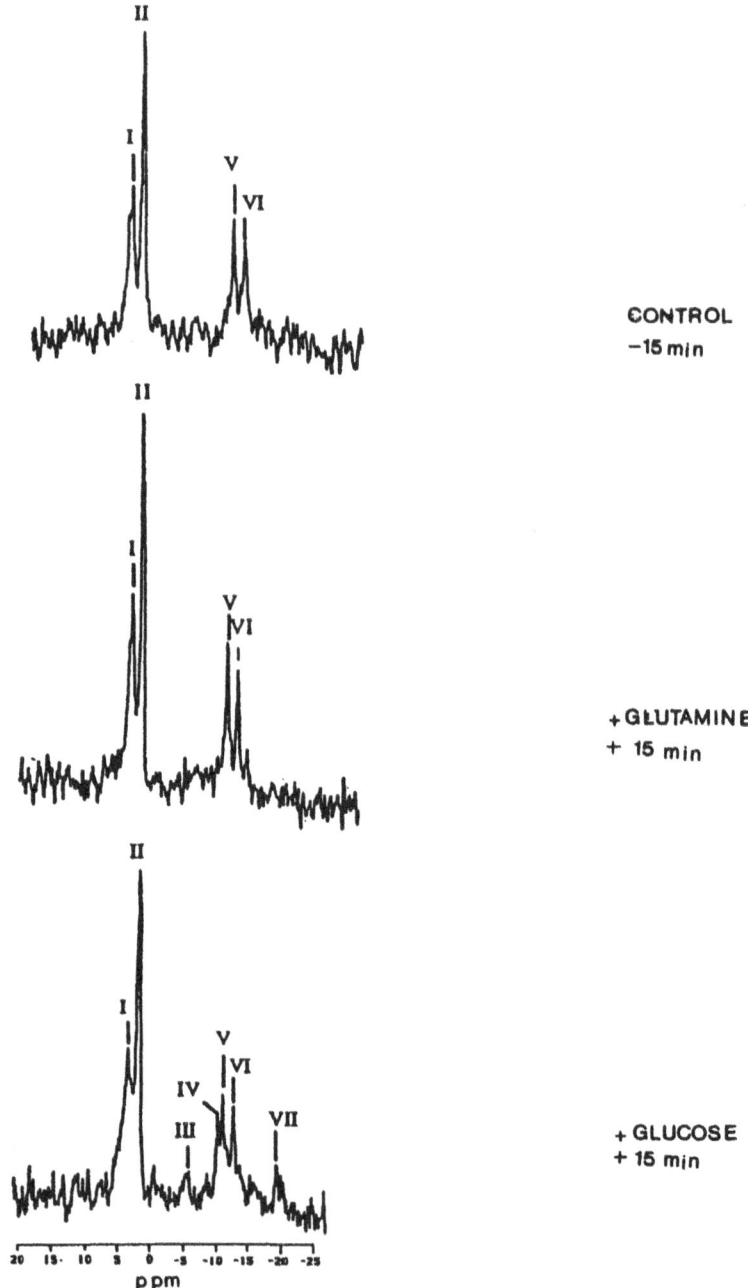

Fig. 4. ^{31}P spectra of Molt 4 cancer cells obtained at 121.5 MHz (1800 scans) before and after the addition of glutamine (final concentration 4 mM) and glucose (final concentration 10 mM). Peak assignments were as follows: I, phosphomonoesters; II Pi; III γATP; IV αADP and αATP; V Nicotinamide adenine dinucleotide phosphate (NADP); VI Uridine diphosphoglucose (UDPG); VI βATP.

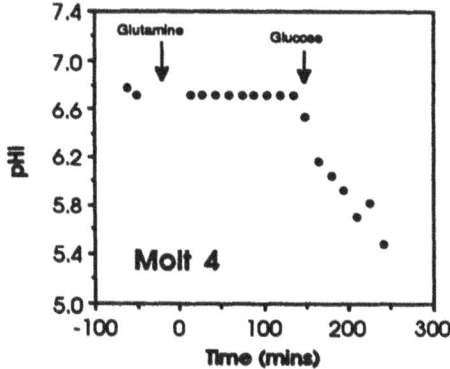

Fig. 5. Alterations in intracellular pH in Molt 4
cells over a time period following the
addition of glutamine (final concentration
4 mM and glucose (final concentration 10
mM).   After glucose addition, pHi
decreased approximately 1.23 pH units.

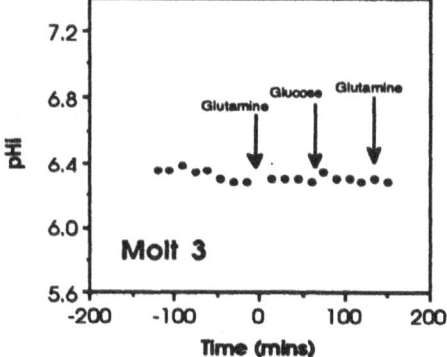

Fig. 6. Alterations in intracellular Ph in Molt 3
cells over a time period following the
addition of glutamine (final concentration
4 Mm), glucose (final concentration 10
Mm) and glutamine (final concentration 4
Mm).

Table 1. Respiration Rates and Glutamine Uptake Rates

	Molt 3	Molt 4	Normal Lymphocytes
Respiration rate ($\mu l\ O_2/h/10^7$ cells)	11.87 ±3.06	1.09 ±0.92	4.91 (bovine)
Preferred substrate	glutamine	glutamine	glutamine
K_m for O_2 (mM)	5.05 x 10^{-5}	3.50 x 10^{-4}	1.78 x 10^{-4}
V_{max} for O_2 ($\mu l\ O_2/hr/10^7$ cells)	0.19	0.11	0.01
Glutamine uptake (nmol gln/min/mg protein)	101 + 1.5	23 + 4.5	1.45 + 5.9 (human)
Crabtree	+	+	-

^{31}P-NMR Since glycolysis generates lactate as a major end product, comparison of transient pH changes brought about by addition glucose to cells could provide information on the glycolytic versus oxidative activities of cells or tissues. To this end we used ^{31}P-NMR to follow changes in pHi associated with the addition of glucose to suspensions of Molt 3, Molt 4 or normal lymphocytes (Figs 2 to 4). Assuming that a fall in pHi correlates with an increase in intracellular lactate (McIntosh et al., 1987), lactate concentration can be quantified. The addition of glucose to a suspension of Molt 4 cells caused a shift in pHi of up to 1.23 pH units (Fig 5), whereas with Molt 3 no such major effect was observed (Fig 6). The observed ΔpH values of 1.23 and 0.28 correspond to 43 mM and 9.8 mM lactate respectively, values which are in the same range as for Ascites tumour cells (Thomas et al., 1979). Intracellular free magnesium levels in the two cell lines were determined from the ^{31}P-NMR chemical shift differences between the α- and β-peaks of ATP. The values estimated for Molt 3 and Molt 4 were 0.4 ±0.09 and 0.25 ±0.06 mM [Mg^{2+}] respectively, which was not statistically significant (students' T-test). A drop in the Molt 4 cell had no significant effect on free magnesium concentration.

CONCLUSIONS

Since the Molt 3 and Molt 4 lines originated from the same patient at the same time, it would be expected that they should display the same gross metabolic characteristics. It is quite evident from this study that they do not, which raised the possibility of heterogeneity of leukemic lymphocytes within the patient from whom the isolation was conducted, or of gross metabolic adaptation by the cell lines in cultures.

REFERENCES

Gupta, R.K., Gupta, P., Yushok, W.D., and Rose, Z.B., 1983, *Physio. Chem. Phys. Med. NMR*, 15: 265-280.
Heldt, H.W., 1980, *Meth. Enzymol.*, 69: 604-613.
Hore, D.J., 1983, *J. Magn. Reson.*, 55: 293-300.
Longmuir, I.S., 1954, *Biochem.*, 57: 81-86.
McCabe, M., Nowak, M., Maguire, D., and Robertson, P., 1984, *Aust. J. Exp. Biol. Med. Sci.*, 62: 539-545.

SHOCK AND WOUND HEALING

WHOLE BODY AND REGIONAL O$_2$ UPTAKE/DELIVERY AND LACTATE FLUX

IN ENDOTOXIC DOGS

S. M. Cain and S. E. Curtis

Departments of Physiology and Biophysics and
Pediatrics, University of Alabama at Birmingham
Birmingham, Alabama 35294, U.S.A.

INTRODUCTION

In many patients critically ill with Adult Respiratory
Distress Syndrome (ARDS) or sepsis, O$_2$ uptake increases if
supply is increased. This occurs even when O$_2$ delivery rates
may exceed O$_2$ demand by more than threefold. Bihari et al.
(1987) suggested that this abnormal O$_2$ supply dependency was
indicative of an occult O$_2$ debt that may be responsible for
the multiorgan failure that so often occurs in these
patients. Additional evidence in favor of this suggestion
was supplied in studies in which arterial blood lactate
levels were measured together with O$_2$ delivery and uptake in
patients with ARDS or sepsis (Astiz et al., 1987; Gilbert et
al., 1986; Haupt et al., 1985; Kaufman et al., 1984). In
all cases in which O$_2$ uptake was observed to increase with O$_2$
delivery, usually by fluid loading, blood lactate levels
decreased. Initial lactate levels were elevated in spite of
the fact that O$_2$ delivery was higher in all cases than the
critical level of 330 ml/min·m^2 reported by Shibutani et al.
(1983) for patients that did not have ARDS or sepsis. The
general assumption was that the elevated lactate was due to
anaerobic metabolism and that it decreased in response to
increased O$_2$ delivery to hypoxic tissues.

Another feature observed in all of the above studies
was that O$_2$ extraction by peripheral tissues appeared to be
unable to increase in spite of apparent tissue hypoxia.
Generally, O$_2$ extraction ratios remained about 30% even while
blood lactate levels were elevated and did not respond to
changes in O$_2$ delivery. This apparent defect in O$_2$ extraction has been suggested to have a multifactorial origin
(Cain, 1986). Prominent among the suggested mechanisms was
an increased diffusion barrier for O$_2$ between the capillary
and the interior of cells. This could result from tissue
injury with consequent endothelial dysfunction and mal-

distribution of blood flow relative to regional O_2 demands.

In the present study, we infused endotoxin into anesthetized dogs to emulate the septic state in patients. Our goals were: 1) to measure O_2 delivery/uptake relationships and lactate fluxes in gut and muscle in context with whole body events to see if there were regional differences; 2) to ascertain whether any regional differences in lactate flux correlated with regional O_2 uptake; and 3) to challenge the ability of tissues to extract O_2 with a lowered inspired O_2 fraction. In all cases, we aggressively treated with donor red blood cells and dextran to maintain cardiac output at or above the levels measured prior to endotoxin infusion. Our goal was not to study the consequences of hypoperfusion in septic shock. Instead, we wished to see whether there was any indication of tissue hypoxia in the face of an adequate O_2 supply or evidence of impaired O_2 extraction.

METHODS

Dogs (n=6) were anesthetized (30 mg/kg pentobarbital sodium iv, supplemented as necessary), paralyzed (30 mg succinylcholine chloride im + 0.1 mg/min iv), and pump-ventilated to maintain PCO_2 between 30 and 35 torr. Catheters were placed in carotid and pulmonary arteries and in the right femoral vein. Venous outflows from the left hindlimb muscles and from a segment of ileum were isolated as previously described (Stork et al., 1989). Both areas were autoperfused and innervated. Regional blood flows were measured every 15 min at the time that arterial and venous blood was sampled. Regional O_2 delivery and uptake were calculated from the results. Whole body O_2 uptake was calculated from analyses and measurement of expired gas volume. Cardiac output was calculated from O_2 uptake and the arteriovenous O_2 content difference. Arterial and regional venous lactate concentrations were measured by electrode (YSI Model 23). Lactate flux across muscle or gut was calculated as the venous-arterial difference times the blood flow.

When all preparations were complete and the animal was stable with respect to blood pressure, cardiac output, and whole body O_2 uptake, the experimental protocol was begun. One set of samples and measurements was taken before beginning infusion of endotoxin (Diffco *E. coli* lipopolysaccharide) at the rate of 33.3 μg/kg·min (for 60 min total dose of 2 mg/kg). After infusion of endotoxin was begun, donor red blood cells and dextran were given in sufficient quantity to maintain cardiac output at or above normal levels with unchanged hematocrit. Sampling continued at 15-min intervals for 2 hr after the infusion was completed but at 1 hr, the inspired gas was changed to a mixture of 12% O_2, balance N_2. After a 30-min period of hypoxia, the inspired gas was restored to room air and measurements continued for another 30 min of recovery when the experiment was concluded. The amounts of perfused muscle and gut were weighed and all values are reported per kg of tissue.

Results were analyzed by repeated measures ANOVA and significant differences were identified by Newman-Keuls' multiple range test. Significance was accepted at $p<0.05$.

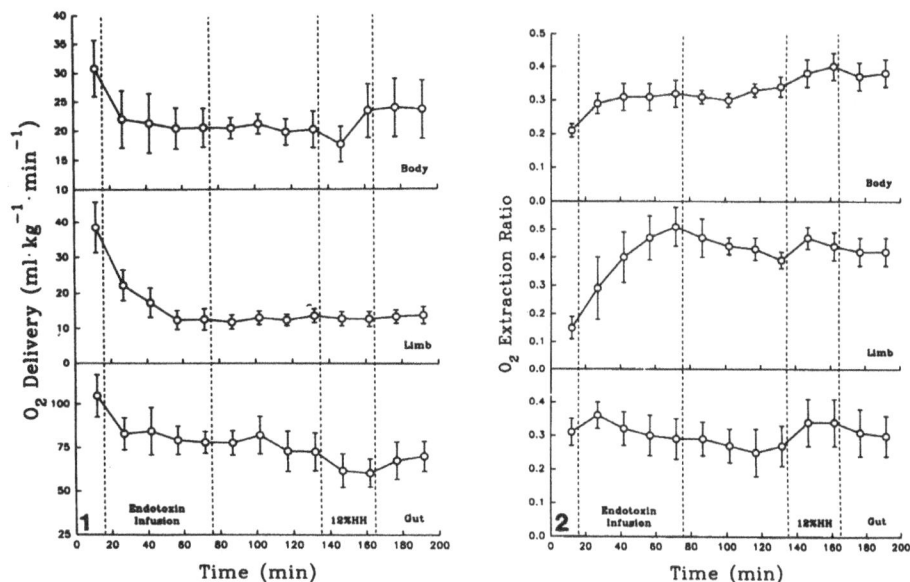

<u>Fig. 1.</u> Total O_2 delivery to whole body, limb, and gut per unit of body weight or organ weight (mean±SE).

<u>Fig. 2.</u> O_2 extraction ratio in whole body, limb, and gut (mean±SE).

RESULTS

We maintained cardiac output at 120 ml·kg^{-1}·min^{-1} or more throughout the experiment. As a result, total O_2 delivery to the whole body (Fig. 1) showed no significant change. This was also true in the gut whereas O_2 delivery per unit weight of muscle did decrease significantly from an initial value that was extraordinarily high. However, muscle O_2 delivery was always greater than 10 ml·kg^{-1}·min^{-1}, even during the period of hypoxic hypoxia (12% HH in Fig. 1) when arterial PO_2 was reduced to 38 torr from its normoxic range of 89 to 61 torr. A significant decrease in arterial O_2 content from 15.4 ml/dl to 13.4 ml/dl during HH was evidently offset by small but important increases in cardiac output and organ blood flows. Average O_2 extraction ratios (Fig. 2), therefore, stayed between 30% and 40% in whole body and gut throughout the experiment and never rose above 50% even in muscle.

Whole body O_2 uptake (Fig. 3) showed a significant upward trend during and after endotoxin infusion. Of interest is the fact that a highly significant increase took place toward the end of the HH period and O_2 uptake then remained elevated, compared to the prehypoxic period, during the last half-hour of normoxic recovery. Limb muscle showed no significant changes in O_2 uptake at any time but gut O_2

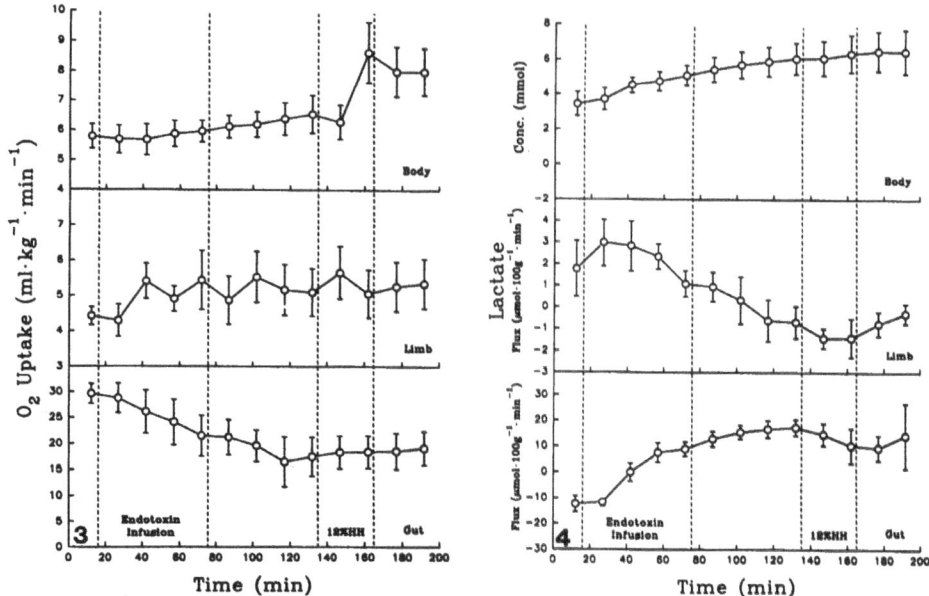

Fig. 3. Whole body O$_2$ uptake per unit body weight and organ O$_2$ uptake per unit organ weight (mean±SE).

Fig. 4. Arterial lactate concentration and lactate fluxes across limb muscle and gut (means±SE).

uptake decreased significantly until the end of the postinfusion period and showed no further change thereafter even with hypoxia.

During the 2-hr period beginning with endotoxin infusion, arterial lactate levels (Fig. 4) significantly increased from 3.4±0.7 to 6.1±0.9 mM/liter. No change was seen with hypoxia or during the subsequent recovery period. Lactate flux across limb muscle decreased during the same period that arterial levels increased but gut converted from taking up lactate to significant output. It too showed no further change with hypoxia.

DISCUSSION

In their study of the effects of vasodilation upon O$_2$ delivery and uptake in critically ill patients, Bihari et al. (1987) concluded that the increase in O$_2$ uptake that occurred with an increase in O$_2$ delivery denoted a substantial O$_2$ debt in 13 patients who subsequently died. All of them had suspected or confirmed sepsis and their death was suggested to be the result of multiple organ failure to which inadequate tissue oxygenation contributed sustantially. Both Haupt et al. (1985) and Gilbert et al. (1986) had earlier found that lactacidosis was a common feature in similar patients who also showed an abnormal dependency of O_2 uptake upon supply. Furthermore, whenever O_2 uptake was raised by raising O_2 delivery, arterial lactate levels were observed to fall. This inverse relationship led to the assumption that elevated lactate levels in septic patients were a sign of tissue O_2 debt and anaerobic metabolism.

We were able to produce arterial lactacidosis in experimental animals infused with endotoxin. The levels that were achieved were consistent with severe tissue hypoxia as judged by other experiments on anesthetized dogs ventilated with hypoxic gas mixtures (Cain, 1969). Based on the relationship described in that paper for arterial lactate increase and net O_2 deficit, the animals infused with endotoxin had an increase in arterial lactate equivalent to a net O_2 deficit of ~40 ml/kg, if the lactate had indeed accrued as a result of tissue hypoxia. Other information, however, raises considerable doubt that lactate increased in arterial blood from that cause.

In their experiments on dogs treated with bolus injections of endotoxin, Nelson et al. (1988) found that whole body O_2 uptake did not become limited until systemic O_2 delivery was lowered to 12.8 $ml \cdot kg^{-1} \cdot min^{-1}$. An even lower critical systemic O_2 delivery of 11.4 $ml \cdot kg^{-1} \cdot min^{-1}$ was found by the same group in another series of experiments (Samsel et al., 1988). Admittedly, these were higher critical O_2 delivery values than those required by control dogs that were not endotoxemic but both were considerably less than the average systemic O_2 delivery of 20 $ml \cdot kg^{-1} \cdot min^{-1}$ that prevailed for most of the experiment in our animals (Fig. 1). Similarly, the O_2 extraction ratio at the critical O_2 delivery was 0.54 in the first study and 0.61 in the second. Average O_2 extraction ratios for the whole body in our experiments never went above 0.40 even with superimposed hypoxia. Based on these results, we conclude that O_2 supply was probably not limiting systemic O_2 uptake in our experiments and, therefore, could not have been responsible for a generalized production of lactate by stimulation of anaerobic metabolism.

Similar information is available for the organ systems that were studied. Bredle et al. (1989) found that the critical O_2 delivery to muscle in endotoxemic dogs did not occur until a value of 8.3 $ml \cdot kg^{-1} \cdot min^{-1}$ was reached or a critical O_2 extraction ratio of 0.60. Muscle O_2 delivery averaged about 12 $ml \cdot kg^{-1} \cdot min^{-1}$ for most of the period of our experiments (Fig. 1) and the average extraction ratio never exceeded 0.50 (Fig. 2). Nelson et al. (1988) found that intestinal O_2 uptake did not become limited by O_2 supply until intestinal O_2 delivery was lowered to 30 $ml \cdot kg^{-1} \cdot min^{-1}$ in their endotoxemic dogs at which point O_2 extraction was 0.47. In our experiments gut O_2 delivery was more than twice that value and O_2 extraction ratio in the gut was generally about 0.30. Again, based on the determination of critical values for O_2 delivery and O_2 extraction ratio made by others in endotoxemic dogs, neither gut nor muscle should have experienced significant tissue hypoxia under the conditions of our experiment. This should have held true even if the tissue's ability to extract O_2 was compromised by the actions of endotoxin. There was no reason to suspect that our animals, which were given a smaller total dose of endotoxin than used in these other studies, should have been in any worse condition.

In the case of muscle, during the hour after the end of endotoxin infusion limb muscle was actually taking up lactate. Average O_2 delivery during this time was, if anything, less than during the infusion when the venoarterial difference in lactate was positive across the limb. Muscle O_2 uptake did not change throughout the experiment. These facts certainly speak against any significant tissue hypoxia in resting skeletal muscle. The picture is less clear in the case of gut.

In spite of well maintained O_2 delivery, gut O_2 uptake fell significantly during the 2-hr period after beginning infusion of endotoxin. During this same time, the gut converted from uptake of lactate to significant production. However, neither O_2 uptake nor lactate output were affected by hypoxic hypoxia in which 5 of the 6 animals experienced a decrease in O_2 delivery to the gut. The decrease in gut O_2 uptake by almost 100% that occurred over the 2-hr period after beginning endotoxin infusion bespeaks a mechanism that may be more specific in gut than in other organ systems. In their discussion of the pathogenesis of nonocclusive ischemic colitis, Bailey et al. (1986) point out the remarkable sensitivity of gut to the renin-angiotensin system. It is likely that the shock phase of endotoxin infusion in which mean arterial blood pressure fell from 140 mmHg to 60 mmHg strongly stimulated the renin-angiotensis system. If the action of angiotensin II on the gut was selective in that flow to the mucosa was reduced even though total flow was maintained, then the well known tendency for mucosal sloughing in canine gut could have taken place. Mucosal necrosis was a notable feature in a porcine model of nonocclusive mesenteric injury (Bailey et al., 1987). We speculate that this occurred and that it is a defense mechanism for the gut by which O_2 demand is lessened. That would still not explain an increase in gut lactate output if the viable tissue was receiving adequate O_2 supply.

Lactate elevation in sepsis is probably the result of a metabolic defect in the interconversion of pyruvate dehydrogenase (PDH) which exists as a complex of active and inactive forms. Sepsis and endotoxin have been shown to cause conversion of PDH to the inactive state with consequent diversion of glucose metabolism into lactate by blocking entry into the tricarboxylic acid cycle (Vary et al., 1986). This can be partially reversed by the action of dichloroacetate so that plasma and tissue lactate concentrations are lowered in sepsis (Vary et al., 1988). Since the only relevant action of dichloroacetate is to activate more PDH, this clearly seems to be a principal reason for lactate elevation with sepsis or endotoxin infusion. Vary et al. (1986) reported a lesser effect of sepsis on inactivation of PDH in skeletal muscle which would explain why we saw no apparent interference with lactate uptake in that tissue in our experiments.

Finally, the use of hypoxic hypoxia to challenge what was supposed to be a marginal delivery of O_2 to the periphery provided some additional useful information. Not only was O_2

delivery apparently adequate, lowering arterial PO$_2$ to 38 torr had no further effect to reduce O$_2$ delivery. There was a significant decrease in arterial O$_2$ content as expected but there was also a compensatory increase in cardiac output that more than offset it. We postulate that the lowered arterial PO$_2$ stimulated sympathetic activity via the carotid chemoreceptors. The increase in cardiac output toward the end of the 30-min period of hypoxia corresponds with similar observations made in dogs that were not infused with endotoxin but were made hypoxic (Cain and Chapler, 1980). The fact that O$_2$ delivery and O$_2$ uptake were maintained even while arterial PO$_2$ was significantly lower would tend to rule against any real increase in diffusion barriers in endotoxemic dogs. Such mechanisms as increased tissue diffusion distance with perfusion heterogeneity and tissue edema have been suggested to explain the apparent defect in the ability of peripheral tissues to extract O$_2$ in septic patients and endotoxin treated animals (Bredle et al., 1989; Haupt et al., 1985; Nelson et al., 1988). If O$_2$ uptake was more diffusion limited after endotoxin infusion, lowering the diffusion gradient for O$_2$ should have had some effect to lower O$_2$ uptake but it did not.

SUMMARY

We infused endotoxin into anesthetized dogs while maintaining cardiac output. Whole body O$_2$ uptake and arterial lactate concentration were measured at the same time with regional O$_2$ uptake and lactate fluxes in muscle and gut. Even though whole body O$_2$ uptake increased, so did arterial lactate levels. The different behavior of organ systems was marked by the fact that muscle took up lactate while its O$_2$ uptake didn't change whereas gut produced lactate as it decreased its O$_2$ uptake. Comparison with critical levels of O$_2$ delivery and O$_2$ extraction ratio for whole body and the two regions opened considerable doubt that generalized and significant hypoxia explained the rise in arterial lactate. Addition of mild hypoxic hypoxia also neither caused O$_2$ uptake to fall nor lactate to rise which reinforced our conclusion that there was little or no tissue hypoxia in endotoxin treated animals given adequate resuscitation with red blood cells and colloid solutions.

ACKNOWLEDGEMENTS

The authors thank W. E. Bradley for his excellent technical assistance in the conduct of these experiments. Funds were furnished by grants from National Institutes of Health (Grant #HL 26927) and from Fisons Pharmaceuticals plc.

REFERENCES

Astiz, M. E., Rackow, E.C., Falk, J. L., Kaufman, B.S., and Weil, M. H., 1987, Oxygen delivery and consumption in patients with hyperdynamic septic shock, Crit. Care Med. 15:26.
Bailey, R.W., Hamilton, H.R., Morris, J.B., Bulkley, G.B., and Smith, G.W., 1986, Pathogenesis of nonocclusive ischemic colitis, Ann. Surg. 203:590.

Bailey, R.W., Bulkley, G.B., Hamilton, S.R., Morris, J.B., and Haglund, U.H., 1987, Protection of the small intestine from nonocclusive mesenteric ischemic injury due to cardiogenic shock, Am. J. Surg. 153:108.

Bihari, D., Smithies, M., Gimson, A., and Tinker, J., 1987, The effects of vasodilation with prostacyclin on oxygen delivery and uptake in critically ill patients, N. Eng. J. Med. 317:397.

Bredle, D. L., Samsel, R. W., Schumacher, P.T., and Cain, S. M., 1989, Critical O_2 delivery to skeletal muscle at high and low PO_2 in endotoxemic dogs, J. Appl. Physiol. 66:2553.

Cain, S. M., 1969, Diminution of lactate rise during hypoxia by PCO_2 and β-adrenergic blockade, Am. J. Physiol. 217:110.

Cain, S. M., 1986, Assessment of tissue oxygenation, Crit. Care Clin. 2:537.

Cain, S.M. and Chapler, C.K., 1980, O_2 extraction by canine hindlimb during α-adrenergic blockade and hypoxic hypoxia, J. Appl. Physiol. 48:630.

Gilbert, E.M., Haupt, M.T., Mandanas, R.Y., Huaringa, A.J., and Carlson, R.W., 1986, The effect of fluid loading, blood transfusion, and catecholamine infusion on oxygen delivery and consumption in patients with sepsis, Am. Rev. Resp. Dis. 134:873.

Haupt, M.T., Gilbert, E.M., and Carlson, R.W., 1985, Fluid loading increases oxygen consumption in septic patients with lactic acidosis, Am. Rev. Resp. Dis. 131:912.

Kaufman, B.S., Rackow, E.C., and Falk, J.L., 1984, The relationship between oxygen delivery and consumption during fluid resuscitation of hypovolemic and septic shock, Chest 85:336.

Nelson, D.P., Samsel, R.W., Wood, L.D.H., and Schumacker, P.T., 1988, Pathological supply dependence of systemic and intestinal O_2 uptake during endotoxemia, J. Appl. Physiol. 64:2410.

Samsel, R.W., Nelson, D.P., Sanders, W.M., Wood, L.D.H., and Schumacker, P.T., 1988, Effect of endotoxin on systemic and skeletal muscle O_2 extraction, J. Appl. Physiol. 65:1377.

Shibutani, K., Komatsu, T., Kubal, K., Sanchala, V., Kumar, V., and Bizzari, D.V., 1983, Critical level of oxygen delivery in anesthetized man, Crit. Care Med. 11:640.

Stork, R.L., Dodd, S.L., Chapler, C.K., and Cain, S.M., 1989, Regional hemodynamic responses to hypoxia and hypermetabolism in polycythemic dogs, J. Appl. Physiol. 67:96.

Vary, T.C., Siegel, J.H., Nakatani, T., Sato, T., and Aoyama, H., 1986, Effect of sepsis on activity of pyruvate dehydrogenase complex in skeletal muscle and liver, Am. J. Physiol. 250 (Endocrinol. Metab. 13):E364.

Vary, T.C., Siegel, J.H., Tall, B.D., and Morris, J.G., 1988, Metabolic effects of partial reversal of pyruvate dehydrogenase activity by dichloroacetate in sepsis, Circ. Shock 24:3.

RESUSCITATION FLUIDS AND OXYGEN TRANSPORT IN HAEMORRHAGIC SHOCK

M.Y. Rady, R.A. Little, E. Kirkman, and
*N.S. Faithfull (Corresponding author)

Northwestern Injury Research Centre,
University of Manchester
*Delta Biotechnology Ltd., Nottingham, U.K.

Haemorrhagic shock may be treated with infusion of blood, simple crystalloid solutions or a number of colloid preparations. Crucial to successful outcome is restoration of cardiovascular function and maintenance of oxygen transport/consumption relationships (Shoemaker et al., 1988). This study examined these effects using Human Albumin and Gelatin solutions in a porcine haemorrhagic shock model.

METHODS

Immature Large White pigs, (body weight 17 to 25 kg) anaesthetised with intramuscular Ketamine hydrochloride (20 mg/kg) were ventilated with 50 percent oxygen, nitrous oxide and isoflurane to maintain arterial carbon dioxide tensions between 30 and 35 mm Hg. Body temperature was maintained at 38.5°C (pulmonary artery) by a heating lamp, a heated operating table and pre- warmed infusion fluids.

Aortic pressure was monitored via a cannula advanced from the left femoral artery. Central venous pressure, pulmonary arterial and pulmonary capillary wedge pressures were obtained using a 7 French gauge Swan Ganz catheter inserted via the right femoral vein. Pressures were transduced using Gould Statham P23BB pressure transducers and displayed on a 6 channel M19 paper recorder (Devices, England). Cardiac output was obtained from an American Edwards COM1 cardiac output computer using thermodilution with 10 ml of 5% dextrose at 4°C. Heart rates were derived from the electrocardiogram obtained from thoracic cutaneous electrodes.

Oxygen Transport to Tissue XIII, Edited by T.K. Goldstick *et al.*
Plenum Press, New York, 1992

Arterial and mixed venous blood gas and acid base status were measured using a Radiometer ABL 330 blood gas analyser; haemoglobin and oxyhaemoglobin saturations were obtained from a Radiometer OSM3 Hemoximeter. Haematocrit was measured in capillary tubes centrifuged at 10,000 G for 6 minutes.

The right femoral artery was canulated and blood was withdrawn at a rate of 1 percent of estimated blood volume per minute for 40 minutes; blood volume was estimated at 75 ml/kg (Engelhardt, 1966). After thirty minutes of haemorrhagic shock the animals were resuscitated with either 4.5% Human Albumin Solution (Immuno) or 4% Modified Fluid Gelatin (Hausman) infused at a rate of 2 ml/kg/min to a total volume of 45 ml/kg. The extra volume, amounting to 20% of blood volume was administered to mimic the clinical situation and to "drive" the cardiac output in order to obtain optimal oxygen transport dynamics.

Derived cardiovascular and oxygen transport variables were calculated at control time (before the onset of haemorrhage), the end of haemorrhage, 30 minutes after the end of haemorrhage, at the completion of resuscitation and 30, 60, 90 and 120 minutes thereafter. Statistical analysis was performed on untransformed data using paired or unpaired Student t tests as appropriate. The nul hypothesis was rejected at a p value of <0.05.

RESULTS

All animals survived to two hours after the end of resuscitation. During shock mean systemic arterial pressure (MAP) was 33 percent reduced (Fig 1) and cardiac index (CI - in ml/kg) decreased by more than 50 percent (Fig 2). Oxygen delivery (DO2), calculated as the product of arterial oxygen content and CI, decreased by 50 percent (Fig 3). Oxygen consumption (VO2) was not decreased (Table 1).

Gelatin initially restored MAP (Fig 1) but two hours after the end of resuscitation it was significantly lower than before bleeding. Albumin restored and maintained MAP during the whole of the post resuscitation protocol. Pressures were significantly greater than in the Gelatin group at 60 and 90 minutes post resuscitation. At 120 minutes pressures in the Albumin group appeared to be decreasing.

TABLE 1. Changes in Oxygen Consumption Index during Shock and
Resuscitation

	Cont	B Sho	E Sho	E Res	30 min	60 min	90 min	120 min
ALBUMIN	6.1	7.2	7.1	6.0	5.8	5.9	5.8	6.0
	0.39	0.31	0.42	0.23	0.30	0.38	0.27	0.39
	(7)	(7)	(7)	(7)	(7)	(7)	(7)	(7)
GELATIN	6.6	5.9	6.7	6.2	6.3	6.4	6.2	6.0
	0.43	0.91	0.62	0.32	0.35	0.31	0.37	0.46
	(7)	(7)	(7)	(7)	(7)	(7)	(7)	(7)

Values for Oxygen Consumption Index at control time (Cont), at the beginning of
shock after removal of 40% of blood volume (B Sho), after 30 minutes of shock
(E Sho), at the end of resuscitation (E Res) and 30, 60, 90 and 120 minutes
thereafter. Figures given are means and standard error. The number of
measurements is in parenthesis.

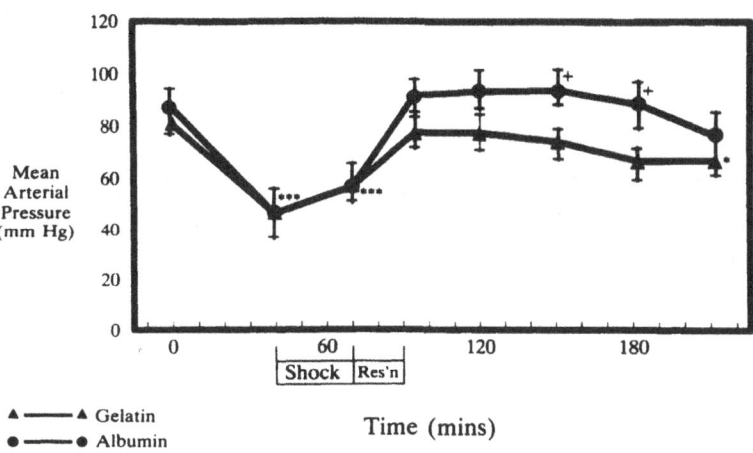

Fig. 1. Effects of haemorrhage and resuscitation on Mean Arterial Pressure in
animals resusciatated with Albumin (circles) or gelatin (triangles).
Significant differences from pre haemorrhagic values are indicated by stars: *
= p <0.05; ** = p <0.01; *** = p <0.001. Significant differences between the
groups is indicated by crosses: + = p <0.05; ++ = p <0.01;.

Both Gelatin and Albumin increased CI 60 to 70 percent above control values (Fig 2). In the Gelatin group CI rapidly decreased and by one hour post-resuscitation it was no longer significantly raised above pre-haemorrhage levels. In contrast Albumin resuscitation maintained CI significantly above control levels and output was significantly greater than in the Gelatin group at 90 and 120 minutes post resuscitation.

The effects of resuscitation on DO2 were strikingly different with the two fluids. Gelatin never restored DO2 to control values whereas at the end of

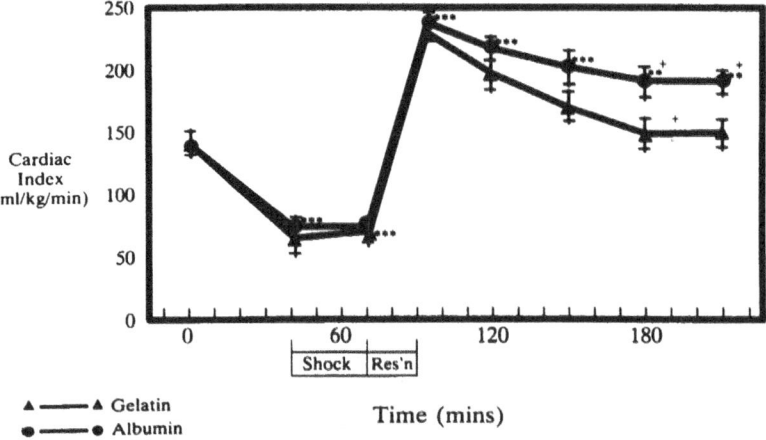

Fig. 2. Effects of haemorrhage and resuscitation on Cardiac Index in animals resuscitated with Albumin (circles) or gelatin (triangles). Significant differences from pre haemorrhagic values are indicated by stars: * = p <0.05; ** = p <0.01; *** = p <0.001. Significant differences between the groups is indicated by crosses: + = p <0.05; ++ = p <0.01.

resuscitation Albumin had restored DO2 to normal. Oxygen transport subsequently fell in both groups but Albumin values were always the higher - significantly so at 30, 90 and 120 minutes. At the end of the protocol DO2 was 56 percent of control in the Gelatin group, whereas in the Albumin group it was at 72 percent of pre-haemorrhage values.

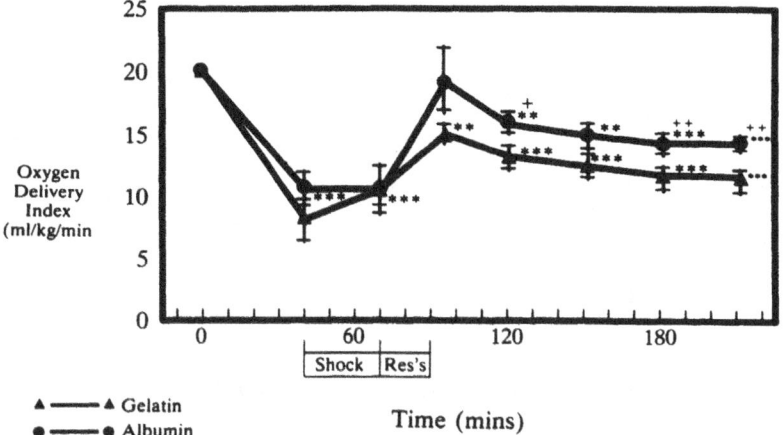

Oxygen Delivery Index (ml/kg/min)

▲ ———— ▲ Gelatin
● ———— ● Albumin

Time (mins)

Fig. 3. **Effects of haemorrhage and resuscitation on Oxygen Delivery in animals resuscitated with Albumin (circles) or Gelatin (triangles). Significant differences from pre haemorrhagic values are indicated by stars: * = p <0.05; ** = p <0.01; *** = p <0.001. Significant differences between the groups is indicated by crosses + = p <0.05; ++ = p <0.01;.**

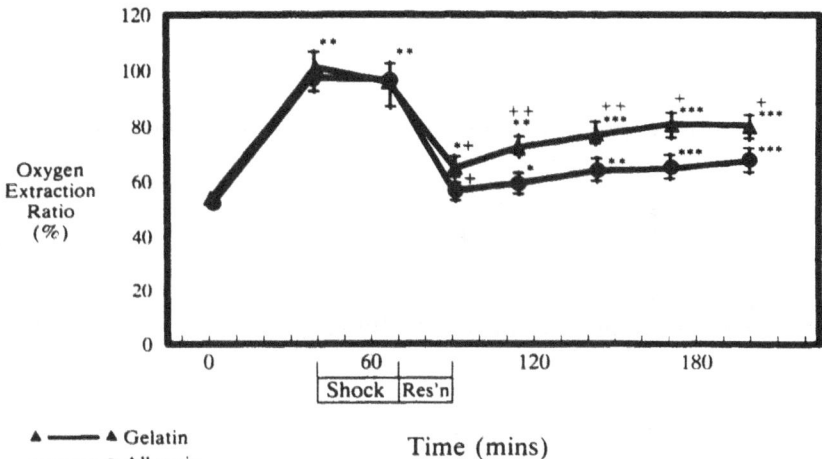

Oxygen Extraction Ratio (%)

▲ ———— ▲ Gelatin
● ———— ● Albumin

Time (mins)

Fig. 4. Effects of haemorrhage and resuscitation on Oxygen Extraction Ratios in animals resuscitated with Albumin (circles) or Gelatin (triangles). Significant differences from pre haemorrhagic values are indicated by stars: * = p <0.05; ** = p <0.01; *** = p <0.001. Significant differences between the groups is indicated by crosses: + = p <0.05; ++ = p <0.01;.

413

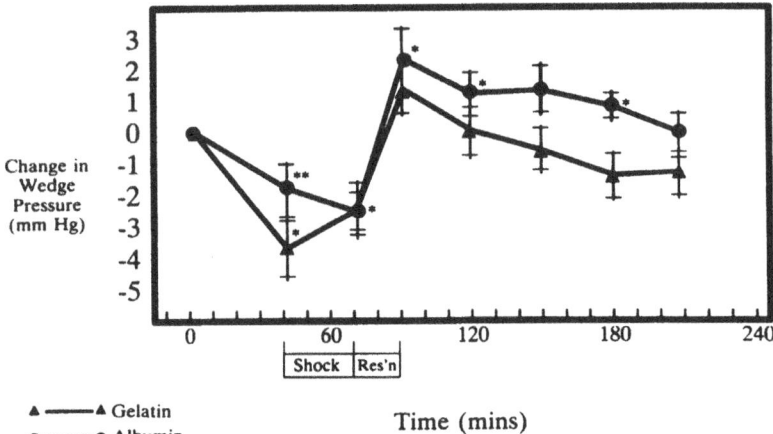

Fig. 5. Effects of haemorrhage and resuscitation on change in Pulmonary
Capillary Wedge Pressure in animals resuscitated with Albumin (circles)
or Gelatin (triangles). Significant differences from pre haemorrhagic
values are indicated by stars: * = p <0.05; ** = p <0.01; *** = p
<0.001.

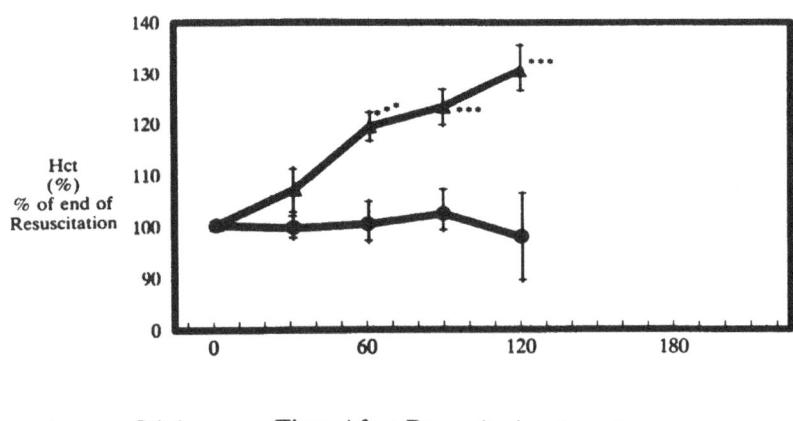

Fig 6. Haematocrit values expressed as a percentage of values at the end of
resuscitation in animals resuscitated with Albumin (circles) or Gelatin
(triangles). Significant differences from end of resuscitation are
indicated by stars: * = p <0.05; ** = p <0.01; *** = p <0.001.

Oxygen Extraction Ratios (OER) rose from 30 percent to 65 percent during haemorrhagic shock (Fig 4). Gelatin resuscitation never returned values to normal; following resuscitation with Albumin they were normal; though they subsequently rose they remained lower than in the Gelatin group. At the end of resuscitation mean values of OER in the latter group were 10 percent higher than in the Albumin group.

DISCUSSION

It is plain from the above results that, following resuscitation from haemorrhagic shock using non oxygen-transporting plasma substitutes, maintenance of adequate DO2 and acceptable levels of OER are intimately linked to the ability of the body to maintain increased CI.

Assuming the maintenance of normal myocardial contactility, the CI is determined by the preload and afterload imposed on the left ventricle - the pulmonary capillary wedge pressure and systemic vascular resistance respectively (SVR). There were no statistically significant differences in SVR between the groups at any time with the exception of a significantly lower SVR in the Gelatin group at the end of resuscitation. This was reflected by a significantly lower Haematocrit (Hct) in this group at this time and may hence be explained in terms of altered blood viscosity.

Changes in wedge pressure from control valves (dWP) are plotted in Fig 5. Both Gelatine and Albumin produced increases in wedge pressure, although the increase achieved statistical significance only in the Albumin treated group.

Following resuscitation with either Albumin or Gelatin there was a significant increase in cardiac index, presumably as a result of the increase in blood volume, due firstly to the administered fluid and additionally, in the case of Gelatin, to the sequestration of fluid from the extravascular compartment (Hint, 1968). Subsequently CI gradually decreased in both groups of animals. However, this post-resuscitation decrease in CI was seen to be more rapid in the Gelatin group than in the Albumin group. Why is this so?

Fig 6 shows Hct as a percentage of that existing at the end of resuscitation. There were no changes in Hct in the Albumin group over the 2 hours following resuscitation, but Hct was significantly increased in the Gelatin group by 60 minutes after resuscitation; Hct had increased

30.5% by 2 hours. Assuming no loss of red cells from the circulation, this represents a 23.4% decrease in blood volume; 60 percent of blood volume had been administered as Gelatin solution. Blood loss was unlikely to have been significant due to meticulous haemostasis and lack of heparinisation. Assuming linear elimination we have calculated its effective half life in this model (based on calculation for each animal) as 3 hours 1 minute (range 1 hour 54 minutes to 4 hours 11 minutes). The mean figure obtained is less that the figure of 4 hours in humans given by Lundsgaard-Hansen and Tschirren (1981) and agrees better with that of 2-3 hours given by Doenicke et al (1977).

Given the Hct values at the beginning and end of resuscitation we can calculate the change in blood volume taking place during resuscitation. The percentage change in blood volume was taken to be:

$$((Hctb/Hcte) \times 100) - 100$$

where Hctb is Hct at the beginning of resuscitation and Hcte is Hct at the end of resuscitation. The change in blood volume on resuscitation in the Gelatin group (154 percent - 7.5 SEM) was significantly more than in the Albumin group (127 percent - 10.0 SEM). However, despite the greater effect of Gelatin on blood volume, it was not able to achieve as great an increase in CI and hence oxygen delivery as that achieved by Albumin.

During severe haemorrhagic shock selective vasoconstriction occurs with sparing of blood flow to myocardial and cerebral tissue and severe reduction to flow to 'non-vital organs' in the splanchnic area. Constriction is most pronounced in the precapillary arterioles and this leads to stasis (Messmer and Kreimeier, 1989). Granulocytes adhere to the endothelial cells and, during resuscitation, oxygen radicals may be formed damaging arteriolar reactivity and decreasing microvascular tone.

It has been suggested that albumin can be used therapeutically to bind inflammatory toxins and as a scavenger of free radicals (Emerson, 1989). In support of this polymorphonuclear luminol-dependant chemiluminescence is decreased by more than 80 percent in the presence of 4 percent Albumin, indicating a marked degree of oxygen radical scavenging (Holt et al, 1984). Wasil et al (1989) confirmed this anti-oxidant effect and demonstrated Albumin protection against myeloperoxidase-derived hypochlorous acid inactivation of the anti-elastase activity of human alpha-1-antiprotease. Albumin can also decrease myocardial hydrogen peroxide levels and increase myocardial contractility in an isolated rat heart reperfusion model (Brown et al, 1989).

We may serendipidously be seeing antioxidant effects in the shock model used in these studies. When used for resuscitation of severe haemorrhagic shock Albumin may, by reason of its antioxidant activity, be able to prevent loss of microvascular reactivity and tone. Whereas the increase in blood volume produced by Gelatin is largely sequestrated in the damaged microvasculation, that provided by Albumin is available for increasing cardiac output.

In conclusion we have shown that resuscitation of haemorrhagic shock with Albumin can better maintain CI and DO2 than can Gelatin. This results in higher OERs when Gelatin is used. This is reflected in lower mixed venous oxygen tensions and will result in lower tissue oxygen tensions, particularly in those organ system with a high oxygen consumption. We have suggested that oxygen radical generation may be partially responsible for these effects.

SUMMARY

This study examined the cardiovascular and oxygen transport/consumption relationships in haemorrhagic shock and resuscitation using Human Albumin and Gelatin solutions in a porcine haemorrhagic shock model. Immature pigs (17-25kg) were anaesthetised and ventilated with nitrous oxide, oxygen and isoflurane. Aortic pressure monitoring and monitoring of pulmonary pressures and cardiac output via a Swan-Ganz catheter were instituted. A controlled haemorrhage of 40% of blood volume at 1% blood volume per minute was followed 30 minutes later by resuscitation with 60% of blood volume of 4.5% Human Albumin (Immuno) or 4% Modified Fluid Gelatin (Gelofusine).

Haemorrhage caused marked decreases in (CI), mean arterial pressure and oxygen delivery (DO2). Oxygen consumption (VO2) remained unchanged. Resuscitation with Albumin and Gelatin raised CI to values 60% to 70% above control though only Albumin restored DO2. Deterioration was more rapid in the Gelatin group due to the estimated half a life of 3 hours in this model. Oxygen extraction was 10% higher in the Gelatin group 2 hours after resuscitation. The role of oxygen radicals in haemorrhagic shock is discussed.

Acknowledgment

Shace, M.Y. Rady is a research fellow funded by Delta Biotechnology Ltd.

REFERENCES

Brown, J.M. Beehler, C.J., Berger, E.M., Grosso, M.A., Whitman, G.J., Terada, L.S., Leff, J.A., Hraken, A.H. and Repine, J.E., 1989. Albumin Decreases Hydrogen Peroxide and Reperfusion Injury in Isolated Rat Hearts. Inflammation, 13:583.

Doenicke, A., Grote, B. and Lorenz, W., 1977. Blood and Blood Substitutes. Br. J. Anaesth, 49:681.

Englehardt, W., 1966. Swine : Cardiovascular Physiology - a Review. In: Swine Biomedical Research. Bustad, L.K., McClellan, R.O., eds. Frayn, Seattle: 307.

Hint, H., 1968. The Pharmacology of Dextran and the Physiological Background for the Clinical Use of Rheomacrodex and Macrodex. Acta Anaesth Belg. 2:119.

Holt, M.E., Ryall, M.E.T., and Campbell, A.K., 1984. Albumin Inhibits Human Polymorphonuclear Leucocyte Luminol-dependant Chemiluminescence: Evidence for Oxygen Radical Scavenging. Br. J. Exp. Path., 65: 231.

Lundsgaard-Hansen, P. and Tschirren, B., 1981. Clinical Experience with 120,000 Units of Modified Fluid Gelatin. Develop biol Standard, 48:251. S. Karger, Basel.

Messmer, K. and Kreimeier, U., 1989. Microciculatory Therapy in Shock. Resuscitation, 18, Suppl: S51.

Shoemaker, W.C., Appel, P.L., Kram, H.B., Waxman, K., Lee, T., 1988. Prospective Trial of Supranormal Values of Survivors as Therapuetic Goals in High Risk Surgical Patients. Chest, 94:1176.

Wasil, M., Halliwell, B., Hutchinson, D.C.S. and Baum, H., 1987. The Antioxidant Action of Human Extracellular Fluids. Biochem. J., 243:219.

IMPACT OF ISCHEMIA ON TISSUE OXYGENATION AND WOUND HEALING:

IMPROVEMENT BY VASOACTIVE MEDICATION

M. Kamler, H.A. Lehr, R.K. Saetzler, T. J. Galla, K. Messmer

Department of Experimental Surgery,
University of Heidelberg, FRG

INTRODUCTION

Wound healing is a complicated process, accomplished by a variety of interrelated metabolic pathways. Disturbances of any of these aspects will result in impaired wound healing. Besides generalized diseases like diabetes mellitus and immunodeficiencies, systemic or local infections can jeopardize the normal wound healing process. Finally, chronic ischemia due to inadequate tissue perfusion and oxygenation can have protracting or deletary effects on the process of wound healing. Inadequate tissue oxygenation results from a discrepancy between oxygen demand and oxygen supply to the impaired tissue. These pathologic conditions include both acute states of ischemia following trauma as well as chronic ischemia in peripheral arterial disease and chronic venous insufficiency. A variety of therapeutic interventions are used to improve the perfusion and oxygenation of the endangered tissue. Elimination of risk factors like cigarette smoking and hypercholesteremia, physical exercise, surgical interventions (bypass, sympathectomy etc.) and finally pharmacological approaches have been studied, with the aim to achieve increased tissue perfusion and improvement of the rheological properties of the blood.

We have developed an animal model which allows to investigate several aspects of wound healing in normal and ischemic tissue and to evaluate the effect of vasoactive drugs on tissue oxygenation and wound healing in ischemic tissue.

MATERIALS AND METHODS

Animal model

Six to twelve weeks old hairless mice, fed standard laboratory chow and water ad libitum, were used for this study (Erikson et al. 1980, Barker et al. 1989). Due to genetic determination (homozygous hr/hr) these animals loose their hair approximately 10 days after birth. This facilitates intravital microscopic investigation of the skin microcirculation. The mice's ears were choosen for investigation since they have a uniform nutritive perfusion by 3 main vessel bundels entering the ear base.

Intravital micoscroscopy

For intravital microscopy the awake animals were immobilized on a plexiglas restrainer and the ears were exposed on a cover slip by three loops of polypropylene thread (fig.1).

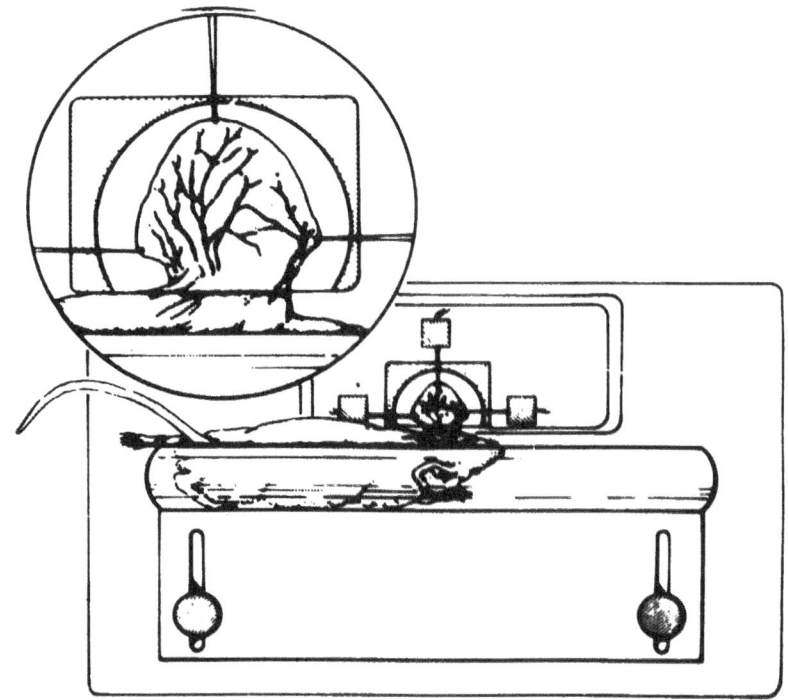

Figure 1. Awake hairless mouse on plexiglas restrainer

For contrast enhancement, the plasma was stained in vivo with fluorescein-isothiocyanate-conjugated dextran (M_r 150000), injected into the tail veins of the animals. The restrainer was then placed on a microscope stage controlled by two computer assisted stepping motors. This allowed repeated measurements on identical observation sites at consecutive time points over the entire experimental period. A saltwater immersion objective (x25) was used for the determination of the microvascular parameters vessel diameter, red blood cell velocity and microvessel density. The microscopic pictures (x560) were recorded on video tape and analysed off-line using a computer assisted micocirculation analysis system (Zeintl et al. 1985). Blood flow in arterioles surrounding the wound was calcutated from red blood cell velocity and vessel diameter according to Baker and Wayland (1974). Baseline measurements of microhemo-dynamic parameters were performed in 8 animals prior to induction of ischemia and follow-up measurements were repeated at 3 day intervals starting 2 days after ischemia induction until the end of the experiment.

Tissue ischemia

Tissue ischemia was induced by ligating 2 of the 3 main vessel bundels; the reduction of blood flow was controlled by direct transillumination under a photomacroscope. To verify ischemia, $tcpO_2$ was measured with a membrane-covered multiwire PO_2 electrode. For that purpose the ears were heated up to 42^0 C. The electrode was moved on the tissue 12-14 times in order to obtain a total of 90-120 single PO_2-values (Kessler, 1981) within the tissue area of app. 10-15mm^2 which later was used for wound creation. Baseline $tcpO_2$ was assessed prior to induction of ischemia and at 3 day intervals starting 2 days after ligation until the end of the experiment.

Wound healing

Circular wounds (ϕ2.5mm) were created in the ears of 20 animals by sharp excision of skin and subcutaneous tissue using microsurgical instruments and an operation microscope. Wound surface area was assessed by digital planimetry at 3 day intervals following wound creation. Concomittantly the microhemodynamic parameters vessel diameter, red blood cell velocity and microvessel density were assessed in the circular edge of the wounds.

Vasoactive medication

All experiments were performed on untreated control animals and on animals, treated with the vasoactive drug Buflomedil (kindly obtained from Abbott/ Wiesbaden, FRG) at an i.v. dose of 3 mg/kg/day given 2 days after induction of ischemia or 1 hour after wound creation, respectively. This dose of Buflomedil elicited no changes of macro- and micro-hemodynamic parameters (Barker et al. 1988).

Statistics

The data were tested for normal distribution using the w-test of Shapiro and Wilk, the hypothesis of normally distributed data being rejected at a level of $p < 0.01$. The results of every experiment showed skewed distribution. They are expressed as the median and upper/lower quartiles. Differences between the groups were tested using the non-parametric Wilcoxon signed rank test. In all cases, statistical significance was accepted at a level of $p < 0.05$.

RESULTS

Verification of ischemia

Two days after induction of ischemia arterial blood flow was reduced to 51% (Q_1 29%, Q_3 73%) of preischemic baseline values and subsequently remained at that level throughout the entire follow-up period. $TcpO_2$ fell from 24 mmHg to 6

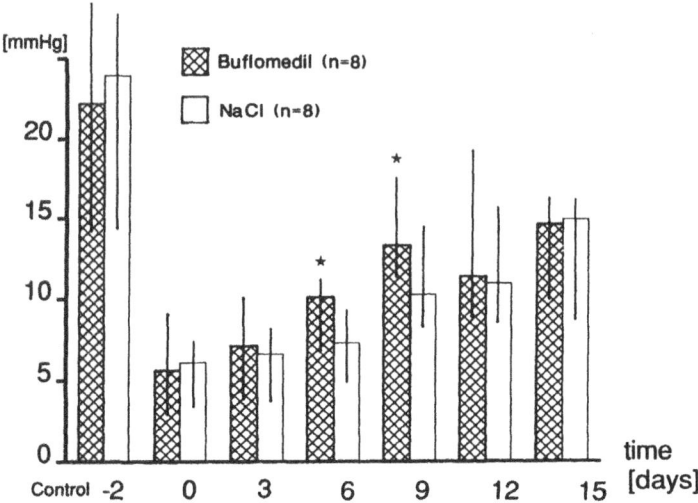

Figure 2. tcpO2 in ischemic tissue (med/Q1,Q3; *p<0.05)

mmHg and slowly recovered to reach 12 mmHg at day 12 after vessel ligation (fig. 2). While no effect of Buflomedil was seen on the initial drop, $_tcpO_2$ at days 6 and 9 days was significantly higher in Buflomedil treated animals. Thereafter no difference was observed between the groups.

Wound healing in ischemic tissue

Immediately after creation, the wounds measured approximately 5mm². In ischemic ears the wound surface area was constantly reduced during the healing process by epithelisation and complete closure was observed 15 days after wound creation (fig. 3). This time for total wound closure was significantly prolonged when compared to wound closure in non-ischemic control ears (Bondar et al. 1990). In Buflomedil treated animals wound healing was accelerated in ischemic tissue 6 days after wound creation. As a consequence, the wounds in Buflomedil treated animals were closed already on day 12 after wound creation. This correlates well with the recovery of tissue oxygenation (fig 2). No correlation was seen between wound healing and vessel diameters or red cell velocities, which remained unaffected by Buflomedil therapy. In particular, no changes were found in arterial blood flow or microvessel density in the circular edge of the wound.

DISCUSSION

Animal model

The developed model allows for quantitative evaluation of the effect of ischemia on oxygenation of the ear tissue of

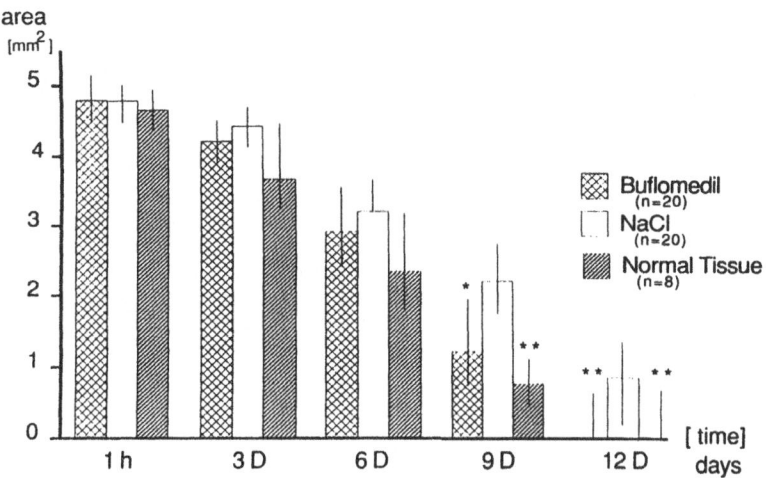

Figure 3. Wound surface area (med/Q1,Q3; *p<0.05 **p<0.01)

the hairless mouse, as well as on healing of standardized wounds. Oxygen tension fell from 24 mmHg before induction of ischemia to 6 mmHg at day 2 after ligation (fig.2). During the course of the observation period, the oxygen pressure in the ear tissue slowly recovered to 15 mmHg. This effect can be ascribed to the formation of collaterals between the main vascular supply routes spread over the entire ear.

After wound creation, wounds were allowed to reepithelize and close without any further manipulation. This fact constitutes one of the main advantages of this model. Any effects of foreign bodies and tissue traumatization in the vicinity of the wound are excluded.

Vasoactive medication

After the initial drop in $tcpO_2$ 2 days after induction of ischemia, $tcpO_2$ values slowly recovered in the course of the experiment. In Buflomedil treated animals, the increase in $tcpO_2$ 6-9 days after induction of ischemia was significantly enhanced as compared to control animals (fig.2). Consecutively, wound healing was accelerated in Buflomedil treated animals (fig.3). These changes were not paralleled by changes in arteriolar blood flow, which was left unaffected by Buflomedil. These results are in accordance with two known mechanisms of action of Buflomedil. Buflomedil improves the formation and perfusion of collaterals by decreasing the capillary resistance and was shown to reduce the oxygen demand, thereby optimizing the supply/demand ratio (Glissold 1987)). Both mechanisms would result in improved tissue oxygenation without affecting microhemodynamic parameters and therefore are in agreement with the observation made in our experiment.

SUMMARY

The influence of vasoactive mediaction on tissue oxygenation and wound healing was investigated in the ear model
of the hairless mouse. Ischemia was induced to the ears by
ligating 2 of the 3 main vessel bundels and verified by
measurements of $tcpO_2$. Reduced tissue oxygenation was followed by a prolongation of the time required for complete
healing of standardized wounds. Treatment with the vasoactive drug Buflomedil (3mg/kg/day iv.) resulted in enhanced
recovery of the tissue from reduced oxygenation and likewise reversed the adverse effects of ischemia on wound
healing. These results warrant the use of the drug in
patients suffering from delayed wound healing due to peripheral arterial disease.

REFERENCES

Baker M., Wayland H., 1974, On-line volumetric flow rate and
velocity profile measurements for blood in microvessels,
Microvasc.Res., 7:131-143.

Barker J.H., Menger M.D., Sack F.U., Messmer K., 1988,
Nutritive Hautdurchblutung bei partieller Ischämie: Wirkung
verschiedener vasoaktiver Substanzen. Vasa, 17:37-41.

Barker J.H., Hammersen F., Bondar I., Uhl E., Galla T.J.,
Menger M.D. Messmer K., 1989, The hairless mouse ear for in
vivo studies of skin microcirculation, Plast.Reconstr.Surg.,
83:948-959.

Bondar I., Barker J.H., Galla T.J., Uhl E., Hammersen F.,
Messmer K., A novel model for quantitative analysis of
microcirculatory changes throughout the process of wound
healing, (submitted)

Clissold S.P., Lynch S., Sorkin E.M., 1987, Buflomedil - A
review on its pharmacological and therapeutical efficacy in
peripheral and cerebral vascular diseases, Drugs 33:430-460.

Erikson E., Boykin J.V., Pittman R.N., 1980, Method for in
vivo microscopy of the cutaneous microcirculation of the
hairless mouse ear, Microvasc.Res., 19:374-379.

Kessler M., 1981, Grundlegende Prinzipien der Sauerstoff-
versorgung des Gewebes, in: "Mikrozirkulation und arterielle
Verschlußkrankheiten", K. Messmer, ed., Karger Verlag,Basel.

Zeintl H., Tompkins, W.R., Messmer, K., Intaglietta, M.,
1986, Static and dynamic microcirculatory video image
analysis applied to clinical investigations, Prog. Appl.
Microcirc. 11:1-10.

OTHER ORGANS AND TISSUES

PAF-ACETHER INDUCED ARTERIAL THROMBOSIS AND THE EFFECT OF

SPECIFIC ANTAGONISTS

R. H. Bourgain[1], R. Andries[1], A. Esanu[2] and P. Braquet[2]

[1]Laboratory of Physiology and Physiopathology, University of Brussels VUB, Laarbeeklaan 103, B-1090 Brussel, Belgium
[2]Henri Beaufour Institute, 17 avenue Descartes, F-92350 Le Plessis Robinson, France

ABSTRACT

Platelet-vessel wall interactions and local thrombosis are investigated *in vivo* in a branch of the mesenteric artery of the guinea pig, using optoelectronic registration and ultrastructural control. Following an electrical challenge resulting in changes of cell membrane polarization, subsequent superfusion by PAF-acether or a stable analogue, (1-O-alkyl-2-N-methylcarbamyl-sn-glycero-3-phosphocholine, 10^{-8} M focal concentration (f.c.)) for a restricted period results in endothelial cell retraction and bleb formation followed by platelet adhesion and the development of a thrombus which over time becomes invaded by leukocytes and eventually occludes the vascular lumen. It was demonstrated in a previous investigation that these phenomena are triggered by the generation of endogenous PAF-acether by the endothelial cells.

Specific PAF-acether-antagonists, such as BN 52021 a ginkgolide, but also synthetic molecules, derivatives of the triazolo-pyridino-diazepine group (BN 50727, BN 50755 and BN 50789), significantly inhibit platelet-vessel wall interactions and thrombosis, but not the formation of blebs in the endothelial cells. Hydrogen peroxide (10^{-5} M f.c.) not only primes the effect of PAF-acether, but is by itself capable of inducing thrombosis through a PAF-acether-mediated mechanism.

Inhibition of acetyl hydrolase by PMSF (phenyl-methyl-sulfonyl-fluoride, 10^{-5} M f.c.) invariably results in a significant enhancement of thrombosis, while conversely, inhibition of acetyl transferase by 27584 RP (4-(naphtylvinyl)pyridine hydrochloride, 10^{-6} M f.c.) inhibits thromboformation indicating that the remodeling pathway is involved.

INTRODUCTION

In western society, the atherosclerosis-thrombosis connection and its closely related pathologies such as tissue hypoxia, ischemia and embolization kill and disable far more people than any other type of disease. Research on the fundamental aspects of both atherogenesis and platelet activation remains therefore urgent and highly important, while at the same time numerous molecules are screened for their pharmacological effects on these phenomena. In its early stage, an arterial thrombus is a platelet mass, originating from a complex series of simultaneous and/or sequential events comprising (i) platelet activation and adhesion onto an injured segment of an artery, (ii) aggregation of these platelets into a sponge-like mass, mediated by the release of platelet content, by the secretion of endothelial cell-derived substances such as PAF-acether and at a later stage by leukocyte products.

A substantial amount of data on the endothelial and platelet metabolism and their interrelation in normal and pathological conditions has been obtained from either *in vitro* or *ex vivo* experimentation, using elaborated cell culture techniques or vessel wall explants. Recently, in inflammation and thrombosis, a key role has been attributed to PAF-acether.

PAF-acether (1-O-alkyl-2-acetyl-sn-glycero-3-phosphocholine) is an autacoid involved in many inflammatory and allergic reactions (Demopoulos et al., 1979; Vargaftig et al., 1981). It was demonstrated that this mediator exerts direct effects on T and B lymphocytes, NK cells, macrophages, keratinocytes and astrocytes. *In vivo* investigation (Desquand et al., 1986) furthermore demonstrated that acute anaphylaxis induced in the lungs of the guinea pig is also PAF-acether-mediated; indeed, it was shown that animals passively sensitized to ovalbumin, responded to antigen stimulation by the appearance of platelet thrombi in the pulmonary circulation, intravascular leukocyte margination and diapedesis and at a later stage disruption of the endothelial lining. Eosinophils and free eosinophylic granules were found scattered in the submucosa of the bronchi near platelet aggregates. These changes mimick closely those observed following the administration of PAF-acether and adduce evidence for the involvement of the autacoid in this type of pulmonary reactivity.

A characteristic feature observed in this response is the diapedesis of platelets through the damaged endothelium and the formation of extravascular aggregates of degranulated and swollen thrombocytes. One hour following the challenge by the antigen, damage of the respiratory epithelium becomes obvious and numerous inflammatory cells (mastocytes, plasmocytes, degranulated eosinophils and macrophages) have penetrated into the bronchial mucosa and submucosa. These observations are clearly indicative for the role of the autacoid in *in vivo* inflammatory and allergic conditions and highlight the involvement of platelets in PAF-acether-mediated reactions.

Using an optoelectronic device developed by Bourgain et al. (1984) to detect platelet-vessel wall interaction and thromboformation, it is possible to apply different pharmacological probes at well defined time intervals. With these experimental facilities available, it was deemed interesting to investigate the reaction pattern following challenges onto the arterial wall in standardized conditions.

PAF-acether (10^{-8} M f.c.) applied topically during a short interval onto a mesenteric arteriole of the guinea pig is known to induce the autogeneration of endogenous PAF-acether by endothelial cells, invariably followed by platelet adhesion and thromboformation (Bourgain et al., 1985). The application of an electrical current (30-40 μA, 1 min) onto the vessel wall, by inducing polarity changes of the endothelial cell membrane, was equally followed by platelet-vessel wall interaction and thromboformation comparable to the phenomena induced by PAF-acether. When the intensity of the current was lowered to 1 μA, no thrombotic effect occurred. At a concentration of 10^{-9} M PAF-acether, no platelet-vessel wall interaction could be detected; however, superfusing the autacoid at this molarity after the application of the sub-threshold electrical challenge triggered the thrombotic phenomenon. Experimental evidence was adduced (Bourgain et al., 1985) that all these reactions are PAF-acether-mediated and induce alterations of the endothelium comprising retraction of the cellular margins and exposure of subintimal structures followed by formation of blebs. Platelet adhesion onto the exposed sub-endothelium is followed by aggregation, which over time results in the formation of a thrombus. Shape change and pseudopod formation of the aggregating thrombocytes are observed, the discoid shape of the newly arriving platelets stands in sharp contrast to the spheroid form of the activated ones. Leukocyte recruitment by the thrombus occurs at a later stage.

Transmission electron microscopy confirms the degranulation of the involved platelets, while within minutes, the thrombus undergoes extensive changes such as an increase in compactness and invasion by leukocytes through crevices in its structure. The ultrastructural data also confirm the activated status of the leukocytes; mostly neutrophils but also monocytes and eosinophils are observed. The platelets, particularly those surrounding leukocytes, undergo extensive swelling at complete degranulation, an aspect identical to the one described in extravascular platelet aggregates during the anaphylactic response in the lung.

These ultrastructural, as well as optoelectronic observations, clearly indicate that PAF-acether is a major mediator involved in platelet-vessel wall interaction and subsequent local thrombosis through a feedback mechanism of autogeneration originating in the vessel wall. PAF-acether antagonists, calcium chelating-agents as well as inhibitors of phospholipase A_2, significantly decreased thromboformation.

Experimental evidence was adduced by Lewis et al. (1988) in *in vitro* models using cultures of bovine and human endothelial cells, that hydrogen peroxide induced the generation of PAF-acether and promoted the adhesion of leukocytes onto the cells.

In this paper, we demonstrate that (i) hydrogen peroxide triggers PAF-acether-mediated lesions in the endothelial cells, that (ii) it primes the effect of PAF-acether, that (iii) a stable analogue of PAF-acether induces alterations comparable to the ones induced by PAF-acether itself and that (iiii) the down-regulation of enzymes such as acetyl transferase and acetyl hydrolase affect the activation phenomena by PAF-acether to a major extent.

MATERIALS AND TECHNIQUES

In order to induce arterial thrombosis, the general principle of the method described originally by Duval et al. (1970) has been followed. The thrombus-inducing procedure described by these authors consisted in the application of an electrical current to a branch of the mesenteric artery of the white Wistar rat which results in extensive lesions affecting the endothelium, the elastic internal lamina and the smooth muscle cells. Furthermore, vasospastic phenomena frequently occurred and interfered to a major extent with the development of the thrombus.

In order to avoid these complications, a current of weak intensity (0.01-5 µA) is applied for a duration of one minute while the polarity is inversed every 5 seconds. In these conditions, a restricted zone of approximately 30 to 50 adjoining endothelial cells is affected.

In dark room (Figure 1) the image of the arterial segment is projected onto a matrix of two columns of 15 LDRs (Light Depending Resistances) fixed on a carrier and the white thrombus is detected as a change in light intensity and recorded as voltage signals. Indeed, the appearance of a white thrombus will, with regard to the blood stream, induce a change in light intensity and consequently variations in the resistance values of the LDRs covered by the projected image of the thrombus. These are measured as voltage differences by Wheatstone bridges containing the LDRs. Each voltage value is a measure for the platelet content along a defined projection line in the thrombus. Two position LDRs, also fixed onto the carrier on each side of its small axis, positioned at equal distance of the center, detect the image of the wall of the artery. Indeed, small movements of the artery result in displacements of its image on the LDR matrix. In order to correct errors due to these phenomena, a translation mechanism with a stepping motor and a nut-and-screw device is adapted in the LDR housing. Transversal displacements of the arterial segment are detected by the position LDRs. The here from derived out-of-balance signal is fed back through a driver circuit to the stepping motor which activates the screw, so that the nut, on which the LDR carrier is fixed, is translated in the direction of the movement in order to adjust the LDR matrix to the projection of the arterial segment. When this is accomplished, the position LDRs are back in their symmetric position with regard to the arterial walls and no further unbalance signal is generated. Longitudinal movements are not corrected for as they are of minor importance.

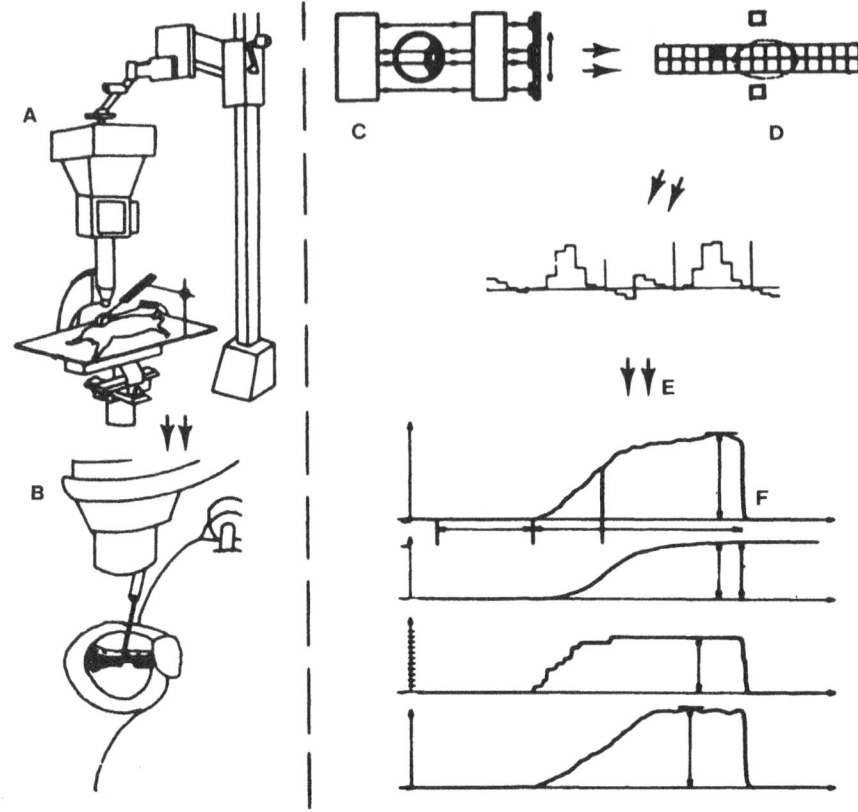

Figure 1. Optoelectronic device for the detection and registration of thrombotic parameters in an arterial segment *in vivo*.
The left part demonstrates the general set-up. The mesenteric preparation is continuously superfused (A) by Ringer solution (37 °C, 1 ml/min). The dissected arterial segment is projected (B) in a dark room unit onto a detection device (C) consisting of a matrix (D, 2x15) of light depending resistances (LDR) which convert the changes in light intensities into potential variations and which are monitored on an oscilloscope (E). The right bottom part (F) represents the thrombus parameters of a thrombus, computed from the variations in electrical potentials.

Monitoring and processing of the derived thrombus image are preceded by multiplexing : periodically (30 times per second) the LDR values are read out sequentially with a sample period of 1 ms. This multiplexed signal is displayed on an oscilloscope triggered by a synchronization signal. By taking a particular LDR as a reference and by measuring all LDR values relative to this one, most errors are eliminated. As a reference an LDR is chosen as close as possible to the thrombus projection, but not covered by it. Evidently, negative voltages are meaningless and these voltage differences between the reference LDR and the other LDRs are electronically set to zero. The LDRs which are not covered by the thrombus image are set to zero using a counting circuit which limits the measuring area on the LDR matrix along the long axis ("head-and-tail adjustment"). Local disturbances such as air bubbles which carry no information on thrombus development can thus be avoided. The corrected multiplexed signal is also displayed on the oscilloscope. Starting from this signal discriminating parameters are derived on-line. On a 5 channel recorder 4 principal parameters together with the heart rate are recorded. On the first channel appears the instantaneous or momentary thrombus value (T(t)), superposed by the "start", "lag" and "stop" pulses, indicating the status of the superfusing sequence. This curve is the sum of the voltages of all LDRs covered by the thrombus and is electronically obtained by the use of a low pass filter which eliminates high frequency variations due to multiplexing, heart rate, and preserves low frequency changes such as variations due to thrombus formation and degradation or embolization phenomena. As such, the T(t) curve is a measure for the the platelet content of the thrombus. The following parameters can be derived from this curve : the lag period $(t_{(l)})$ or the time required to detect the first signs of platelet-vessel wall interaction; the time to reach 500 mV from the onset of platelet-vessel wall interaction $(t_{(0.5V)})$ and the maximal value $(m_{(T)})$. On the second channel, the O(t) curve expresses the instantaneous thrombus area, recorded by electronic circuit and which represents the number of LDRs which are covered by the thrombus image with its maximal value as the $m_{(O)}$ parameter. On the third channel. the instantaneous thickness of the thrombus (D(t) curve) with its maximal value $m_{(D)}$ is defined as the highest LDR voltage, while on the fourth channel the integral of the over time recorded T(t) curve is represented by the TTV(t) curve. Its maximal and end value defines the TTV parameter, while the TVM parameter represents the integrated T(t) curve from the onset of platelet-vessel wall interaction up to the point where $m_{(T)}$ is reached. The integrator is reset at the beginning of each experiment by the "start" pulse. In figure 1 are also represented the different parameters together with the oscilloscope image of a thrombus.

The statistical analysis is perfomed by the analysis of variance (anova) on the rank values in all groups of investigated parameters.

PAF-acether-induced Thrombosis

Thrombus induction by topical superfusion can be performed by using a constant concentration of the mediator or by a gradient method. Local arterial thrombosis induced by PAF-acether follows a typical pattern of recurrent thrombosis. Indeed, once thrombosis by PAF-acether has been triggered, it is not halted, even when the topical superfusion by the agonist is discontinued and isotonic Ringer is applied. Furthermore, although some embolization by fragmentation, or even complete embolization eventually occurs, PAF-acether-induced thrombogenesis is a continuously recurring process leading ultimately to complete obstruction of the vascular lumen. In relation to this observation, experimental evidence was adduced that endogenous PAF-acether generated by the endothelial cells maintains the thrombotic state.

The mesenteric arterial branches are carefully chosen in relation to their anatomy in order to get comparable conditions for thrombus induction. In a previous control study performed in 15 randomly chosen guinea pigs, 3 arteries were investigated successively at 80 minutes intervals; no statistically significant differences were observed between the thrombus parameters of the 3 groups of arteries. Evidently, if the animal is to serve as its own control, as is mostly required in the testing of the effect of a pharmacon, two or more arteries have to be submitted to the thrombus-inducing treatment. The first artery serves to establish the control thrombus parameters, whereas the following arteries are investigated to test the pharmacon in different modes of application.

The experimental protocol for PAF-acether induced thrombosis by the gradient method is given here in detail :

First Artery (control). A concentrated solution of PAF-acether (10^{-6}Moles/l) is infused (at 1 ml/minute) into the first (contents 25 ml) of a double fluid container, allowing free exchange of fluids between both containers under mixing with a continuous supply of air bubbles. The contents of the second container (contents 12.5 ml) is superfused over the arterial preparation (at 1 ml/minute), following heating to 37 °C. The concentration of PAF-acether, applied by topical superfusion over the arterial segment, increases continuously in function of time approximating the following equation:

$$C_2(t) = \frac{C_0}{(V_1 - V_2)}[V_1(1 - e^{-\frac{Q}{V_1}t}) - V_2(1 - e^{-\frac{Q}{V_2}t})]$$

where: C_2 = superfusing concentration of PAF-acether (Moles/l)
C_0 = starting concentration of PAF-acether (10^{-6} Moles/l)
V_1 = volume of the first fluid container (25 ml)
V_2 = volume of the second fluid container (12.5 ml)
Q = flow rate (0.0167 ml/s)
t = time (s)

At the first signs of thromboformation (end of the "lag" period), the concentration of PAF-acether is kept constant and the superfusion is continued by a PAF-acether solution at this molarity. When the T(t) curve reaches 500 mV, PAF-acether superfusion is discontinued, and isotonic Ringer is applied. Registration of the thrombus parameters is then continued for another 15 minutes.

Second Artery (Drug applied "post"). The same protocol is repeated, but when the T(t) curve reaches 500 mV, the drug is injected intravenously or administered in topical superfusion at the desired molarity.

Third Artery (Drug applied "init"). The same protocol for the induction of thrombosis is followed, but 3 minutes prior to the electrical challenge, the drug is injected intravenously or continuously administered in topical superfusion at the desired molarity.

Hydrogen Peroxide-induced Thrombosis

The experiments with hydrogen peroxide are performed according to an identical protocol. The application of an electrical challenge is followed by superfusion of the peroxide using the gradient method (where $C_0 = 10^{-3}$ M). Drugs can be applied either in "post" or "init" mode to evaluate their effect on H_2O_2-induced thrombosis.

When for the first artery of some guinea pigs no platelet-vessel wall interaction occurs after 30 minutes of peroxide application, the molarity present at that time (4.9 10^{-6} M f.c.) is kept constant for another 15 minutes followed by 15 minutes of isotonic Ringer superfusion. The normal PAF-acether superfusion procedure is then started (priming experiments).

As anti-PAF-acether agents we used : the synthetic compounds BN 50727 (Tetrahydro-4,7,8,10 methyl-1 (chloro-2 phenyl)-6 (methoxy-4 phenylcarbamoyl)-9 pyrido [4',3'-4,5]), BN 50755 (Tetrahydro-4,7,8,10 methyl-1 (chloro-2 phenyl)-6 (trifluoromethyl-3 phenyl thiocarbamoyl)-9 pyrido), BN 50789 (Tetrahydro-4,7,8,10 methyl-1 (chloro-2 phenyl)-6 t-butylcarbamoyl-9 pyrido [4',3'-4,5] thieno [3,2-f]) and the natural compound derived from the Ginkgo Biloba BN 52021 (all gifts from H. Beaufour Institute, France). 27584 RP was a gift from Rhône Polenc Santé, France, while PAF-acether and Carbamyl-PAF-acether were purchased from Novabiochem, Switzerland and PMSF was from SIGMA, USA.

RESULTS

1. Hydrogen Peroxide-induced Platelet-Vessel Wall Interaction

In tables 1 and 2 are represented the data obtained in 6 animals using H_2O_2 as the inductor of thrombosis following the depolarizing electrical challenge. The different discriminating registered parameters are explained in the technical introduction. In table 1 are represented the control values and in table 2 the results after administration of the antagonist of PAF-acether according to the "post" mode.

Table 1. Thromboformation induced by H_2O_2. Control values

Arterial Code	H_2O_2 10^{-4} M	$t_{(0.5V)}$ s	$m_{(T)}$ mV	$m_{(D)}$ mV	TVM V.s	TTV V.s
P90410A	24.2	32.0	6280.0	1272.0	1936.8	2088.0
P90411A	14.1	68.0	10000.0	2020.0	1862.4	1870.4
P90419A	1.8	92.0	2240.0	708.0	880.8	1513.6
K90406A	4.9	94.0	8880.0	1420.0	3803.0	4680.8
K90412A	48.9	127.0	5711.0	933.0	1196.0	1967.0
K90418A	1.3	131.0	2570.0	608.0	919.0	983.0
MEAN	15.9	90.7	5946.8	1160.2	1766.3	2183.8
SEM	7.5	15.2	1295.4	214.5	447.7	525.4

Control values are obtained following 15 minutes of Ringer superfusion from 0.5 V at the T(t) curve.

The statistical analysis performed on these data using the anova technique, clearly indicates, as made evident by a highly significant ($p < 0.01$) effect on the respective parameters that the administration of the PAF-acether antagonist BN 50727 results in a decrease of vascular thrombogenesis. The results observed with BN 50755 and BN 50789 are equivalent to those observed with BN 50727. No untoward effects are observed during and after the intravenous administration of the synthetic compounds; the hemodynamic parameters remained unchanged during the whole course of observation.

Table 2. Thromboformation induced by H_2O_2. Effect of BN 50727

Arterial Code	H_2O_2 10^{-4} M	$t_{(0.5V)}$ s	$m_{(T)}$ mV	$m_{(D)}$ mV	TVM V.s	TTV V.s
P90410B	46.2	132.0	712.0	255.0	34.4	82.4
P90411B	32.6	140.0	784.0	74.0	130.4	204.0
P90419B	4.9	116.0	896.0	385.0	104.8	214.4
K90406B	4.9	313.0	1269.0	318.0	159.0	237.8
K90412B	48.9	108.0	883.0	318.0	78.7	129.3
K90418B	4.9	190.0	1092.0	142.0	115.7	286.0
MEAN	23.7	166.5	939.3	248.7	103.8	192.3
SEM	8.7	31.6	84.2	48.3	114.3	30.3

Values are obtained following superfusion of Ringer solution from 0.5 V at the T(t) curve and simultaneous injection of BN 50727 (0.1 mg/kg b.w.).

Table 3. Thromboformation induced by PAF-acether. Control values

Arterial Code	PAF	t$_{(0.5V)}$	Parameters m$_{(T)}$	m$_{(D)}$	TVM	TTV
	10^{-8} M	s	mV	mV	V.s	V.s
P90414B	7.4	136.0	2560.0	775.0	689.6	902.4
P90417B	30.5	140.0	2120.0	430.0	305.6	948.8
P90418B	16.8	52.0	6000.0	1112.0	1344.8	1530.4
P90424C	23.0	268.0	7680.0	808.0	1053.6	1804.8
P90426B	44.6	600.0	2620.0	984.0	1339.2	1510.4
K90419C	9.2	152.0	1820.0	438.0	533.6	689.6
MEAN	21.9	224.7	3800.0	757.8	877.7	1231.1
SEM	5.7	80.2	992.8	113.9	177.3	180.5

Control values are obtained following 15 minutes of Ringer superfusion from 0.5 V at the T(t) curve.

Table 4. Thromboformation induced by PAF-acether following H$_2$O$_2$-superfusion

Arterial Code	PAF	t$_{(0.5V)}$	Parameters m$_{(T)}$	m$_{(D)}$	TVM	TTV
	10^{-8} M	s	mV	mV	V.s	V.s
P90414A	37.6	96.0	5760.0	798.0	1812.0	2729.6
P90417A	16.5	24.0	1760.0	191.0	1092.8	1227.2
P90418A	38.1	136.0	4640.0	1464.0	755.2	1232.8
P90424A	32.1	76.0	6640.0	1048.0	2552.8	3741.6
P90426A	33.7	36.0	5200.0	1760.0	1952.8	2064.8
K90419B	12.7	53.0	8292.0	829.0	1719.0	1816.0
MEAN	28.4	70.2	5382.0	1015.0	1647.4	2135.3
SEM	4.5	17.0	892.5	225.0	261.3	394.9

PAF-acether gradient was started 15 minutes following the end of the peroxide application. Values are obtained following 15 minutes of Ringer superfusion from 0.5 V at the T(t) curve.

No alterations in the vasomotricity were observed during the superfusion interval using hydrogen peroxide as the superfusing agent.

2. Priming Effect of Hydrogen Peroxide on PAF-acether

In tables 3 and table 4 are represented the results of the paired data recorded in 6 animals. In each animal as indicated by its code, on the first artery the effect PAF-acether was examined following superfusion by H_2O_2 (table 3) and on the second artery the effect of PAF-acether alone (table 4) was studied. Statistical analysis (anova) demonstrates that there is a marked ($p < 0.01$) difference between the two groups represented respectively in table 3 and in table 4. As such, even though the superfusion of hydrogen peroxide in the group of animals represented in table 4 does not by itself induce thrombosis, the subsequent superfusion by PAF-acether demonstrates a marked enhancement, when compared to the control PAF-acether superfusion (Table 3). This enhancement results from the priming effect of the peroxide.

Table 5. Thromboformation induced by Carbamyl-PAF-acether. Control values

Arterial Code	C-PAF	Parameters $t_{(0.5V)}$	$m_{(T)}$	$m_{(D)}$	TVM	TTV
	10^{-8} M	s	mV	mV	V.s	V.s
K00404B	35.8	157.0	3537.0	808.0	720.0	884.1
K00406A	37.1	101.0	5956.0	1032.0	2217.1	2217.1
K00409A	25.6	445.0	5681.0	1056.0	779.0	1406.0
K00410A	45.8	148.0	3208.0	674.0	670.8	1390.3
K00412A	30.5	118.0	3129.0	1014.0	601.0	944.0
W00427A	43.8	244.0	4300.0	775.0	1093.8	1298.5
MEAN	36.5	202.2	4301.8	893.2	1013.6	1356.7
SEM	3.1	52.6	509.7	65.7	250.6	195.0

Control values are obtained following 15 minutes of Ringer superfusion from 0.5 V at the T(t) curve.

Table 6. Thromboformation induced by Carbamyl-PAF-acether. Effect of BN 50727

Arterial Code	C-PAF	Parameters $t_{(0.5V)}$	$m_{(T)}$	$m_{(D)}$	TVM	TTV
	10^{-8} M	s	mV	mV	V.s	V.s
K00327C	5.3	152.0	569.0	214.0	35.7	40.0
W00629B	23.1	184.0	1010.0	180.0	117.6	213.0
W00702B	34.4	160.0	1725.0	245.0	299.7	460.5
W00703B	42.6	120.0	1140.0	300.0	130.8	200.4
K00627B	16.9	80.0	1412.0	329.0	64.0	141.9
K00628B	14.5	192.0	961.0	432.0	139.2	232.0
MEAN	22.8	148.0	1136.2	283.3	131.2	214.6
SEM	5.6	17.1	162.4	37.2	37.5	56.8

Values are obtained following 15 minutes of Ringer superfusion from 0.5 V at the T(t) curve and the simultaneous injection of BN 50727 (0.1 mg/kg b.w.).

3. Effect of Carbamyl-PAF-acether on platelet-vessel wall interaction

The PAF-acether-analogue-analogue is applied in topical superfusion in identical conditions as PAF-acether itself. The results of the application are given in table 5 and the effect of the PAF-acether antagonist BN 50727 is represented in table 6. The analogue demonstrates properties in inducing arterial thrombosis comparable to PAF-acether itself; furthermore the inhibiting effect of the antagonist is highly significant ($p < 0.01$).

4. Effect of the down-regulation of acetyl hydrolase

The effect of the down-regulation of the enzyme acetyl hydrolase by PMSF is demonstrated in tables 7 and 8. A significant increase in thrombus parameters ($p < 0.01$) is obtained when PMSF in topical superfusion (10^{-5} M f.c.) is applied.

5. Effect of the down-regulation of acetyl transferase

The down-regulation of the rate limiting enzyme acetyl transferase had already been investigated previously through the administration of ketotifen, an anti-asthmatic drug (Bourgain et al., 1988). A synthetic compound 27584 RP was described (Stern, 1971) as being potentially active in relation to the down-regulation of the enzyme. The results obtained are summarized in tables 9 and 10. As can be concluded from these findings, a definite inhibiting effect ($p < 0.01$) of the drug when applied in topical superfusion (10^{-6} M f.c.) could be made evident. In these experiments, a constant concentration of PAF-acether (10^{-9} M f.c) is used.

Table 7. Thromboformation induced by PAF-acether. Control values

Arterial Code	PAF 10^{-8} M	$t_{(0.5V)}$ s	$m_{(T)}$ mV	$m_{(D)}$ mV	TVM V.s	TTV V.s
R80713A	31.9	200.0	4009.0	1007.0	788.5	1086.8
R80715A	32.2	324.0	4900.0	1723.0	1043.1	1646.0
R80719A	24.4	176.0	2253.0	409.0	1077.9	1300.0
R80726A	23.6	256.0	3303.0	1105.0	589.8	1943.0
R80727A	17.3	279.0	3705.0	786.0	1400.0	1575.8
R80728A	25.6	140.0	5500.0	1040.0	1276.5	1820.0
W00322A	19.2	228.0	4025.0	960.0	889.9	1097.6
K00327A	13.3	296.0	3478.0	716.0	671.7	1244.4
K00328A	30.2	135.0	2307.0	571.0	830.3	927.0
MEAN	24.2	226.0	3720.0	924.1	952.0	1404.5
SEM	2.2	22.7	355.9	126.1	90.2	118.4

Control values are obtained following 15 minutes of Ringer superfusion from 0.5 V at the T(t) curve.

Table 8. Thromboformation induced by PAF-acether. Effect of PMSF

Arterial Code	PAF 10^{-8} M	$t_{(0.5V)}$ s	$m_{(T)}$ mV	$m_{(D)}$ mV	TVM V.s	TTV V.s
R80713B	12.5	112.0	8799.0	1542.0	1059.0	2240.0
R80715B	4.9	68.0	10350.0	2480.0	1564.5	2815.0
R80719B	21.0	96.0	4225.0	737.0	2842.0	4053.8
R80726B	9.7	80.0	5223.0	1135.0	1316.5	1784.3
R80727B	1.0	56.0	9964.0	1900.0	2650.0	2800.0
R80728B	4.9	92.0	6995.0	1100.0	2041.7	2316.2
W00322C	8.3	148.0	2800.0	615.0	377.9	1013.7
MEAN	8.9	93.1	6908.0	1358.4	1693.1	2431.9
SEM	2.5	0.7	1109.3	250.7	332.6	358.4

Values are obtained following PMSF superfusion (10^{-5} M f.c.) 3 minutes prior to the application of PAF-acether and continued throughout the experiment.

Table 9. Thromboformation induced by PAF-acether. Control values

Arterial Code	$t_{(0.5V)}$ s	$m_{(T)}$ mV	$m_{(D)}$ mV	TVM V.s	TTV V.s
K70708A	812.0	2527.0	687.0	793.4	698.0
K70714A	188.0	4995.0	382.0	707.9	2064.2
K70715A	360.0	1620.0	463.0	506.3	1080.6
K70716A	569.0	4525.0	632.0	456.9	964.7
K70716B	308.0	1800.0	810.0	345.9	1320.0
R80311B	394.0	2496.0	412.0	523.2	1179.0
MEAN	438.5	2993.8	564.3	555.6	1217.8
SEM	90.3	581.0	69.9	67.6	189.9

Control values are obtained following Ringer superfusion from 0.5 V at the T(t) curve.

DISCUSSION

It was already made evident by our group that PAF-acether plays a major role in platelet-vessel wall interaction and subsequent thrombosis, the major cause of ischemia and tissue anoxia.   Indeed, a challenge to the arterial wall such as changes in polarity at the

Table 10. Thromboformation induced by PAF-acether. Effect of 27584 RP

Code	$t_{(0.5V)}$	Arterial $m_{(T)}$	$m_{(D)}$	TVM	TTV
	s	mV	mV	V.s	V.s
K70708B	558.0	1777.0	312.0	636.1	972.7
K70714B	60.0	209.0	62.0	10.0	13.2
K70715B	635.0	1086.0	647.0	471.9	741.0
K70716C	804.0	2250.0	412.0	1316.4	2067.0
R80311C	2400.0	0.0	0.0	0.0	0.0
MEAN	891.4	1064.4	286.6	486.7	758.8
SEM	397.0	434.7	118.1	242.4	380.1

Values are obtained following 27584 RP superfusion (10^{-6} M f.c.) throughout the experiment.

level of the endothelium, the topical superfusion by PAF-acether or the application of a sub-threshold electrical current together with the inactive precursor lyso-PAF-acether, all resulted in the induction of PAF-acether-mediated platelet-vessel wall interaction, starting with the formation of blebs in the endothelial cells, later complicated by platelet thrombosis. In all these conditions, evidence was adduced that PAF-acether was generated by the damaged endothelium and as such became an attractor for platelets an an inductor of platelet activation.

In the light of the data described by Lewis et al. (1988) in *in vitro* cell cultures of endothelium, it was indeed interesting to examine whether the challenge of the arterial wall *in vivo* by hydrogen peroxide resulted equally in changes mediated by PAF-acether. The observed results confirmed the idea; indeed, the peroxide in topical administration constantly induced the characteristic generation of blebs followed by platelet adhesion and thromboformation. The administration of PAF-acether-antagonists according to the "post" mode, invariably down-regulated thromboformation as indicated in the respective tables. Evidently, these observations confirm that the vascular reactivity is mediated by PAF-acether. Administration of the pharmaca in the "init" mode (data not documented here) resulted in the complete absence of platelet-vessel wall interaction which confirmed our previous observations (Bourgain et al., 1985) that PAF-acether is generated by the endothelial cells.

The ultrastructural data furthermore indicated that the blebs in the endothelium and the subsequent thromboformation induced by hydrogen peroxide were quite comparable to the alterations induced by PAF-acether itself. The effect of a stable analogue of PAF-acether equally resulted in the activation of platelet-vessel wall interaction and thromboformation in a way quite comparable to the one induced by the mediator itself or by electrical challenge. These effects were equally inhibited by specific anti-PAF-acether agents. The down-regulation of acetyl hydrolase by PMSF was invariably accompanied by a significant increase of platelet-vessel wall interaction and thrombosis. In these conditions, as was to be expected, the enzyme involved in the breakdown of the mediator being no longer fully active, the PAF-acether-induced effects, such as its autogeneration, its activating effect on platelets and leukocytes as well as on the vessel wall, followed a positive feedback route.

The important question as to the pathway followed in the biosynthesis of PAF-acether has been answered by the data observed when the rate-limiting enzyme acetyl transferase was inhibited. The administration in topical superfusion of the synthetic compound 27584 RP, being an inhibitor of this enzyme, was followed by a significant decrease in platelet-vessel wall interaction and subsequent thrombosis.

In previous experiments, we had demonstrated already that ketotifen, an anti-asthmatic drug interfered likewise with acetyl transferase activity, results which were in accordance with the data published by the group of Joly et al. (1987).

Our investigations clearly demonstrate that: (i) hydrogen peroxide in *in vivo* conditions induces PAF-acether-mediated effects and primes the effect of PAF-acether; (ii) a stable analogue of PAF-acether is equally capable of triggering endothelial cell activation and the generation of endogenous PAF-acether; (iii) the inhibition of acetyl transferase results in a marked enhancement of the effect of the mediator; (iiii) the down-regulation of acetyl transferase is followed by a marked inhibition of the PAF-acether-induced effects clearly demonstrating that the remodeling pathway is followed in the synthesis of the mediator. The involvement of PAF-acether in these phenomena will lead to the occlusion of the arterial lumen and will ultimately result in tissue anoxia through the mechanism of ischemia.

ACKNOWLEDGEMENTS

This investigation was made possible through the skilled technical assistance of Decuyper K. and Vansteenkiste W., and was partially supported by F.G.W.O. Contract 3.0017.88 (Fund for Medical Scientific Research).

REFERENCES

Bourgain, R. H., Vermariën, H., Andries, R., Vereecke, F., Jacqueloot, J., Rennies, J., Blockeel, E., and Six, F., 1984, A standardized 'in vivo' model for the study of experimental arterial thrombosis. Description of a method, Adv. exp. Med. Biol., 180: 635.

Bourgain, R. H., Maes, L., Braquet, P., Andries, R., Touqui, L., Braquet, M., 1985, The effect of 1-O-alkyl-2-acetyl-*sn*-glycero-3-phosphocholine (PAF-acether) on the arterial wall, Prostaglandins, 30: 185.

Bourgain, R. H., Andries, R., Braquet, P., Xie, Z. W., Sédivy, P., Decuyper, K., and Vargaftig, B. B., 1988, PAF-acether autogenerated in the guinea pig arterial wall may account for local thromboformation, Prostaglandins, 35: 809.

Demopoulos, C. A., Pinckard, R. N., and Hanahan, D. J., 1979, Platelet-activating factor. Evidence for 1-O-alkyl-2-acetyl-*sn*-glycerol-3-phosphorylcholine as the active component (a new class of lipid mediators), J. Biol. Chem., 254: 9355.

Desquand, S., Touvay, C., Randon, J., Lagente, V., Maridonneau-Parini, I., Etienne, A., Lefort, J., Braquet, P., and Vargaftig, B. B., 1986, Interference of BN 52021 with the bronchopneumonial effects of PAF-acether in the guinea pig, Eur. J. Pharmacol., 127: 83.

Duval, D. L., Didisheim, P., Titus, J. L., Spitell, J. A., and Owen, C. A., 1970, Experimental arterial thrombosis: description of a method, Mayo Clin. Proc., 45: 388.

Joly, F., Bessou, G., Benveniste, J., and Ninio, E., 1987, Ketotifen inhibits PAF-acether biosynthesis and β-hexosaminidase release in mouse mast cells stimulated with antigen, Eur. J. Paharmacol., 144: 133.

Lewis, M. S., Whatley, R. E., Cain, P., McIntyre, T. M., Prescott, S. M., and Zimmerman, G. A., 1988, Hydrogen peroxide stimulates the synthesis of platelet-activating factor by endothelium and induces endothelium cell-dependent neutrophil adhesion, J. Clin. Invest., 82: 2045.

Stern, P., 1971, Pharmakologische analyse eines specifischem cholinacetylase-inhibitors, Arzneim. Forsch., 7: 991.

Vargaftig, B. B., Chignard, M., Benveniste, J., Lefort, J., and Wal, F., 1981, Background and present status of research on platelet-activating factor (PAF-acether), Ann. N.Y. Acad. Sci., 370: 119.

DEVELOPMENT OF AN *IN VIVO* PERFUSION SYSTEM FOR

BOVINE FETAL SMALL INTESTINE

Howard D. Tyler[1], Lloyd P. Tate, Jr.[3],
Harold A. Ramsey[1] and Ian S. Longmuir[2]

College of Agriculture and Life Sciences
Departments of Animal Science[1] and Biochemistry[2]
and
College of Veterinary Medicine
Department of Food Animal and Equine Medicine[3]
North Carolina State University
Raleigh, North Carolina USA 27695

INTRODUCTION

Previous work at this institution has focused on the role of oxygen availability on the development of bovine small intestine during the perinatal period. In particular, the change in arterial oxygen tension associated with the conversion from placental to pulmonary respiration at birth was hypothesized to initiate an alteration in intestinal macromolecular permeability during the first 24 hours of life that is characteristic of the bovine neonate. The first model used to test this hypothesis was the hypoxic postnatal calf (Tyler and Ramsey, 1989). By providing the newborn calf with a 90:10 mixture of $N_2:O_2$, arterial P_{O2} was maintained at a level similar to that of the fetal calf. The results of this study were inconclusive, however, which may have been due to other changes occurring at birth that influence intestinal development.

In a postnatal *in vivo* model, there appears to be no acceptable way to isolate the neonatal small intestinal system. Moreover, the potential for an *in vitro* model is limited by the fragility of the newborn intestinal tissue, thus giving a high degree of inaccuracy in absorptive studies lasting more than a few hours. A better approach, therefore, may to be to increase oxygen availability to the small intestine of the fetus, thereby initiating oxygen-induced changes which normally occur at birth. This approach has several distinct advantages over a postnatal model. Fetal calves are relatively stable metabolically and endocrinologically compared to newborns. Additionally, changes that normally occur in the prepartum period, while they may have a role in the cessation of macromolecular permeability of the small intestine in the postnatal period, do not normally induce this change prenatally. Although a surgical approach is required, the intestinal tissue itself is not involved, and any effect of surgical manipulation would be reflected in the control animals.

Chronic catheterization of fetal vessels as a technique for studying fetal metabolism has become a relatively common procedure. There are, however, some inherent limitations in this technique. Catheters are subject to blockage and prone to inaccuracies, such as dislodging of position, which can be difficult to detect and impractical to correct. Additionally, chronic alteration of arterial oxygen tension in the fetus to any significant extent is difficult to accomplish by any previously established techniques. To further complicate matters, mixing of oxygenated and deoxygenated blood occurs at several locations in the fetus between the point where oxygenated blood enters the fetus in the umbilical vein and the cranial mesenteric artery. This effectively eliminates the possibility of manipulating either uMwilical blood or maternal blood to effectively control P_{O2} of the blood supply of the small intestine. Also, increasing oxygen tension in the entire fetus may introduce additional sources of error into the experimental design.

The development of our technique was intended to overcome these drawbacks in an effective manner and allow manipulation of a single organ or system while minimizing effects on the fetus as a whole. The approach presently being tested involves the implantation of a circulatory extension in the cranial mesenteric artery of the bovine fetus at approximately day 268 of gestation. The goals of this technique are fourfold: 1) to implant an extension in the cranial mesenteric artery and maintain continuous blood flow through the extension post-operatively, 2) complete post-operative recovery for the dam and her fetus culminating in parturition at the appropriate time, 3) to maintain catheter patency through parturition and into the postnatal period, and 4) to generate precise, accurate and meaningful data throughout the experimental period.

The purpose of this discourse is to discuss the model under development, the application of this model in terms of oxygen manipulation, and potential applications under consideration for later study. In light of this, and since the technique is still being modified, surgical approaches will be discussed only in general terms. With respect to the current study which necessitated development of this model, the objective is to test the hypothesis that increasing mesenteric arterial oxygen availability may stimulate development of small intestinal tissue prior to parturition.

MATERIALS AND METHODS

Catheter Preparation

The goal of treatment is to alter oxygen tension in flowing blood with minimal effects on peripheral resistance and flow rate. Our approach is to utilize the gas permeability characteristics of Silastic brand tubing[1] as an oxygenator. However, the same characteristic that makes Silastic ideal as a treatment catheter in this model makes it unacceptable elsewhere in the system. Although equilibration with the air occurs in the exteriorized section of the tubing, equilibration with tissue would be occurring in other parts of the catheter, thereby creating significant inaccuracies. Silastic is also a pliable tubing with a propensity for kinking and/or collapsing under external pressure. Therefore,

---

[1] Silastic brand medical-grade tubing. Dow Corning Corporation. Medical Products. Midland, Michigan, USA. 48640.

polyethylene tubing[2] is used internally as it is relatively gas impermeable and rigid.

For control animals, a single length of polyethylene tubing (3.17mm x 3.99mm x 6.75m) is used. Three-way valves are spliced in the tubing 2.5m from either end. For treated animals, a 1.75m piece of Silastic brand tubing (2.94mm x 4.08mm) is spliced between two sections of TDMAC-heparin treated polyethylene (3.17mm x 3.99mm x 2.5m) using three-way valves at the splice junctions. Therefore, the treatment determines the type of catheter utilized. Silastic cuffs are used to connect polyethylene catheters to three-way valves. The free ends of both polyethylene cannulas are encased in Silastic tubing to provide a better surface for anchoring within the artery. Much of the polyethylene is similarly sheathed to improve suturing characteristics at exteriorization sites. Sheathing in this manner also facilitates healing at these sites. Silastic sheaths are prepared by swelling in toluene (1 h) prior to slipping over the polyethylene. Evaporation of the toluene shrinks the Silastic, providing a tight-fitting sheath (Huntington *et al.*, 1989). Polyethylene cannulas are treated with 2% TDMAC-heparin complex. All catheters are gas sterilized prior to surgery.

Surgery

Pregnant cows are obtained on approximately day 268 of gestation. Feed is withheld for 24 hours prior to surgery. The cow is anesthetized and placed in a dorsal position. The abdomen is opened via a left paramedian incision immediately dorsal to the mammary vein. The pregnant uterine horn is isolated and opened in a hypovascular area, and the fetus is delivered caudally to cranially, exposing only the hindlegs and abdomen as far as the sternum. The head of the calf remains *in utero* to minimize external stimulations and prevent the possibility of spontaneous initiation of breathing. The fetal abdomen is then opened and the cranial mesenteric artery isolated. The artery is clamped both distally and proximally and severed between the clamps. The tubing is then inserted upstream in the proximal severed artery while simultaneously removing the clamp and securing the tubing inside. Blood is allowed to flow to the end of the tubing, which is then inserted into the distal portion of the severed artery as the clamp is removed, again securing the tubing. This reestablishes blood flow to the small intestine.

The cannula now serves as an artificial circulatory extension (ACE). The tubing is then carefully sutured as it exits the fetus. The tubing exits the uterus through the incision, exits the cow through a stab wound on the flank and the exteriorized section is secured to the body of the cow with surgical tape, still allowing access to the three-way valves for blood sampling.

Sampling and Analysis

Heparinized 1 ml blood samples are drawn twice daily from both the upstream and downstream ends of the Silastic tubing, reflecting pre- and post-treatment values, respectively. Samples are analyzed on an Instrumentation Laboratories System 1302 Blood Gas System and subsequently on an Instrumentation Laboratories 482 Co-oximeter System. Sampling frequency will increase in the immediate periparturient period (parturition $\pm$ 1 d), and then return to schedule.

---

[2] Intramedic polyethylene tubing. Clay Adams Division of Becton Dickinson and Company. Parsippany, New Jersey, USA. 07054.

Additional samples are drawn daily for both serum and plasma and analyzed for several metabolites, hormones and IgG. These provide information as to effect of treatment as well as providing some reference values that are lacking in the fetal calf.

RESULTS

Exteriorization of the ACE in this manner allows direct measurements of blood flow, access for manipulation of both blood flow and blood constituents, and ease of sampling. The use of Silastic tubing allows manipulation of oxygen tension in the blood. Mixing of this highly oxygenated blood with the less oxygenated blood from the umbilical vein at the ductus venosus should result in only small changes to the P_{O2} of the blood supplying the rest of the fetus. Thus, this system effectively isolates treatment effects to small intestinal tissue in an *in vivo* system.

Prenatal perfusion of fetal intestinal tissue with highly oxygenated blood should allow developmental changes to occur prenatally similar to those occurring in newborn calves. If cessation of immunoglobulin transport in the small intestine is initiated by the change in oxygen tension associated with birth, treated calves should absorb no immunoglobulins after colostrum ingestion postnatally while control calves should absorb immunoglobulins in a normal manner.

In the two surgeries that have been performed, the first goal of the four outlined in the introduction has been achieved. However, kinking of the ACE and postsurgical complications in the cow has precluded complete success in this endeavor. In particular, the extended duration of surgery and position of the cow throughout surgery appear to be contributing factors to a postsurgical rumen bloat, probably due to a delayed return of rumen motility. The bloating rumen physically prevents blood flow through the catheter by compressing it against the body wall. New methods are being tested to alleviate these problems. Additional modifications in both technique and postsurgical procedures should allow for attainment of the final three goals of this model.

One successful aspect of this project has been the use of Silastic as an oxygen exchanger (Table 1).
It should be noted that these results are from a longer length of Silastic tubing than is proposed in the protocol. The P_{O2} values obtained after equilibration with air are elevated beyond the P_{O2} in air. This may have been due to slight acidification of the blood samples by heparin within the sealed syringes. The pH of the heparin used was 5.8, which may have altered the pH of the blood enough to dislodge some of the oxygen from hemoglobin, especially at 100% saturation. Even a small increase in unbound oxygen would translate into relatively large increases in oxygen tension. For future samples, heparin pH will be adjusted to 7.4 to alleviate these inaccuracies.

The altered acid-base status of the fetus due to decreased P_{CO2} is an obvious concern. If shortening the length of Silastic required does not appreciably alleviate the situation to acceptable levels, a method for increasing the CO_2 content of the air around the catheter will be necessary. This requires knowledge of acceptable levels for fetal calves that have yet to be determined.

DISCUSSION

The results from two attempts to create an exteriorized arterial extension in fetal calves have been described. Both attempts were only partially successful, as frequently happens in

Table 1. Arterial blood gas and acid-base values in fetal calves for blood flowing through a 1.75m length of Silastic catheter.

	pH	P_{O2} (mm Hg)	P_{CO2} (mm Hg)	HCO_3^- (mEq/L)
initial values	7.176	27	45.8	17.1
post-treatment	7.713	179	6.6	8.5

the development of any new procedure. The major technical problems associated with this procedure appear to be post-surgical bloat and preventing the catheter from kinking due to movement of the fetus. Techniques to alleviate these problems need to be perfected for success in such an experiment.

These results indicate that blood flowing through a short length of exteriorized Silastic tubing equilibrates quickly with air. The exact length of tubing required for equilibration will depend on diameter of the tubing and velocity of the blood. Therefore, the length recommended in the protocol is an estimate based on preliminary observations and will doubtless require further adjustment.

Other research applications for this technique include metabolite and mesenteric blood flow response to different gut peptides and/or metabolite and hormonal responses to restricted mesenteric blood flow. Implantation of an additional ACE into the portal vein of the fetus would allow direct measurement of substrate utilization and/or production by small intestinal tissue *in utero*. However, this greatly increases the complexity of the surgery and the postsurgical complications. Therefore, the perfusion technique will need to be perfected prior to attempting any additional cannulations.

REFERENCES

Huntington, G.B., C.K. Reynolds and B.H. Stroud. 1989. Techniques for measuring blood flow in splanchnic tissues of cattle. J. Dairy Sci. 72:1583

Tyler, H.D. and H.A. Ramsey. 1989. Effect of hypoxia on intestinal transport of immunoglobulins in newborn calves. J. Dairy Sci. 72 (Suppl. 1):316

A NON-LINEAR CALCULATION METHOD FOR IN VIVO ESTIMATION OF

SINUSOIDAL OXYGEN SATURATION IN THE LIVER OF RATS

Takashi Matsunaga, Shigetoshi Okumura, Hiroshi Eguchi
Hiroyuki Fukui, Nobuhiro Sato and Takenobu Kamada

Department of Medicine, Osaka University
1-1-50 Fukushima, Osaka, Japan, 553

INTRODUCTION

Microcirculatory disturbance occurs in a variety of liver
pathological situations and may cause hypoxia in some areas
of the liver. In this regard it is important to determine
hepatic oxygenation level in situ.

We reported a spectro-photometric method to monitor
hemoglobin oxygen saturation(SO2) of the region of
interest(ROI) in the liver of rats(Eguchi et al. 1988). This
method separates the absorption spectra of the liver into
two parts, the spectra of oxy-hemoglobin and deoxy-
hemoglobin. This allows us to calculate the ratio of oxyhemoglobin to total hemoglobin(i.e. SO2).

It works for a larger ROI(87μ m in diameter) since the .
To evaluate SO2 in a smaller ROI(14μ m in diameter,
corresponding to the size of the liver sinusoid), we have to
divide the absorption spectra into three parts, including
the absorption spectra of liver pigments without
hemoglobin(i.e. perfused liver). Because in a smaller ROI
the interference of absorption of liver pigments is not
negligible unlike in a larger ROI.

Furthermore we cannot assume Lambert-Beer's law in case
of a smaller ROI. In view of this, we have developed a new
calculation method for the determination of SO2 and carried
out the validation of the method.

MATERIALS AND METHODS

System Features

Our system is described in Figure 1. We made observations
using a Leitz microscope(ORTHOLUX) with a water-immersion
objective(ULTRAPAK 23x). A xenon lamp illuminated an
object(a microcuvette or a rat liver). An optic fiber(400μ m
in diameter) was placed at the focusing plane so that the
transmitted light was guided to a spectrophotometer(USP501,

UNISOKU, Japan). Dividing the diameter of the fiber(400μ m) by the magnification of the objective(23x) and optvar factor(1.25) gave the size of the ROI(400μ m/23x/1.25=14μ m). The ROI can be placed at the center of the field. The absorption spectra in the 530nm-590nm range was recorded on the disk of a personal computer(PC-9801VX, NEC, Japan) coupled with the spectrophotometer. These data were able to be analyzed easily with the same computer. The microscopic image was videotaped simultaneously during the measurement of the spectra.

Calculation Procedures

The principal idea is that the observed absorption spectrum is comprised of three components.
The idea is expressed in the following equation.

$$m (x) = c (x , p) + o (x , q) + d (x , r) \quad (1)$$

m(x)   :the observed optical density(OD) at wavelength x.
c(x,p):the OD of liver pigments at wavelength x
o(x,q):the OD of oxygenated blood at wavelength x
d(x,r):the OD of deoxygenated blood at wavelength x
p   :the amount of liver pigments
q   :the amount of oxygenated blood
r   :the amount of deoxygenated blood
   c(x,p) ,o(x,q) and d(x,r) were obtained from in vitro experiments(described below). The coefficient a,b and c were chosen to minimize

$$\sum_{x} \{m(x) - c(x,p) - o(x,q) - d(x,r)\}^2 \quad (2),$$

where x's are 534, 540, 546, 552, 558, 564, 570, 576, 582, 588nm. This is a non-linear least square method. SO2 is given by the equation

$$SO2 = q / (q + r) \quad (3)$$

We used Gauss-Newton's method to calculate p,q and r. The program was coded in Quick BASIC(Microsoft). Gauss-Newton's method requires initial values for p,q and r. To get the initial values we assumed Lambert-Beer's law temporarily.

$$c(x,p) = pc(x) \quad (4)$$
$$o(x,q) = qo(x) \quad (5)$$
$$d(x,r) = rd(x) \quad (6)$$

Substituting (4)(5)(6) into (1) yields

$$m(x) = pc(x) + qo(x)+ rd(x) \quad (7)$$

The initial values are the values that minimize

$$\sum_{x} \{m(x) - pc(x) - qo(x) - rd(x)\}^2 \quad (8)$$

This is a linear least square method. That was able to be easily solved by using QR method. Then we carried out Gauss-Newton's method, assuming Lambert-Beer's law no longer.

In Vitro Experiment

Determination of c(x,p)   The perfused rat liver(an open system) was prepared with standard procedures from Sprague-Dawley(S-D) rats weighing 100-150g. The liver was perfused with oxygen rich saline(37° C) and placed on the microscope

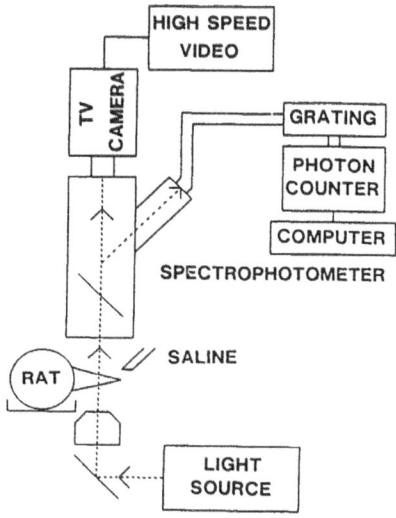

Figure 1. Block diagram of our system for the determination of oxygen saturation of blood in sinusoidal capillaries.

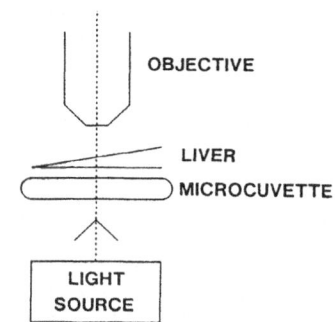

Figure 2. In vivo simulation with the perfused rat liver placed on the microcuvette containing blood solutions.

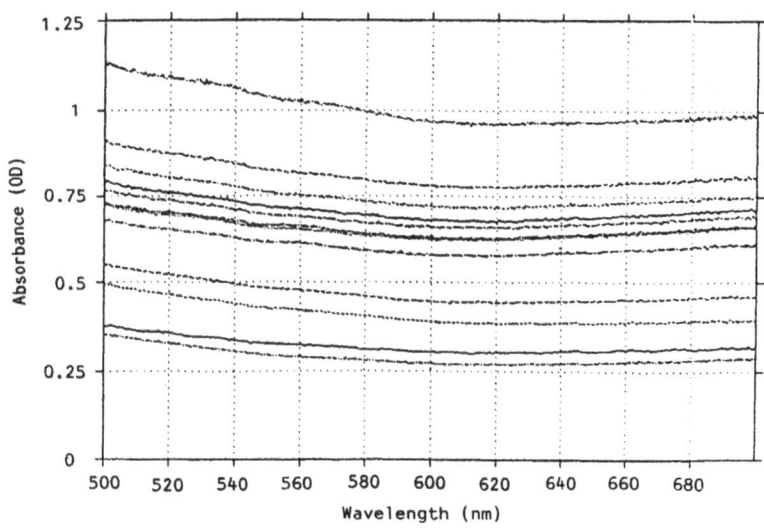

Figure 3. Absorption spectra for the perfused rat liver with ROIs set at various positions.

stage. ROIs were set at various positions 0-200μ m distant from the the liver edge so that various amounts of liver pigments existed in the ROI.

Determination of o(x,q) and d(x,r)   Blood solutions with various hematocrits were prepared. The blood was collected from S-D rats ,heparinized and diluted with phosphate buffered saline(37° C). The solutions were oxygen saturated with oxygen gas or completely deoxygenated by adding 20μ l of dihydrosulfite(60mg/ml). Then the solutions were placed in a glass-microcuvette(pathlength:300μ m) and the absorption spectra for them were obtained.

Simulation for in Vivo Situation   A perfused rat liver was placed on the microcuvette containing an oxygen saturated rat blood solution(Figure 2). Then the absorption spectra were obtained as the function of the distance between the ROI and the liver edge. We made the distance less than 200μ m, because we usually observe the area less than 200μ m from the liver edge in vivo. From these spectra we estimated the SO2 of the blood in the microcuvette with the procedures described above.

Calibration   Blood solutions with various oxygen saturations were prepared by adding various amounts of dihydrosulfite to oxygen saturated blood solutions. Then the solutions were placed in the microcuvette for the determination of the absorption spectra. The PO2 of the blood solutions was measured with an auto gas analyzer(ABL-330 RADIOMETER, Denmark). Then SO2 was determined from the dissociation curve for hemoglobin. We compared SO2 obtained from spectral analysis to that obtained from gas analysis.

In Vivo Experiments

Animals were prepared as described by Sato et al.(1986). In brief, S-D rats weighing 100-150g were anesthetized with pentobarbital(35mg/kg) and underwent laparotomy. The liver edge was placed on a specially designed stage for intravital microscopy. The ROIs were set along the sinusoids from periportal region to pericentral region. The distance between the adjacent ROIs were 30-40μ m. The absorption spectra were recorded every 10 min for 60 min.

RESULTS

In Vitro Experiments

Determination of c(x,p), o(x,q) and d(x,r)   Figures 3-5 show the absorption spectra of a perfused rat liver, oxygen saturated blood and oxygen free blood respectively. From these spectra we were able to get c(x,p), o(x,q) and d(x,r). For the calculation convenience c(588,p) versus c(534,p)/c(588,p), o(588,q) versus o(534,q)/o(588,q) and d(588,r) versus d(534,r)/d(588,r) were plotted(Figures 6-8), indicating they do not obey Lambert-Beer's law.

Simulation for in Vivo Situation   Figure 9 shows the estimated SO2 of the blood in the microcuvette placed under the perfused rat liver. They are the function of the distance from the liver edge. Open circles show estimated

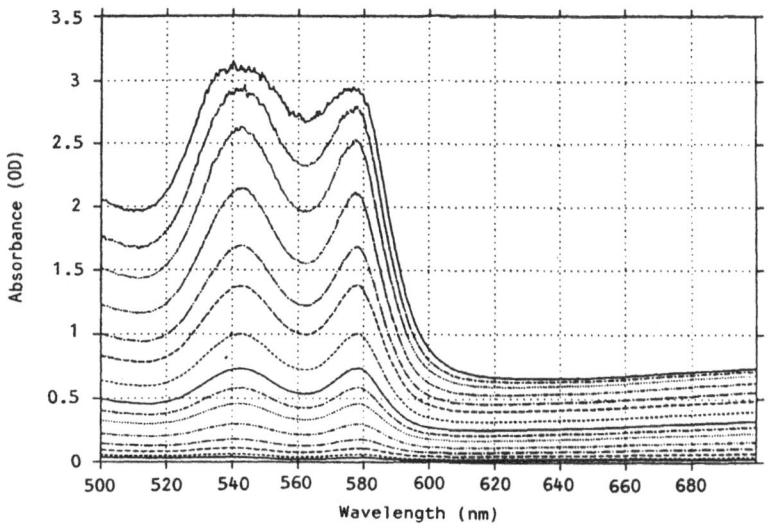

Figure 4. Absorption spectra for oxygenated blood solutions with various hematocrits.

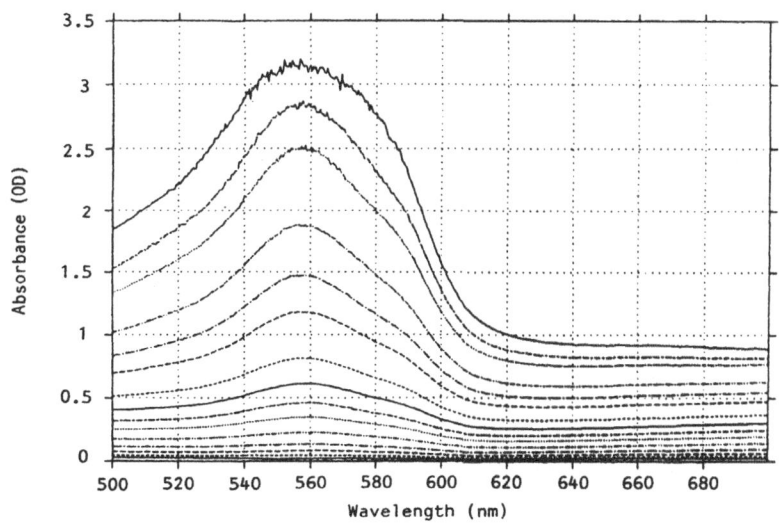

Figure 5. Absorption spectra for deoxygenated blood solutions with various hematocrits.

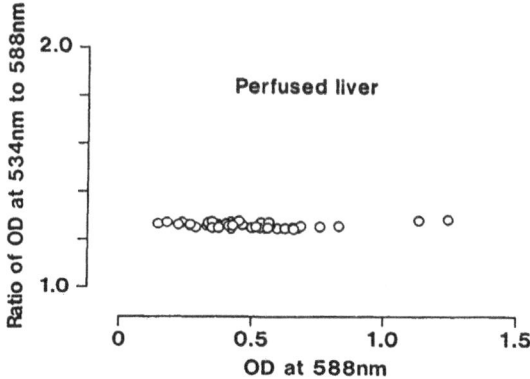

Figure 6. Ratio of OD for the perfused rat liver at 534nm to OD at 588nm.

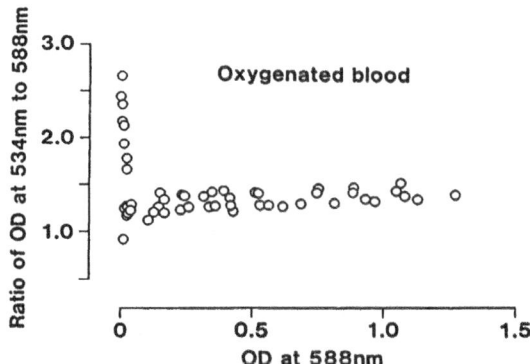

Figure 7. Ratio of OD for oxygenated blood solutions at 534nm to OD at 588nm.

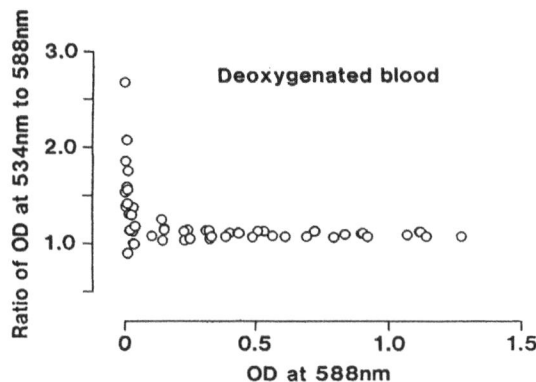

Figure 8. Ratio of OD for deoxygenated blood solutions at 534nm to OD at 588nm.

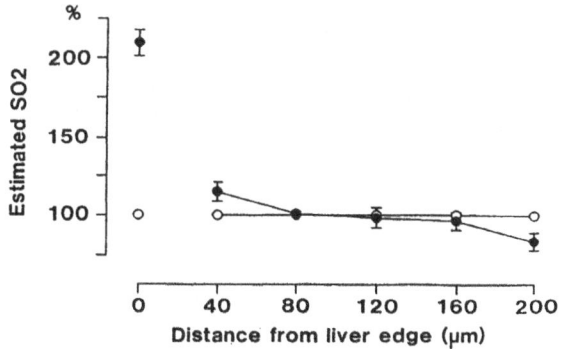

Figure 9. Estimated oxygen saturation in the microcuvette as the function of the distance from the liver edge. Open circles:our method. Closed circles:the method ignoring liver pigments.

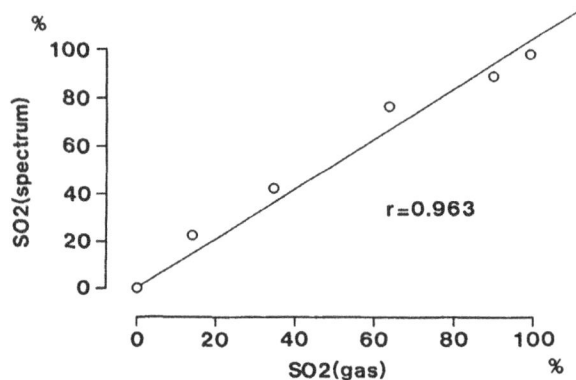

Figure 10. Correlation between spectrophotometrically determined SO2 and SO2 determined by the auto-gas analyzer

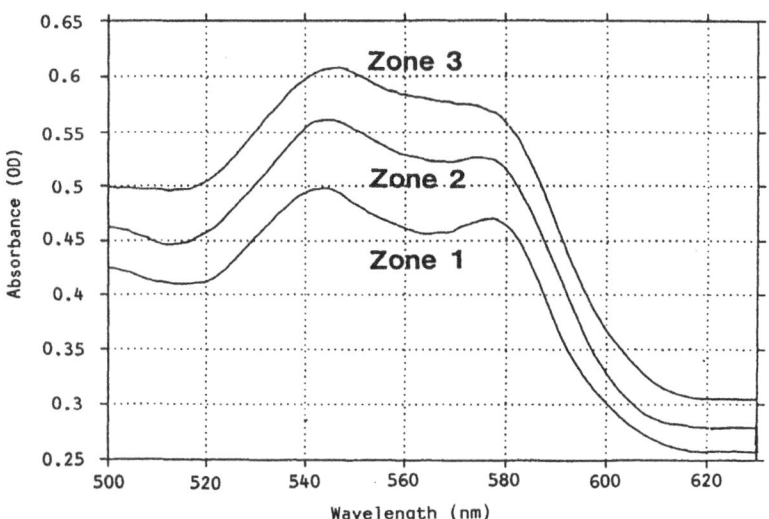

Figure 11. Absorption spectra for the region of interest in the rat liver _in vivo_

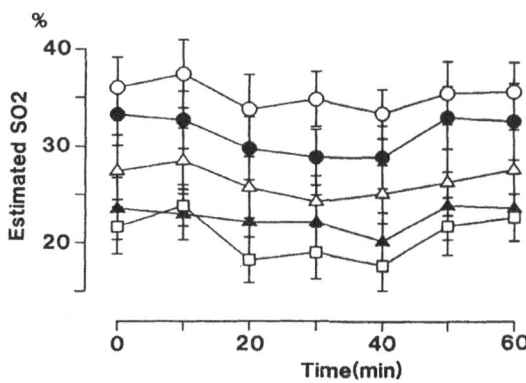

Figure 12 . Time-course of estimated oxygen
saturation of blood in capillaries in vivo.

SO2 calculated with our method. SO2 was about 100% at the
place 0-200μ m distant from the liver edge. Closed circles
show the results obtained from the method ignoring the liver
pigments. range. SO2 was overestimated in the range 0-40μ m
and underestimated in the 120-200μ m range.

Calibration   Spectrophotometrically determined SO2 for
the blood solutions in the microcuvette and independently
determined SO2 with the auto-gas analyzer were well
correlated(Figure 10).

In Vivo Experiments

The absorption spectra in vivo were shown in Figure 11.
Figure 12 shows SO2 for the ROI set along the sinusoid.
Apparent oxygen gradient exists along the sinusoid. SO2 was
relatively constant with some fluctuation for 60min(Figure
13).

DISCUSSION

Differences in the spectra of oxy- and deoxy-hemoglobin
have been used for the determination of SO2 in red blood
cell hemoglobin(Anderson et al. 1965).
Several investigators reported on spectrophotometric
methods to determine SO2 or hematocrit in microvessels(
Pittman et al. 1975, Pries et al. 1983). Recently Ellsworth
et al.(1987) described the method for the determination of
the SO2 of blood within capillaries of the hamster cheek
pouch retractor muscle. Unfortunately we could not apply
their methods to the rat liver because of pigments in the
liver other than hemoglobin.
Eguchi et al. (1988) presented a spectrophotometric
method to determine the SO2 for ROIs(87μ m in diameter) in
the hepatic lobule in rats. However, the ROI is not small
enough to measure the SO2 of blood within capillaries in the
hepatic sinusoid.

Difficulties exist in the fact that the liver contains various pigments other than hemoglobin. For a smaller ROI(14μ m in diameter) that correspond to the size of capillaries in the sinusoid, the amount of the liver pigments varies depending on the diameter of the capillary and the position of the ROI. In view of this we have to estimate the amount of liver pigments in the ROI. To do this we have obtained the OD of liver pigments by recording the absorption spectra of the perfused rat liver.

In addition we can no longer assume Lambert-Beer's law as indicated in Figures 6-8. Our non-linear least square method do not require this condition.

Our in vivo simulation indicated our method well canceled the influence of the liver pigments on the estimated SO2. Our calibration allows us to measure SO2 of blood in sinusoidal capillaries with this spectrophotometric method.

In summary we have presented a computer-assisted spectrophotometric method for the estimation of SO2 of blood in hepatic sinusoidal capillaries. Our method can be extended to the tissue which contains various amounts of mitochondrial and microsomal cytochromes, if the perfused system of the tissue is available.

REFERENCES

Anderson, N.M., and Sekelj, P., 1965, Studies on the light transmission of nonhemolyzed whole blood. Determination of oxygen saturation, J. Lab. Clin. Med., 65: 153-166.

Eguchi, H., Sato,N., Matsumura, T., Kawano, S., and Kamada, T., 1988, In vivo estimation of oxygen saturation of hemoglobin in hepatic lobules in rats, in: "Oxygen transport to tissue X", Mochizuki, M., Honig, C.R., Koyama, T., Goldstick, T.K., and Bruley, D.F., Plenum Publishing Corp., New York.

Ellsworth, M.L., Pittman, R.N., and Ellis, C.G., 1987, Measurement of hemoglobin oxygen saturation in capillaries, Am. J. Physiol., 252: H1031-H1040

Kessler, M., Lang, H., Sinagowitz, E., Rink, R., and Hoeper, J., 1973, Homeostasis of oxygen supply in liver and kidney, in: "Oxygen Transport to tissue," Bicher, H.I., and Bruley, D.F., eds., Plenum Press, New York.

Pitmann, R.N., and Duling, B.R., 1975, A new method for the measurement of percent oxyhemoglobin, J. Appl. Physiol. 38: 315-320.

Pries, A.R., Kanzow, G., and Gaehtgens, P., 1983, Microphotometric determination of hematocrit in small vessels, Am. J. Physiol., 245: H167-H177

Sato, N., Eguchi,H., Inoue ,A., Matsumura, T., Kawano, S., and Kamada, T., 1986, Hepatic microcirculation in Zucker fatty rats, in :"Oxygen transport to tissue VIII", Longmir, I.S., ed., Plenum Publishing Corp., New York.

A COMPARISON OF PARAMETERS USED TO STANDARDIZE RESULTS FROM *IN VITRO*

PERFUSIONS OF HUMAN PLACENTAE

D.J. Maguire, R.S. Addison, T.J. Harvey, R.H. Mortimer[2]
and G.R. Cannell[2]

Division of Science and Technology
Griffith University, Nathan, Queensland, Australia
[2]Conjoint Internal Medicine Laboratory
Royal Brisbane Hospital, Herston, Queensland
Australia

INTRODUCTION

The placenta is a relatively short-lived organ which is required to mimic some of the complex functions of highly specialized organs. Some examples of this include;

(i)   Trans-placental gaseous exchanges (oxygen and carbon dioxide) which must occur at rates approaching those exhibited by alveolar tissue in lungs.

(ii)   Detoxification of foetal waste-products, in particular various amines associated with developing tissues, thereby mimicking one of the important roles of hepatic tissue.

(iii)   Synthesis of a number of pregnancy-specific peptide hormones, such as prolactin and gonadotrophin, and transformation of steroid hormones of maternal origin. These "endocrine-like" activities are essential to survival of the developing foetus and are, indeed, used clinically to monitor foeto-placental function and detect foetal distress.

(iv)   Trans-placental solute exchange. Although this exchange does not occur against a concentration gradient, this activity is similar to some aspects of renal function.

Despite the importance of these placental functions, detailed physiological and biochemical study on placental tissues have been largely neglected. Difficulties confront researchers who choose to study this abundant and readily available tissue. It is difficult to compare data obtained from perfusions of placentae from different individuals or even from lobules within the same placenta. Most homogenized placental preparations used in biochemical investigations will contain contaminating maternal and foetal blood. Placental lobules prepared for *in vitro* perfusions are contaminated with varying amounts of non-functional (non-contributing) tissue. In the work reported here, the transfer of a number of solutes which might potentially be used for standardizing between perfusion preparations have been compared. Two of these have previously been proposed as placental markers of either rapid transfer (antipyrine; Challier. 1985) or slow transfer (creatinine; Eaton *et al*, 1985). Inulin and antipyrine transfer have also been studied by Brandes *et al* (1983).

Table 1. Typical results for an individual six hour perfusion (maternal to foetal compartments, N=6, mean, ml per minute per kg.)

TIME (HOURS)	0	2	4	6
OXYGEN DELIVERY	17.9	18.2	18.2	17.8
OXYGEN TRANSFER	7.9	7.3	7.3	6.8
OXYGEN CONSUMPTION	0.33	0.36	O.49	0.42

EXPERIMENTAL

The placental perfusion technique was investigated in some detail, in order to assess the clinical significance of novel steroid-transforming activities which we have recently reported (Addison et al, 1991). The perfusate system illustrated (Fig. 1) is adapted from that described by Miller et al (1985) in order to incorporate a novel second circuit. This enables us to minimize perfusate dead volume during transfer experiments. The perfusate was tissue culture medium M199, obtained from Difco (Detroit, MI, U.S.A.) and supplemented with heparin (25 IU/ml) and gentamicin (100 mg/l), glucose (2g/l), sodium bicarbonate (2.9 g/l) and dextran (fetal 29 g/l; maternal 7.5 g/l). Term human placentae were obtained at caesarian section within 5 min of delivery from healthy women with no significant drug history and transported immediately in cold, oxygenated perfusate to an adjacent laboratory. An artery-vein pair to an intact peripheral lobule was cannulated for fetal circuit perfusion with maternal perfusion being established by insertion of two cannulae approx. 0.8 cm through the decidual plate. The volume of perfusate in the maternal and fetal circuits was 150 and 100 ml, respectively. The pH was maintained at 7.35 to 7.45 by the addition of sodium bicarbonate. Perfusion was maintained for 6 h in a dual recirculating mode with maternal and fetal flow rates of 25 and 3 ml/min, respectively. Tissue temperature was maintained at 37°C and fetal circuit pressure below 40 mm Hg.

A number of different methods have been applied in this and other studies to monitor the viability of placental function. Oxygen consumption is one such marker.

The oxygen content of perfusate delivery and return lines were assessed by collecting discrete samples throughout the perfusion and assaying those samples in a blood gas analyser (ABL300, Radiometer, Copenhagen, Denmark). From these measurements, it was apparent that various oxygen parameters are relatively constant for an individual placenta during the period of perfusion (Table 1). However, quite wide variations in these parameters were found when comparisons were made between different placentae (Table 2).

Table 2. Oxygen parameters measured or calculated (n=6, mean +/- S.E.M., ml per minute per kg (micromoles per minute per gram) wet weight.

OXYGEN DELIVERY	24.5	+/- 7.1 (1.093 +/- 0.317)
OXYGEN TRANSFER	9.7	+/- 2.8 (0.433 +/- 0.125)
OXYGEN CONSUMPTION	0.52	+/- 0.23(0.023 +/- 0.01)

Table 3. Transfer parameters for solutes examined. Clearance was calculated as described in the text. Clearance index was calculated by assigning a value of unity to clearance of antipyrine.

Compound	C (ml/min)	Index	Time to equilibrium
antipyrine	1.99	1.00	2 - 2.5 hr
creatinine	0.43	0.22	4 - 4.5 hr
inulin	0.28	0.14	>6 hr

Another way of assessing placental function in an *in-vitro* situation is to measure one of the "organ-like" functions described above. We have investigated the clearance of several molecules from the maternal perfusate (Fig. 2). Clearance from the maternal circulation is defined as;

$$\text{clearance (C)} = \frac{(FV - FA)QF}{MA}$$

MA = solute concentration in maternal arterial perfusate
FV = solute concentration in foetal venous perfusate
FA = solute concentration in foetal arterial perfusate
QF = fetal flow rate

In order to assess clearances, it is practical to either measure clearance only at the beginning of a reperfusion experiment or to introduce a "single pass" technique. This latter approach has the advantage that it is possible to reassess clearance during the perfusion period, at least for substances with moderate clearance rates. An alternative approach is to simulate the transfer by mathematical modelling techniques. This latter approach was adopted in the work reported here, and involved the development of a two-compartment model.

Concentrations of solutes compared in this study were estimated as follows: Prednisolone, inulin and antipyrene concentrations were assessed using radioactive monitoring techniques. [3H]-prednisolone (7-10 μCi, HPLC purified) and non-radiolabelled prednisolone (150 μg) were added directly to the maternal resevoir, as too were [14C]-antipyrine (5-10 μCi) or [3H]-inulin (10μCi). All radioisotopes were from Amersham Int. (Bucks., U.K.). Perfusates were sampled at times shown during perfusions, added to scintillation fluid (Ready Safe) obtained from Beckman Instruments Inc. (Fullerton, CA, U.S.A.) and counted in a scintillation counter (Tricarb 2000CA, Packard, U.S.A.). Creatinine was assayed by the Jaffe reaction on an RA1000 analyser (Technicon, New York, U.S.A.). Gamma glutamyl transferase activity was assayed using a kinetic spectrophotometric method on a SMAC II analyser (Technicon, New York, U.S.A.).

Typical transfer curves for the solutes are shown (Fig. 2) from which clearances were calculated (Table 3). Clearances were not calculated for prednisolone, since the metabolism of this compound makes it unsuitable as a standard for comparing between different placental preparations. Gamma glutamyl transferase is not transferred and thus its clearance cannot be estimated. It is however a useful indicator of integrity of individual preparations.

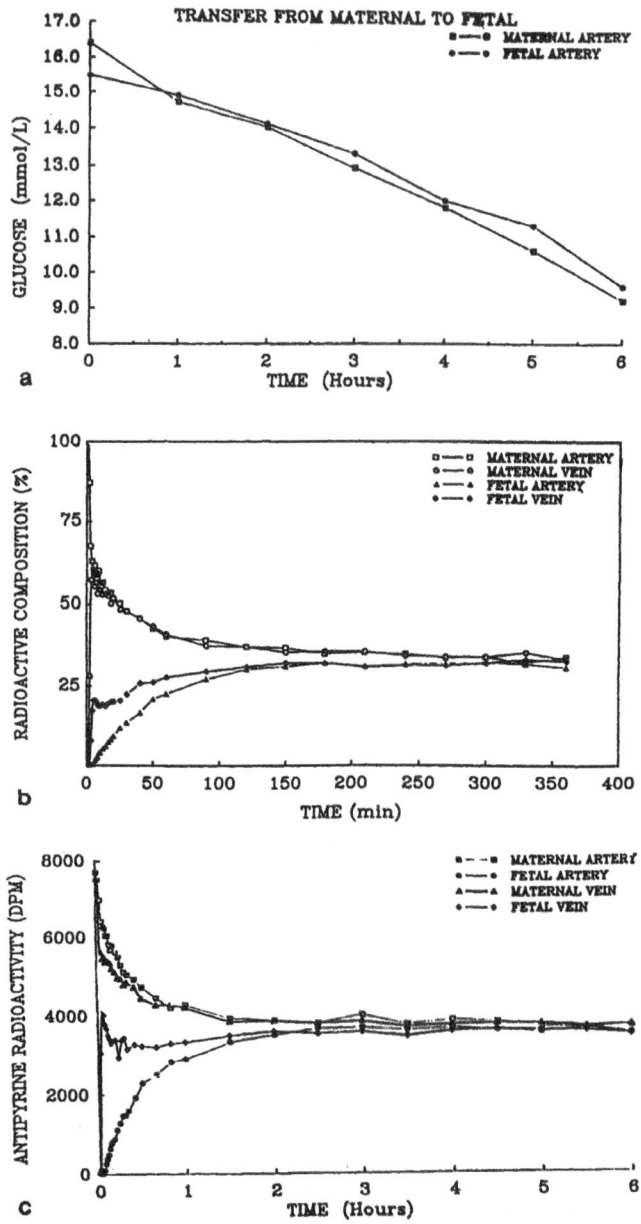

Fig. 2. Examples of transfer curves of solutes: (a) glucose, (b) prednisolone, (c) antipyrine

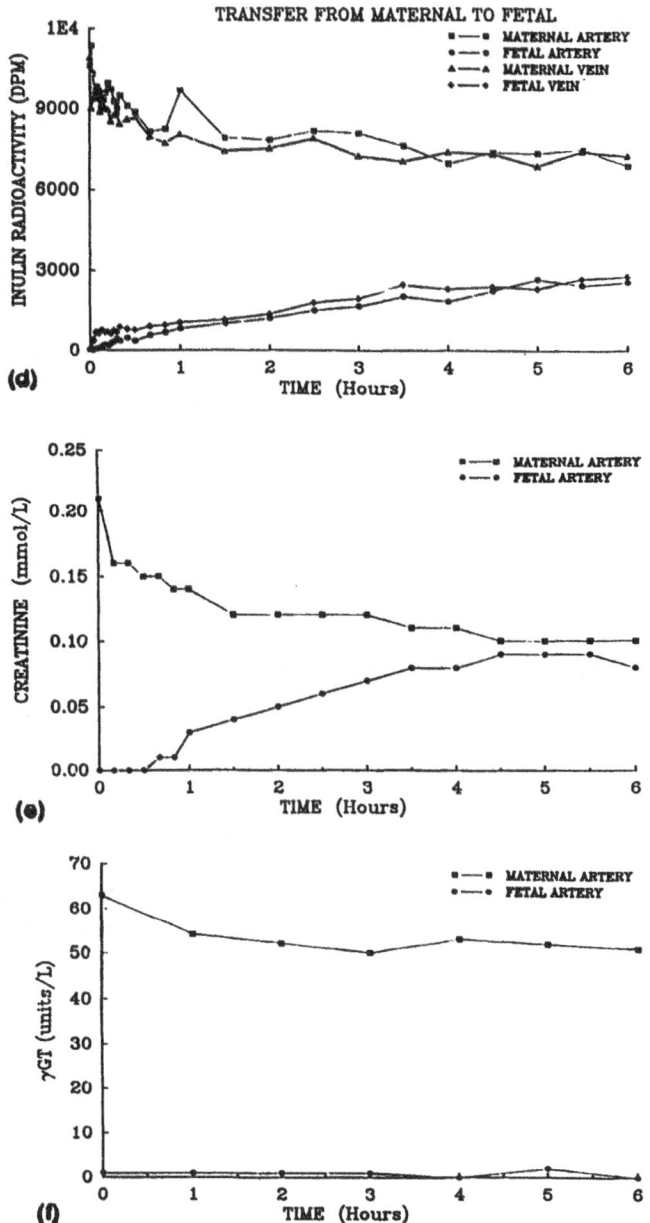

Fig. 2. Examples of transfer curves of solutes:
(d) inulin (e) creatinine (f) gamma glutamyl transferase

CONCLUSIONS

It was found that the oxygen consumption rate for an individual perfused placenta is relatively constant during the perfusion period, provided conditions remain constant for the tissue under investigation. However, oxygen consumption rate is quite variable when comparisons are made between different placentae. This makes it difficult to use oxygen consumption rate as a standard upon which to base comparisons of other metabolic parameters.

A survey was made of other indicators which might serve as standards and in particular the clearance of various solutes. Of the solutes investigated, the order of clearance was shown to be:

> antipyrine>(prednisolone)>creatinine>inulin>>glutamyl transferase.
>   (or metabolites)

By comparison with oxygen consumption rates there was less variability between placentae for these parameters. It is suggested that, although oxygen consumption rate is an important indicator of viability, it should not be used to standardize results obtained for other metabolic measurements when making comparisons between placentae. As a means of monitoring viability of individual placental lobules throughout a perfusion experiment, the measurement of oxygen consumption rates remains an important parameter.

REFERENCES

Addison, R.S., Maguire, D.J., Mortimer, R.H. and Cannell, G.R., 1991, Metabolism of prednisolone by the isolated perfused human placental lobule, J Steroid Biochem. Molec. Biol., 39: 83-90.

Brandes, J.M., Tavalioni, M., Potter, B.J., Sarkozi, L., Shepard, M.D. and Berk, P.D., 1983, A new recycling technique for human placental cotyledon perfusion application: application to studies of the feto/maternal transfer of glucose, inulin and antipyrine, Am. J. Obstet. Gynec., 146: 800-806.

Challier, J.C., 1985, Criteria for evaluating perfusion experiments and presentation of results, Contr. Gynec. Obstet. 13: 32-39.

Eaton, B.M., Browne, M.J. and Contractor, S.F., 1985, Placenta, 6: 341-346.

Miller, R.K., Weir, P.J., Maulik, D. and Di Sant'agnese, P.A., 1985, Human placenta *in vitro*. characterization during 12 hours of dual perfusion, Contr. Gynec. Obstet. 13: 77-84.

HUMAN PLACENTAL OXYGEN METABOLISM

D. J. Maguire, V. Voroteliak, D. Cowley[1] and G. R. Cannell[2]

Division of Science and Technology, Griffith University, Nathan
Queensland, Australia
[1]Department of Chemical Pathology, Mater Hospital
Wooloongabba, Queensland, Australia
[2]Conjoint Internal Medicine Laboratory, Royal Brisbane Hospital
Herston, Queensland, Australia

INTRODUCTION

The highly-vascularized capillary network of the human placenta provides an adequate structure for gaseous exchange between foetal and maternal circulations, although it may not be the most efficient such structure. By comparison with other mammals, it has been suggested that guinea pigs, rats and rabbits may provide better conditions for growth in-utero than do humans (Bartels, 1970).

It has been estimated that half of the oxygen is depleted from maternal blood during its circulation through the placenta and much of this oxygen is used by the tissues of the placenta. The fate of that oxygen has not been fully documented in biochemical terms. The highly oxidative metabolism of placenta is further supported by the presence of a significant mitochondrial population in placental tissue (Plate 1). Respiratory demands (involving the electron transport pathway) might reasonably be expected to account for a large proportion of the oxygen consumed in this tissue. This contrasts with reports of relatively high rates of lactate production and glucose utilization during *in vitro* perfusions of isolated placental lobules (Hauguel *et al*, 1983).

A number of other oxygen-requiring activities have also been claimed to be localized in placental tissues, including some of those catalysed by enzymes involved in the detoxification of waste fetal products and in steroid transformations.

In the present work, an estimate has been made of the maximal contribution to placental oxygen consumption by the enzyme diamine oxidase (E.C. 1.4.3.6), involved in one of these processes. This enzyme catalyses the oxidative deamination of a number of amines which are present during pregnancy, including monoamines such as histamine and diamines such as putrescine.

$$RCH_2NH_2 + H_2O + O_2 \longrightarrow RCHO + H_2O_2 + NH_3$$

These amines are believed to play an essential, though as yet poorly-defined, role in the rapid cell division which accompanies foetal development and the concomitant growth of the placenta.

Oxygen Transport to Tissue XIII, Edited by T.K. Goldstick *et al.*
Plenum Press, New York, 1992

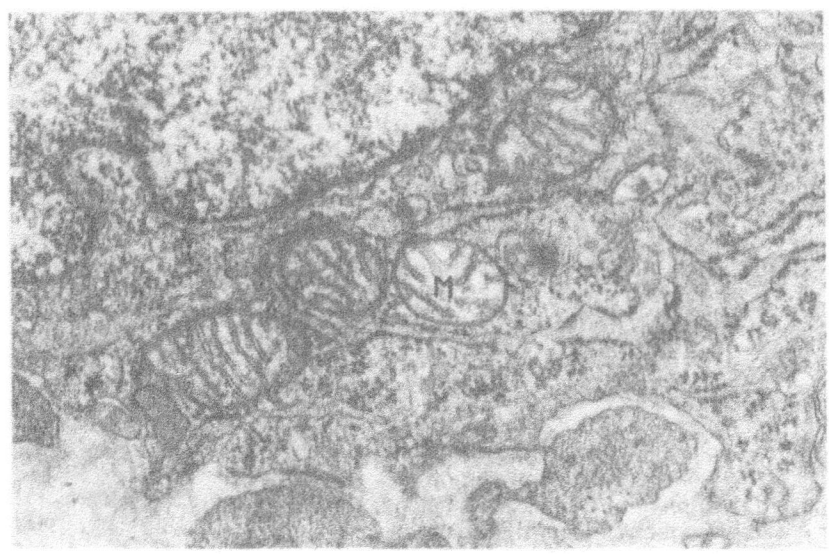

Plate 1. Transmission electron micrograph of fresh placental specimen, showing mitochondria (M).

EXPERIMENTAL

Isolated placental lobules were collected and prepared for perfusion as described in the previous paper. The pO_2 (oxygen partial pressure) in discrete samples of arterial and venous perfusates (foetal circulation) and affluent and effluent perfusates (maternal circulation) were measured using a blood gas analyser (ABL300, Radiometer, Copenhagen).

Glucose concentrations in perfusates were also determined by discrete sampling and analysis using a hexokinase method on a Technicon RA1000 analyser. Lactate was measured using a lactate dehydrogenase spectrophotometric kit method (Behring, Germany)

From these measurements, oxygen consumption rates, glucose utilization rates and lactate production rates shown in Table 1 were calculated as described by Cannell et al (1987). Results from this calculation are in good agreement with, though slightly higher than, estimates of rates of oxygen consumption, glucose utilization, and lactate production in the isolated perfused placental lobule made by Hauguel et al, (1983).

Using these results, two limiting cases can be defined. Firstly, if it is assumed that all the lactate is produced by anaerobic glycolysis (two moles of lactate produced per mole of glucose consumed) then lactate production will account for a glucose utilization rate of 0.25 +/- 0.07 µmol/g/min (90% of the value calculated in Table 1). In the second case, if all oxygen consumption is due to the aerobic metabolism of glucose (six mole of oxygen per mole of glucose), oxygen would be consumed at a rate of 1.68 +/- 0.42 µmol/g/min. compared to the value of 0.28 +/- 0.07 µmol/g/min shown (Table 1).

Calculation of the maximal consumption of oxygen by the enzyme, diamine oxidase, was based on experiments in which putrescine was used as the substrate. In this calculation, it was assumed that the enzyme could function at Vmax under conditions which exist in intact tissues. This dictates that substrate is present at saturating concentrations and that diffusion is not a limiting factor. Clearly, neither of these assumptions prevail but they permit the definition of an upper limit for the purpose of the present exercise.

Table 1. Energy parameters (rates) calculated in the present work.

GLUCOSE UTILIZATION	0.28 +/- 0.07 µmol/g/min	(n=9, mean +/- S.E.M.)
LACTATE PRODUCTION	0.50 +/- 0.15 µmol/g/min	(n=9, mean +/- S.E.M.)
OXYGEN CONSUMPTION	0.02 +/- 0.01 µmol/g/min	(n=9, mean +/- S.E.M.)

Estimates of the activity of diamine oxidase in homogenized, rinsed placental tissue are from our own (unpublished) experiments. These are in good agreement with previous reports in which the enzyme has been partially purified(Southren et al, 1985). Activity was measured using a modification of the assay described by Okuyama and Kobayashi (1961) and involves incubation of tissue homogenates with $[^{14}C]$-putrescine. 500µl of homogenized tissue was added to 1.5 ml of 0.067M phosphate buffer, pH 7.4. 100µl of labelled substrate (0.5 mM, 70 nCi) was added and the mixture was incubated at 37°C for 2hr. At the end of the incubation, 200µl of 10mM aminoguanidine and 3 ml of organic scintillation fluid were added to each tube. Product was extracted into the organic phase by vigorous shaking and this phase was removed for scintillation counting after freezing the aqueous phase. Enzyme activity was expressed as µmol substrate oxidized per minute per ml of homogenate, and converted to activity per placenta (Table 2). For ease of reference, activity is also expressed in the units used by Southren. et al (1966). The oxidation of one mole of substrate involves the consumption of one half-mole of oxygen and thus it is possible to calculate oxygen consumption attributable to this enzyme reaction.

The value obtained is compared to the total placental oxygen consumption (Table 2). Also included in this table is an estimate of the contribution to oxygen consumption by the exposure of plasma diamine oxidase to substrate during the passage of blood through placenta. For this calculation, a (maternal) blood flow rate of 500 ml/min is used and an average value for the enzyme activity in pregnancy plasma at term (Forget et al, 1986). For this calculation, it is also assumed that saturating levels of substrate(s) are available.

Table 2. Comparison of theoretical maximal oxygen consumption by diamine oxidase with actual placental oxygen consumption.

Activity of DAO (homogenate)	0.004	µmol/min/g
	(2.02 x10^6 units/g)	
Oxygen consumption (due to DAO)	0.002	µmol/min/g
Oxygen consumption (by placenta)	0.023	µmol/min/g
Oxygen consumption (per placenta)	11.6	µmol/min
Serum DAO	0.6	nmol/min/ml
Oxygen consumption (due to serum DAO)	0.3	nmol/min/ml

From the results of other investigations we have performed, it has been revealed that different forms of this enzyme exist. Diamine oxidase has been extracted from placental tissue, pooled maternal plasma and pooled amniotic fluid. Although pregnancy plasma and amniotic fluid can generally be obtained free of contamination from other body tissues or fluids, any preparation of placental tissue is undoubtedly still contaminated with some foetal and maternal blood. We have partially characterized the enzyme in pregnancy plasma and amniotic fluid. Results obtained using chaotropic electrophoresis (unpublished results) are consistent with differences in the subunit molecular weight of the enzyme from these two sources. Polyclonal antibodies have been prepared against the purified enzymes and these have been used to provide further evidence of differences between the two forms. Further studies are underway to purify the placental form of this enzyme and to describe its physicochemical properties and substrate specificity. The results obtained from that study may then necessitate adjustments in our estimate of the contribution of this enzyme to placental oxygen consumption. Another phenomenon which may impact upon this discussion is the observed alteration in activity and affinity of the enzyme for some substrates upon binding to surfaces (Stevanato et al, 1989)

CONCLUSIONS

Estimations of the maximal contribution by diamine oxidase to placental oxygen consumption have been made, based on experiments using putrescine as substrate. From these results it is possible to calculate a theoretical value for the maximum oxygen consumption in the prescence of substrates with higher affinities. The contribution is negligible when compared with the total placental oxygen consumption rates.

The differences observed between the amniotic fluid and pregnancy plasma forms of this enzyme warrant further investigation, and justify further studies of the putative placental enzyme, aimed at better understanding of substrate specificity. It is unlikely that the contribution to oxygen consumption by this enzyme will be a significant proportion of the total placental oxygen consumption, even if excess amounts of its natural substrate were present or the observed kinetic parameters were significantly altered during extraction from.a membrane bound environment.

REFERENCES

Bartels, H., 1970, "Prenatal Respiration," North-Holland Publishing Company, Amsterdam.

Cannell, G.R., Kluck, R.M., Hooper, W.D. and Dickinson, R.G., 1987, Calculation of viability parameters for the isolated perfused human placental preparation, Intelligent Instruments and Computers, 5: 109-113.

Forget, P., De Curtis, M., Senterre, J., Serum diamine oxidase in the neonate, Biol. Neonate. 50: 1-5.

Hauguel, S., Challier, S., Cedard, L. and Olive, G., 1983, Metabolism of the human placenta perfused in vitro: glucose transfer and utilization, O2 consumption, lactate and ammonia production, Pediatr. Res., 17: 729-732.

Okuyama, T. and Kobayashi,Y., 1961, Determination of diamine oxidase activity by liquid scintillation counting, Arch. Biochem. Biophys., 95: 242-250.

Southren, A.L., Kobayashi, Y., Jung, W.,Carmody, N.C. and Weingold, A.B., 1966, In Vitro production of diamine oxidase by the perfused human placenta, J. Clin. Endocrin. 16: 1005-1009.

Stevanato, R., Porchia, M., Befani, O., Mondovi, B. and Rigo, A., 1989, Characterization of free and immobilized amine oxidases, Biotechnol. and App. Biochem., 11: 266-272.

THE MEASUREMENT OF THE DIFFUSION COEFFICIENT OF OXYGEN THORUGH SMALL
VOLUMES OF VISCOUS SOLUTION: IMPLICATIONS FOR THE FLUX OF OXYGEN
THROUGH TISSUES

M. McCabe and D.J. Maguire*

Department of Chemistry and Biochemistry, James Cook
University of North Queensland, Townsville Qld. 4811
Australia and * Division of Science and Technology
Griffith University, Nathan Qld. 4111 Australia

INTRODUCTION

There is great confusion in the literature concerning the diffusion
of oxygen through solutions of high viscosity. Some results suggesting
that the Stokes-Einstein relationship does not apply to such small
diffusing species. Various attempts at explanation for this have
included suggestions of an anomalously increased chemical activity of
oxygen in highly concentrated solutions, and/or the possibility of "slip"
of the diffusing species in contact with the solvent medium.

Davies and Brink (1942) first considered the use of a recessed type
of polarographic electrode for measuring diffusion coefficients of
electro-reducible substances, and the application of such an electrode
system to a study of diffusion of oxygen through solutions of non-
Newtonian fluids was also suggested by Longmuir and Pistel (1960).
However accurate and reproducible results have proved elusive in practice,
due in part to corrosion of the mercury surface at the cathode (unlike
the dropping mercury electrode the recessed electrode is not regularly
nor automatically renewed). An additional and serious source of error
seems to be consequent on the electrocapillary effect which causes a jump
of the cathode mercury surface when the electrical circuit is completed,
which in turn induces stirring of the solution at the cathode/solution
interface.

It is important to obtain reliable measurements of oxygen diffusion
through stationary non-Newtonian solutions since these can be valuable
models for biological systems. Measurements of oxygen flux through
living tissues are bedevilled with problems of non-simple architecture
and heterogeneity as well as of oxygen consumption within tissues and so
it would be useful to study model systems of variable and controllable
complexity, and one major variant of interest is the microscopic viscosity
of the tissue fluids.

APPARATUS DESIGN

An apparatus has been designed to overcome the experimental problems
associated with a simple recessed type electrode. The design allows for

convenient and regular renewal of the cathode surface between each reading while the electro-capillary jump is prevented by the inclusion of a
teflon tap just below the cathode and its electrical contact. Finally
for added convenience the anode is incorporated as a part of the apparatus
and is made with a large surface area to avoid the possibility of anode
polarisation. The apparatus is shown diagramatically in Figure 1. Teflon
was the material of choice for the tap since it could be used without the
need for any type of lubricating grease which avoided the problem of
contamination of the mercury.

EXPERIMENTAL

All experiments and manipulations were conducted in a constant
temperature room at 23°C. All solutions contained 0.1M NaCl as a carrier
electrolyte.

The viscosity of solutions was altered by dissolving appropriate
concentrations of sucrose, and the viscosities of these sucrose solutions
were measured using a simple Ostwald type capillary (flow time) viscometer. The addition of the sucrose also diminished the solubility of
oxygen. Oxygen solubility coefficients were measured by a chemical
method due to Winkler.

Equilibration of the sucrose solutions with air was performed by
sparging the sucrose solutions for 2 hours with air which had previously
passed through water to almost saturate it with water vapour. Despite
this some minor evaporation of the sucrose solutions did occur, and this
was corrected by readjusting them back to their initial volume after the
first 1½ hours of sparging.

The radius of the polarographic capillary was obtained by measuring
the length of a weighed thread of mercury within the capillary, and from
this radius a value for the surface area of the cathode was obtained.

RESULTS

The decay of current following the completion of the electrical
circuit of the polarograph was recorded with a pen recorder with an
appropriate fast response. The results with solutions of sucrose in
saline are shown in Figure 2. Figure 3 shows a series of derived graphs
corresponding to the i_t versus t curves obtained experimentally. From
Figure 3, appropriate diffusion coefficients for oxygen were calculated.

The O_2 content of the series of solutions is shown in Figure 4.
The relationship between viscosities and diffusion coefficients for
oxygen is shown in Table 1.

In one series of experiments the sucrose solutions within the
capillary were replaced with solutions of hyaluronic acid (also previously equilibrated with air and containing NaCl). The results are shown
in the table while the corresponding values for oxygen diffusion coefficients and hyaluronate concentration are shown in Table 1.

DISCUSSION

Using the apparatus it was possible to show that a relationship of
the form $i_t = nFCA(D/t)^{\frac{1}{2}}$ is obeyed where: n is the number of electrons
involved in the cathode oxygen reaction at the applied potential (approx-

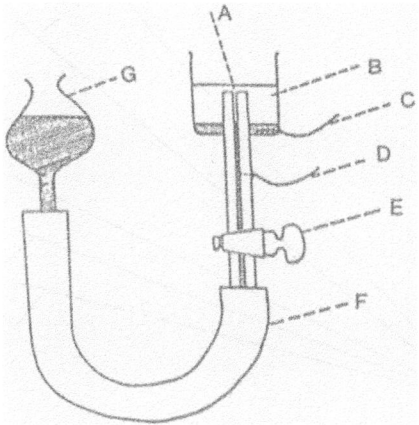

Fig. 1.   A, capillary containing viscous solution;
          B, carrier electrolyte solution; C, Pt
          wire passing into mercury anode pool;
          D, Pt wire connection to mercury cathode
          within the capillary; E, teflon tap;
          F, thick walled plastic tubing; G, mercury
          store.

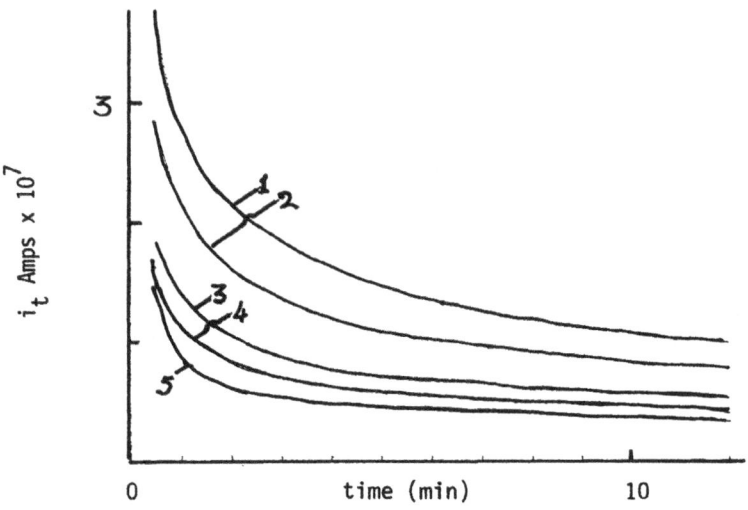

Fig. 2.   The decay of current with time following
          the completion of the electrical circuit
          of the polarograph.

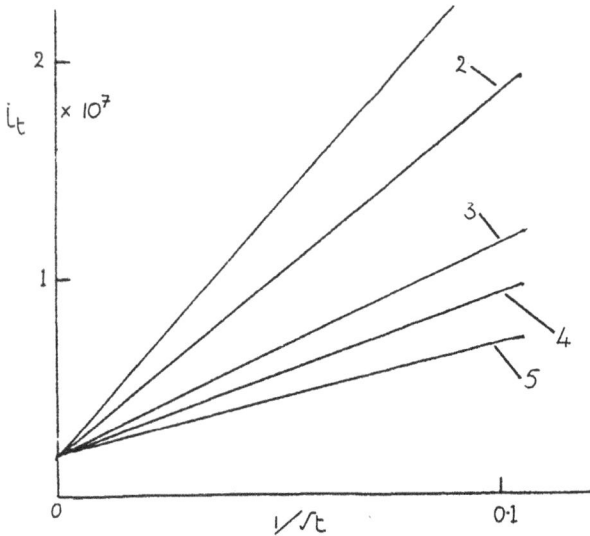

Fig. 3. Relationship between the oxygen
limited diffusion current and $1/\sqrt{t}$
for a series of sucrose solutions
previously equilibrated with air.

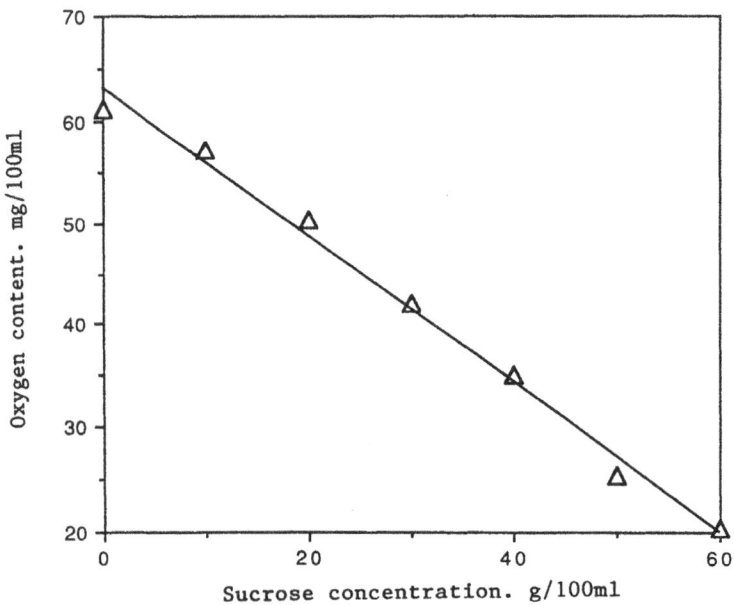

Fig. 4. Oxygen content of a series of sucrose solutions
equilibrated with air at 23^0, containing 0.1
NaCL.

Table 1. Showing diminution on oxygen diffusion coefficient with increasing solution viscosity, and diffusion coefficients expressed in activity rather than concentration.

Solution	Viscosity	Conventional Diffusion Coefficient $D_F.cm^2s^{-1}$	Activity Diffusion Coefficient D_A ie D_F/f^2
Saline	105	2.2×10^{-5}	2.07×10^{-5}
12% Sucrose	150	1.65×10^{-5}	1.12×10^{-5}
31% Sucrose	317	1.10×10^{-5}	0.39×10^{-5}
37% Sucrose	462	0.73×10^{-5}	0.23×10^{-5}
51% Sucrose	1573	0.51×10^{-5}	0.12×10^{-5}

Table 2. Values for S (slope of derived graph); oxygen concentration and Fick diffusion coefficients for O_2 in solutions of hyaluronic acid, and in a series of dextran solutions.

Solution	Slope of Derived Graph	Concn. of O_2 in Soln.	$D_F \times 10^7 cm^2 s^{-1}$
0.75 M KCl	146	1.97	2.2
0.75 M KCl with 2 mg/ml hyaluronate	133	2.04	1.71
0.75 M KCl with 4 mg/ml hyaluronate	154	1.84	1.27
KCl soln. with Dextran 15 10 mg/ml	221	1.96	3.74
KCl soln. with Dextran 60C 10 mg/ml	190	2.06	3.43
KCl soln. with Dextran 200C 10 mg/ml	174	2.01	2.96
KCl soln. with Dextran 500C 10 mg/ml	171	2.17	2.48
KCl soln. with Dextran 2000C 10 mg/ml	167	2.18	2.34
KCl soln. with Dextran 2000C 50 mg/ml	154	1.71	1.93

imately 4); F is the Faraday; C the current in amps at zero time; A is the surface area of the cathode; D is the diffusion coefficient of the electro-reducible substance (oxygen in this case); and i_t is the current t sec after completion of the polarographic electrical circuit. From the derived graph it is possible to calculate a value for D, the diffusion coefficient of oxygen through the solution. It can be seen that D varies inversely with the gross macroscopic viscosity of the solutions as measured using a simple capillary viscometer.

Effectively the above set of operations have the effect of calibrating the apparatus in its use as a viscometer and it becomes possible to use the apparatus to find the viscosity of small samples of solution placed inside the capillary of the apparatus by measuring the oxygen diffusion coefficient through the solution. Where this has been done for solutions of hyaluronic acid, the results have allowed an estimate of the viscosity of the hyaluronic acid. Most importantly, the viscosity measurements were obtained on the solutions when they were stationary (only the oxygen molecules move as probes through the solutions), and hence the apparatus can supply values for oxygen diffusion coefficients and hence of microscopic viscosities of solutions which are non-Newtonian when the solution is completely stationary.

Many, if not all biological systems can be (under some circumstances at least) non-Newtonian, due to the common presence of often severely assymetric and interacting polymer species. Until now the microscopic viscosities of non-Newtonian solutions have been notoriously difficult to obtain since they require an extrapolation of results obtained on sheared solutions, down to zero shear. It has been commonly observed that it is in the regions of the lowest shear that the viscosity changes are greatest (Young, 1958). The results obtained here with hyaluronate solutions suggest that viscosities of these non-Newtonian fluids are also anomalously high at the zero shear rate prevailing in these experiments.

The data obtained with hyaluronic acid solutions suggest that viscosities of similar and non stationary non-Newtonian fluids may also be high with correspondingly low values for oxygen diffusion. These results are therefore in accord with the low values previously found for the diffusion of oxygen (and nitrogen and tritiated water) through gels containing hyaluronic acid (McCabe and Laurent, 1975), and for other uncharged and low molecular weight species (McCabe, 1972). Such values cannot be adequately explained by theories based on a simple obstructive effect of the polymer on the diffusing species (McCabe, 1967). Theories for non-Newtonian viscosities need to explain adequately the greatly increased microscopic rigidity at zero shear rate. The implication of the results is that a considerable fraction of the water within the hyaluronate solutions is unavailable for oxygen diffusion.

The results obtained here are also in keeping with the data obtained some time ago by Zander and his colleagues (Zander, 1976) who found anomalously low values for oxygen solubilities within connective tissues.

The results may provide another physiological reason for the presence of hyaluronate in such high concentrations in many connective tissues, for example in Whartons jelly surrounding the unbilical vessels, the effect would be to diminish the losses of oxygen during the transport of blood from placenta to foetus. Additionally the greatly increased flux of oxygen observed across skin warmed to 43°C (and which is the basis for the success of transcutaneous blood gas measurements), may be due in part to vibrations and movements induced in the connective tissues of the dermis consequent on the arterialisation of the skin with the accompanying induced pulse.

In one set of experiments the sucrose solutions within the capillary were replaced with solutions of dextrans. Results are shown in Table 2.

The results with the dextrans suggests that diffusion coefficients are strongly dependent on the degree of polymerisation as well as on the dry weight content of the solutions, while low concentrations of low molecular weight dectrans may actually enhance oxygen diffusion coefficients.

REFERENCES

Davies, P.W., and Brink, F., 1942, Rev. Sci. Instr., 13:524.

Longmuir, I.S. and Milesi, J., 1960, Measurement of the diffusion of oxygen through samples of viscous liquids, J. Polarog. Soc., 6:18.

McCabe, M., 1967, Diffusion coefficients in polymer solutions, Biochem. J., 104:8.

McCabe, M., 1972, The diffusion coefficient of caffein through agar gels containing a hyaluronic acid-protein complex; a model system for the study of the permeability of connective tissues, Biochem. J., 127: 249.

McCabe, M. and Laurent, T.C., 1975, The diffusion of oxygen, nitrogen and and water in hyaluronate solutions, Biochim et Biophys. Acta, 399: 131.

Young, J.T., 1958, Determination of the intrinsic viscosity of rigid particle at zero rate of shear, J. Phys. Chem., 62:894.

Young, J.T., 1958, Non Newtonian viscosity of poly-γ-benzyl-L-glutamate solutions, J. Amer. Chem. Soc., 80:1783.

STUDIES OF LYMPH AND LYMPHOCYTE RESPIRATION TO ESTIMATE TISSUE OXYGEN

PRESSURES AND OXYGEN PERMEABILITY OF THE LYMPH DUCT

M. McCabe and D. Maguire*

Department of Biochemistry, James Cook University of North
Queensland, Townsville Qld. 4811 Australia and *Division
of Science and Technology, Griffith University, Nathan Qld.
4111 Australia

INTRODUCTION

It is well established that a mixed peripheral blood lymphocyte
population, when subjected to a mitogenic stimulus in vitro, will show a
significant increase in glucose consumption which is matched by lactate
output. This increased glycolysis has been ascribed to a direct effect of
the mitogen upon glucose transport. Less convincing are the reports on
the stimulation of oxygen uptake following mitogenic stimuli, however it
seems that those reports which indicate an absence of an oxygen uptake
stimulation following in vitro mitogenesis (Weidemann and Kolbuch, 1974
and Culvenor and Weidemann, 1976) are the consequence of failure to
follow the respiration rate for prolonged periods after the stimulation.
It seems that in vitro stimulation of lymphocytes does generate a significant increase in oxygen uptake which is not maximal until about day 3
(Roos and Loos, 1973).

In vivo studies of the pO_2 of lymph have been attempted by collecting
lymph from cannulated ducts. Because of the slow flow rates and the large
volumes that have been demanded for O_2 measurements, the collected lymph
is inevitably partly re-equilibrated with air or oxygen containing
solutions during its collection. A more satisfactory approach is by the
insertion of oxygen microelectrodes into the lymph duct. A particularly
convenient electrode has been designed by Hagihara et al. (1981). It can
be made small enough to insert into a suitable lymph duct of any moderate
sized animal without causing a blockage of the duct. Using this electrode
the physiological pO_2 within the lymph duct has been measured.

By presenting the animal with an immune challenge in the region of
the cannulated lymph duct, and observing the subsequent changes in lymph
pO_2 and the respiration rate of the lymph suspension, it is possible to
calculate the rate of oxygen flux and hence the permeability to oxygen
of the lymph duct, and to estimate the average tissue pO_2 in the region
of the duct.

EXPERIMENTAL

The polarographic cathode of Hagihara et al. (1981) was used without significant modification. That electrode system consists of a Pt cathode protected at its tip with a coating of cellulose diacetate (Hagihara et al., 1978) into which is incorporated heparin, which slowly leaks out and discourages clot formation around the tip when the electrode is used in vivo. Electrical continuity to a silver/silver chloride anode is normally maintained by a stream of saline which flows from anode to cathode down the electrode catheter (and which also contains heparin). When inserted into the lymph vessel, the direction of the flow inside the electrode catheter was reversed so that the lymph flowed from the lymph duct into the catheter, first past the cathode and finally the anode before it was collected. When used in this way the electrode system relies upon the lymph fluid electrolyte components for its electrical continuity. To minimise the sensitivity of the electrode to variations in flow rates, the electrode tip was given an extra coating of the cellulose diacetate. While this diminished flow rate dependence it also diminished the sensitivity of the system and increased its response time. By trial and error an optimum thickness was determined which permitted accurate readings to be completed before any clots started to form around the electrode tip.

Antigenic challenge in vivo was performed by injecting a solid antigen of Salmonella muenchen organisms (2 mg of acetone extracted freeze dried organisms in 0.2 ml of saline). The injection was subcutaneous into the region drained by the cannulated lymph node. Readings on pO_2 were taken over prolonged periods both before and after the antigenic challenge.

RESULTS AND DISCUSSION

The oxygen partial pressure of the efferent lymph was measured before and after the node was challenged. Measurements were made during the 2 days preceding and for three days after the challenge. Figure 1 shows the alterations in lymph pO_2 consequent on the antigenic challenge. It can be seen that there is a significant increase in the lymph pO_2 which occurs within 10 hours, and this is followed by a drop in the pO_2 which becomes most pronounced approximately 80 hours after the challenge. These changes can be correlated with concomitant changes in the flow rates and the total cell count appearing in the cell mix. The minimal value of lymph pO_2 coincided exactly with the time for maximum cell output into the lymph fluid, while the transient high pO_2 also coincided exactly with the time of the disappearance of almost all of the cells from the lymph fluid (see Figure 2).

The larger lymph vessels resemble veins histologically, the chief difference being that the walls of lymphatics are thinner. Oxygen is applied to the interior of the lymphatic by diffusion from surrounding tissues, and the question has arisen as to whether the very significant glycolysis which is observed in stimulated lymphocytes is a consequence of a relative hypoxia within the lymphatic vessel consequent on the immunological challenge to the animal.

It can be seen from the results of Figure 2 that the pO_2 within the lymph duct never fell below approximately 30 mm Hg making it extremely unlikely that there is ever a hypoxic stress on lymphocytes at least when they are liberated into the duct from the node.

476

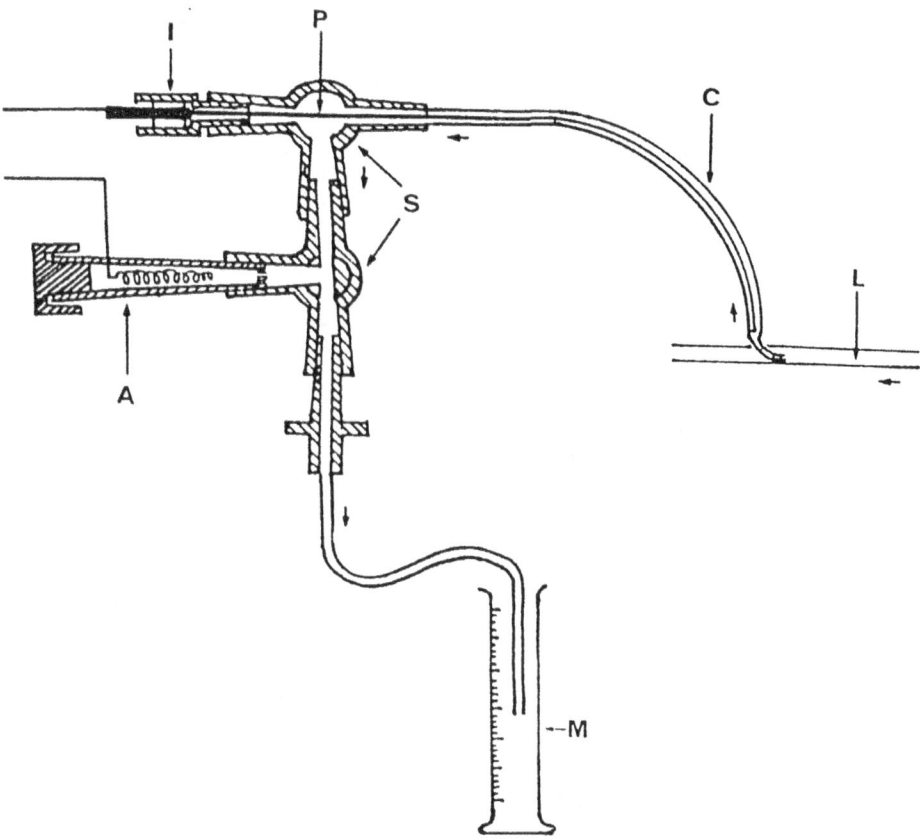

Fig. 1. Showing the insertion of the intravascular cathode into a cannulated lymph duct, with the collection of the lymph for measuring cell densities, and respiration rates *in vitro*, and for measuring flow rates of lymph.

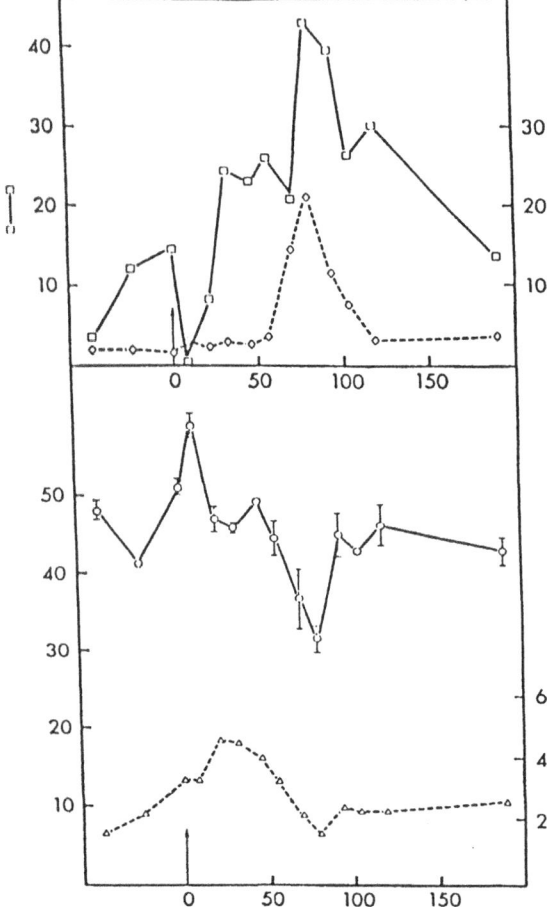

Fig. 2. The changes in intravascular pO_2 and concurrent changes in the lymph flow rate and cell numbers.

Intralymph pO_2 (mm Hg) O–O; flow rate of lymph fluid past the cathode Δ----Δ; total cell numbers in the lymph □——□; and the percentage of large (blast) cells Δ----Δ

The results also show that the pO_2 within the lymph duct could reach up to approximately 60 mm Hg when the cell concentration became very low, and presumably this is close to the value for pO_2 of the surrounding interstitial fluid.

Use of lymph pO_2 values during immunological challenge to calculate oxygen permeability through the lymph duct wall. The increase and decrease in lymph fluid pO_2 consequent on the dramatic changes in cell numbers (and hence the respiratory load within the lymph duct) can be used to give approximate measurements of the flux of oxygen across the lymph vessel wall. The total flux across the wall must exactly balance the total consumption inside, once the system is in oxygen equilibrium. Consider the flux of oxygen across L cm of length of the lymph vessel, then

$$\text{total flux of oxygen} = J2\pi rL$$

where J is the oxygen flux per unit area (in cm^2), r is the radius of lymph duct (in cms). Also

$$J = P(c_2 - c_1)$$

where P is the permeability to oxygen (in cm sec^{-1}), and c_1 and c_2 are the oxygen partial pressures when there is effectively zero respiration inside the lymph vessel, and at maximum respiration respectively.

This flux of oxygen is balanced by the consumption, U, within L cm of the length of the lymph vessel, i.e.

$$U = Q\pi r^2 L$$

where Q is the oxygen uptake of lymph fluid (at c_1 atmospheres), expressed as g O_2 per ml lymph fluid per second, i.e.

$$P(c_2 - c_1)2\pi rL = Q\pi r^2 L$$

$$P = \frac{Qr}{2(c_2 - c_1)}$$

Substitution of the corresponding values for the respiration rate, the radius of the lymph duct and the maximum and minimum values for C, gave a value for oxygen permeability (P) = 2.4×10^{-7} cm sec^{-1}.

REFERENCES

Culvenor, J.G. and Weidemann, M.J., 1976, PHA stimulation of rat thymus lymphocyte glycolysis, Biochim. Biophys. Acta, 437:354-363.

Hagihara, B., Ishibashi, F., Sato, N., Minami, T., Okada, Y., and Sugimoto, T., 1981, Intravascular oxygen monitoring with a polaro-graphic oxygen cathode, J. Biomed. Eugur., 3:263-270.

Roos, D. and Loos, J.A., 1973, Changes in the carbohydrate metabolism of mitogenically stimulated human peripheral lymphocytes, Exp. Cell. Res., 77:127-135.

Weidemann, M.J. and Kolbuch, M.E., 1974, Effect of concanavalin A on the oxidative metabolism of rat thymus lymphocytes, Proc. Aust. Biochem. Soc., 7:27.

GROUP PHOTO

AUTHOR INDEX

SUBJECT INDEX*

*The page numbers given are the first pages of the papers in which the subjects listed are covered.

Neonate
  pig, 341
NIR
  imaging, 155, 173
  spectrophotometry, 119, 125, 131,
      137, 155, 163, 173
  spectroscopy, 143
Nuclear magnetic resonance (NMR)
  spectroscopy, 187
    deuterium, 373
    ^{31}P, 393

Oleic acid, 299
Optical pathlength, 143
Oxidative stress, 211, 223
Oxidative tissue damage, 239
Oxygen
  availability, 441
  consumption, 203
    adaptation to hypoxia, 41
    mitochondria, 279
    on-line measurement, 195
    placenta, 457, 463
    skin, 49
  delivery, 203, 401, 409
    placental, 457
  dissociation curve, *see* Oxygen
      hemoglobin equilibrium curve
  electrode, *see* PO_2 electrode
  free radicals, 223, 253, 409
    spatial distribution, 211
  hemoglobin equilibrium curve, 11
  metabolism
    placenta, 463
    tissue, 187
  radicals, *see* Oxygen free radicals
  release, 285
  tension, *see* PO_2
  uptake
    effect of delivery, 401
    lung, 195, 203, 285
    measurement, 195, 203
    spatial distribution, 401
Oxyhemoglobin saturation measure-
      ment, 447

PAF-acether, 427
Paralysis of vascular reactivity, 299
PCO_2, arterial blood, 11
Perfluorocarbon emulsions, 119, 125, 131
pH, arterial blood, 11
Phospholipid bilayer, *see* Membrane

Phosphorescence
  lifetime, 279
  quenching by O_2, 179, 279, 341
  probe, 179
Phosphorimeter, 179
Phosphorylation state, 21
Photometry, picosecond, 131
Pig
  brain
    oxygenation, 341
    PO_2, 179
  hemorrhagic shock, 409
  lung oxygen uptake, 203
  newborn, 341
Placenta, human
  clearance of tracer, 457
  O_2 consumption, 457, 463
Plasma expander, 409
Platelet adhesion inhibition, 427
Pneumocystis Carinii pneumonia (PCP),
      293
Pneumonia, 293, 319
PO_2
  alveolar-arterial difference, 3
  arterial, continuous monitor, 71, 75
  blood, 341
  brain, 341
  electrode, 61, 71, 75, 85, 285, 475
    for diffusion coefficient measure-
        ment, 467
  gradient, 21
  intracellular, 61
  intravascular, 71, 75
  lymph duct, 475
  surface, 61
  retina, 113
  tissue, 61, 85, 103, 155
  tumor, 361
  vitreous humor, 113
Prostacyclin (PGI_2), 299
Pulmonary artery
  mean pressure, 299
  wedge pressure, 299

Rabbit
  blood, 11
  brain, 85, 103
  heart, 271
  lung, 311
  respiratory control, 347
Rat
  brain, 119, 125, 131

The manufacturer's authorised representative in the EU is Springer
Nature Customer Service Centre GmbH, Europaplatz 3, 69115 Heidelberg,
Germany. If you have any concerns regarding our products, please
contact ProductSafety@springernature.com

Printed and bound by CPI Group (UK) Ltd, Croydon, CR0 4YY
23/04/2026
02095623-0012